Very High Energy Cosmic Gamma Radiation

A Crucial Window on the Extreme Universe

Very High Energy Cosmic Gamma Radiation

A Crucial Window on the Extreme Universe

F. A. Aharonian

Max-Planck-Institut für Kernphysik
Heidelberg, Germany

NEW JERSEY · LONDON · SINGAPORE · BEIJING · SHANGHAI · HONG KONG · TAIPEI · CHENNAI

Published by

World Scientific Publishing Co. Pte. Ltd.
5 Toh Tuck Link, Singapore 596224
USA office: Suite 202, 1060 Main Street, River Edge, NJ 07661
UK office: 57 Shelton Street, Covent Garden, London WC2H 9HE

British Library Cataloguing-in-Publication Data
A catalogue record for this book is available from the British Library.

Front cover: X-ray image of the Crab Nebula, from the Chandra Photo Album; courtesy of NASA/CXC/ASU/J. Hester *et al.*

VERY HIGH ENERGY COSMIC GAMMA RADIATION: A CRUCIAL WINDOW ON THE EXTREME UNIVERSE

Copyright © 2004 by World Scientific Publishing Co. Pte. Ltd.

All rights reserved. This book, or parts thereof, may not be reproduced in any form or by any means, electronic or mechanical, including photocopying, recording or any information storage and retrieval system now known or to be invented, without written permission from the Publisher.

For photocopying of material in this volume, please pay a copying fee through the Copyright Clearance Center, Inc., 222 Rosewood Drive, Danvers, MA 01923, USA. In this case permission to photocopy is not required from the publisher.

ISBN 981-02-4573-4

Printed in Singapore by World Scientific Printers (S) Pte Ltd

Preface

The branch of high energy astrophysics that studies the sky in energetic γ-ray photons – *gamma-ray astronomy* – is destined to play a crucial role in the exploration of non-thermal phenomena in the Universe in their most extreme and violent forms. The great potential of the discipline allows an impressive coverage of a diverse range of "hot topics" in modern astrophysics and cosmology, in particular (i) the origin of galactic and extragalactic Cosmic Rays, (ii) acceleration and radiation processes in extreme astrophysical conditions, e.g. in pulsar magnetospheres, in the vicinity of accreting black holes, in relativistic outflows like the quasar jets and the pulsar winds; (iii) the nature of enigmatic transient phenomena like the γ-ray bursts (GRBs); (iv) cosmological issues connected with the diffuse background radiation and intergalactic magnetic fields; the search for dark matter in the form of WIMPs through their characteristic annihilation radiation, and tests of non-acceleration ('top-down') scenarios for the production of the highest energy particles observed in Cosmic Rays, *etc*.

The results from the Compton Gamma Ray Observatory (GRO) have confirmed a number of these prime motivations of gamma-ray astronomy. Many classical representatives of different galactic and extragalactic source populations, e.g. pulsars, supernova remnants, giant molecular clouds, quasars, which were predicted as potential MeV/GeV γ-ray emitters, are now among the almost 300 γ-ray sources detected by EGRET, and approximately 30 sources detected by COMPTEL. The nature of most of these sources remains, however, unknown. Moreover, the origin of γ-radiation from even firmly identified objects is poorly understood. This clearly justifies future gamma ray missions with new generation detectors like the Gamma-ray Large Area Space Telescope (GLAST). GLAST, with its advanced performance, has been carefully designed for deep surveys of the sky in γ-rays with an ambitious aim of providing *"γ-ray astronomy with thousands of sources"* in the energy region from tens of MeV to 10 GeV. Also, since most EGRET sources do not exhibit spectral cutoffs in the 1-10 GeV region, the extension of their study into the unexplored region beyond 10 GeV is another important issue for the GLAST. Meanwhile, the area limitations of space-borne detectors compels the study of Very High Energy (VHE) photons above 100 GeV to remain (except for the specific topic related to the diffuse extragalactic γ-ray background) the domain of ground-based gamma-ray astronomy.

The recent exciting observational results and theoretical predictions supply a strong rationale for the systematic study of primary γ-radiation in the VHE domain. Further improvement of the detection technique will be linked to stereoscopic observations of air showers with imaging Cherenkov telescope arrays with energy thresholds as low as 10 GeV, angular resolutions better than a few arcminutes, and flux sensitivities approaching to $10^{-13}\,\mathrm{erg/cm^2 s}$. This will elevate the status of the field, which currently can be characterised as an *"astronomy with a few sources"*, to the level of truly *observational* discipline.

The further study of the sky in high energy γ-rays promises a new path towards understanding of the non-thermal phenomena in the Universe. It is expected that with forthcoming powerful space-borne and ground-based detectors, gamma-ray astronomy will enter a new era with an objective of providing crucial insight into a number of fundamental problems of astrophysics and cosmology. This necessitates a comprehensive discussion of major motivations and objectives of this rapidly developing field.

When writing this book, I tried to highlight the principal objectives of the field, as well as to demonstrate its relevance and links to other branches of Astronomy and Cosmology. Preference has been given to three topical areas - the *Origin of Cosmic Rays*, the *Physics and Astrophysics of Relativistic Jets*, and *Observational Cosmology*. One chapter of the book is devoted to the discussion of principal γ-ray production and absorption mechanisms, with emphasis upon the processes that play dominant roles in the high and very high energy domains.

The chosen topics are among the scientific interests of the author. Also, a substantial part of the book is based on my own studies performed in close collaboration with my colleagues. Many results and conclusions reflect, to a large extent, my understanding of the subject in general, and my assessment of the achievements, as well as existing difficulties, ambiguities and "nasty problems" of the field. Therefore I cannot exclude a somewhat subjective (but hopefully not completely wrong) character of some parts of the book concerning both the interpretation of observations and the preference given to certain methods and approaches in phenomenological and theoretical studies.

This book would have not been possible without intensive collaboration with my co-workers. Also, I have profited and learned a lot from discussions with numerous colleagues working in the field of high energy astrophysics. Special thanks must go to Armen Atoyan and Paolo Coppi for many years of fruitful collaboration. I am indebted to a large number of friends and colleagues for their invaluable contributions to the different aspects of our joint projects in several areas of high energy astrophysics: C. Akerlof, S.Bogovalov, J. Cronin, L. Costamante, E. Derishev, L. Drury, G. Heinzelmann, W. Hofmann, D. Horns, T. Kifune, S. Kelner, J. Kirk, H. Krawczynsky, A. Plyasheshnikov, G. Rowell, V. Sahakian, D. Schramm (deceased), R.A. Sunyaev, T. Takahashi, A. Timokhin, Y. Uchiyama, V. Vardanian (deceased), H.J. Völk. Finally, I am grateful to Phil Edwards for his careful reading of the manuscript and important comments.

Contents

Preface v

1. Introduction 1
   1.1 "The Last Electromagnetic Window" 1
   1.2 Energy Domains of Gamma Ray Astronomy 4
   1.3 Gamma Ray Astronomy: A Discipline in Its
       Own Right 7

2. Status of the Field 23
   2.1 Low Energy Gamma Ray Sources 24
       2.1.1 The COMPTEL source catalog 25
   2.2 High Energy Gamma Ray Sources 31
       2.2.1 GeV blazars 34
       2.2.2 GeV pulsars 36
       2.2.3 Unidentified EGRET sources 39
   2.3 The Status of Ground-Based Gamma Ray Astronomy . . 42
       2.3.1 Brief historical review 42
       2.3.2 Reported TeV sources 46
             2.3.2.1 The Crab Nebula 47
             2.3.2.2 Other plerions 51
             2.3.2.3 Gamma ray pulsars 54
             2.3.2.4 Gamma rays from supernova remnants . . 54
             2.3.2.5 Other galactic sources 63
             2.3.2.6 TeV blazars 67
             2.3.2.7 Other extragalactic objects 79
       2.3.3 Next generation of IACT arrays 83

		2.3.3.1	Atmospheric Cherenkov radition	83
		2.3.3.2	Stereoscopic detection of Cherenkov images	84
		2.3.3.3	IACT arrays	87
		2.3.3.4	Sub-10 GeV ground based detectors?	90
		2.3.3.5	Large field-of-view detectors	95
		2.3.3.6	IACT arrays for probing PeV γ-rays	96

3. Gamma Ray Production and Absorption Mechanisms 99

    3.1 Interactions with Matter 100
        3.1.1 Electron bremsstrahlung and pair-production . . . 101
        3.1.2 Electron-positron annihilation 105
        3.1.3 Gamma rays produced by relativistic protons . . . 106
            3.1.3.1 π^0-decay gamma rays 106
            3.1.3.2 Nuclear gamma ray line emission 110
    3.2 Interactions with Photon Fields 112
        3.2.1 Inverse Compton scattering 113
        3.2.2 Photon-photon pair production 117
        3.2.3 Interactions of hadrons with radiation fields 121
    3.3 Interactions with Magnetic Fields 123
        3.3.1 Synchrotron radiation and pair-production 123
        3.3.2 Synchrotron radiation of protons 126
    3.4 Relativistic Electron-Photon Cascades 130

4. Gamma Rays and Origin of Galactic Cosmic Rays 135

    4.1 Origin of Galactic Cosmic Rays: General Remarks 135
        4.1.1 What do we know about Cosmic Rays? 135
        4.1.2 What we do not know about Cosmic Rays? 137
        4.1.3 Common beliefs and "nasty" problems 140
        4.1.4 Searching for sites of production of GCRs 143
    4.2 Giant Molecular Clouds as Tracers of Cosmic Ray 147
        4.2.1 Proton fluxes in the ISM near the accelerator . . . 149
            4.2.1.1 Impulsive source 149
            4.2.1.2 Continuous source 151
            4.2.1.3 The case of dense gas regions 154
        4.2.2 Gamma rays from a cloud near the accelerator . . 155
        4.2.3 Accelerator inside the cloud 159

		4.2.4	On the level of the "sea" of galactic cosmic rays 161

    4.3 Probing the Sources of VHE CR Electrons 165
          4.3.1 Distributions of VHE electrons 166
          4.3.2 Extended regions of IC gamma radiation 168
    4.4 Diffuse Radiation from the Galactic Disk 173
          4.4.1 CR spectra in the inner Galaxy 174
          4.4.2 Diffuse radiation associated with cosmic ray electrons 177
                  4.4.2.1 IC gamma rays 177
                  4.4.2.2 Electron bremsstrahlung 181
                  4.4.2.3 Annihilation of CR positrons in flight . . 182
          4.4.3 Gamma rays of nucleonic origin 183
          4.4.4 Overall gamma ray fluxes 185
          4.4.5 Probing the diffuse γ-ray background on small scales 189
          4.4.6 Concluding remarks 195

5. Gamma Ray Visibility of Supernova Remnants 199

    5.1 Gamma Rays as a Diagnostic Tool 199
    5.2 Inverse Compton Versus π^0-Decay Gamma Rays 205
    5.3 Synchrotron X-ray Emission of SNRs 208
    5.4 TeV Gamma Radiation of SN 1006 and Similar SNRs . . 209
          5.4.1 Inverse Compton models of TeV emission 209
          5.4.2 Hadronic origin of TeV emission? 215
          5.4.3 Distinct features of electronic and hadronic models 217
          5.4.4 Concluding remarks 219
    5.5 Molecular Clouds Overtaken by SNRs 220
          5.5.1 Bremsstrahlung X-rays from γ Cygni 223
          5.5.2 The case of RX J1713.7-3946 225
    5.6 A Special Case: Gamma Rays from Cassiopeia A 231
    5.7 "PeV SNRs" 238

6. Pulsars, Pulsar Winds, Plerions 243

    6.1 Magnetospheric Gamma Rays 244
          6.1.1 Polar cap versus outer gap models 247
          6.1.2 Magnetospheric TeV gamma rays? 251

6.2 Gamma Rays from Unshocked Pulsar Winds 253
   6.2.1 Characteristics of the KED wind 254
   6.2.2 The ejection rate and the particle spectrum 255
   6.2.3 IC Radiation of the pulsar wind in Crab 256
   6.2.4 Gamma rays from winds of PSR B1706-44 and Vela? 261
   6.2.5 IC γ-rays from the binary pulsar PSR B1259-63 265
6.3 Gamma Rays from Pulsar Driven Nebulae 268
   6.3.1 Broad-band nonthermal radiation of the Crab Nebula 268
      6.3.1.1 Synchrotron and IC radiation 270
      6.3.1.2 Second High Energy Synchrotron Component 273
      6.3.1.3 Bremsstrahlung and π^0-decay gamma rays? 275
      6.3.1.4 The objectives of future gamma ray studies 278
6.4 High Energy Gamma Rays from Other Plerions 281
   6.4.1 Time-evolution of electrons 283
   6.4.2 Target photon fields 284
   6.4.3 Effects of B-field, electron energy, and pulsar age . 286
   6.4.4 Synchrotron and IC nebulae around PSR B1706-44 289

7. Gamma Rays Expected from Microquasars 293
   7.1 Do We Expect Gamma Rays from X-Ray Binaries? . . . 293
   7.2 Nonthermal Phenomena in Microquasars 295
   7.3 Modelling of Radio Flares of GRS 1915+105 301
   7.4 Expected Gamma Ray Fluxes 304
   7.5 Searching for Gamma Ray Signals from Microquasars . . 309
   7.6 The Case of Microblazars 311
   7.7 Ultraluminous Sources as Microblazars? 313
   7.8 Persistent Gamma Ray Emission from Extended Lobes 317

8. Large Scale Jets of Radio Galaxies and Quasars 321
   8.1 Synchrotron and IC Models of Large Scale AGN Jets . . 323

	8.2	Ultra High Energy Protons in Jets	327
		8.2.1 Secondary electrons	327
		8.2.2 Synchrotron radiation of protons	330
		8.2.3 Pictor A, PKS 0637-752, and 3C 120	332
		8.2.4 The case of 3C 273	339
	8.3	Large Scale Jets Powered by Gamma Rays	347
	8.4	Concluding Remarks	354
9.	Nonthermal Phenomena in Clusters of Galaxies	359
	9.1	Nonthermal Particles and Magnetic Fields	359
	9.2	Inverse Compton and Bremsstrahlung Models	364
		9.2.1 Inverse Compton models	364
		9.2.2 Nonrelativistic bremsstrahlung	368
	9.3	Synchrotron X- and γ-rays of "Photonic" Origin?	369
	9.4	Nonthermal Radiation Components Associated with Very High and Extremely High Energy Protons	378
		9.4.1 High energy radiation from cores of clusters	379
		9.4.1.1 Detectability of gamma rays	384
		9.4.1.2 Detectability of X-rays	386
		9.4.2 Nonthermal radiation beyond the cluster cores	388
		9.4.2.1 Weak magnetic field	389
		9.4.2.2 Intermediate magnetic field	390
		9.4.2.3 Strong magnetic field	390
10.	TeV Blazars and Cosmic Background Radiation	393
	10.1	Cosmic Infrared Background Radiation	394
	10.2	Intergalactic Absorption of Gamma Rays	397
	10.3	TeV Blazars	399
	10.4	Leptonic Models of TeV Blazars	402
		10.4.1 Constraints on the SSC parameter space	404
		10.4.2 Time-dependent SSC treatment	409
		10.4.3 The case of 1ES 1426+428	415
	10.5	Hadronic Models	417
		10.5.1 Mass-loaded hadronic jet models	418
		10.5.2 Photo-pion and synchrotron losses of protons	419
		10.5.3 Proton synchrotron models	422
		10.5.3.1 Fitting the TeV spectrum of Mkn 501	423
		10.5.3.2 X-rays from secondary electrons	425

     10.5.3.3 Broad band SED of 1ES 1426+428
         within the proton-synchrotron model . . . 427
  10.6 "IR background–TeV Gamma Ray Crisis"? 432

11. **High Energy Gamma Rays — Carriers of Unique Cosmological Information** 437

  11.1 Probing DEBRA Through γ-Ray Absorption Features . . 437
  11.2 The Effect of Cascading in the CIB 447
  11.3 Pair Halos as Unique Cosmological Candles 450
  11.4 Diffuse Extragalactic Background as Calorimetric
     Measure of the VHE Emissivity of the Universe 460

Appendix A Spherically symmetric diffusion from a
       single source 463

Appendix B Evolution of relativistic electrons
       in an expanding magnetised medium 465

  B.1 Kinetic equation 465
  B.2 Time-independent energy losses 467
  B.3 Expanding cloud 470

Bibliography 473

Index 493

Chapter 1

Introduction

1.1 "The Last Electromagnetic Window"

Many review articles and books related to gamma-ray astronomy start with a statement that the γ-ray domain of cosmic radiation is the last of the electromagnetic windows to be opened. This concerns not only the first papers written several decades ago by the pioneers of gamma-ray astronomy, but also the recent assessments of the status of the field. Actually, a gamma-ray astronomer may argue that this window is already opened, at least in the sense that we already have a map of the sky in γ-rays (see Fig. 1.1). Gamma-ray astronomy has indeed entered the main stream of modern astrophysics with more than 300 reported sources, and reliable detection techniques with a potential for further significant improvement. Moreover, many "hot" topics of modern astronomy like the physics and astrophysics of relativistic jets in Active Galactic Nuclei (AGN) essentially rely on γ-ray observations in the MeV (10^6 eV), GeV (10^9 eV), and TeV (10^{12} eV) energy regions. On the other hand, it is difficult to object to the most critical representatives of the advanced areas of astronomy who argue that the performance of current gamma-ray detectors (in particular the flux sensitivities and angular resolution) needs to be improved significantly in order to match the performance of Radio, Optical and X-ray telescopes. The fact that a major fraction of more than 270 sources detected by the Energetic Gamma Ray Experiment Telescope (EGRET) aboard the Compton Gamma Ray Observatory (GRO) remains unidentified, supports, to a large extent, such a critical view. The ratio of identified-to-unidentified sources is significantly better in the TeV region – only one unidentified source from more than 10 reported objects (see Fig. 1.1). However, in the TeV regime presently we deal with an astronomy with a very limited number of detected sources. Thus, one may conclude that the γ-ray sky remains a largely unexplored frontier representing one of the last energy bands of the electromagnetic spectrum to be explored with detectors of an adequate sensitivity.

With the new generation of space-based and ground-based detectors, the energy flux sensitivity will be significantly improved approaching to the level between 10^{-13} and 10^{-12} erg/cm^{2}s over a broad energy range from 100 MeV to 10 TeV (see Fig. 1.2). For example, the flux sensitivity at GeV energies will be improved (for point-like sources) by the Gamma-ray Large Area Space Telescope (GLAST) by two orders of magnitude. It is expected that with GLAST we will enter an era of "gamma-ray astronomy with thousands of sources". Dramatic improvements are expected also in the TeV regime. There is a confidence that the new generation of stereoscopic arrays of Imaging Atmospheric Cherenkov Telescopes (IACTs) with energy

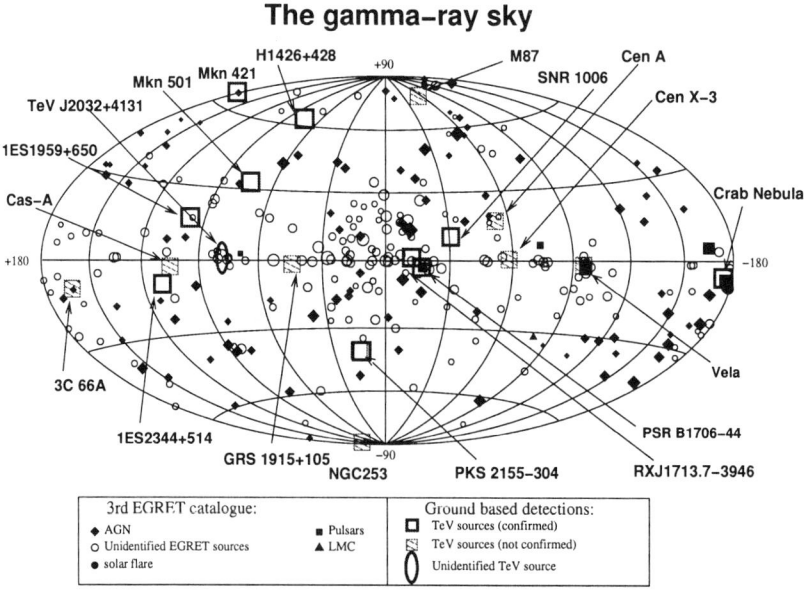

Fig. 1.1 The reported MeV/GeV (EGRET) and TeV gamma-ray sources. Note that the presentation of the locations of EGRET sources by symbols with variable size is chosen to illustrate the level of γ-ray fluxes detected from individual sources (Hartman et al., 1999). The locations of TeV sources are shown by larger symbols in order to distinguish them from EGRET sources, but they do not correlate with the reported flux or source angular size.

threshold around 100 GeV will bring many important results and discoveries relevant to various aspects of high energy astrophysics and cosmology.

Another high priority objective of future instrumental developments will be an attempt of exploration of the energy interval between 10 and 100 GeV. The interest to this relatively narrow energy band is motivated not only by the natural desire to enter a new domain which remains a *terra incognita*, but also because it provides a bridge between the high and very high astronomies, and thus may allow key inspections of the current concepts concerning both the GeV and TeV regimes. Moreover, this energy region is crucial for proper understanding of a number of astrophysical and cosmological phenomena related to the physics and astrophysics of AGN and Gamma-Ray Bursts (GRBs).

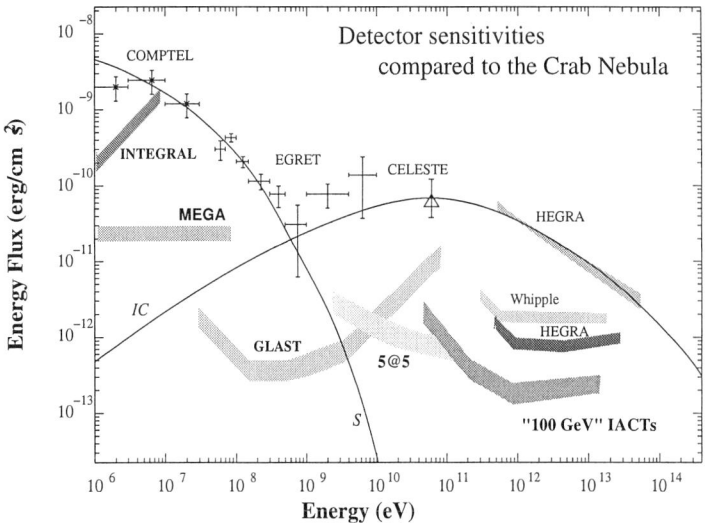

Fig. 1.2 Energy flux sensitivities of the future low-energy (INTEGRAL, MEGA) and High Energy (GLAST) space-based detectors shown together with flux sensitivities of the current (Whipple, HEGRA), upcoming "100 GeV" threshold (e.g. H.E.S.S.) and future "sub-10 GeV" threshold (e.g. 5@5) arrays of Imaging Atmospheric Cherenkov Telescopes. For comparison, the predicted synchrotron (S) and inverse Compton (IC) fluxes, as well as the reported γ-ray fluxes (COMPTEL, EGRET, CELESTE, HEGRA) from the Crab Nebula are shown.

In spite of lack of information about γ-ray sources in this energy region, the extension of a hard diffuse extragalactic γ-radiation detected by EGRET up to 100 GeV (see Fig. 1.3), which represents the γ-ray emissivity of the entire Universe, is a clear indication of the ongoing nonthermal activity in the Universe with significant release of power in this energy band. To a large extent, this general statement does not depend on the specific origin of this radiation, in particular on the question whether it is mainly contributed by discrete unresolved sources or it is a product of truly diffuse processes that take place in the intergalactic medium. Different aspects of gamma-ray astronomy in this intriguing energy interval will be be effectively covered by GLAST and, hopefully, also by arrays of low energy-threshold Cherenkov telescopes (see Fig. 1.2).

1.2 Energy Domains of Gamma Ray Astronomy

Returning back to the question of the "last window", we notice that it covers at least 14 decades in frequency ! The energy domain of gamma-ray astronomy spans from approximately $E = m_e c^2 \simeq 0.5 \times 10^6$ eV to $\geq 10^{20}$ eV. The lower bound characterises the region of nuclear γ-ray lines, as well as the electron-positron annihilation line. The second bound characterises the energy of highest energy particles observed in cosmic rays. This limit can be also interpreted as the maximum energy of protons and nuclei that can be achieved in conventional cosmic ray accelerators. The interactions of protons with surrounding gas or radiation fields should unavoidably lead to production of γ-ray photons of comparable energy. In fact, much higher (by two or three orders of magnitude) energy γ-rays can be in principle produced due to the collapse or decay of hypothetical very massive relic particles.

This enormous energy band of cosmic electromagnetic radiation is covered rather inhomogeneously in the sense of essentially different detection methods and flux sensitivities of space- and ground-based instruments. Therefore it is convenient to introduce several sub-divisions, taking into account the specific astrophysical objectives and detection methods relevant to different energy bands. Generally, the observational gamma-ray astronomy can be divided into 6 areas - *low* (LE: below 30 MeV), *high* (HE: 30 MeV - 30 GeV), *very high* (VHE: 30 GeV - 30 TeV), *ultra high* (UHE: 30 TeV - 30 PeV), and finally *extremely high* (EHE: above 30 PeV) energies. The "keV ... EeV" symbols characterise the most commonly used energy units in high energy astrophysics and imply: $1 \text{keV} = 10^3$ eV, $1 \text{MeV} = 10^6$ eV, $1 \text{GeV} = 10^9$ eV, $1 \text{TeV} = 10^{12}$ eV, $1 \text{PeV} = 10^{15}$ eV, $1 \text{EeV} = 10^{18}$ eV. While observations in the first two energy bands are covered by satellite-

or balloon-borne detectors, the last three energy intervals can be best addressed with ground-based instruments. It should be noticed, however, that the term "astronomy" presently can be adequately applied, strictly speaking, to the activities in the LE, HE, and VHE energy domains. The significant efforts in the past unfortunately did not not result in detection of UHE and EHE γ-ray sources, although presently we cannot exclude that a fraction of highest energy cosmic rays detected beyond the so-called GZK cutoff may have photonic origin.

Only a limited fraction of the Universe is available for γ-ray observations in the VHE, UHE and EHE domains. The γ-ray horizon of the Universe is strongly energy-dependent (see Fig. 1.4). It is determined by interactions of γ-rays with the diffuse extragalactic photon fields. For example, the mean free path of γ-rays less than several GeV exceeds the Hubble size of the Universe, whereas the free path of 1 PeV γ-rays is only 8.5 kpc. This implies that all objects beyond our Galaxy are not visible in PeV γ-rays. The visibility range increases up to several 100 Mpc for TeV γ-rays and ≈ 10 Mpc for 100 EeV. Only by reducing the energy threshold of detectors down to 100 GeV can we approach cosmological distances corresponding to redshifts $z \sim 1$. And finally, in order to detect γ-rays from the most distant

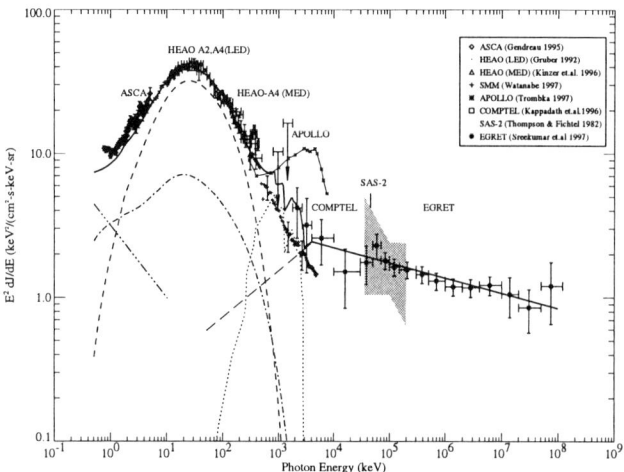

Fig. 1.3 Diffuse extragalactic X- and γ-ray backgrounds (from Sreekumar *et al.*, 1998).

quasars with redshifts $z \geq 3$ the energy threshold of detectors should be reduced to 10 GeV and lower.

Thus, cosmological epochs cannot be directly probed with TeV and higher energy γ-rays. When a high energy γ-ray photon is absorbed, its energy however is not lost. Their interactions with extragalactic radiation fields initiate electromagnetic cascades which lead to formation of huge (≥ 10 Mpc) nonthermal structures like hypothetical (not yet detected) ultrarelativistic electron-positron Pair Halos. Actually, the entire Universe is a scene of continuous creation of electromagnetic cascades. All cosmic γ-rays above several GeV have a similar fate. Sooner or later they terminate on Hubble scales. The superposition of contributions of γ-rays from these cascades should constitute a significant fraction of the diffuse background shown in Fig. 1.3.

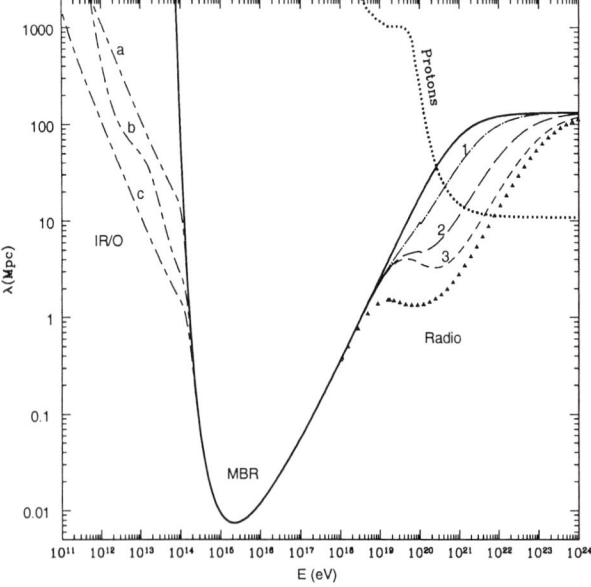

Fig. 1.4 Mean free paths of gamma-rays in the intergalactic medium at redshifts $z \ll 1$. Below 10^{14} eV γ-rays interact with infrared and optical photons, above 10^{19} eV - with low frequency radio emission. Large uncertainties in predicted mean free paths are the result of poorly known fluxes of the extragalactic diffuse background radiation at these wavelengths. Between 10^{14} eV and 10^{19} eV, γ-rays interact mainly with the 2.7 K CMBR, therefore the mean free paths can be predicted with very high accuracy (from Coppi and Aharonian, 1997).

1.3 Gamma Ray Astronomy: A Discipline in Its Own Right

High energy γ-rays combine three characteristics that make these energetic photons ideal carriers of information about nonthermal relativistic processes in astrophysical settings: (i) copious production in many galactic and extragalactic objects due to effective acceleration of charged particles and their subsequent interactions with the ambient gas, low frequency radiation, and magnetic fields; (ii) free propagation in space without deflection in the interstellar and intergalactic magnetic fields; (iii) effective detection by space-borne and ground based instruments. Therefore it is commonly believed that very high energy gamma-ray astronomy is destined to play a crucial role in exploration of nonthermal phenomena in the Universe in their most extreme and violent forms. The major driving motivations of this field conditionally can be grouped in three topical areas: (i) *Origin of Cosmic Rays*, (ii) *Physics and Astrophysics of Relativistic Outflows*, and (iii) *Observational Gamma Ray Cosmology*.

Origin of Cosmic Rays

Galactic Cosmic Rays. For more than 40 years, ideas have circulated about the crucial role of gamma-ray astronomy in solving the problem of origin of galactic cosmic rays (CRs). The realization of this seminal prediction recognised by pioneers of the field in the 1950's and 1960's is still considered as one of the major goals of γ-ray astronomy. The basic idea is simple and concerns both the acceleration and propagation aspects of CRs. Namely, while the localised γ-ray sources exhibit the sites of production/acceleration of CRs, the angular and spectral distributions of the diffuse galactic γ-ray emission provide unique information about the character of propagation of CRs in galactic magnetic fields. The prime objective of this activity is the decisive test of the hypothesis that supernova remnants (SNRs) are responsible for the bulk of observed CRs up to 10^{15} eV. Detection of TeV γ-rays from shell-type SNRs would be the first straightforward proof of the widely believed model of diffusive shock acceleration of CR protons in these objects. Conservative phenomenological estimates show that a certain number of 10^3 to 10^4 yr old SNRs should be visible in γ-rays. Although 3 shell type SNRs already have been reported as TeV emitters, the limited information about both the spectral and spatial distributions of detected signals does not allow definite conclusions concerning the nature of TeV emission, also because the latter could be substantially "contaminated" by γ-rays of leptonic (inverse Compton) origin.

The failure to detect positive γ-ray signals from several selected SNRs would impose a strong constraint on the total energy in the accelerated protons, $W_{\rm p} \leq 10^{49}$ erg. Contrary to current belief this would indicate the inability of SNRs to explain the bulk of the observed cosmic ray fluxes. SNRs are still only one of the plausible sites of CR acceleration. It is quite possible that different galactic source populations, e.g. pulsars and microquasars, contribute comparably to the observed cosmic ray flux. This

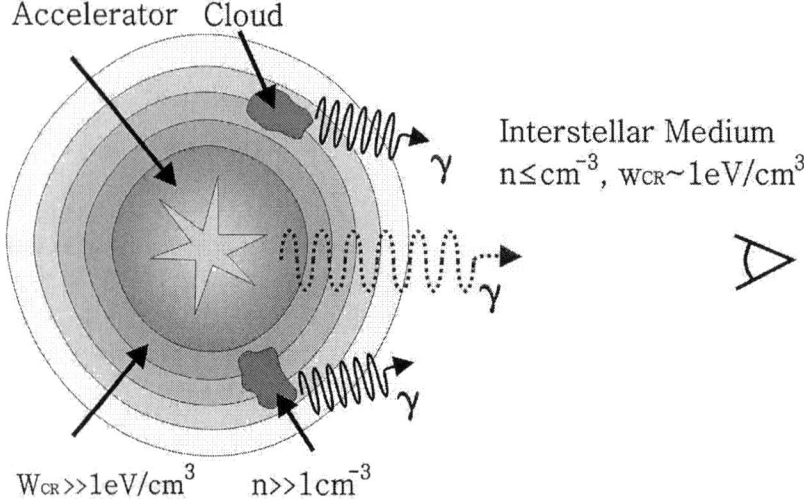

Fig. 1.5 Gamma-rays radiated by dense molecular clouds located in the vicinity of a young ($\leq 10^4$ yr old) proton accelerator, where the density of relativistic particles significantly exceeds the average level of the "sea" of galactic CRs (≈ 1 eV/cm^3) determined by the mixture of contributions from all individual sources during the CR propagation in the Galactic Disk on timescales $\sim 10^7$ yr. The particle accelerator itself is a source of γ-rays, but its intensity could be quite low due to the lack of sufficiently dense target material inside compared to the average density of the interstellar medium (≈ 1 cm^{-3}). Also, the γ-ray spectrum could be suppressed at very high energies due to the energy-dependent escape of particles from the accelerator.

makes the problem rather complicated. Only direct identification of these sources as particle accelerators through their characteristic γ-ray emission can help to elucidate the origin of galactic CRs. The existence of a particle *accelerator* by itself is not enough for efficient γ-ray production; one needs the second component - the *target*. The so called giant molecular clouds (GMCs) with diffuse masses $\sim 10^4$ to 10^6 $M_\odot$, seem to be ideal objects to play that role (see Fig. 1.5). These objects are intimately connected with star formation regions that are strongly believed to be the most probable locations (with or without SNRs) of cosmic ray production in our Galaxy. The search for TeV γ-rays from GMCs is important to ascertain the possible existence of nearby high energy proton accelerators.

Extragalactic Cosmic Rays. Although there is little doubt that the highest energy particles observed in cosmic rays, with $E \sim 10^{20}$ eV, are produced outside of our Galaxy, the sites and relevant acceleration mechanisms continue to be a mystery. Powerful extragalactic objects like radiogalaxies, AGN, clusters of galaxies, the enigmatic gamma-ray burst sources, have all been (phenomenologically) suggested as possible acceleration sites. However boosting particles to such energies is a serious theoretical challenge. Even if the electrodynamical system accelerating particles had an infinite lifetime and the acceleration proceeds at the maximum possible rate of about $\sim qBc$, the attainable energy is limited by two conditions: (1) by the confinement - particles can stay in the acceleration region as long as their gyroradius remains smaller of the characteristic linear size of the accelerator; (2) by synchrotron energy losses. Even assuming that the particles are moving along smooth field lines, there is still curvature radiation which limits the maximum attainable energy. These two conditions allow us to derive a robust lower limit for the total electromagnetic energy, W, stored in the acceleration region, and thus estimate the size, l, and the magnetic field strength, B, that are optimal with respect to minimisation of the electromagnetic energy.

In large scale ($\gg 1$ kpc) structures the "confinement condition" is more critical. In this case, the radio lobes in radiogalaxies, hot spots in AGN jets, and clusters of galaxies seem the most likely sites of particle acceleration (see Fig. 1.6a). In many cases these objects may contain enough target material in the form of gas and photon fields to convert a substantial fraction of these particles into detectable VHE radiation. Moreover, even if these particles do not spend much time in their production region, but leave the source, they unavoidably collide with 2.7 CMBR photons at relatively small distance from the source (especially for objects at large redshifts). The secondary electrons and photons – the products of photohadronic interactions – initiate two-stage pair cascades in the 2.7 K CMBR and optical/infrared extragalactic radiation fields. The typical angular size of the active region of cascade development does not exceed 1 degree for a source located at a distance ~1 Gpc. The resulting γ-radiation at energies below 100 GeV can be detectable by GLAST and the new generation of low-threshold IACT arrays.

The relativistic bulk motion reduces the energy requirements significantly, but requires compact objects and large magnetic fields (see Fig. 1.6b). For example, for a bulk motion Lorentz factor $\Gamma \sim 10$, the optimal size and magnetic field are $R \sim 10^{14}$ eV and $B \sim 300$ G, respectively. This corresponds to a quite reasonable total electromagnetic energy $\sim 3 \times 10^{47}$ erg for the inner jets of blazars. A reasonable combination of model parameters is obtained also for GRBs for which a bulk motion Lorenz factor of about 300 is more typical.

Energy losses, due to either synchrotron or curvature radiation, play an increasing role in the energy balance of accelerated particles, with decreasing size (and accordingly increasing magnetic field) of the accelerator. In compact objects like small scale AGN jets or GRBs, the radiative losses become the dominant factor that limits the maximum attainable energy. This implies that proton acceleration in such objects should be always accompanied by hard synchrotron (or curvature) radiation extending to $\sim \alpha_f^{-1} m_p c^2 \Gamma \sim 1(\Gamma/10)$ TeV (in the observer's frame). The search for such a characteristic radiation can be used to probe the potential accelerators of highest energy cosmic rays. The protons can interact effectively also with ambient dense photon fields through photomeson production. Due to the hadronic and electromagnetic cascades initiated by these interactions, a substantial part of the primary energy can be effectively transported through neutrons and γ-rays to very large distances from the central engine. Moreover, the multiple conversions of nucleons from charged to neutral state and back may significantly increase the acceleration rate compared to the standard diffusive shock acceleration mechanism, and thus make more effective the production of 10^{20} eV protons. This mechanism provides a very effective conversion of the kinetic energy of bulk relativistic flows in GRBs and inner jets of blazars to the accompanying high energy γ-ray and neutrino emission.

Top-Down scenarios. The above constraints on the parameter space of 10^{20} eV particle accelerators set by classical electrodynamics obviously cannot be applied to the so-called "top-down" models of cosmic rays which make use of quantum effects and non-electromagnetic interactions. The hypothesis of the non-acceleration or "top-down" scenario as an alternative to the ordinary acceleration ("bottom-up") scenario is motivated by difficulties of current theoretical models to provide adequate acceleration rates that could boost the particles to energies $\geq 10^{20}$ eV. In the "top-down" models the cosmic rays are the result of decays of the so-called topological defects or relic super-heavy particles. These models, however, have a

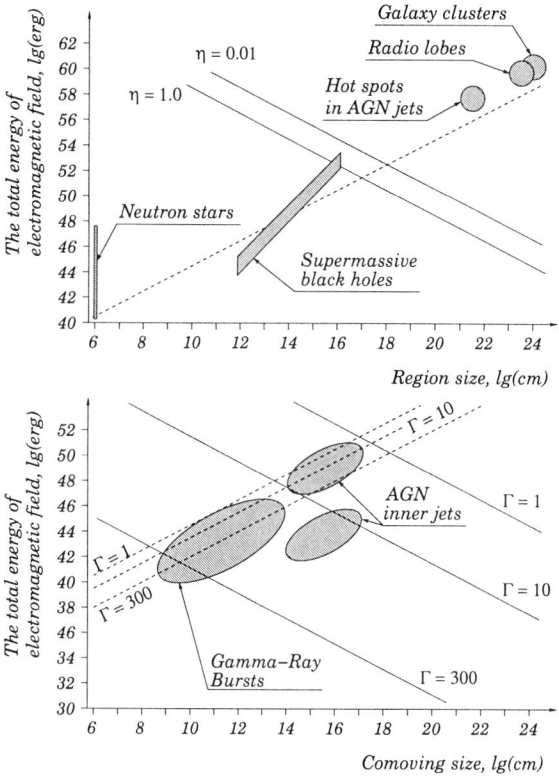

Fig. 1.6 The minimum energy requirement to the potential accelerators of 10^{20} eV protons as a function of the characteristic linear size of the accelerator. The dashed lines corresponds to the condition of particle confinement in large scale magnetic field or by the difference in electric field potential (generalised Hillas criterion), the solid lines correspond to the limits set by radiative (synchrotron or curvature) energy losses: **a** (top panel) accelerators at rest; for two values of the acceleration efficiency η given by $\dot{\varepsilon}_{\rm acc} = \eta e B c$; **b** (bottom panel) accelerators moving with relativistic speed; for $\eta = 1$, and 3 different values of the bulk motion Lorentz factor Γ. The characteristic ranges of different source populations (as potential cosmic ray accelerators) on the (W, l) plane are also shown. For the inner AGN jets, the upper zone corresponds to hadronic models of γ-ray emission, the lower zone - to leptonic models (from Aharonian et al., 2002b).

major problem. The electromagnetic cascades initiated by secondary electrons and photons lead to fluxes of ≤ 100 GeV γ-rays that exceed the observed flux of diffuse extragalactic γ-ray background shown in Fig. 1.3. This inconsistency can be avoided if one associates these exotic parents of cosmic rays to the halo of our Galaxy. Independent of the type of GUT particles, their decays lead to an excess of pions (and thus photons) over nucleons at production, and consequently to a high photon-to-proton ratio in cosmic rays at energies above 10^{19} eV. Thus the photon-to-proton ratio can be used as a diagnostic tool for the "top-down" model of highest energy cosmic rays. The detection of an unusually large content of γ-rays at $E \sim 10^{20}$ eV may become the first astrophysically meaningful result in the EHE gamma-ray domain.

Physics and Astrophysics of Relativistic Flows

Relativistic flows in astrophysics in the forms of winds and jets are common in many astrophysical settings. Most nonthermal phenomena observed from pulsars, microquasars, AGN and Gamma Ray Bursts are linked in one way or another to relativistically moving plasmas. The relativistic outflows are tightly coupled with compact relativistic objects - neutron stars and black holes. The theory of relativistic collimated outflows is very complex and not yet properly understood. It deals with magnetohydrodynamics, electrodynamics, strong shock waves and related with them particle acceleration. Each of these aspects challenges many uncertainties and problems. For example, many fundamental questions do not have yet definite answers concerning the origin (electromagnetic or gas dynamical ?) and content (electron-positron or electron-proton ?) of jets in AGN, as well as the processes which determine the jet power and support its propagation over distances up to several hundred kpc. There is an unsolved fundamental problem also in the theory of pulsar winds. It is believed that the spin-down power of a pulsar is carried away by a MHD wind in which the energy originally (closer to the magnetosphere) is dominated by Poynting flux. On the other hand, observations of the Crab Nebula show that when the wind approaches the inner edge of the visible synchrotron nebula, which is believed to be the site of wind termination, most of the energy must be in the form of kinetic energy of ultrarelativistic flow with Lorentz factor $\Gamma \sim 10^6$. Apparently somewhere between the pulsar and the termination shock the Poynting flux is converted to kinetic energy. How and where the acceleration of the wind takes place, remains a theoretical challenge despite recent intensive efforts in this direction.

The distinct feature of relativistic outflows is the effective acceleration of particles at different stages of its development - close to the central engine, during the propagation on large scales, and at the termination shock fronts. Nonthermal radio emission observed from different regions associated with relativistic jets and winds, e.g. from inner jets of blazars, large scale structures (knots, hot spots) of jets of powerful radiogalaxies and quasars, from compact expanding plasmons in microquasars, from pulsar-driven nebulae (plerions), *etc.*, carry information about relativistic electrons accelerated to relatively modest (GeV) energies. Remarkably, in many cases the synchrotron emission extends to the optical and X-ray bands. This implies that the electrons are accelerated to TeV energies. The inverse Compton scattering of the same electrons as well as interactions of the hadronic component of accelerated particles with the the surrounding radiation and magnetic fields, result in very high energy γ-rays. Both the acceleration and radiation processes with participation of ultrarelativistic electrons and protons may proceed with very high efficiency. Therefore it is believed that high energy γ-ray emission should provides us with important, in some cases crucial, information about nonthermal processes in jets and winds.

Small and Large scale AGN jets. Active Galactic Nuclei represent a large population of compact extragalactic objects characterised with extremely luminous electromagnetic radiation produced in very compact volumes. Although this source population consists of several classes of galaxies with substantially different characteristics, the prevailing concept of structure of AGN assumes that the differences are basically due to the strongly anisotropic radiation patterns. Consequently, the current classification schemes are dominated by random pointing directions rather than by intrinsic physical properties (Urry and Padovani, 1995). The presently most popular picture of physical structure of AGN is illustrated in Fig. 1.7.

AGN with relativistic jets close to the line of sight (so-called blazars) are very effective TeV γ-ray emitters. The dramatically enhanced fluxes of the Doppler-boosted radiation ($\propto D_j^4$) coupled with the fortuitous orientation of the jets towards the observer, make these objects ideal laboratories to study the underlying physics of AGN jets through multi-wavelength observations of temporal and spectral characteristics of radiation from radio to very high energy γ-rays. First of all this concerns the BL Lac objects, a sub-population of AGN of which several nearby representatives are already established as TeV γ-ray emitters. The TeV radiation not only tells us

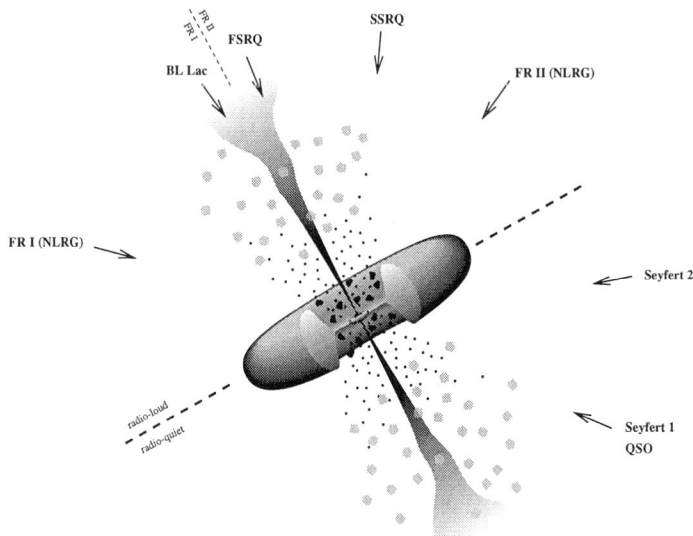

Fig. 1.7 Schematic illustration of the current paradigm of radio-laud AGN. At the center of the galaxy there is a supermassive black hole ($\sim 10^6$ to $\sim 10^{10}$ M$_\odot$) the gravitational potential energy of which is the ultimate source of power of the system released in different forms – through the thermal emission of the accretion disk, as well as through nonthermal processes in the relativistic jets that emanate perpendicular to the plane of the accretion disc. Particle acceleration takes place throughout the entire jet extending up to 10^{24} cm, i.e. well beyond the host galaxy. These particles interact with the ambient photon and magnetic fields, and thus result in nonthermal (synchrotron and inverse Compton) emission components observed on different (sub-pc, kpc, and multi-hundred kpc) scales. Broad emission lines are produced in clouds orbiting above the accretion disc. They are located typically within the zone between 0.01 to 0.1 pc. The accretion disk and the broad-line region is surrounded by a thick dusty torus. Narrow emission lines are produced in clouds located much farther from the central engine, typically between 0.3 and 30 pc. (from Urry and Padovani, 1995).

that particles in these objects are accelerated to very high energies, but also provides the strongest evidence in favour of the commonly accepted paradigm that the nonthermal radiation is produced in relativistic outflows (jets) with Doppler factors $D_j \geq 10$.

Presently, the leptonic (basically, inverse Compton) models of TeV emission represent the preferred concept for TeV blazars. These models have two attractive features: (i) capability of the relatively well developed model of shock waves to accelerate electrons to multi-TeV energies, (ii) effective production of tightly correlated X-ray and TeV emission components via synchrotron and inverse Compton channels. However, the very fact of strong X/TeV correlations does not yet exclude the hadronic models.

Effective acceleration of electrons to very high energies takes place also in large scale structures of AGN jets. Although there is no alternative to the nonthermal origin of large scale (up to several 100 kpc) jets of radiogalaxies, it remains a theoretical challenge to explain the variety of morphological and spectroscopic peculiarities observed from these objects. This concerns, first of all, the X-ray data. The recent exciting discoveries by the Chandra X-ray Observatory added much to our knowledge of X-ray structures of large scale jets in quasars and radiogalaxies. However, these results did not solve the old problems, and, in fact, brought new puzzles. The standard models that relate X-ray emission of distinct jet features to the synchrotron radiation or inverse Compton scattering of directly accelerated electrons face certain problems. The synchrotron mechanism is "over-efficient" in the sense that the TeV electrons, due to severe radiative losses, have very short propagation lengths, and thus hardly can form *diffuse* X-ray structures on kpc scales. The inverse Compton models require low energy (~ 1 GeV) electrons, and, therefore, are free of this problem. On the other hand, in many cases this mechanism appears not sufficiently efficient to provide the observed X-ray fluxes.

The synchrotron radiation of protons could be an alternative interpretation of X-ray emission. For a certain combination of parameters characterising the acceleration, propagation and radiation of very high energy protons, this model can provide effective cooling of protons via synchrotron radiation on quite comfortable timescales of about $10^7 - 10^8$ yr. This allows effective propagation of protons in the jet over kpc scales, and thus production of extended X-ray structures. Yet, the model allows high radiation efficiencies, and demands quite reasonable proton acceleration rates to explain the observed X-ray fluxes from typical representatives of this source population. Although these rates are comparable with the electron acceleration rates required in the electron-synchrotron models, the proton-synchrotron model implies much higher energy densities in the form of nonthermal particles and magnetic fields. The success of the proton-synchrotron model largely relies on 3 principal assumptions: (*i*) acceleration of protons to energies of at least $E_{\rm p} = 10^{18}$ eV; (*ii*) a strong ambient magnetic field, $B \geq 1$ mG; (*iii*) slow propagation of protons in the knots. Any observational evidence in favour of the proton-synchrotron origin of the large-scale X-structures would imply acceleration of protons to extremely high energies. This may have a direct link to another puzzle of the high energy astrophysics - the origin of extremely high energy cosmic rays (EHECRs) observed up to 10^{20} eV. An interesting consequence of this model is that it predicts significant γ-ray emission from the surrounding cluster environments initiated by interactions of runaway protons with the 2.7 K CMBR. The spectral and angular characteristics of this radiation component strongly depend on the ambient magnetic field, thus they carry important information about both the total power of acceleration of the highest energy particles in the jet, and the strength and structure of intra-cluster magnetic fields.

An interesting alternative to the conventional interpretations of X-ray features of AGN jets is a scenario in which the jet is powered by external γ-rays through their interactions with local low-frequency radiation fields and by subsequent development of electromagnetic cascades penetrating through the regular and random fields in the MHD jet. Generally, the nonthermal γ-ray phenomena are associated with *accelerated* particles. In this scenario, we have exactly opposite picture when the external γ-radiation is the primary substance. Consequently, there is no need for particle acceleration immediately in the jet. At the same time, this hypothesis contains certain components of standard MHD jets. Because of its prime motivation to explain the diffuse X-ray structures of large scale jets by electron synchrotron radiation, this model can be considered "leptonic". On the other hand, this scenario might be called "hadronic", because the primary γ-rays of energy $10^{15} - 10^{19}$ eV can be produced only in hadronic interactions, most likely in the vicinity of the central engine. The high energy γ-ray beam provides a direct link between the large scale jet and the central engine, thus it can be considered as an alternative to the Poynting flux assumed in the standard AGN models for extraction of energy from rotating black holes. While the transformation of Poynting flux into kinetic energy in the outflow, and eventually (through termination shocks) into relativistic particles remains an unsolved theoretical problem, the transformation of γ-rays to relativistic electrons can be realized effectively through the photon-photon pair production. The ultrahigh energy γ-rays may have broader implications. In particular, the γ-ray beams can power the surrounding intergalactic medium resulting in diffuse synchrotron X- and γ-ray emission from clusters of galaxies.

Pulsar winds and nebulae. High energy γ-rays emitted by rotation powered pulsars can be produced in three physically distinct regions: the pulsar magnetosphere, the unshocked relativistic wind, and the synchrotron nebula (Fig. 1.8). High energy γ-ray observations have a great potential to test current theoretical concepts concerning the production of an ultra-relativistic pulsar wind in the vicinity of the star and its interactions with the ambient medium.

The energy spectra and the structure of light curves in the region from several GeV to 30 GeV carry key information about the location of the γ-ray production region in the magnetosphere.

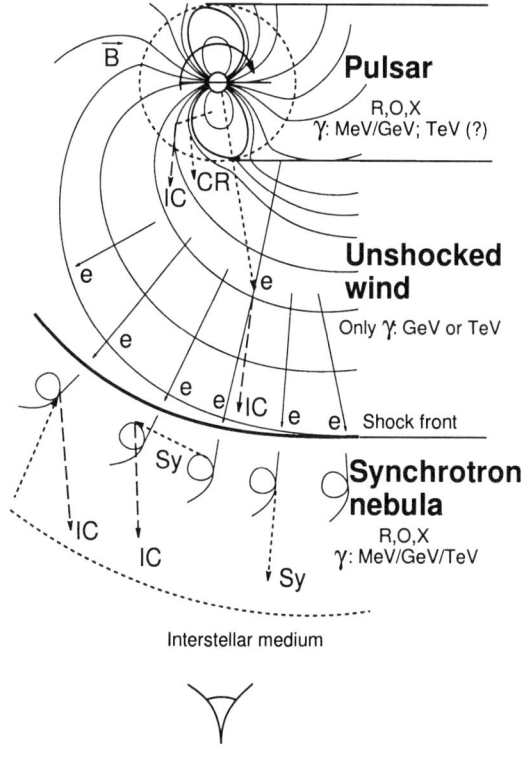

Fig. 1.8 Three regions of nonthermal radiation associated with a rotation powered pulsar: *pulsar* – magnetospheric pulsed γ-ray emission produced within the light cylinder due to the curvature, synchrotron, and inverse Compton processes; *unshocked wind* – gamma-radiation of the cold wind at GeV and TeV energies through the relativistic bulk-motion Comptonization; *synchrotron nebula* – broad-band, synchrotron and IC emission of the nonthermal nebulae (from Aharonian and Bogovalov, 2003).

The energy interval from 10 GeV to 1 TeV is the most informative region to search for γ-ray emission from the unshocked pulsar wind. Detection and identification of a specific ("asymmetric-line") type emission would provide direct information about the Lorentz factor and the site of acceleration of the wind i.e. the sites where the transformation from a Poynting flux dominated state to a kinetic energy dominated state occur. Finally, combined with X-ray observations, the spectral properties and morphology of 10 GeV to TeV γ-rays provide unambiguous information about the distributions of electrons and the magnetic fields in pulsar nebulae produced by wind termination shocks.

Microquasars. X-ray binaries are traditionally treated as *thermal* sources that transform the gravitational energy of accretion onto a compact object (a neutron star or a black hole) into X-ray emission radiated by the hot accreting plasma. However, since the discovery of galactic sources with relativistic jets – called *microquasars* – the basic concepts on X-ray binaries have been significantly revised. It is established that the non-thermal power of synchrotron radio jets (in the form of accelerated electrons and kinetic energy of the relativistic outflow) during strong radio flares can be comparable to or even exceed the thermal radiation luminosity of these accretion-driven objects. The discovery of microquasars opened new possibilities to study the phenomenon of relativistic jets common elsewhere in AGN. Because of their proximity, microquasars offer an opportunity for monitoring jets on much smaller spatial and temporal timescales.

Synchrotron and infrared emission observed from microquasars implies the existence of electrons up to energies ~ 10 GeV. If the electron acceleration proceeds at a sufficiently high rate, the synchrotron spectrum can extend to hard X-rays. In addition, the high density photon fields produced by the jet itself, as well as coming from the accretion disk and the companion normal star, create favourable conditions for effective production of inverse Compton γ-radiation. Apart from this episodic component of radiation associated with strong radio flares, one may expect persistent X- and high energy γ-ray emissions components from extended regions caused by synchrotron and inverse Compton radiation of ultrarelativistic electrons accelerated at the interface between the relativistic jet and the interstellar medium. Termination of the jets in the interstellar medium may result also in effective acceleration of protons. In this regard, microquasars are potential sites for the production of galactic cosmic rays.

Observational Gamma-Ray Cosmology

The fact that the spectra of extragalactic sources extend beyond 100 GeV opens a unique path for realization of exciting cosmological aspects of very high energy (VHE) gamma-ray astronomy. The promise here is connected with the energy dependent absorption of γ-rays interacting with diffuse extragalactic photon fields. The photon-photon absorption features, expected in the spectra of high energy γ-rays arriving from distant extragalactic objects depends on the spectrum and absolute flux of the diffuse extragalactic background at infrared and optical wavelengths. The detection and identification of these features should provide important information about the epochs of galaxy formation and their evolution in the past. Obviously this method of extracting information about the diffuse extragalactic background requires (i) TeV γ-ray beams emitted from extragalactic objects located at different distances between 100 and 1000 Mpc; (2) good gamma-ray spectrometry, and (3) good understanding of the intrinsic spectra of γ-rays (i.e. before their deformation in the intergalactic medium). Blazars do provide us with intense TeV beams, and the IACT arrays do allow an adequate γ-ray spectrometry based on an energy resolution as good as 10 per cent and large γ-ray photon statistics. A serious obstacle in practical realization of this interesting method is our poor knowledge about the primary γ-ray spectra produced in the source. The recent remarkable progress in well coordinated observations of several TeV blazars in different energy bands gives a certain optimism that eventually the gamma-ray astronomers will be able to identify the principal radiation mechanisms, fix/constrain the relevant model parameter space, and reconstruct robustly the intrinsic γ-ray spectra based on multiwavelength studies of the spectral and temporal characteristics of blazars. This should allow reliable estimates of the intergalactic absorption effect, and consequently derivation of the flux and spectrum of diffuse extragalactic background between 1 and 100 μm.

Strictly speaking the intergalactic absorption features contain information about the product of the diffuse extragalactic background radiation density u_r and the Hubble constant H_0. In principle, it would be possible to decouple u_r and H_0 by studying the spectral and angular characteristics of VHE γ radiation from hypothetical electron-positron Pair Halos surrounding powerful nonthermal extragalactic objects. These giant but light (electron-positron) structures which are unavoidably formed around any extragalactic VHE source due to development of pair cascades initiated by interactions of primary multi-TeV photons with the extragalactic

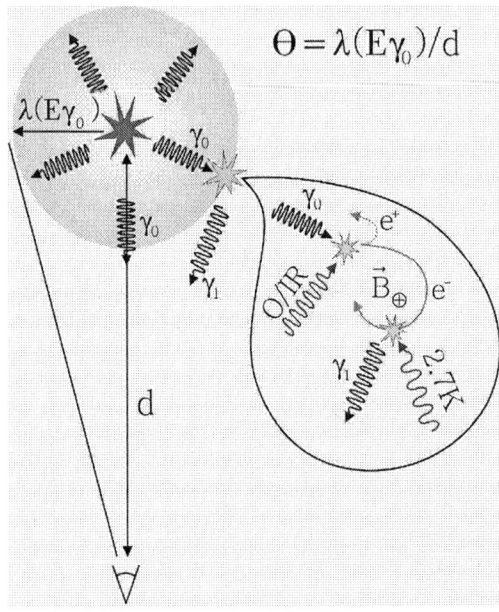

Fig. 1.9 Formation and radiation of a Pair Halo.

background photon fields, may serve as unique cosmological candles. The formation and radiation of a Pair Halo is illustrated in Fig. 1.9.

The radiation of a Pair Halo can be recognised by its distinct variation in spectrum and intensity with angular distance from the halo centre. This variation depends weakly on the details of the central source, for example on the orientation and beaming/opening angle of a possible emitting jet, but depends on u_r and H_0. Thus detection of a halo would give us two observables – angular and spectral distributions of γ-radiation – that might make it possible to disentangle u_r and H_0. Since the angular size of a halo around a source with known redshift z_0 is determined by the density of the background radiation in the vicinity of the source (i.e. at the epoch z_0), observing pair halo radiation from sources at different redshifts should provide an important probe of cosmological evolution of the background radiation.

The extended character of γ-ray emission emitted by isotropic Pair Halos (up to several degrees) makes their mapping and spectroscopy a difficult measurement that must wait future sensitive IACT arrays. But in any case, the existence of strong extragalactic TeV sources sustains the hope that the

Pair Halos will be eventually discovered. For formation of isotropic Pair Halos the intergalactic field should be sufficiently strong (larger than 10^{-12} G) In fact, both the current observations and cosmological concepts do not exclude the possibility that in some regions with typical linear scales of about 100 Mpc, the intergalactic magnetic field could be arbitrarily small. If so, instead of detecting extended and persistent isotropic Pair Halos, we should expect cascade radiation penetrating almost rectilinearly from the source to the observer. As far as cosmological distances are concerned, even very small deflections of the cascade electrons in the intergalactic magnetic fields should lead to significant delays of arriving cascade radiation. If such delays can be distinguished from the intrinsic time structure of radiation of a γ-ray source, it would be possible to probe the primordial magnetic fields of the Universe at the level down to 10^{-18} G.

Independent of details concerning the structure and strength of intergalactic magnetic fields, as well as the flux and spectrum of the extragalactic radiation fields, the ensemble of all VHE source in the Universe produce isotropic cascade γ-radiation which serves as a calorimetric measure for the integrated power of the Universe in the form of any phenomenon accompanied by radiation of VHE γ-rays. These γ-rays may have quite different natures. They are copiously produced by AGN jets, radiogalaxies and galaxy clusters – the most powerful extragalactic objects in the Universe. They are also contributed by less powerful but more populous sources like Pulsars and SNRs – the most active sites of γ-ray production in ordinary galaxies. Gamma-rays may appear during grandiose processes like the formation of large-scale structures in the Universe and at brief solitary events like Gamma Ray Bursts. We may also expect quite intense γ-ray production of "non-acceleration" origin related to decays of hypothetical cosmological relics like the topological defects as well as to annihilation of non-baryonic Dark Matter indexDark Matter. Independent of their origin, all high energy γ-rays have a common fate – due to interactions with the extragalactic radiation fields they inevitably terminate on Hubble (space and time) scales, and thus make the entire Universe an active scene of continuous creation and development of electromagnetic cascades in the intergalactic medium. The superposition of these cascades should not exceed the observed flux of the diffuse extragalactic γ-ray background. This provides robust constraints on the overall VHE γ-ray luminosity of the Universe.

Finally, the search for hypothetical γ-ray emission from Dark Matter Halos in our Galaxy and other nearby galaxies by GLAST and by forthcoming powerful ground-based instruments should provide a very deep and meaningful probe for the existence of non-baryonic Dark Matter in the Universe. The detection of positive signals would reveal the nature of Dark Matter with extremely important implications for both Cosmology and Particle Physics.

Chapter 2

Status of the Field

This Chapter presents a brief overview of the achievements of observational gamma-ray astronomy in different energy bands. The status of space-based gamma-ray astronomy in the low energy (LE) and high energy (HE) regimes are discussed in Sections 2.1 and 2.2. Comprehensive description of the results at MeV and GeV energies obtained basically with BATSE, OSSE, COMPTEL and EGRET detectors aboard NASA's Compton Gamma Ray Observatory can be found in the proceedings of the Fourth (Dermer et al., 1997) and Fifth (McConnell and Ryan, 2000) Compton Symposia, as well as in the proceedings of the Gamma-Ray Astrophysics-2001 symposium (Ritz et al., 2001). The methods of detection of cosmic MeV and GeV γ-rays are described in great details in the book by Fichtel and Trombka (1997). Therefore, Sections 2.1 and 2.2 will be rather short. Only the major observational results, with brief commentaries on their astrophysical implications, will be highlighted.

Several review articles on VHE gamma-ray astronomy have been written over the last decade – see e.g. papers by Cronin et al. (1993), Aharonian and Akerlof (1997), Ong (1998), Hoffman et al. (1999), Catanese and Weekes (1999), as well as the proceedings of the recent two symposia on "High Energy Gamma-Ray Astronomy" (Aharonian and Völk, 2001) and "The Universe Viewed in Gamma-rays" (Enomoto et al., 2003). It should be noticed, however, that currently the field is developing so fast that every year brings new discoveries and surprises, and, consequently, new ideas and models (as well as puzzles), thus review articles quickly become "old" on timescales shorter than the typical time needed for their publication in regular journals or conference proceedings. Most likely (hopefully !) the same will happen with this Chapter.

2.1 Low Energy Gamma Ray Sources

This energy regime is uniquely related to several areas of high energy astrophysics, in particular to the phenomena of Gamma Ray Bursts, nucleosynthesis of heavy elements in the Universe, origin of low energy (subrelativistic) cosmic rays, solar flares, *etc.* Also, this energy regime provides key observations for understanding of high energy processes in stellar black-holes in our Galaxy, as well as the massive black-holes believed to exist in the centers of AGN. In particular, many popular ideas and models that assume formation of relativistic, both thermal and nonthermal, plasmas around these objects, can be tested via studies of characteristic electron-positron annihilation radiation.

At the same time, this energy region has been and unfortunately remains a challenge for design and construction of adequately sensitive γ-ray detectors that would rise the low-energy gamma-ray astronomy to the level of its immediate neighbours - X ray astronomy and high-energy gamma-ray astronomy. There are several reasons for slow development in instrumentation in this energy region: (i) the small photon interaction cross-section in the transition region from the Compton scattering to pair production; (ii) small energy deposits and large mean free paths of secondary products; (iii) large uncertainties in reconstruction of full kinematics of the first interaction, and correspondingly rather limited angular resolution; (iv) large local backgrounds, especially at MeV energies due to excitation of surrounding materials by cosmic rays. The combination of these factors – low γ-ray detection efficiency, modest angular resolution and high background – severely limits sensitivities of γ-ray detectors operating in this energy region. The minimum detectable energy fluxes at hard X-ray and low-energy γ-rays are, indeed, not very impressive, and even after significant improvements by next generation instruments, they unfortunately will remain relatively modest compared to sensitivities expected in the foreseeable future in the HE and VHE domains (see Fig. 1.2).

Even so, low-energy γ-rays contain invaluable astrophysical information that cannot be obtained by other means. Therefore, any further improvement of detector performance would lead to exciting results in several areas of astrophysics and cosmology.

2.1.1 The COMPTEL source catalog

The imaging Compton telescope COMPTEL provided the first all sky survey in the MeV γ-ray domain (Schönfelder et al., 2000). This instrument was designed to operate in the energy range from 0.75 to 30 MeV. Within its large FoV of about 1 steradian, the source location accuracy was of about 1°. Typically COMPTEL could resolve different sources, provided that they were about 3 to 5 degrees away from each other. With 5 to 10 percent energy resolution, COMPTEL was able to detect and identified several γ-ray lines of extra-solar origin - 1.1809 MeV (^{26}Al), 1.157 (^{44}Ti), 0.847 and 1.238 MeV (^{56}Co), as well as 2.223 MeV (deuterium or neutron-capture line). The sensitivity of COMPTEL was significantly limited due to the instrumental background. Also, for identification of sources in the galactic disk, detailed modelling of the diffuse galactic emission was essential for this experiment. On average, for a 2-week observation period, the source detection threshold was an order of magnitude below the Crab flux, i.e. at the level of $\approx 2 - 3 \times 10^{-10}$ erg/cm^{2}s (see Fig. 1.2). Unfortunately, the INTEGRAL mission cannot offer better sensitivity above 1 MeV for continuum emission. But INTEGRAL will indeed greatly improve the detection sensitivity for line emission.

The first COMPTEL source catalog includes 32 persistent sources and 31 GRBs reported at $\geq 3\sigma$ statistical significance level (Schönfelder et al., 2000). The persistent sources of continuum emission belong to 3 types of source populations. In addition γ-ray line emission has been detected from 7 objects. And finally, 9 sources remain unidentified.

- **Spin-Down pulsars** - Crab, Vela, and PSR1509-58.

While the Crab and Vela pulsars are established as prominent γ-ray emitters detected at MeV/GeV energies, PSR 1509-58 remains up to now as a "MeV" γ-ray pulsar (see Fig 2.1). COMPTEL has detected both pulsed and continuum emission components from the direction of the Crab. The continuum component is presumably associated with the nebula, i.e. has a synchrotron origin. This interpretation agrees with the multiwavelength data obtained from the Crab Nebula (see Fig. 1.2).

- **Stellar Black-Hole Candidates** - Cyg X-1, Nova Persei 1992 (GRO J0422+32).

In addition to these galactic black-hole candidates with spectra extending beyond 1 MeV, OSSE – another hard X-ray/low energy γ-ray instrument aboard Compton GRO - has detected hard tails of radiation from several other similar objects, presumably microquasars, extending to 1 MeV (Grove et al., 1997). The observations of Cyg X-1 by OSSE and COMPTEL show significant variations in the MeV region between the so-called hard and soft spectral states, that are discovered and well studied in X-rays.

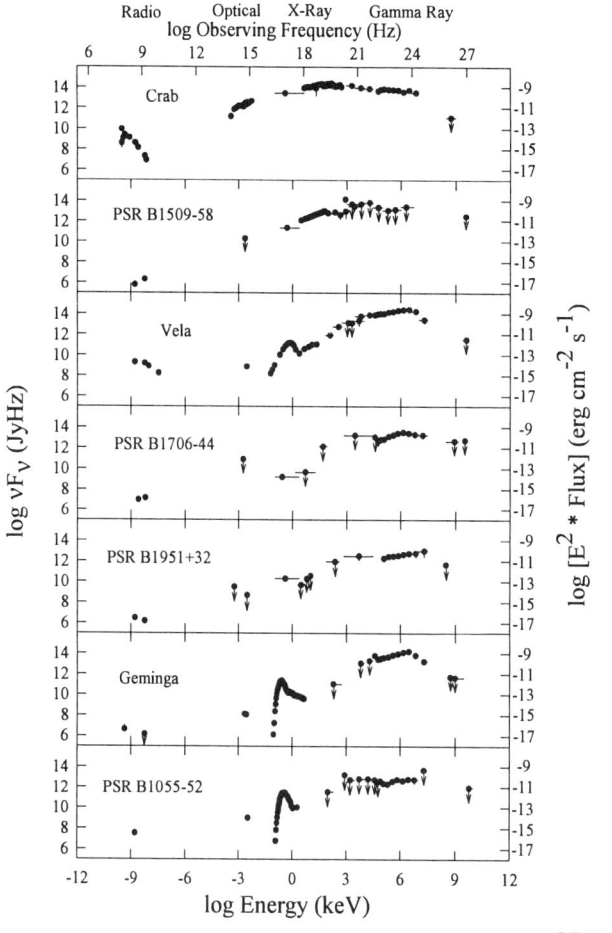

Fig. 2.1 Broad-band spectral energy distributions of pulsed emission of γ-ray pulsars (from Thompson, 1999). The MeV data are from COMPTEL, GeV data - from EGRET, and the TeV upper limits are from observations with the Whipple (pulsars in the Northern hemisphere) and CANGAROO/Durham (pulsars in the Southern hemisphere) telescopes.

The MeV spectra in these two states are shown in Fig. 2.2 together with model curves calculated within the so-called hybrid thermal/nonthermal Comptonization model (McConnell et al., 2002). This model assumes that γ-rays are produced by a non-thermal population of electrons accelerated in the accretion plasma around the black hole. This radiation can originate also in the synchrotron jet recently discovered in several representatives of this source population, including Cyg X-1. MeV radiation from synchrotron radio jets can be result of inverse Compton (Georganopoulos et al., 2002) or synchrotron radiation (Aharonian and Atoyan, 1998; Markoff et al., 2001). In either case, the radiation should be of nonthermal origin associated with relativistic electrons accelerated in the jet.

Until now high energy γ-rays above 100 MeV have not been convincingly detected from a black-hole candidate with hard X-ray/soft gamma-ray spectra. Nevertheless, it is possible that some of the unidentified γ-ray sources are from the same source population (see Sec.2.2).

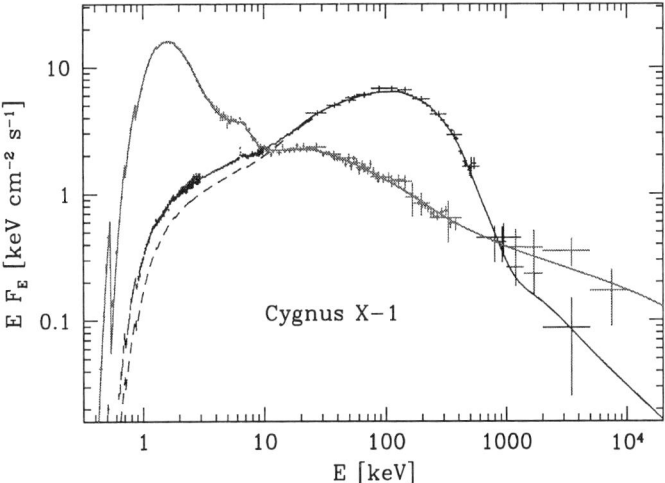

Fig. 2.2 X- and gamma-ray spectra of the black-hole candidate Cyg X-1 in soft ("high") and hard ("low") spectral states as measured by COMPTEL (MeV), OSSE (sub-MeV) and BeppoSAX (X-ray) instruments. The spectral fits are obtained within the hybrid thermal/nonthermal Comtonization model. (From McConnel et al., 2002).

- **AGN** - CTA 102, 3C 454.3, PKS 0528+134, GRO J0516-609, PKS 0208-512, 3C 273, PKS 1222+216, 3C 279, Centaurus A, PKS 1622-297.

Except for the radiogalaxy Centaurus A, all these objects are blazars - highly variable AGN with relativistic jets close to the line of sight. Because the spectral energy distributions of high energy branches of these objects peak at MeV energies, they are called *MeV blazars*, The famous quasar 3C 273 is a prominent representative of this class of objects (Lichti *et al.*, 1995). The broad-band spectral energy distribution of 3C 273 is shown in Fig 2.3.

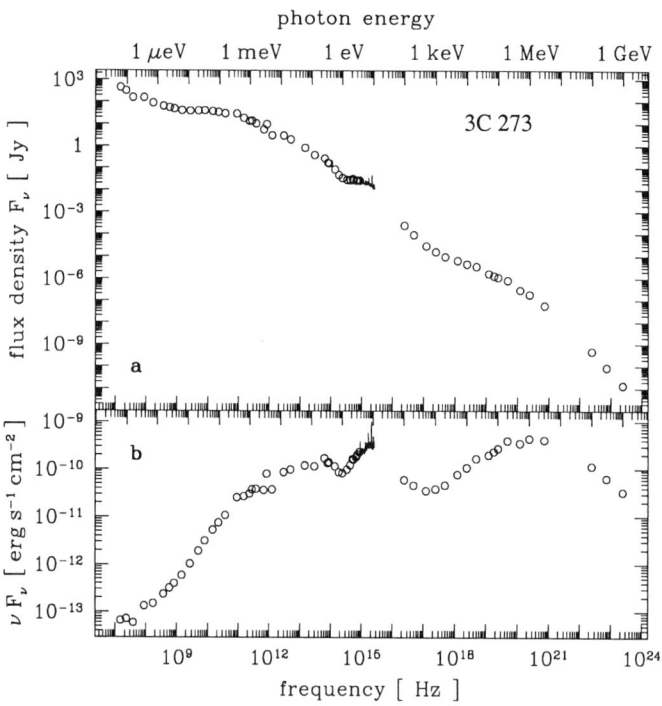

Fig. 2.3 Flux density and spectral energy distribution of the quasar 3C 273 (from Courvoisier, 1998).

MeV blazars have quite steep spectra beyond MeV energies (with photon index greater than 2), in contrast to the so-called GeV blazars with flat γ-ray spectra extending to GeV energies. But, most probably, there is no fundamental difference between MeV and GeV blazars. It is believed that both classes represent the same AGN population - flat spectrum radio quasars (FSRQs). The difference between MeV and GeV blazars can be explained within the so-called external Compton model, assuming that

GeV flat spectra originate in the broad emission line (BEL) regions, where their production is dominated by Comptonization of optical-UV emission lines, whereas the spectra of MeV blazars are formed at distances where the target photons are supplied by hot dust. Moreover, it is possible that the MeV and GeV blazar phenomena can appear interchangeably within the same object like in PKS 0208-512 (Sikora et al., 2002).

- **Unidentified Sources**

The first official COMPTEL catalog contains nine unidentified sources. Five of them are located at low galactic latitudes ($|b| \leq 10°$), and four above the galactic plane ($|b| \geq 10°$). Three of the five unidentified high-latitude sources are not point-like. Either they have diffuse origin or are result of superposition of several faint objects. Two of the four low-latitude sources possibly coincide with the EGRET unidentified sources 2EG 2227+61 and 2EG0241+6119 (discovered initially by COS B). The nature of all unidentified COMPTEL sources remains highly unknown. One cannot exclude the possibility that some of them are caused by statistical fluctuations.

- **Gamma Ray Line Sources**

COMPTEL has demonstrated that a variety of astrophysical objects may produce γ-ray line emission. A noticeable success in this regard was generation of a maximum entropy COMPTEL map at 1.809 MeV. It is widely believed that this γ-ray line from ^{26}Al is mainly produced in Wolf-Rayet stars, therefore it serves as a unique tracer of star formation in our Galaxy over the last several millions of years. The all-sky map in the 1.809 MeV line revealed bright extended regions, in particular in the inner Galaxy, as well as in the Vela, Cygnus and Aquila regions. An excess emission was reported also from the direction of Carina, but in this case the source seems to be point-like, i.e. with an angular extension that does not significantly exceed 1 degree.

Gamma-ray line emission was detected from three other point-like sources, although in different lines. The line 1.157 MeV from ^{44}Ti has been detected from Cas A and, perhaps, also from RX J0852-4621 (a supernova discovered recently in the Vela region). A tentative detection of two coupled γ-ray lines at 0.847 and 1.238 MeV that are associated with ^{56}Co was reported from SN 1991T. If confirmed, these results would have great impact on the theory of nucleosynthesis providing invaluable probes into the inner layers of exploding stars.

Finally, an excess of 2.2 MeV deuterium line emission has been found with $\simeq 3.7\sigma$ statistical significance from a region that does not show any remarkable activity at other wavelengths. If true, this would imply a very effective source of neutron production, most likely through spallation of nuclei in a very hot, sub-relativistic plasma or by energetic nonthermal particles interacting with the ambient cold gas. In either case, the deuterium line production region, which generally could be separated from the neutron production region, should be very dense, $n \geq 10^{16}$ cm^{-3}, otherwise the neutrons would decay before being captured by thermal protons. This could happen, for example, in a binary system – the neutrons can be produced in the two-temperature accretion plasma around the compact object, a black hole or a neutron star, "evaporate" from the accretion disk, and be captured by a dense atmosphere of the normal star. This scenario proposed by Aharonian and Sunyaev (1984), later has been discussed and developed by many authors (e.g. Guessoum and Kazanas, 1990; Guessoum and Jean, 2002). Due to relatively narrow width of the detected 2.2 MeV line, the latter cannot be explained by neutron capture in the hot accretion plasma. It should be noted, however, that production of free neutrons, and therefore also the 2.2 MeV line is a relatively inefficient mechanism, which makes this (in fact, any) interpretation of the reported 2.2 MeV flux quite problematic, especially given the lack of a prominent object in the error box of COMPTEL.

Unfortunately, the energy threshold of COMPTEL of about 0.75 MeV did not allow studies of the 0.511 MeV electron-positron annihilation line. This line is the longest-known and the most intense extra-solar γ-ray line. The first evidence of positron annihilation radiation from the direction of the Galactic Center was obtained in early 1970s, but it took several years until the signal was confirmed and unambiguously identified, by a high-resolution germanium balloon experiment (Leventhal *et al.*, 1978), with the positron annihilation line.

The history of annihilation line studies over the past 30 years contains many controversial issues like the the uncertainty in the size of the line production region, or claims about an existence of a variable component of radiation. If confirmed, the variable emission would appear to imply production and annihilation of positrons in a compact object(s). However, the observations of recent years, in particular by OSSE, did not succeed in finding any evidence for a point-like source on top of the diffuse 0.511 MeV emission. Also, no evidence has been found also for variability of emission above or below the diffuse flux (Harris, 1997). On the other hand, the OSSE observations brought new puzzles, like a high-latitude, asymmetrically extended feature of annihilation radiation that was claimed to be larger than the Galactic bulge feature (Purcell et al., 1997). However, the history of the extra-solar γ-ray line observations tells us to be cautious about this as well as with all the γ-ray line results mentioned above, given the inadequate angular resolution and the limited capability of the current low-energy γ-ray detectors to suppress and identify reliably the high instrumental background.

A breakthrough is expected in this regard from the new INTEGRAL mission. The significant improvement in the γ-ray line sensitivity, by a factor of five or so, over the Compton GRO detectors, should allow INTEGRAL not only to confirm the results discussed above, but, more importantly, will provide deeper insight into the nuclear processes that take place in different environments in our Galaxy and, hopefully, also beyond. On the other hand, INTEGRAL unfortunately cannot provide deeper γ-ray probes of the continuum component of radiation above 1 MeV. This unfortunate fact, as well as the urgent need for new instruments in this important energy region with more than factor of 10 improvement over the COMPTEL sensitivity, is recognised by the gamma-ray astronomical community. There is a hope that MEGA (Bloser et al., 2002) – likely the first representative of the new generation Compton telescopes – will dramatically change the status of low energy γ-ray astronomy.

2.2 High Energy Gamma Ray Sources

At energies above 30 MeV, detection of cosmic γ-rays becomes significantly easier. The detection principle is based on conversion of the primary photon to an electron-positron pair, and on subsequent measurements of the tracks of secondary electrons with tracking detectors and their energy with a total-absorption calorimeter. This technique, originally developed for particle accelerator experiments, has a great potential for cosmic γ-ray studies. It allows reconstruction of the arrival direction and energy of primary γ-rays on an event-by-event basis. The energy resolution is basically determined by fluctuations in the electromagnetic cascade that develops in the calorimeter, as well as, especially at higher energies, by the absorbing capability of the calorimeter. The best energy resolution is achieved at GeV energies because of relatively small fluctuations and, at the same time, due to high efficiency of the total confinement of the cascade products in the calorimeter. The energy resolution could be as good as a few per cent, although it would require quite massive calorimeter. Below several 100 MeV, the energy of the electron and positron can be determined also by analysing the multiple Coulomb scattering angles in converters. However, generally this is not considered as a prime priority in current designs of γ-ray telescopes. In fact, the multiple Coulomb scattering of electrons significantly limits the accuracy of determination of the arrival direction of primary photons. Therefore, as a basic element, in high energy γ-ray telescopes a multi-layer tracker with thin converters is used. The thickness and number of converters is determined from the trade-off between the overall efficiency of photon conversion and minimisation of the Coulomb scattering effect. Actually, there is another effect that limits the angular resolution. It is induced by the "invisible" recoil momentum at pair production that prevents full (unambiguous) reconstruction of kinematics of the process, and thus significantly limits the angular resolution of telescopes, especially at sub-GeV energies. The uncertainty in the angular resolution induced by this intrinsic effect is estimated $\approx (m_e c^2/E_\gamma)\ln(E_\gamma/m_e c^2)$, e.g. $\sim 1.5°$ at 100 MeV, and $\sim 0.2°$ at 1 GeV.

Angular resolution is one of the key parameters characterising performance of high energy detectors. In addition to accurate determination of the size and location of a γ-ray source, good angular resolution improves, to a certain extent, the minimum detectable flux (sensitivity) from point sources through reduction of γ-ray backgrounds - both of local and astronomical (diffuse galactic and extragalactic) origins. The background caused by charged cosmic rays can be removed with very high efficiency using an active anti-coincidence shield consisting of thin scintillation counters. The shield is "transparent" for γ-rays but provides an effective veto against charged particles. The flux sensitivity of a γ-ray telescope is determined by the residual background rate and the effective detection area – the physical area of the detector multiplied by the (energy-dependent) detection efficiency. In the space-based experiments the physical area of telescopes is limited, and cannot significantly exceed several m^2.

The first meaningful observational results of γ-ray astronomy appeared in the 1970s, basically due to two successful space missions called SAS-2 (Fichtel *et al.*, 1975) and COS B (e.g. Bignami and Hermsen, 1983). Four point sources were detected by SAS-2, including the Crab and Vela pulsars, as well as the X-ray binary source Cyg X-3, although the identification of the excess emission with this object remains rather controversial. The origin of the fourth source which later was called Geminga (Bignami *et al.*, 1983), remained a mystery over almost 20 years until it was identified with a X-ray pulsar.

The COS-B mission increased the number of γ-ray sources to 25, although most of these sources were not identified. An undisputed success of COS-B was a detailed study of the spatial distribution of the galactic diffuse γ-ray emission and discovery of several "hot spots" in some active star formation regions and giant molecular clouds like the Orion complex. Also, COS-B discovered the first extragalactic γ-ray source which was promptly identified with the quasar 3C 273 – the first representative of the most populous high-energy γ-ray source population associated with blazars, as revealed by EGRET a decade later.

EGRET, as a part of the Compton GRO mission, provided a deep study of the high-energy γ-ray sky during nine very successful and exciting years, from 1991 to 2000. It is often said that EGRET brought gamma-ray astronomy to maturity. This instrument with an effective energy threshold around 50 MeV and FoV of about 0.5 sr, had best performance at energies around 1 GeV: (i) effective detection area ~ 0.1 m^2, (ii) angular resolution $\sim 1.5°$, (iii) energy resolution $\sim 10\%$, (iv) minimum detectable photon flux $J(\geq 1 \text{ GeV}) \simeq 10^{-8}$ ph/cm^{2}s or energy flux 2×10^{-11} erg/cm^{2}s from a point source during a 1 year all-sky survey.

The current list of high-energy γ-ray sources released by the EGRET team in the form of the 3rd EGRET Catalog (Hartman *et al.*, 1999) consists of 271 sources detected above 100 MeV. The catalog includes 66 high-confidence and 27 lower confidence identifications with AGN, five pulsars, a nearby dwarf galaxy - the Large Magellanic Cloud, a nearby radiogalaxy - Centaurus A, as well as a single Solar flare detected in 1991. Finally, the catalog contains 170 sources not yet firmly identified with known objects. After the release of the 3rd EGRET Catalog, a number of papers have been published suggesting possible identifications of many γ-ray sources with individual objects representing several source populations – pulsars, supernova remnants, microquasars, molecular clouds, plerions, clusters of galaxies, *etc.* The distribution of high energy γ-ray sources from the the 3rd EGRET Catalog in galactic coordinates is shown in Fig. 1.1.

Along with discrete sources, γ-rays of diffuse origin, i.e. photons produced by interactions of cosmic rays with ambient gas and photon fields, are expected. Actually, diffuse emission from the galactic disk dominates over the contribution of resolved sources. After removal of all, identified and unidentified, objects the diffuse emission appears with several interesting spatial and spectral features which reflect distributions of cosmic rays and the interstellar gas in the galactic disk (see Chapter 4). The fluxes of diffuse gamma-radiation from the inner Galaxy detected by EGRET between 20 MeV and 10 GeV are shown in Fig. 4.20, together with fluxes at lower energies measured by COMPTEL and by OSSE.

Diffuse high energy γ-ray emission is observed also at large galactic latitudes. The bulk of the detected flux is believed to be of extragalactic origin. Almost surely, a significant fraction of this component comes from superposition of faint unresolved γ-ray sources, first of all from blazars, and possibly also from more extended structures like galaxy clusters. There may also be significant contributions from electromagnetic cascades in the intergalactic medium triggered by interactions of very high energy γ-rays from discrete sources with extragalactic photon fields (see Chapter 11). In either case, the diffuse extragalactic radiation detected by EGRET up to 100 GeV shown in Fig. 1.3 contains unique cosmological information.

Below we briefly discuss three major high-energy source populations detected by EGRET.

2.2.1 GeV blazars

Almost all AGN detected by EGRET belong to the blazar population - objects characterised by nonthermal continuum emission with high radio and optical polarizations and short timescale variability observed at all wavelengths. In addition, EGRET has found evidence of γ-ray emission from Centaurus A, a nearby radiogalaxy also detected by COMPTEL at MeV energies.

The main contributors to the EGRET list of blazars (see e.g. Mukherjee *et al.*, 1997) reported with a high degree of confidence (at least 4σ detection for high galactic latitudes, and 5σ detection for $|b| \leq 10°$) are the so-called flat-spectrum radio quasars (FSRQs). Sixteen objects are identified with BL Lac objects which are characterised by stronger polarization and weaker optical lines than FSRQs. BL Lac objects are relatively closer and have lower luminosities than FSRQs.

A significant fraction of GeV blazars, in particular 3C 273 and 3C 279, exhibit apparent superluminal motion (Jorstad *et al.*, 2001) detected by VLBI radio observations. Many EGRET blazars show variability on timescales of months. For many EGRET blazars, the study of short timescale variability is limited by the γ-ray photon statistics. Nevertheless, short flares on timescales less than 10 h have been detected from very strong objects like PKS 1622-297 and 3C 279.

The spectra of EGRET blazars are well-described by a simple power-law over the energy region from 30 MeV to 10 GeV, with an average photon index ~ 2.2. There is no evidence for spectral cutoffs, at least at energies below 10 GeV. Also, there is no apparent correlation between the photon indices and the source redshift (distance), despite a very strong luminosity-redshift correlation, especially at low z. In Fig. 2.4 the spectra of two EGRET blazars representing the FSRQ (3C 279 at z=0.538) and BL Lac (1219+285 at z=0.102) AGN populations are shown. Remarkably, despite almost 3 order of magnitude difference in apparent γ-ray luminosities, the spectra of these object are quite similar.

The strongest GeV Blazars do not show TeV emission. Only three EGRET blazars have been detected at TeV energies - Mkn 421, Mkn 501, and PKS 2155-304, all three being low-redshift X-ray selected BL Lac objects. At the same time all these three objects are weak GeV γ-ray emitters. Actually this GeV-TeV anti-correlation agrees with expectation.

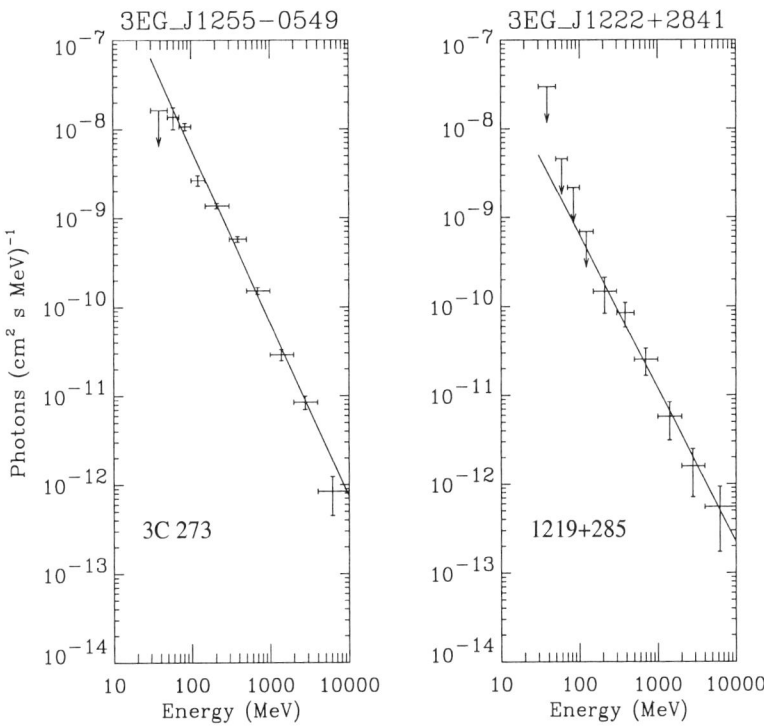

Fig. 2.4 Gamma-ray spectra of blazars 3C 279 and 1219+285 as representatives of FSRQ and BL Lac source populations (from R. Mukherjee, 2001).

The strong EGRET blazars are located at large distances, and therefore TeV γ-rays emitted by these objects suffer severe intergalactic absorption due to interactions with extragalactic diffuse infrared radiation (see Chapter 10). Moreover, there are other reasons for the TeV-GeV anti-correlation associated, for example, with essentially different conditions for particle acceleration, as well as for the production and absorption of γ-rays in the jets of radio-loud quasars and BL Lac objects. The significantly higher densities of infrared and optical photons in quasars not only provide effective γ-ray production in these objects through inverse Compton scattering, but also limit the maximum energy of accelerated electrons due to the same process. As a result, one may expect a strong shift of both synchrotron and inverse Compton peaks in the spectral energy distributions of powerful blazars towards lower frequencies. Such a tendency of "becoming redder" with increasing source bolometric luminosity is indeed observed (see Chapter 10 and Fig. 10.4), although the picture could be, of course, more complex and sophisticated compared to this simple interpretation. It is important to note in this regard that the two brightest TeV blazars, Mkn 421 and Mkn 501, have sub-luminal parsec-scale jets, in contrast to the apparently superluminal jets of majority of GeV blazars detected by EGRET (Edwards and Piner, 2002).

Unfortunately, the low TeV *source* statistics and low GeV *photon* statistics do not allow any detailed quantitative studies of the links between the TeV and GeV blazars. This issue could be properly addressed only after having more information about both source populations, especially at the intermediate energies around 100 GeV. Such information will be available in the foreseeable future with GLAST (the Gamma-ray Large Area Space Telescope) and the new generation of 100 GeV threshold Cherenkov telescope arrays.

2.2.2 GeV pulsars

Pulsars - single neutron stars powered by fast rotation - are effective high energy γ-ray emitters. Two prominent representatives of this source population, the Crab and Vela pulsars, were the first astronomical objects discovered in high energy γ-rays. Together with the Geminga pulsar, they are the brightest persistent γ-ray sources on the GeV sky. At least six EGRET sources (with another 3 possible candidates) are identified with pulsars. The multiwavelength spectral energy distributions of these objects are shown in Fig. 2.1. Only in the case of the Crab pulsar does the luminosity peak at sub-MeV energies. For the other EGRET pulsars the dominant power is released in the γ-ray band; in the case of PSR B1951+32 – beyond 10 GeV. Although the γ-ray spectra of all pulsars are very hard with photon index ≤ 2, at higher energies one expects significant steepening or a spectral cutoff. In some cases this can be directly seen in the highest energy bins of EGRET data, or is implied from upper limits obtained at TeV energies. A break around 10 GeV in the Vela spectrum is clearly seen in Fig. 2.5.

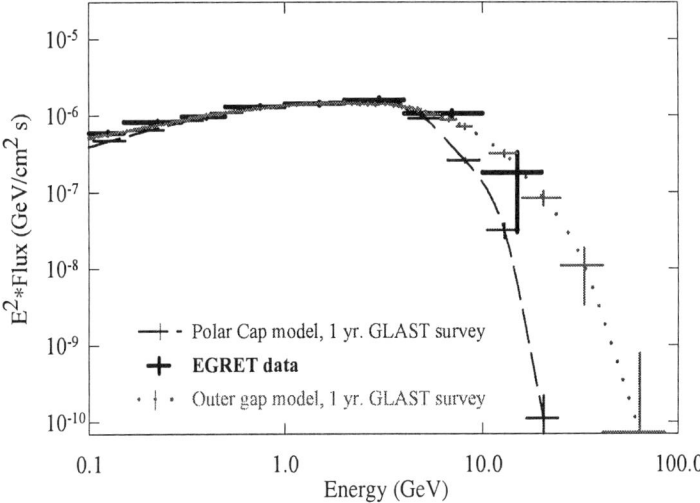

Fig. 2.5 High energy spectrum of the Vela pulsar. Heavy error bars - EGRET data. Dotted line - prediction of the outer gap model (Romani, 1996), dashed line - prediction of the polar cap model (Daugherty and Harding, 1996). The error bars shown at the model curves are those expected from the 1 year survey of GLAST. (From Thompson, 2001).

Pulsars are characterised by their so-called light curves - indicators of the time structure of emission of these astronomical clocks. The light curves of all EGRET pulsars show a double-peak structure. The Crab light curve above 100 MeV is rather similar to the light curves seen at other wavelengths, in both the pulse shape and phase. At the same time, the light curve of Vela does not resemble the radio light curve. At present, these differences, as well as some other peculiarities of GeV pulsars do not have a convincing theoretical interpretation. The current two basic concepts based on the polar cap and outer gap models (see Chapter 6), can explain certain, but not all, features of individual pulsars. Therefore perhaps these two models should be taken as only a first step in the direction of development of a self-consistent theory of nonthermal radiation of radio pulsars.

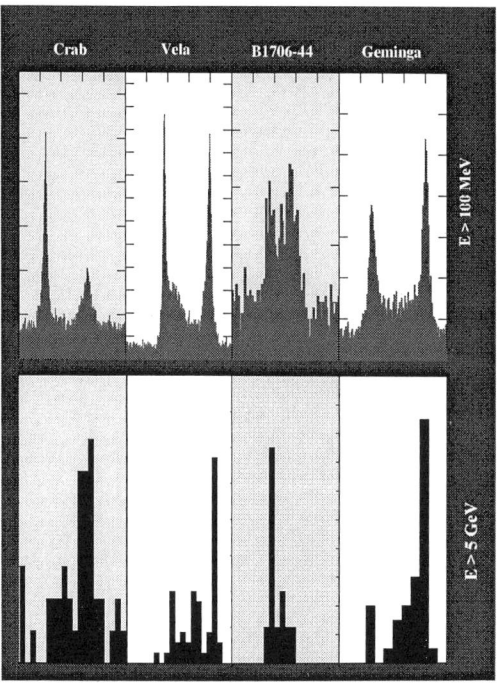

Fig. 2.6 Light curves of four γ-ray pulsars at energies above 100 MeV and 5 GeV (from Thompson, 2001).

Both the poor source statistics of the γ-ray pulsar population and poor photon statistics of individual γ-ray pulsars do not allow conclusive tests in favour of, or against, the polar cap and outer gap models. In Fig. 2.5 the energy spectra predicted by these two models for the Vela pulsar are shown. While at energies below several GeV these models predict similar γ-ray spectra, both being in good agreement with EGRET measurements, at energies above 10 GeV the theoretical spectra are dramatically different. Unfortunately, large statistical uncertainties in the EGRET data in this crucial energy region do not allow us to give preference to either of these models (note that the last spectral point in Fig. 2.5 is based on only 4 detected photons). GLAST (Thompson, 2001) and future low-energy threshold ground-based instruments like 5@5 (Aharonian and Bogovalov, 2003) will be able to discriminate between these two models. Moreover, the statistics of detected γ-rays should be sufficient to probe the light curves at different energies. The EGRET data show evidence of change of structure of light curves with energy. The light curves of four EGRET pulsars with reasonable statistics in two energy bands, above 100 MeV and above 5 GeV, are shown in Fig. 2.6. The multi-GeV light curves are dominated by one of the two pulses seen at lower energies (Thompson, 2001). There is little doubt how crucial will be for the pulsar physics a confirmation of this trend with much better photon statistics, and an extension of these studies to even higher energies. Also, high photon statistics is a necessary condition for probes of possible irregular and regular (i.e. as a function of phase) time-variations of γ-ray spectra. While GLAST can provide adequate detection rates above 100 MeV, it might run out of ≥ 10 GeV photons at given narrow phase intervals. Studies of the light curves at different energies, and the energy spectra at different phase intervals can be effectively covered by sub-10 GeV threshold Cherenkov telescope arrays (Aharonian and Bogovalov, 2003).

The high detector sensitivity is an important factor allowing effective searches for periodic signals from γ-ray sources without relying on observations at other energy band. Because pulsars are not perfect clocks (their periods increase with time due to rotational losses), it is important to accumulate photon statistics adequate for search for periodic signals during rather short periods (say less than 1 day), thus any change of a signal's phase can be ignored. The superior sensitivities of GLAST and 5@5 should allow to increase significantly the number of γ-ray pulsars, as well as to reveal the pulsed emission component from a number of unidentified EGRET sources, if they have indeed pulsar origin.

2.2.3 *Unidentified EGRET sources*

Almost 2/3 of sources from the 3rd EGRET Catalog are not yet identified with known astrophysical objects. Although a number of potential identifications have been suggested, the origin of the major fraction of the high energy γ-ray sources remains a mystery. Although it is quite possible that a part of these sources constitute a new class of astrophysical objects that shine mainly in γ-rays, the poor angular resolution of EGRET is likely to be the major reason for such a large fraction of unidentified objects. The main hope of solving the puzzle of unidentified EGRET sources is the GLAST mission, and perhaps also future low-energy threshold ground based γ-ray detectors. On the other hand, statistical studies of the properties of these objects, as well as continuation of multiwavelength probes of environments surrounding unidentified γ-ray sources ("source by source analysis") is an important approach to be continued during the next several years – before GLAST comes on line. The multiwavelength studies of unidentified EGRET sources will have also an important impact on preparation of the observation programs of GLAST and ground-based detectors.

The statistical studies allow differentiation of characteristics of unidentified sources that might give a hint for a possible links of the EGRET sources to certain source populations. Such a study (Gehrels et al., 2000) shows that indeed the stable (time-independent) EGRET sources are grouped in two populations when characterised by the spectral indices and the logN-logS distributions. Namely, it is found that brighter sources with harder energy spectra are concentrated at low galactic latitudes ($|b| \leq 5°$), while fainter and softer ones are located at medium galactic latitudes ($5° \leq |b| \leq 30°$).

A rather smooth longitude profile of about 50 low-latitude sources sets a limit to their distance of a few kpc (Grenier, 2001). Most of these sources do not show signs of variability. Although it is not possible to link these objects to a single class of galactic sources, their positions seem to correlate with objects which are believed to be tracers of active star formations regions in our Galaxy - HII regions, pulsars, SNRs, OB associations. It has been argued (Romero et al., 1999) that 22 of these sources can be associated with SNRs and 26 with OB association, with ten sources coincident with both SNRs and OB association – regions called SNOBs (Montmerle, 1979). Using the known distances of counterparts, the luminosities of γ-ray sources are estimated between $10^{34} - 10^{35}$ erg/s, in a good agreement with the luminosity estimates inferred from the global spatial distribution (Grenier, 2001). There is a very large dispersion in photon indices of these sources ranging from 1.7 to 3.1. At first glance, this contradicts the assumption that all these objects belong the same source population. However, such a dispersion can be readily explained, provided that γ-rays are produced by interactions of cosmic rays with molecular clouds, and that the particle accelerator is separated from the target/cloud (Aharonian, 2001a).

Actually, the large γ-ray luminosities up to 10^{35} erg/s require a combination of a powerful accelerator (e.g. a relatively young SNR shell or a pulsar or a microquasar) coupled with a nearby dense gas target. On the other hand, the accelerators hardly can operate effectively in dense environments. Therefore, both the dispersion of photon indices and large luminosities provide an argument that γ-rays are produced in systems like SNRs interacting with dense molecular clouds. Motivated by this argument, Torres et al. (2003) looked at the positional coincidences between unidentified EGRET sources and the regions around SNRs, namely, "enlarging" the size of the SNR by a half degree. They found approximately 30 coincidences of this kind. Interestingly, the associations of unidentified EGRET sources with SNRs are strongest for remnants close to molecular clouds. However, it should be noticed that no supernova remnant has yet been firmly detected in high energy γ-rays. The probability for chance alignment is quite high, exceeding 0.1 per cent (Grenier, 2001).

Representatives of some other galactic source populations like colliding winds in Wolf-Rayet binaries (Benaglia and Romero, 2002), plerions (Roberts et al., 2001) and microquasars (Parades et al., 2000; Kaufman Bernado et al., 2002) have been suggested as possible counterparts of low-latitude EGRET sources. In particular, recent searches of X-ray regions by ASCA in the fields containing bright sources of GeV emission, resulted in discovery of several hard X-ray sources, presumably pulsar-driven nebulae, positionally coincident with unidentified sources of GeV γ-ray emission (Roberts et al., 2001).

Although the EGRET observations did not reveal γ-ray fluxes from the most prominent microquasars like GRS 1915+105, galactic X-ray binaries with relativistic jets have been proposed as new potential candidates. This hypothesis recently received interesting support based on the positional coincidence of the microquasar LS 5039 with an unidentified GeV source (Paredes et al., 2000). If confirmed, this would perhaps require reconsideration of a general sceptical view (see however, Mori et al., 1997 and Vestrand et al., 1997) that EGRET detections of GeV γ-rays from directions of the X-ray binaries Cyg X-3 (presumably a microquasar, or even microblazar) and Cen X-3 are results of random coincidence. Unlike shell type SNRs, Giant Molecular Clouds (GMCs) and plerions, galactic jet sources are variable objects on timescales down to 1 day or less. Therefore at present they seem to offer one of a few possibilities for interpretation of variable low-latitude unidentified γ-ray sources.

Whereas there is no shortage in candidates as counterparts for low-latitude unidentified EGRET sources, away from the galactic plane the situation is reversed. The error boxes of many of unidentified high-latitude EGRET sources are "empty". Approximately half of the $\approx$ 130 sources above $|b| = 2.5°$ are variable. The large scale height of variable sources, as well as their spectral shapes indicate that this subset of high latitude sources is likely to consist largely of blazars. On the other hand, the distribution of the persistent sources closely follows the curved lane of a local (within a few 100 pc) structure at medium galactic latitudes called the Gould Belt (Gehrels et al., 2000, Grenier, 2001). Like in the galactic plane, several type of objects could be counterparts of EGRET sources. Among such sources are massive stars in OB associations with highly supersonic winds which can supply relativistic particles, e.g. through terminal shocks, for further production of γ-rays at interactions with dense ambient gas or photon fields. Pulsars born in the Gould Belt during the last 3 million years (Grenier and Perrot, 1999) are currently discussed as another promising candidate. Harding and Zhang (2001) have recently noticed that off-beam γ-ray pulsars in the Gould Belt, i.e. those viewed at large angles to the neutron star magnetic pole, can qualitatively match both the detected γ-ray fluxes and the number of EGRET sources at medium galactic latitudes. If so, GLAST and 5@5 type ground-based instruments will be able, as argued above, to detect γ-ray pulsations from most of these sources.

2.3 The Status of Ground-Based Gamma Ray Astronomy

2.3.1 Brief historical review

Gamma-rays interacting with the Earth's atmosphere initiate electromagnetic cascades. At sufficiently high energies the number of cascade particles is sufficient to obtain adequate information about the energy, direction and type of primary particles based on the study of spatial and temporal properties of secondary cascade products. Therefore the arrays of particle (electron, muon, hadron) detectors used in the traditional cosmic ray experiments can serve as effective tools also for a search for sources of very high energy γ-rays. In the 1980s, trying to pursue this technique, several cosmic ray groups reported the detection of excess events over the isotropic cosmic ray background from the direction of famous X-ray binaries Cygnus X-3 (see e.g. Samorski and Stamm, 1983; Lloyd-Evans et al., 1983), and Her X-1 (Dingus et al., 1988). Actually, claims of detection of γ-rays from Cygnus X-3 at lower, TeV energies were first made in the mid 1970s (by the Crimean group) and continued through the mid 1980s (by the Whipple, Durham, Haleakala and some other groups). This controversial episode in gamma-ray astronomy is described in a review article by Weekes (1992).

These exciting reports initiated new air-shower arrays specifically designed for γ-ray studies, in particular the CASA-MIA (Borione et al., 1994), and HEGRA (Karle et al., 1995) detectors with significantly improved sensitivities, lower energy thresholds, and relatively effective hadron/γ separation capabilities. The all-sky surveys by these detectors did not, however, reveal point sources of γ-rays (Cronin et al., 1993) down to flux levels of $\sim 10^{-14}$ ph/cm^{2}s above 100 TeV. To a certain extent, this cannot be interpreted as a big surprise. The production of such energetic photons requires charged parent particles of energy exceeding several times $\sim 10^{14}$ eV. Although the spectrum of cosmic rays extends up to 10^{20} eV, it is quite possible that the acceleration efficiency of protons in the galactic sources, in particular in SNRs, drops at 100 TeV/amu (see Chapter 5). Thus even in the presence of dense target material, the π^0-decay γ-ray emission above 10 TeV is expected to be strongly suppressed. The problem of the high-energy cutoff exists, actually even more seriously, also for the second important channel of γ-ray production through the inverse Compton scattering. The severe synchrotron losses and reduction of the cross-section of the Compton scattering due to the Klein-Nishina effect make this mechanism at such high energies much less efficient than in the TeV region.

From this point of view, nonthermal extragalactic sources like the jets in powerful radiogalaxies and quasars, and the rich galaxy clusters that can accelerate protons well beyond 10^{15} eV, are certainly more promising objects for 100 TeV observations. But, unfortunately, because of absorption in the extragalactic radiation fields, only a small part of the local intergalactic space within a few Mpc is transparent for ≥ 100 TeV γ-rays.

Thus, by reduction of the energy threshold of detection methods down to 10 TeV and below one may hope to boost the chances of discovery of VHE γ-ray sources both in and beyond of our Galaxy. Such sources have been indeed detected, due to the successful realization of the the so-called Imaging Atmospheric Cherenkov Telescope (IACT) technique.

A remarkable feature of this technique is its high detection rate capability, a consequence of the large integration area of air showers. Even a simple device consisting of a fast (nanosecond) detector of optical radiation (a photomultiplier) in the focal plane of a modest both in quality and size (≈ 10 m^2) reflector, can provide huge, as large as 3×10^8 m^2, area for detection of 1 TeV γ-rays. However, the goal cannot be achieved without an effective method of suppression of several heavy backgrounds of different origin. For example, such a simple device cannot distinguish between electromagnetic and hadronic showers, and thus works also as an effective collector of cosmic rays, the flux of which exceeds by several orders of magnitude the flux of γ-rays. This obviously limits the sensitivity of the instrument as a γ-ray detector. Another background caused by the integrated light from the night-sky, limits the minimum detectable energy of γ-rays. Fortunately, the imaging technique provides an adequate background rejection power (see below). The reported TeV γ-ray signals from more than 10 astrophysical objects by several instruments installed in both the northern (Whipple, HEGRA, CAT, Telescope Array, CrAO, SHALON, TACTIC) and southern (CANGAROO, Durham) hemispheres basically proved the early theoretical predictions (Hillas, 1995) concerning the potential of this technique.

The first Cherenkov light pulses from atmospheric air showers were registered by Galbraith and Jelley in 1953. The attempt to pursue the detection of γ-rays from astrophysical sources with the first atmospheric Cerenkov telescopes (Jelley and Porter, 1963, Chudakov et al., 1965) resulted in meaningful upper limits that appeared below the optimistic theoretical predictions. Several years later the first positive signal was reported from the Crab Nebula. The result was obtained with a 10-meter-diameter Cherenkov telescope completed in 1968 at Mt. Hopkins in southern Arizona. In contrast to the high mechanical and optical qualities of this telescope, which after more than 30 years still remains one of the best in the field, the focal plane instrumentation was relatively primitive, thus only a marginal signal at a level of $\approx 3\sigma$ was revealed after 150 hours of observations accumulated during 1969-1972 (Fazio et al., 1972).

For the next 10 years or so the field languished. The activity in ground based observations significantly declined, and the interest was shifted to two successful satellite-based experiments, SAS-II and COS B, that opened up the observational gamma-ray astronomy at energies above 100 MeV. But in the mid 1980s the interest in ground-based observation turned back, motivated basically by the above mentioned claimed of unusual signals from Cygnus X-3 and some other X-ray binaries (for a review see Weekes, 1992). Even more astonishment was introduced by the claims of periodic signals from Cyg X-3 by underground experiments originally designed for searches for proton decays. However, despite the number of claimed detections, each individual result did not exceed a few-standard-deviations significance, and as observations improved in sensitivity, the signal from Cyg X-3 appeared to diminish proportionally.

It should be noticed, however, that this disappointing episode in the history of gamma-ray astronomy had also positive impacts. Since the unusual "signals" from Cyg X-3 could not be explained in the framework of conventional physics, it attracted many experienced specialists from other fields, in particular from the high energy physicists community, and initiated a new research area called *Astroparticle Physics* – currently a very popular discipline, albeit with somewhat different (from the 1980s) emphasis on the potential topics and priorities.

Also, as the interest in confirming the existence of TeV signals from Cygnus X-3 began to peak, the Crimean group led by A.A. Stepanian and the Whipple group led by T.C. Weekes started practical steps in the direction of improving the sensitivity of the Cherenkov telescopes by implementing the imaging technique. The idea was that the analysis of the angular distribution of the Cherenkov radiation of air-showers should allow a significant reduction of the cosmic ray background. Hillas (1985) clearly demonstrated that indeed the analysis of the second moments of the Cherenkov images of air showers – as detected by a high quality mirror with a multi-channel imaging camera at its focus – should be able to discriminate between the γ-ray and proton- induced showers, and thus to improve significantly the signal-to-noise ratio. The exploitation of this technique by the Whipple telescope equipped with a 37-photomultiplier camera resulted in the first high-confidence detection of TeV γ-rays from an astrophysical object - a 9σ γ-ray signal from the Crab Nebula (Weekes *et al.*, 1989). The construction of a new 109-channel camera, as well as subsequent improvements in the data analysis technique soon led to new important discoveries with this telescope - the detection of γ-rays from Mkn 421 (Punch *et al.*, al. 1992) and Mkn 501 (Quinn *et al.*, 1996).

With arrival of several new projects in the mid 1990s – the CANGA-ROO 3.8m, Durham (both in Australia), CAT in the French Pyrenees, the HEGRA telescope system on the Canary Island La Palma, the Telescope Array in Utah (USA), G-48 in Crimea and some others, vigorous activity commenced with a hope to increase significantly the number of TeV γ-ray sources. Perhaps one may conclude that, in the sense of number of discovered objects, the hope has been only partly fulfilled. Presently, 6 to 8 objects are firmly established as γ-ray emitters, whereas another 10 or so are considered as likely candidates. However, the achievements of the field, in particular the astrophysical significance of the reported results, cannot be reduced to the number of detected sources. For example, the recent comprehensive studies of spectral and temporal characteristic of TeV emission of Mkn 421 and Mkn 501 by the CAT, HEGRA and Whipple IACTs on timescales down to 1 hour, yielded perhaps the highest experimental quality achieved in gamma-ray astronomy, including the MeV and GeV bands. It is difficult to overestimate the significance of these observations for the current models of nonthermal processes in the relativistic jets of blazars. The same is true for TeV γ-ray emission reported from three shell type supernova remnants, SN 1006, RX J1713.7-3946 (CANGAROO) and Cas A (HEGRA), for understanding of the origin of galactic cosmic rays. Several TeV sources have been detected by the HEGRA stereoscopic telescope system at the flux level as small as 10^{-12} erg/cm^{2}s, and localised within a few arcminutes. This is a remarkable accomplishment that can be achieved at MeV/GeV energies only with the next generation satellite-borne instruments like GLAST. The number of TeV sources is growing rather fast, and hopefully many more sources will be found with the forthcoming IACT arrays CANGAROO-III, H.E.S.S., MAGIC and VERITAS. These detectors will operate at thresholds around 100 GeV, and provide flux sensitivity at TeV energies down to 10^{-13} erg/cm^{2}s. This should lead, hopefully in the near future, to a dramatic growth in the number of VHE sources.

2.3.2 *Reported TeV sources*

The sources reported as TeV γ-ray emitters by different groups are shown in Fig. 1.1. They are referred to two categories of detection – "confirmed" (by the same or an independent group) and "not confirmed". Actually, such a division is rather conditional, and does not fully describe the ambiguity of conclusions concerning the confidence level of the reported results. Some of these sources are detected with very high, 10σ or more, statistical significance and are confirmed by at least two independent groups. Some others have been detected at high confidence level, e.g. with more than 6σ significance, although only by a single group. Finally, several reports claiming detection of new TeV sources have not been confirmed by followup observations by the same or by other groups. Currently, the latter is considered as a key condition to ensure membership of the "VHE Source Club". Although generally well justified, this robust condition should be applied cautiously, especially when it concerns *a priori* or suspected variable objects. On the other hand, one should not overemphasise the claimed very high statistical significance of some detections, because sometimes the large "sigmas" are obtained after optimisation of the image parameter cuts. Also, in some cases the systematic effects are neglected, although they in fact may dominate over the statistical uncertainties. The necessity of independent observations in such cases cannot be questioned.

GALACTIC SOURCES

2.3.2.1 The Crab Nebula

The Crab Nebula, one of the most prominent objects in the sky, is a unique particle accelerator. The undisputed synchrotron nature of the non-thermal radiation from radio to low energy γ-rays (see Fig. 2.7) indicates the existence of relativistic electrons of energies up to 10^{16} eV. Given the large magnetic field in the nebula, $B \geq 100$ μG, this implies an extremely effective acceleration at a rate quite close to the maximum possible rate allowed by classical electrodynamics. The Compton scattering of the same electrons

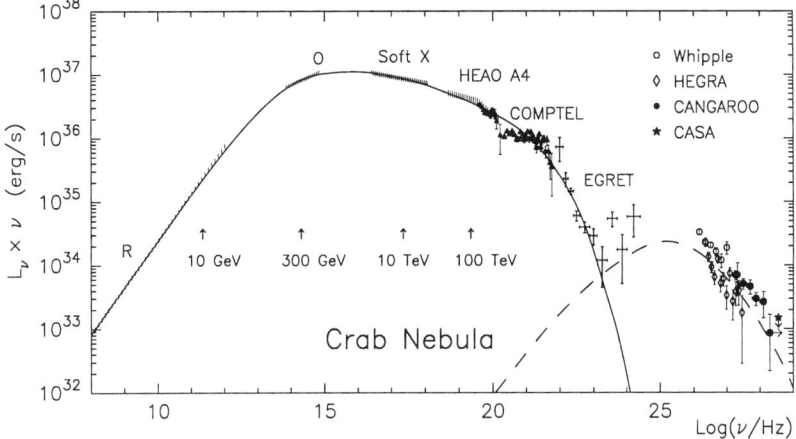

Fig. 2.7 Nonthermal radiation of the Crab Nebula from radio to very high energy γ-rays. The solid and dashed curves correspond to the synchrotron and inverse Compton components of radiation, respectively, calculated in the framework of the spherically symmetric MHD wind model. The vertical arrows indicated the ranges of characteristic frequencies of synchrotron photons emitted by electrons of different energies.

leads to effective TeV γ-ray emission. Despite different approaches and accuracies of calculations performed by many authors in the past, the Crab Nebula was confidently predicted as a strong VHE γ-ray source. Since the first positive report by the Whipple collaboration (Weekes et al., 1999), the Crab Nebula has been detected by more than 10 independent groups using different ground-based techniques. Presently the reported fluxes cover a very broad energy range that extends from 60 GeV (de Naurois et al., 2002) to 20 TeV (Aharonian et al., 2000a) or even to higher energies (Tanimori *et al.*, 1998a; Horns *et al.*, 2003).

As the brightest persistent TeV source seen effectively from both hemispheres, the Crab Nebula has become the standard candle for cross-calibration of different detectors. Currently, this is often treated as the most important aspect of ground-based γ-ray observations of Crab, assuming that the "astrophysical" objectives are already achieved, given the good agreement between the reported fluxes and the theoretical predictions (see Fig. 2.7). However, many details remain unresolved (see Sec. 6.3.1 in Chapter 6), and should be addressed by future observations with significantly improved performance in the entire γ-ray domain.

Probing the magnetic fields and electrons. The most informative frequency band to probe the acceleration site(s) and the character of propagation of electrons is the X-ray domain. Chandra, with its sub-arcsecond imaging capability and excellent spectral resolution, is an ideal instrument for such studies. However, the synchrotron data alone tell us only about the product of the magnetic field strength and the density of relativistic electrons. These parameters can be disentangled using additional information contained in γ-rays. Since the TeV γ-rays are produced by IC scattering of electrons responsible also for the observed X-rays, an estimate of the magnetic field based on keV/TeV data concerns the central $r \sim 0.5\,\mathrm{pc}$ region of the nebula. This corresponds to less than 1 arcmin angular size of the region surrounding the central pulsar. The HEGRA collaboration using its stereoscopic system of 5 IACTs with an angular resolution $\approx 0.1°$, set an interesting limit on the angular size of the TeV emission of about 1.5 arcmin (Aharonian *et al.*, 2000b), confirming that the TeV γ-rays indeed originate in the central part of the nebula. But unfortunately this constraint is not yet sufficient for more definite conclusions about the standard wind termination shock model. The accuracy of the determination of arrival direction of individual γ-rays by future stereoscopic "100 GeV" threshold arrays in the high energy (TeV) domain is expected to be better than a few arcmin. This, combined with large TeV photon statistics, may allow, hopefully, an adequate mapping of the source on ≤ 1 arcmin scales, and thus provide an accurate estimate of the magnetic field in the most interesting sub-pc part of the central region of the nebula.

In all models of the Crab Nebula the calculations of IC fluxes are "controlled" by the observed X-ray flux. The X-ray emission of the Crab Nebula has a distinct axisymmetrical structure (see Fig. 6.14). This is strong evidence that most of the rotational energy of the pulsar is released in the form of a wind which flows out from the pulsar equator. It is natural to expect that the observed TeV fluxes are produced in the same region of X-torus. However, it is difficult to avoid a suspicion that the real picture is more complex. In particular, if the B-field in this region is significantly enhanced and exceeds 0.2 mG, the γ-radiation from the X-torus cannot explain the observed TeV flux. If so, an alternative site for γ-ray production could be regions outside of the X-ray torus, i.e. the parts of the nebula powered by a possible *quasi-spherical* component of the wind. Although the energy budget of this component of the wind should be significantly less than the luminosity of the equatorial wind, the observed flux of TeV radiation can be achieved assuming a smaller magnetic field in this region. The existence of non-equatorial outflow from the pulsar can be examined by future detailed spatial and spectrometric studies in the X- and TeV γ-ray regimes.

Within the IC models of γ-radiation, the magnetic field in outer parts of the (optical) nebula can be best probed by sub-100 GeV IACT arrays in the northern hemisphere. The angular resolution of the VERITAS telescope array at the threshold of 50-100 GeV is expected to be close to 0.1° which should be sufficient for extraction of such important information. The flux sensitivity and angular resolution of GLAST can provide a complementary study at lower energies (see Fig. 1.2). And finally, the γ-ray fluxes above $E \geq 10\,\mathrm{TeV}$ combined with hard X-ray/low energy γ-ray data, should allow determination of the magnetic field in the vicinity of the wind shock front at $r \sim 0.1\,\mathrm{pc}$. This compact region cannot be spatially resolved by γ-ray instruments. Nevertheless an indirect "identification" of this region could be possible by detection of time variability of the highest energy tail of γ-ray spectrum on timescales of several months, expected because of the unsteady structure of the shock and rapid synchrotron losses of 100 TeV electrons.

Searching for gamma-rays of "hadronic" origin. The shape of the IC spectrum is rather stable to the basic parameters of the nebula, and can be predicted with high confidence. While at GeV energies the IC spectrum is very hard with a power-law index $\alpha_\gamma \approx 1.5$, in the VHE region the spectrum gradually steepens from $\alpha_\gamma \approx 2$ at $E \sim 100\,\mathrm{GeV}$ to $\alpha_\gamma \approx 2.5 - 2.6$ at $E \sim 1\,\mathrm{TeV}$, and $\alpha_\gamma \approx 2.7$ at $E \sim 10\,\mathrm{TeV}$. This behaviour, which implies an almost constant slope between 1 and 10 TeV, but significant flattening around 100 GeV, is in general agreement with high energy results (e.g. Hillas *et al.*, 1998, Aharonian *et al.*, 2000a), and with the recently measured flux at 60 GeV (de Naurois *et al.*, 2002) by

CELESTE – a Cherenkov-wave-front detector using a former solar plant at the Themis site in the French Pyrenees (see Fig. 1.2). But for a final conclusion, detailed spectral measurements in the transition region around 100 GeV are needed. It can be done by VERITAS and MAGIC at energies around 100 GeV, and by GLAST at lower energies. Another crucial test can be provided by precise spectrometric measurements at the highest energies well above 10 TeV. In particular, the confirmation of relatively flat, $\propto E^{-2.6}$ type, spectrum extending well beyond 20 TeV as reported by the CANGAROO (Tanimori et al., 1998a) and HEGRA (Horns et al., 2003) groups, would perhaps require an additional radiation mechanism. The γ-rays of π^0 origin seem to be an interesting possibility (see Sec. 6.3.1 in Chapter 6). Because of the limited energy budget determined by the spin-down luminosity of the pulsar, this hypothesis requires an ambient gas density n_{eff} in the γ-ray production region exceeding by an order of magnitude the average density of the nebula, $\bar{n} \approx 5\,\text{cm}^{-3}$. This seems quite unlikely, however cannot not be excluded, e.g. due to possible effective confinement of protons in dense *filaments*. If so, this should unavoidably result also in an enhanced contribution from electron bremsstrahlung, and, as a consequence, in a noticeably higher γ-ray flux at 100 GeV compared to the pure IC flux. The recent CELESTE measurement of a relatively low flux at 60 GeV does not support this hypothesis, but further studies are needed for a final conclusion.

Searching for gamma-ray signatures of the unshocked wind. The Crab Nebula and other plerions are powered through the termination shocks of the cold ultrarelativistic electron-positron wind with a bulk motion Lorentz factor as large as $\Gamma \sim 10^6$. It is believed that the region between the pulsar magnetosphere and the shock, where almost all the rotational energy of the pulsar is somehow released in the form of kinetic energy of the wind, is *invisible* – despite the large Lorentz factor, the wind electrons move together with the magnetic field and thus do not emit synchrotron radiation. Even so, the wind can be directly observed through the bulk motion Comptonization caused by illumination of the wind (Bogovalov and Aharonian, 2000). This radiation has distinct spectral characteristics depending essentially on the position of the "birthplace" of the *particle dominated wind* (i.e. the site where the *Poynting flux dominated* wind undergoes to the regime dominated by the *kinetic energy of particles*), as well as on the Lorentz-factor and the geometry of propagation of the wind (see Sec. 6.2 of Chapter 6). Thus, dedicated searches for such specific radiation components in the Crab spectrum may provide unique (not accessible at other wavelengths) information about the unshocked pulsar wind. In particular, they can "localise" the region of formation of the particle dominated wind, and "measure" the Lorentz factor of the wind.

2.3.2.2 Other plerions

The existence of the bright synchrotron nebula around the Crab pulsar very often is interpreted as a crucial condition for effective production of IC γ-rays. In fact, the *strong* magnetic field in the Crab Nebula produced by the *strong* wind only reduces the γ-ray production efficiency. Indeed, the energy density of the B-field exceeds by more than two orders of magnitude the radiation density, thus only ≤ 1 per cent of the energy of accelerated electrons is converted to the IC γ-rays, the rest being emitted in the form of synchrotron X-rays. In other plerions with significantly weaker winds, the resulting nebular B-fields are more than one order of magnitude smaller, which makes these objects more effective γ-ray emitters. The main target photons for inverse Compton scattering in these objects is contributed by the 2.7 K CMBR, therefore the radiative loss of electrons is shared between synchrotron and IC channels as $L_\gamma/L_X = w_{MBR}/w_B \simeq 1\,(B/3\mu G)^{-2}$. In a plerion with nebular magnetic field less than 30 μG, the γ-ray production efficiency should exceed 1 per cent, given that the cooling time of ≥ 10 TeV electrons ($\approx$ 1000 yr) is less than the age of typical plerions. This is by an order of magnitude more efficient than the γ-ray production in the Crab Nebula. Correspondingly, one may expect that the next generation IACT arrays with sensitivity better than 10 mCrab should be able to probe TeV γ-ray emission from plerions containing pulsars with the so-called "spin-down" energy flux $S_0 = (L_0/10^{37}\,\text{erg/s})(d/1\,\text{kpc})^{-2} \geq 10^{-3}$ (see Sec. 6.4). Tens of pulsars with $S_0 \geq 10^{-3}$ are found in our Galaxy. This provides optimism that IC γ-ray nebulae surrounding some selected pulsars finally will be detected.

In fact, three plerions in the Southern Hemisphere – PSR B1706-44, Vela, and PSR B1508-58 – have been already claimed by the CANGAROO group as TeV emitters. While the statistical significance of the signal from PSR B1706-44 detected by the first 3.8m diameter CANGAROO telescope (Kifune *et al.*, 1995) was quite high, and later claimed to be confirmed both by the Durham group (Chadwick *et al.*, 1998a) and by the new 10m CANGAROO telescope (Kushida *et al.*, 2003), the reports on tentative detection of TeV emission from Vela and PSR B1509-58 with the 3.8m

CANGAROO telescope are not yet confirmed by other measurements.

Although above we argued that one should expect TeV γ-ray emission from some selected plerions, the reported fluxes of both PSR 1706-44 and Vela are too high to be easily accommodated by the conventional synchrotron-inverse Compton models.

PSR B1706-44. Generally, it is reasonable to assume that the unpulsed TeV radiation from PSR B1706-44 is not directly connected with the 102 ms EGRET pulsar, but rather originates in the surrounding IC nebula with a total TeV γ-ray luminosity $L_{\rm TeV} \sim 10^{33}$ erg/s. But the problem is that the X-ray luminosity of the region around the pulsar within 1 arcmin is a factor of 3 below the γ-ray luminosity. This is a rather unexpected result, and puts very tight limits on the parameters characterising the TeV γ-ray production region. Indeed, assuming that the X-rays and TeV γ-rays are produced in the same region, we come to the conclusion of an uncomfortably low (for a *pulsar wind nebula*) magnetic field, $B < 3\mu{\rm G}$. A possible way to avoid the problem of such a low B-field is to assume that the electrons occupy a significantly larger region than the ≤ 1 arcmin (unresolved by ROSAT) synchrotron X-ray nebula. This could be realized if the $\geq 10\,{\rm TeV}$ electrons quickly leave (e.g. due to the diffusive propagation) the pulsar wind nebula with a conventional, e.g. $B \sim 10\,\mu{\rm G}$ field, and enter the interstellar medium with much lower field. There they upscatter the 2.7 K CMBR photons and thus produce the bulk of the observed TeV flux. This scenario, discussed in Sec. 6.4 of Chapter 6, can be inspected through spectral and morphological studies of X- and γ-ray emission components by more powerful detectors like XMM-Newton and H.E.S.S.

Vela. In addition to the high TeV/X-ray flux ratio, the TeV signal of this source contains another puzzle. Surprisingly, the γ-ray production region reported by the CANGAROO group is *offset* from the Vela pulsar position by about 0.13°. It has been claimed that the position of the γ-ray excess coincides with the supposed "birthplace" of the pulsar as determined from its age and proper motion. The total γ-ray luminosity above 1.3 TeV is estimated as $L_\gamma \sim 10^{33}$ erg/s.

Despite a calm reaction from the gamma-ray community to this result ("why not ?"), the most natural inverse Compton interpretation of the observed TeV emission faces very serious difficulties. The simplest solution would be (from the point of view of a theorist) that both the reported flux and the position of the TeV source are not correct. On the other hand, in the case of confirmation of these results, one would need to invoke certain extraordinary assumptions concerning both the origin of the electrons and the strength of the ambient magnetic field.

The X-ray luminosity of the same region seen by ROSAT and ASCA does not significantly exceed 10^{32} erg/s (Harding *et al.*, 1997), an order of magnitude less than the TeV luminosity. Thus, in a one-zone synchrotron-Compton model, the magnetic field cannot exceed a few μG. This implies that the pressure of relativistic electrons in the X-ray hot spot exceeds by two orders of magnitude the pressure of the B-field. Formally we cannot exclude the scenario with an effective escape of electrons, provided that the escape is compensated by continuous particle acceleration. However, since the B-field in the nebula is believed to be much stronger, the escape of electrons should produce an X-ray image with a *hole* rather than a *hot spot* in the γ-ray production region. These difficulties of the hypothesis that we are dealing with long-lived particles left in a trail by the pulsar after it moved from its birthplace (Harding *et al.*, 1997), can be somehow tolerated, if we assume that the TeV γ-rays are of hadronic origin, i.e. they are produced by interactions of "relic" energetic *protons* with the ambient gas (Aharonian, 1999). The required total energy in "relic" protons could be reduced down to a reasonable amount of $W_\mathrm{p} \simeq 10^{49}$ erg if we assume that the ambient gas density in this region is significantly higher than outside, e.g. $n = 10\,\mathrm{cm}^{-3}$. For effective confinement of particles in this region with a size of $R \sim 1\,\mathrm{pc}$ the B-field should be close to its equipartition value, i.e. $B = (6\,W_\mathrm{p}/R^3)^{1/2} \simeq 1\,\mathrm{mG}$. In principle, both the relativistic particles and the B-field could be created by the powerful relativistic wind of the "baby" pulsar. An alternative source could be the kinetic energy released at the supernova explosion. Obviously, the 'relic' electrons could not survive severe synchrotron losses in a such strong magnetic field. At the same time, the X-ray emission of the hot spot still could be explained by the synchrotron radiation of *secondary* ($\pi^{\pm}$-decay) electrons and positrons. The energy released in secondary electrons at p-p interactions depends on the spectrum of protons; for hard proton spectra it could be as large as half of the energy transferred to π^0-decay γ-rays. The secondary TeV electrons quickly lose their energy in the form of synchrotron X-rays. Thus, even within this pure 'hadronic' scenario the TeV radiation should be accompanied with non-negligible nonthermal X-ray emission. Unfortunately, the lack of adequate spectral information in X-ray and TeV regions does not allow more quantitative conclusions about the parameters characterising the TeV source.

2.3.2.3 *Gamma ray pulsars*

The first realistic attempts of using the atmospheric Cherenkov technique for the detection of VHE γ-ray sources in the late 1960s coincided with the great discovery of pulsars. Since then pulsars have been in the highest priority target lists of many Cherenkov groups, although, as we understand now, this idea does not have a solid theoretical background, and there is no much hope to detect magnetospheric TeV γ-ray emission even from pulsars that are bright in GeV γ-rays. Actually, there were several reports of possible detections of pulsed TeV radiation, in particular from the Crab and Geminga pulsars, however follow-up observations with more sensitive imaging telescopes failed to confirm these early claims. The upper limits of TeV radiation from the MeV/GeV pulsars are shown in Fig. 2.1. These agree with the models which predict intrinsic spectral cutoffs beyond 10 GeV. Irrespective of details of these models, it is likely that flat GeV γ-ray spectra should significantly steepen at higher energies due to attenuation caused by pair production in the pulsar magnetic field. The position of the spectral turnover depends essentially on the localisation of the γ-ray production region(s). Thus, spectrometric measurements at energies above 10 GeV by GLAST, and, possibly, also by very low-threshold (sub-20 GeV) ground based detectors may provide a crucial test for different scenarios of particle acceleration in pulsar magnetospheres.

Meanwhile, the outer magnetosphere gap models predict a new hard component of γ-radiation produced due to the inverse Compton mechanism in the outer magnetosphere (Chapter 6). An interesting feature of this radiation is its hard spectrum below 1 TeV, with a sharp cutoff of several TeV. Thus, the most promising energy region for detection of this component seems to be a rather narrow interval around a few TeV (Romani, 1996). In particular, the calculations for the Vela pulsar predict that the energy flux in pulsed TeV γ-rays could exceed 0.1% of the pulsed GeV flux (Romani, 1996). A strong upper limit on the flux from the Vela pulsar reported by the CANGAROO collaboration (Yoshikoshi *et al.*, 1997) at the "right" energies (see Fig. 2.1) is quite close to the predicted flux.

2.3.2.4 *Gamma rays from supernova remnants*

It is believed that cosmic rays below the so-called "knee" around 1 PeV have galactic origin and are produced in shell-type SNRs (see Chapters 4 and 5). The arguments leading to this important statement are, however, indirect and rather circumstantial. If SNRs are the principal sources of galactic cosmic rays providing the bulk of the observed flux, and if the particle acceleration proceeds through the so-called diffusive shock acceleration mechanism, then the relatively young remnants should be visible in γ-rays, first of all at TeV energies. Thus, the detection of positive TeV signals of hadronic origin from SNRs would provide the first straightforward proof of the shock acceleration of cosmic ray protons and nuclei in these objects (see Chapter 5). On the other hand, the failure to detect π^0-decay γ-ray signals from several selected SNRs would impose strong constraints on the overall energy in accelerated protons per SNR, $W_{\rm CR} \leq 10^{49} - 10^{50}\,{\rm erg}$. Contrary to the current belief, this would indicate an inability of the ensemble of galactic SNRs to explain the observed cosmic ray flux. Motivated by a perspective to make an important contribution to the solution of the long-standing problem of origin of galactic cosmic rays, the HEGRA, CANGAROO and Whipple collaborations have conducted extensive programs of observations of young and mid-age galactic shell-type SNRs.

The first attempts led to the flux upper limits from several "good candidate" SNRs. The Whipple collaboration has published the results of observations of six SNRs (Buckley *et al.*, 1998) shown in Fig.2.8, together with the flux upper limits at higher energies obtained by air shower arrays. The low-energy γ-ray measurements by EGRET from the directions close to three of these SNRs, IC 443, γ Cygni, and W44, are also shown.

The SNRs shown in Fig.2.8 were selected as possible TeV candidates based on their nonthermal radio properties, relatively small distance, "right" age (to be in the so-called Sedov phase) and, in some cases, because of possible association with nearby molecular clouds. The solid curves are normalised to the EGRET fluxes at 100 MeV assuming that the EGRET emission is produced by hadronic interactions of accelerated particles with an $E^{-2.1}$ type spectrum. The upper limits derived for IC 433, W44 and γ Cygni lie significantly below the "model" extrapolations represented by solid curves, and therefore initiated quite heated discussions in the γ-ray and cosmic ray communities concerning the possible need for revision of the standard concept of the origin of galactic cosmic rays – "diffusive shock acceleration in shell-type supernova remnants". Although a possible revision of current views cannot be excluded, at least for some specific aspects of the problem, the strong claims based on the above upper limits apparently are premature and to some extent exaggerated. The conclusions essentially rely on the normalisation to the EGRET data which, in turn, implies two assumptions: (a) the EGRET fluxes originate in the supernova remnants, and (b) they are produced entirely by hadronic interactions. Neither of these assumptions are, however, well justified. Indeed,

(i) The localisation of γ-ray regions by EGRET is very poor, especially in the "crowded" regions in the galactic plane where these SNRs are placed.

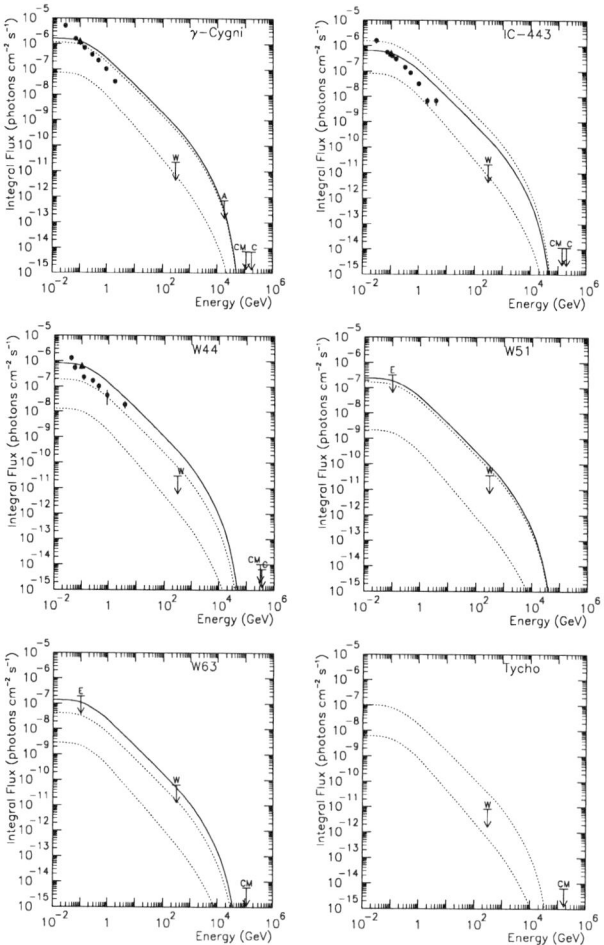

Fig. 2.8 Whipple flux upper limits (W) from 6 SNR (from Buckley *et al.*, 1998). Also are shown the EGRET fluxes (filled circles) or upper limits (E), as well as the flux limits obtained at higher energies by air-shower detectors – CASA-MIA (CM), CYGNUS (C), and AIROBICC (A). The experimental points are compared to extrapolations from the EGRET fluxes (solid curves), as well as to the conservative estimates of the allowable range of fluxes from the diffusive shock acceleration model (dashed curves).

Therefore, the proximity of the EGRET sources to SNRs should be treated as an interesting hint, but not yet as an established physical association.

(ii) Even adopting these associations as being real, one cannot be sure that the low energy γ-rays are the result of cosmic ray interactions. For example, they can be produced by embedded pulsars - a possible scenario in the case of γ Cygni (Brazier et al., 1996).

(iii) Even if the low energy γ-rays are produced at interactions of particles accelerated by SNR shocks with the ambient diffuse gas, the contribution of the electron bremsstrahlung in the EGRET energy region can dominate over the hadronic interactions.

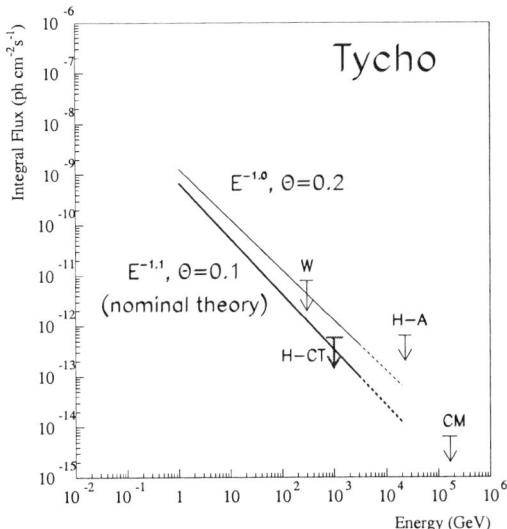

Fig. 2.9 Flux upper limit (indicated as H-CT) from Tycho's SNR obtained with the HEGRA telescope system (Aharonian et al., 2001e). The Whipple (W), HEGRA-AIROBIC (H-A) and CASA-MIA (CM) upper limits are also shown. The thin and thick solid lines correspond to the standard phenomenological predictions assuming two combinations of the power-law index (2.0 and 2.1) of accelerated protons and efficiency (0.1 and 0.2) of conversion of the kinetic energy of SNR explosion to cosmic rays.

Thus, until the arrival of high quality MeV/GeV data, unbiased constraints on the parameter space characterising the diffusive shock acceleration model can be derived only from TeV observations. In this regard, the dashed curves in Fig. 2.8 represent the range of TeV γ-ray fluxes based on the conservative estimates of parameters allowed by the standard shock acceleration model. Note that the typically (especially for SNRs in the galactic plane) poor information about the the nebular gas density and the distance to the source results in a significant (as large as an order of magnitude) combined uncertainty in the predicted TeV fluxes ($F \propto n/d^2$). It is seen from Fig.2.8 that within such large uncertainties the Whipple upper limits are not in conflict with theoretical predictions.

Generally, it is reasonable to expect significant enhancement of γ-ray fluxes caused by the presence of high density environments, e.g. giant molecular clouds in the vicinity of a SNR. On the other hand, expansion of a SNR in a dense environment may prevent acceleration of particles to very high energies. Therefore, it is possible that the best TeV emitters are SNRs in relatively low density environments. Motivated by this argument, the HEGRA collaboration conducted a deep observation of Tycho's supernova remnant that resulted in a flux upper limit at the 0.03 Crab level (Fig. 2.9). It is quite close to the conservative theoretical predictions that follow from the standard diffusive shock acceleration theory. This implies that the next generation of telescope arrays in the Northern Hemisphere and GLAST should be able to detect Tycho at GeV energies, or introduce significant corrections to our current understanding of particle acceleration processes in SNRs.

While the ultimate aim of the next generation IACT arrays will be inspection of the long-standing hypothesis that the population of shell-type SNRs *as a whole* is responsible for the galactic cosmic rays through the diffusive shock acceleration, presently we already have evidence that some *individual* representatives of this source population do accelerate, in one way or another, particles to energies well beyond 10 TeV. Importantly, this belief no longer relies merely on theoretical arguments, but has an observational backing.

SN 1006 and RX J1713.7-3946. The detection of featureless X-ray emission from several relatively young SNRs, of which the shells of SN 1006 and RX J1713.7-3946 are the most convincing examples, is best interpreted as synchrotron radiation of relativistic electrons accelerated to energies $\sim$ 100 TeV (e.g. Koyama, 2001). Motivated by the ASCA X-ray discoveries, the CANGAROO group over the last several years has extensively observed SN 1006 and RX J1713.7-3946 and ... found TeV γ-ray signals from both objects. The original reports of detection of TeV γ-rays from these objects obtained with the 3.8 m telescope (Tanimori *et al.*, 1998b, Muraishi *et al.*, 2000) were later claimed to be confirmed by more sensitive observations with the new 7m and 10 m diameter telescopes (Tanimori, 2003).

The synchrotron origin of X-radiation from these objects implies the existence of $\gg 10$ TeV electrons which should also produce TeV γ-ray emission trough upscattering of the 2.7 K CMBR. This idea was immediately adopted by the TeV and X-ray communities, and continues to dominate as the most likely and natural interpretation of the detected TeV γ-ray emission from both objects. However, this interpretation requires very small magnetic fields in the X- and γ-ray production regions, and might face a serious theoretical challenge connected with the limited acceleration rate allowed by the standard (parallel shock-acceleration) mechanism (Aharonian, 1999). On the other hand, it has been argued (Aharonian, 1999) that in spite of certain problems, the hadronic origin of gamma-radiation should not be discarded based on simplified model calculations.

The question of the origin of the TeV emission from SN 1006 remains rather uncertain and controversial. Decisive tests can be obtained only after comprehensive spectrometric and morphological studies of properties of γ-ray emission. The most exciting upshot of these measurements could be the localisation of the TeV γ-ray production region. In the case of hadronic (π^0-decay) origin of γ-rays we should expect, within the diffusive shock acceleration scenario, a quite compact region in the shell, because of the strong shock-compression of the target material. In contrast, the leptonic (inverse Compton) model predicts that the TeV emission should cover an extended region, because of fast diffusion of the most energetic electrons in the weak (required by this model) ambient magnetic fields (Aharonian and Atoyan, 1999; Berezhko et al., 2002).

Formally, the above arguments concerning the question of the "hadronic versus leptonic" origin of TeV emission can be applied also to the second CANGAROO source, RX J1713.7-3946 (Muraishi et al., 2000). However, the case of this source seems to be even more complicated and knotty. Recently Enomoto et al. (2002) published the spectrum of TeV emission of this source based on newer CANGAROO observations. The spectrum is claimed to be steep with a power-law photon index $\Gamma = 2.8 \pm 0.2$ between 400 GeV and 8 TeV (see Fig. 2.10). They argued that such a steep spectrum is inconsistent with the inverse Compton model, but can be explained by π^0-decay γ-rays. If true, for the standard shock acceleration spectrum with power-law index ≈ 2, this would require an energy cutoff in the proton spectrum around 10 TeV. Immediately after publication of this paper, Reimer and Pohl (2002) and Butt et al. (2002) criticised this interpretation arguing that it violates the γ-ray upper limits set by EGRET at GeV energies. Note that in Fig. 2.10 from Enomoto *et al.* (2002) the EGRET flux upper limit is indicated only at 100 MeV, whereas the flux upper limits above 1 GeV are more restrictive. However, this is not a sufficiently robust argument to discard the hadronic origin of the reported TeV emission (see Sec.5.5.2 in Chapter 5).

On the other hand, the arguments against the IC model should be backed by thorough theoretical studies based on higher quality data from the radio, X-ray and γ-ray domains. The X-ray observations are of particular interest because the synchrotron X-ray fluxes reliably "control" the predictions of IC emission at TeV energies. The recent morphological and spectral studies of the northwest rim of RX J1713.7-3946 (the region where the acceleration of relativistic electrons and presumably also protons takes

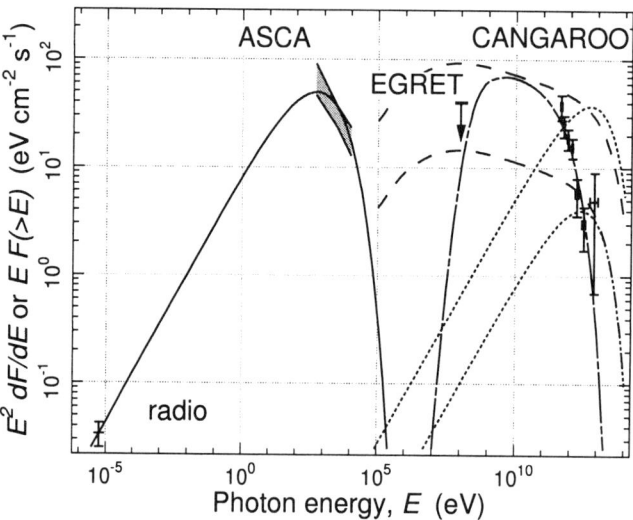

Fig. 2.10 Multiwavelength spectral energy distribution of RX J1713.7-3946 (from Enomoto *et al.*, 2002). The solid, dotted, dashed, and dot-dashed curves correspond to model calculation for synchrotron, inverse Compton, bremsstrahlung, and π^0-decay radiation components. The upper and low curves for the inverse Compton and bremsstrahlung components are calculated for two representative magnetic fields, 3 and 20 μG, respectively. The absolute fluxes of π^0-decay γ-rays are calculated assuming a spectral index of accelerated protons $\Gamma_p = 2.08$, an exponential cutoff at $E_0 = 10$ TeV, and $(W_p/10^{50}\text{ erg})(n_0/300\text{ cm}^{-3})(d/6\text{ kpc})^{-2} = 1$, where W_p is the overall energy in protons, d is the distance to the source, and n is the ambient gas density.

place) with Chandra unexpectedly revealed remarkable structure of this region with bright filaments and hot spots, accompanied by dark voids (Uchiyama et al., 2003). This adds more puzzles to this source which may challenge the perceptions of the standard diffusive shock-acceleration model, and makes new γ-ray studies with forthcoming IACT arrays like H.E.S.S. and CANGAROO-III extremely important Both the high TeV fluxes reported by the CANGAROO collaboration (approximately at the Crab level), and the typical angular dimensions of the relevant regions (from several to tens of arcminutes) perfectly match the performance of these instruments for an adequate spectral and spatial studies in a broad energy region from 100 GeV to 10 TeV.

Cas A. While the origins of TeV emission claimed to be detected from SN 1006 and RX J1713.7-3946 are not yet established, and remain controversial, it is likely that the TeV emission from the young supernova remnant Cas A detected by the HEGRA telescope system can be relatively easy explained by interactions of accelerated protons with the ambient gas. The "brute-force" observation strategy applied by the HEGRA collaboration to one of the most prominent objects in the sky, was eventually rewarded by detection of a tiny flux, $(5.8 \pm 1.2_{\text{stat}} \pm 1.2_{\text{syst}}) \times 10^{-13}$ ph/cm^{2}s above 1 TeV (Aharonian et al., 2001d). More than 200 hours of data accumulated during 3 years from 1997 to 1999, revealed a positive signal with statistical significance of 5σ. The high quality of data obtained by the stereoscopic system of imaging Cherenkov telescopes, as we well as the power of the stereoscopic approach to control the background conditions, make this result highly reliable.

Cas A is the result of the youngest supernova event in our Galaxy that took place around 1680. It is a very bright radio source. The source is bright also in X-rays, a noticeable fraction of which may have nonthermal origin. But, unlike SN 1006 and RX J1713.7-3946, the hard X-ray emission of this source can be interpreted in different ways, therefore we still do not have strong observational evidence of acceleration of electrons in this object to multi-TeV energies. These electrons present a certain interest as contributors to the TeV γ-ray emission through the inverse Compton scattering. However, because of the well established strong magnetic field in the nebula, well over 100 μG, it is likely that the inverse Compton contribution is not very significant, unless we assume multi-zone structures containing regions with a very low magnetic field (see Fig. 2.11, and discussion in Sec. 5.6 of Chapter 5). On the other hand, because of very low reported TeV γ-ray flux and high average density of the nebular gas, a little effort is needed to explain the TeV radiation by interactions of accelerated protons. This interpretation requires $\approx 10^{49}$ erg total energy in accelerated protons (see Fig. 2.11) which exceeds by only a factor of 10 the energy in relativistic electrons derived from radio observations.

Crucial information about the radiation mechanism is contained in the energy spectrum. For example, inverse Compton scattering predicts a differential spectrum steeper than E^{-3}. Therefore, the detection of a steep γ-ray spectrum would be an additional argument in favour of the hadronic origin of TeV emission. The tiny signal detected by HEGRA does not allow the derivation of the energy spectrum. But hopefully this information will be available soon with the forthcoming IACT arrays.

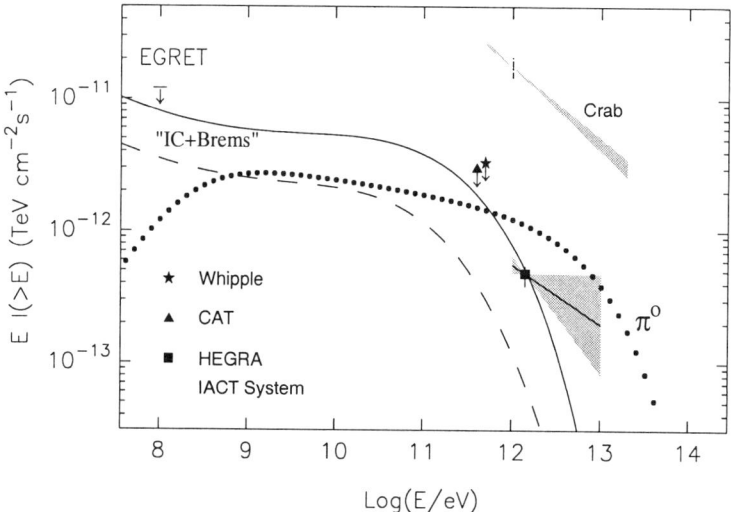

Fig. 2.11 Gamma-rays from Cas A. The shaded area shows the 1σ error range for the fluxes measured by the HEGRA CT-system (Aharonian et al., 2001d). Also indicated are the flux upper limits set by EGRET, Whipple and CAT telescopes. The model predictions are from Atoyan et al. (2000b). The dotted curve represents the fluxes for π^0-decay decay γ-rays calculated for relativistic protons with power-law index $\Gamma = 2.15$ (identical to the spectral index of radio emitting electrons), exponential cutoff at $E_0 = 200$ TeV and total energy 2×10^{49} erg. The density in the shell is assumed $n = 15$ cm^{-3}. The solid and dashed lines correspond to the γ-ray fluxes produced by electrons (IC+Bremsstrahlung) calculated in the framework of a 3-zone model for 2 set of basic parameters discussed in Sec. 5.6 of Chapter 5).

2.3.2.5 Other galactic sources

The high γ-ray detection rate capability of atmospheric Cherenkov telescopes makes them effective tools for searches for possible episodic or burst-like TeV emission from compact objects in our Galaxy. After the failure of the present generation of ground-based instruments to confirm the early optimistic claims of detection of TeV signals from X-ray binaries and cataclysmic variables, these sources have been removed from the highest priority target list of ground-based observations. However, after the discovery of galactic sources with relativistic jets, or the so-called *microquasars*, which unexpectedly revealed that non-thermal high energy processes play an important role in these accretion-powered objects, the X-ray binaries regained popularity in the gamma-ray community. Some models of X-ray binaries with synchrotron radio jets predict that in addition to comptonized thermal X-ray emission formed in the hot accretion disks, one may also expect hard nonthermal X-rays produced in the jets by synchrotron radiation from ultrarelativistic, multi-TeV electrons. Under certain conditions, the same electrons may effectively radiate episodic VHE γ-ray emission through the inverse Compton channel (Chapter 7). Also, microquasars have been recently claimed as possible sources of high energy neutrinos due to photo-meson processes in the inner parts of the jets (Distefano *et al.*, 2002). If true, high energy neutrinos should be accompanied by observable gamma-radiation as well.

In this regard, the reports about tentative (but unfortunately not yet confirmed) detections of transient TeV signals from GRS 1915+105 (Aharonian and Heinzelmann, 1998) and from the accreting X-ray binary Cen X-3 (Chadwick *et al.*, 1998b) represent intriguing evidence that VHE phenomena indeed may take place in these objects. Therefore, it is quite important to monitor these sources with the next generation of IACT arrays. The possible links between relativistic motion and accretion phenomena in these objects may provide a key insight into the nature of not only compact galactic X-ray sources, but also for the engines of AGN and quasars.

The episodic character of VHE γ-ray emission from highly variable sources requires detailed search strategies supported by multiwavelength observations. TeV episodic sources can be revealed also serendipitously during systematic surveys of the galactic plane. Generally, the atmospheric Cherenkov telescope technique is designed for pointing observations. However, the stereoscopic approach coupled with the relatively large field of view of the imaging cameras, provides highly accurate reconstruction of γ-ray arrival directions up to 2 degrees off-axis, thus allowing reasonable surveys of the sky from single tracking positions. The HEGRA collaboration (Aharonian et al., 2002d) has demonstrated the effectiveness of this technique using the system of telescopes equipped with 271 pixel cameras with 4.3° diameter FoV for the first sensitive TeV survey. One quarter of the galactic plane ($-2° \leq l \leq 85°$) has been covered, which included 86 radiopulsars, 63 supernova remnants, and nine GeV sources. In total, 115 hours of high quality data distributed over 92 individual locations were accumulated. For reference and calibration purposes, several observations of the Crab Nebula in different periods were also carried out.

No evidence for TeV emission was found; the flux upper limits range from 0.15 Crab to several Crab units depending on the observation time and angular size of the objects. The ensemble sums over selected pulsar, SNR and the so-called GeV-source sample (Lamb and Macomb, 1997) do not show positive signals either. Nevertheless they allow quite robust upper limits from these potential source populations – 3.6%, 6.7%, and 5.7% of the Crab flux for the pulsar, SNR, and GeV-source subsets, respectively.

The observations followed mainly the galactic equator with a spacing between individual scan positions of 1 degree, but in some regions, in particular in the Cygnus region, the survey covered a larger range in galactic latitude, although at the expense of a reduced overlap between adjacent points. In addition, the Cygnus region has been later observed more extensively due to several pointing observations in search of TeV counterparts for two EGRET GeV sources, GeV J2035+4214 and GeV J2026+4124, as well as from the binary X-ray source Cyg X-3. These observations during 1999, 2000 and 2001 (for a total 113 hours) have serendipitously revealed a signal positionally inside the core of the OB association called Cygnus OB2 (Aharonian et al., 2002b). The sky-map of the excess events of this region obtained by the HEGRA telescope system is shown in Fig. 2.12. The excess significance at the source center of gravity amounts to 5.9σ, which however is reduced to 4.6σ after accounting for the statistical trial factor. The source did not show flux variability, but the marginal signal accumulated over 3 years does not allow a strong statement in this regard. On the other hand, the source seems to be extended with an intrinsic radius of about 5 arcminutes. Another interesting feature of the source is its quite hard spectrum (see Fig. 2.13) that can be approximated as power-law with a photon index $\Gamma = 1.9 \pm 0.3_{stat} \pm 0.3_{sys}$. The integral flux above 1 TeV amounts to 25 mCrab.

To verify the first serendipitously discovered TeV source, labelled

TeV J2032+4130, the same region was monitored again in 2002. The observations over 130 hours confirmed the presence of the TeV source at the same position with the same (originally reported) flux and angular size of γ-ray emission. The excess significance from all data exceeds 7σ (G. Rowell, private communication). The energy spectrum also appeared consistent with the previous result, confirming that the hard spectrum extends to ~ 10 TeV.

Although the TeV source is located in a remarkable place, in the core of Cygnus OB2 (Knödlseder, 200), where the total mechanical energy of young stellar winds exceeds several times 10^{39} erg/s, so far no remarkable counterpart is identified with the TeV source. No luminous star or catalogued X-ray source from the ROSAT survey lies within the TeV error circle. The archival ASCA GIS data give a 99% upper limit of 2×10^{-12} erg cm^{-2} s^{-1} in the interval 2-10 keV (see Fig. 2.13).

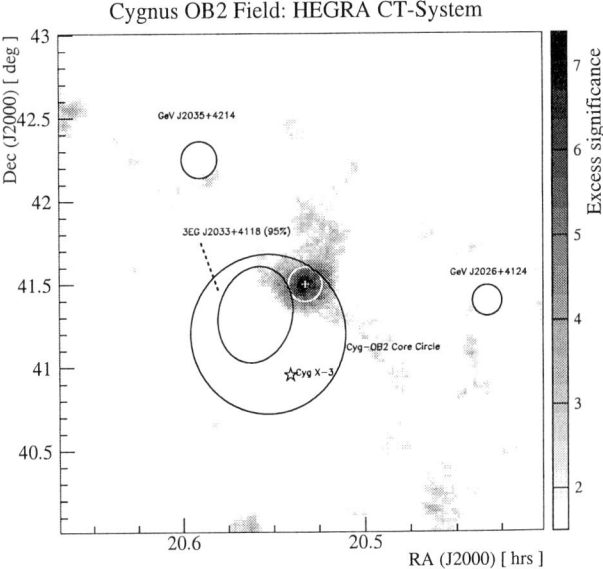

Fig. 2.12 Sky-map of event excess significance for a $3° \times 3°$ region centered on TeV J2032+4130. Only significances larger than 3σ are shown. The locations ellipses (95 % confidence) of 3 EGRET unidentified sources, the nearby X-ray binary Cyg X-3, and the core of Cygnus OB2, are also shown (courtesy of G. Rowell).

Such results may imply that the particle accelerator is not co-located with the TeV source, i.e. the particle acceleration and γ-ray production regions are well separated. This scenario generally favours a hadronic origin of the radiation. While this radiation should peak in the highest density environments, e.g. in the nearby dense molecular clouds, the TeV γ-ray emissivity immediately in the accelerator can be quite modest because of low density of the target material inside, as well as due to fast escape of the highest energy particles from the acceleration region. In this scenario, the hard TeV spectrum of the observed emission can be readily explained by (energy-dependent) diffusion effects characterising propagation of protons from the accelerator to the target (see Chapter 4). Assuming that the TeV source is indeed located in Cygnus OB2, the distance to which is estimated as 1.7 kpc, the TeV γ-ray luminosity of $\sim 6 \times 10^{32}$ erg/s between 1 and 10 TeV requires a reasonable total energy in ≥ 10 TeV protons, $W_{\mathrm{p}} \simeq 10^{47}(n/100 \text{ cm}^{-3})^{-1}$ erg, where n is the gas density in the γ-ray production region.

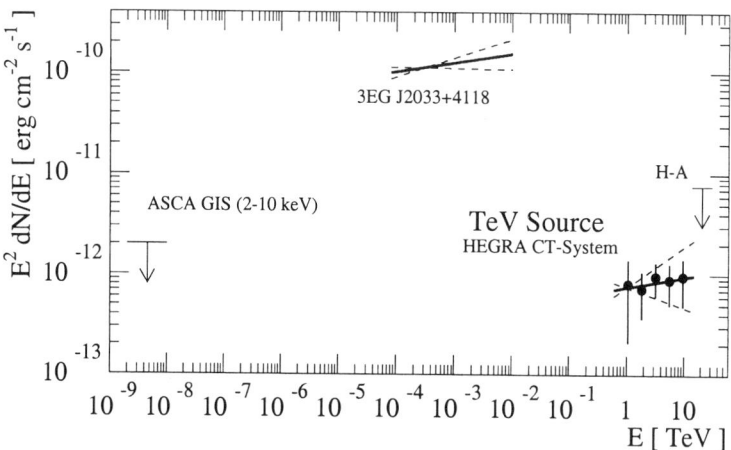

Fig. 2.13 The differential energy spectrum of TeV J2032+4130 (from Aharonian et al., 2002b). The energy flux of the EGRET source 3EG J2033+4118 and the ASCA GIS upper limits are also shown.

An alternative mechanism for extended TeV emission can be the inverse Compton scattering of ultrarelativistic electrons. This mechanism can be realized in two possible scenarios – in a plerion (with possibly a misaligned pulsar) or in a jet-driven termination shock. The second case apparently implies the existence of a nearby microquasar. One may speculate that two EGRET sources in the proximity of the TeV source in Fig. 2.12 (one of them, 3EG J2033+4118, is possibly associated with Cyg X-3; see Mori et al., 1997), could be possible candidates for such a microquasar. Independent of the origin of accelerated electrons, the inverse Compton model requires a rather low magnetic field in the region of γ-ray production, not larger than 10 μG, otherwise the synchrotron radiation of multi-TeV electrons would exceed the ASCA upper limit.

And finally, it is quite possible that the TeV source is a representative of a new source population in our Galaxy. The discovery of this source required more than 100 h observation time with the HEGRA stereoscopic system. Because of its hard spectrum, the source is not accessible, for any reasonable observation time, for other current ground-based instruments. However, with arrival of the next generation of IACT arrays with significantly improved sensitivities at TeV energies (by a factor of 5 or more compared to the HEGRA telescope system at 1 TeV), statistically significant signals from similar sources should appear after several hours of observations. The discovery and detailed study of spectral and angular properties of more TeV objects of this kind seems very promising in the context of planned deep survey of the the galactic plane with the H.E.S.S. IACT array, given its ideal location in the southern hemisphere, effective energy threshold around 100 GeV, superior flux sensitivity $\sim 10^{-12}$ erg/cm^{2}s (above 1 TeV for less than 10 hour observation time, and for relatively compact objects with angular size less than several arcmin), and large, 5-degree FoV.

EXTRAGALACTIC SOURCES

2.3.2.6 *TeV blazars*

Many nonthermal extragalactic objects representing different classes of AGN and located within 1 Gpc are potential TeV sources. First of all this concerns the BL Lac population of blazars of which four representatives, Mkn 421 (at redshift $z = 0.031$), Mkn 501 ($z = 0.034$), 1ES 1959+650 ($z = 0.047$), and 1ES 1426+428 (at $z = 0.129$) are firmly established by independent observations of several groups as TeV γ-ray emitters. The current list of extragalactic TeV sources contains two more BL Lac objects detected at 4 to 6 standard-deviation significance level – 1ES 2344+51 at $z = 0.044$, and PKS 2155-304 at $z = 0.116$, as reported by the Whipple (Catanese et al., 1998) and Durham (Chadwick et al., 1999) groups, respectively. The Crimean group has claimed detection of TeV γ-rays from two other AGN – 3C 66A (Neshpor et al., 1998) and BL Lacertae (Neshport et al., 2001). The reported γ-ray fluxes significantly exceed the upper limits set by other groups with more sensitive instruments. But this cannot be interpreted as contradiction, given the highly variably behaviour of blazars. It should be noted in this regard that BL Lacertae ($z = 0.069$) – the prototypical object for a class of AGN (BL Lacs) that are characterised by strong variability in continuum emission and polarization, but with very weak emission and absorption lines – has not been established as an outstanding MeV/GeV γ-ray emitter until the remarkable optical outbursts in 1997 July, when the follow-up observations with EGRET revealed strong γ-ray flux above 100 MeV (Bloom et al., 1997). On the other hand, the claim about detection of TeV γ-rays from the distant blazar 3C 66A ($z = 0.44$) seems rather unlikely on theoretical grounds, given the fact that only a negligible fraction of TeV γ-rays would survive severe intergalactic absorption due to interactions with the diffuse extragalactic radiation fields (see Chapter 10).

Since the discovery of TeV γ-radiation of Mkn 421 by the Whipple group (Punch et al., 1992), this object has been subject of intensive studies through multiwavelength observations. The source is variable with typical average TeV flux between 30% to 50% that of the Crab Nebula, but with rapid, as short as 0.5 h, flares (Gaidos et al., 1996, Aharonian et al., 2002c). Until the spectacular high state of the source in 2001, the spectral studies were based mainly on the data taken during quiescent or moderately high states. In particular, the spectrum measured by HEGRA during the 1998 "ASCA" campaign (Takahashi et al., 2000) is fitted by a steep power-law with photon index $\Gamma \simeq 3$ (Aharonian et al., 1999d). The exceptionally bright and long-lasting activity of Mkn 421 in 2001 allowed the HEGRA and VERITAS groups (Aharonian et al., 2002c, Krennrich et al., 2002) to derive the time-averaged gamma-ray spectrum of the source in the high state which up to ~ 15 TeV is described as $dN/dE = KE^{-\Gamma} \exp(-E/E_0)$, i.e. by the same canonical "power-law with exponential cutoff" function found earlier by the HEGRA group for the high-state, time-averaged spectrum of Mkn 501 (Aharonian et al., 1999c). The spectra of these sources are not, however, identical. In Fig. 2.14 are shown the spectral energy distributions based on approximately 40,000 and 60,000 (!) TeV γ-ray events detected by the HEGRA telescope system during the exceptionally high states of Mkn 501 in 1997 and Mkn 421 in 2001, respectively. They are described by different $[\Gamma, E_0]$ combinations: $\Gamma = 1.92$ and $E_0 = 6.2$ TeV for Mkn 501 (Aharonian et al., 1999c) and $\Gamma = 2.19$ and $E_0 = 3.6$ TeV

(Aharonian et al., 2002c). The power-law part of the spectrum of Mkn 421 is steeper, and the exponential cutoff starts earlier. The difference is clearly seen from the the lower panel of Fig. 2.14 where the ratio of two energy spectra is shown. Since both sources are located at approximately the same distance, the difference in the cutoff energies can be interpreted as an indication against the hypothesis that attributes the cutoffs to the pure intergalactic-absorption effect (see Chapters 10).

For the typical energy resolution of Cherenkov telescopes of about 20%

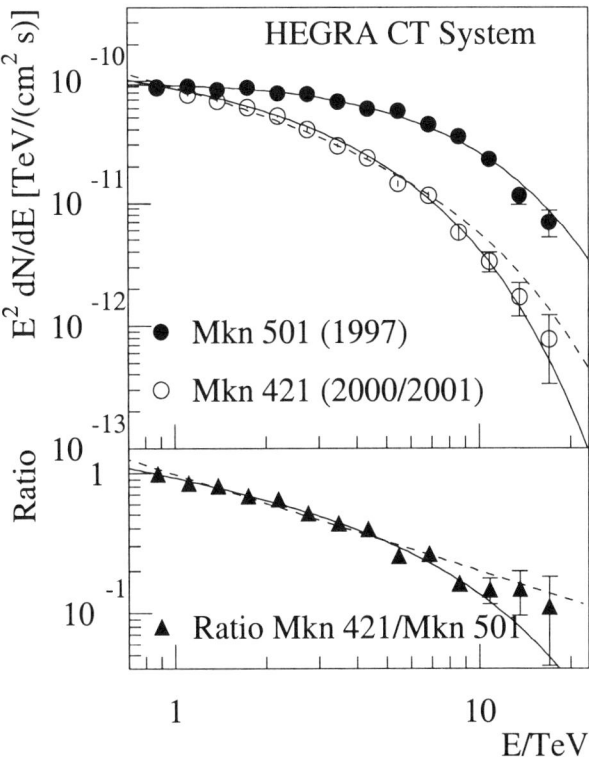

Fig. 2.14 The spectral energy distributions of Mkn 501 and Mkn 421 in high states (from Aharonian et al., 2002c). The observed differential spectra are well described by a power law with exponential cut-off. Both the cut-off energy and the photon index are different for the two energy spectra (see the text). A fit of a fixed cut-off energy at 6.2 TeV, as measured for Mkn 501 but letting the photon index vary, applied to the Mkn 421 data, results in the dashed curve in the upper. To demonstrate the difference in the two energy spectra and to reduce the impact of possible systematic effects, the ratio of the Mkn 421 and Mkn 501 spectra is shown in the lower panel.

or larger, one has to be careful with conclusions concerning the spectral shape at energies well beyond the cutoff energy E_0. Obviously for determination of spectra with sharper (super-exponential) cutoffs, a significantly better energy resolution is required. Motivated by this, the HEGRA collaboration recent re-analysed the "1997 high-state" spectrum of Mkn 501 using an improved method for energy reconstruction with $\simeq 10\%$ resolution achievable for the stereoscopic mode of observation (Hofmann et al., 2000). The new analysis confirms the result of the previous study up to 17 TeV, but provides a stronger constraint on the flux at 21 TeV (Aharonian et al., 2001a). With this new point, the Mkn 501 spectrum at very high energies is better described by a power-law with super-exponential cutoff (see Fig. 10.3).

The low fluxes of Mkn 421 and Mkn 501 detected by EGRET indicate that the "power-low with a (super)exponential cutoff" spectral presentation at TeV energies with photon index ≈ 2 in the high states of these objects, cannot be extrapolated towards low energies. There is little doubt that

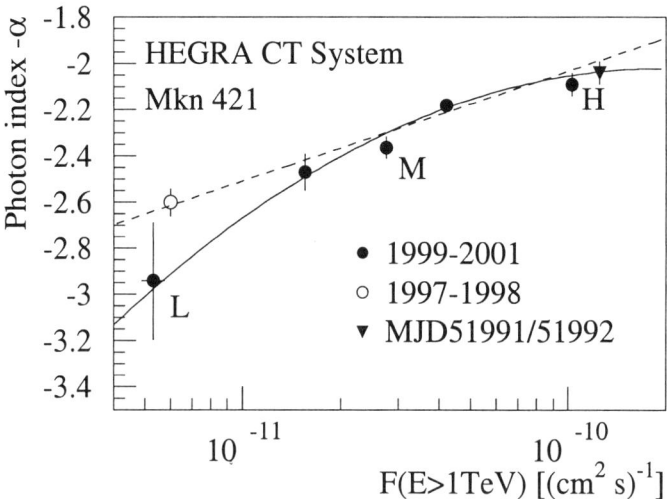

Fig. 2.15 The "photon index - absolute flux" relation based on the TeV measurements of Mkn 421 at different epochs (from Aharonian et al., 2002c). The energy spectra get harder with increasing flux as is seen when splitting data in separate flux intervals ("L", "M", and "H" indicate the levels of low, medium, and high fluxes, respectively) and fitting by a power law with a fixed exponential cut-off energy (3.6 TeV) to the individual energy spectra. Also indicated are the values for the night MJD51991/51992, as well as the average flux from the 1997/1998 observations of Mkn 421.

significant spectral changes should occur in the transition region from GeV to TeV energies. Therefore, the spectral studies at energies around 100 GeV are of a great interest. During the high state of Mkn 421 in 2001, an important measurement at 140 GeV was performed by STACEE – a Cherenkov-wave-front sampling detector using heliostats of the National Solar Thermal Test Facility near Albuquerque, New Mexico (Boone et al., 2002). The temporal evolution of the flux during the March to May period appeared consistent with observations in both the TeV (Whipple) and X-ray (RXTE) bands. The mean energy flux of about 10^{-10} erg/cm^{2}s is comparable with the TeV flux averaged for the same period of activity of the source. However, the lack of information about fluxes derived by STACEE and Whipple for precisely overlapping time intervals, prevents a broad-band compilation of the γ-ray spectral energy distribution, given the variability of the source on timescales of several hours or less.

Despite the huge overall photon statistics, the spectra of both sources above 10 TeV can only be determined using data accumulated over long periods of observations. Obviously, the time-averaged spectra of variable sources obtained in such way can be considered as *astrophysically meaningful* only if the spectral shape is time-independent, i.e. does not strongly correlate with the absolute flux. Remarkably, during the high state in 1997 the shapes of daily spectra of Mkn 501, despite dramatic flux variations, remained essentially stable (Aharonian et al., 1999a), albeit for some specific flares non-negligible spectral variations cannot be excluded. For example, the CAT group (Djannati-Ataï et al., 1999) found evidence for a noticeable hardness-intensity correlation for the 1997 April 7 and 16 flares.

The sensitivity of current ground-based detectors is not sufficient to study the spectral variability of TeV radiation of Mkn 421 and Mkn 501 in quiescent states on timescales less than several days. However, even the time-averaged spectra in low states are of certain interest. The HEGRA observations of Mkn 501 in 1998 and 1999, when the source was at a ≈ 10 times lower flux level compared to the average flux in 1997, revealed a noticeable steepening (by 0.44 ± 0.1 in photon index) of the energy spectrum (Aharonian et al. 2001b). Clear "spectral shape - absolute flux" correlations on large time-scales have been found also for Mkn 421 by the Whipple (Krennrich et al., 2002) and HEGRA (Aharonian et al., 2002c) groups (see Fig. 2.15).

Recently, similar behaviour has been observed from another TeV blazar - 1ES 1959+650 . This is an X-ray selected BL Lac object at z=0.047 with a steep X-ray spectrum. In the quiescent state, the synchrotron peak is located at several eV, but during strong flares it can be significantly shifted towards higher, keV energies. The first evidence of TeV emission from this source was reported by the Telescope Array group (Nishiyama et al., 1999). The source was monitored with the HEGRA system during the July 2000 - October 2001 period. The data, based on 94 h of observation time, revealed a positive ($\approx 5\sigma$) γ-ray signal (Götting et al., 2001; Aharonian et al., 2003a). The energy spectrum relevant to this period of observations with an average flux of the source of about 0.08 Crab is well described by a power-law with a photon index 3.3 ± 0.7 (see Fig. 2.16).

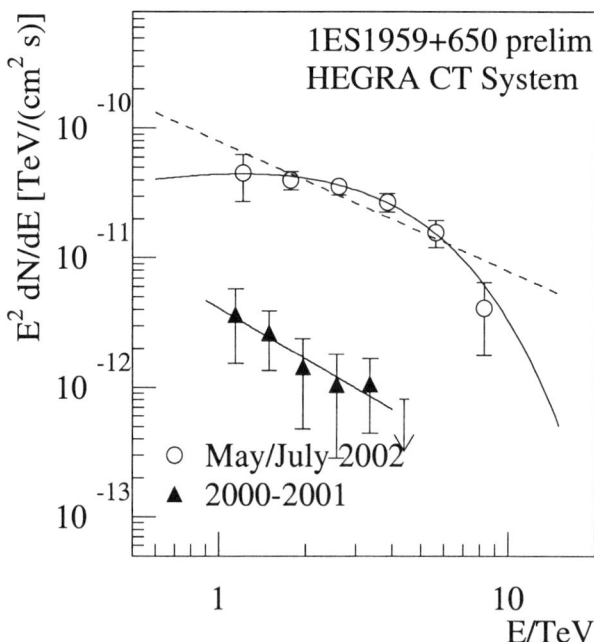

Fig. 2.16 The energy spectra of 1ES1959+650 in the low (2000/2001) and in the high states by combining data of May 18/19, 19/29, July 13 and July 14 flares (from Horns et al., 2003).

In May 2002 the source suddenly entered a remarkably high state that lasted at least 3 months. The high TeV activity noticed by the Whipple collaboration (Dowdall et al., 2002), triggered intensive monitoring of the source also with the CAT and HEGRA telescopes, as well as by the pointed X-ray instruments aboard RXTE. During this period, the variable TeV flux of the source very often exceeded the Crab flux. The energy spectrum of 4 combined flares in May 18/19, 19/29, July 13 and July 14 is shown Fig. 2.16. The function $E^{-1.5}\exp{(-E/E_0)}$ with $E_0 \approx 2.4$ TeV describes the time-averaged spectrum quite well . Unfortunately, the limited spectral points detected by HEGRA cannot tell us whether the index of 1.5 reflects the real power-law photon index (slope) of the spectrum at energies $E \ll E_0$, or whether it should be treated as a formal fit parameter for the region above 1 TeV. In this regard, results from low-energy threshold CAT data, obtained at preferable (smaller) zenith angles (because of the location of this instrument at 42N latitude) is of great interest, in particular for derivation of constraints on the diffuse extragalactic background at near infrared and optical wavelengths (see Chapter 10).

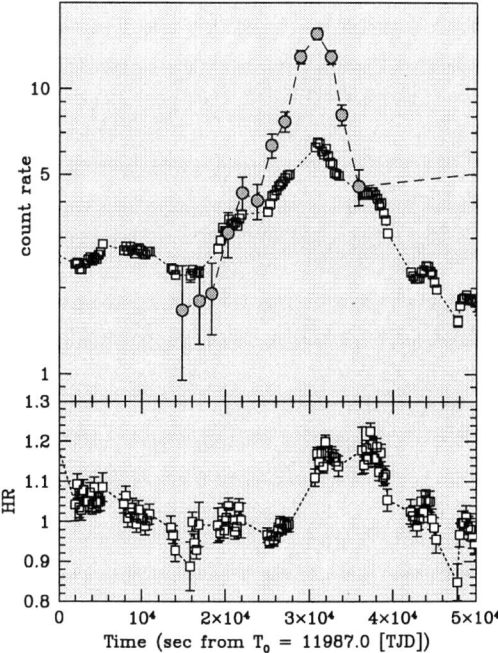

Fig. 2.17 Simultaneous X-ray and γ-ray observations of the 2001 March 19 flare of Mkn 421 by RXTE (squares) and VERITAS (circles). The bottom panel shows the RXTE/PCA "6-11 keV/3.5-4.5 keV" hardness ratio (courtesy of G. Fossati).

The imaging atmospheric Cherenkov telescopes are nicely suited to searches for short signals from TeV blazars in flaring states. In particular, they can follow flares at flux level ≈ 10^{-11} erg/cm^{2}s on timescales less than several hours, and thus are well-matched to the sensitivity and spectral coverage of X-ray detectors of the RXTE, BeppoSAX and XMM satellites for multiwavelength studies of flux variations. This is a key condition which makes simultaneous X- and γ-ray observations meaningful.

The first important results on the correlated TeV and X-ray flares were obtained during the extraordinary activity of Mkn 501 in 1997 (see Figs.10.8 and 10.9 in Chapter 10), as well as during the 1998 multiwavelength campaign of Mkn 421 (Takahashi et al., 2000). The best results so far in this regard became available after the well coordinated multiwavelength campaigns of spectacular flares of Mkn 421 in 2001. On several occasions, truly simultaneous observations by RXTE and TeV instruments with durations up to 6 hours per night were carried out. A nice sample of such events, the 2001 March 19 flare detected by RXTE and VERITAS shows a clear keV/TeV correlation on sub-hour timescales (Fig.2.17).

The detection of a large number of strong flares during the high activity of Mkn 421 in 2001 by the HEGRA telescope system shows that the temporal and spectral properties of individual flares can be very specific and essentially different from each other. This is demonstrated in Fig.2.18 where the time profiles of two flares observed during two subsequent nights, March 21/22 and March 22/23, are shown. Both flares have asymmetric profiles described by a surprisingly simple analytical function consisting of two exponentials that characterise the rise and decay times,

$$F(t) = f_0 + f_1 \cdot \begin{cases} \exp[\tau_{\rm rise} \cdot (t-t_0)] & t \leq t_0 \\ \exp[-\tau_{\rm decay} \cdot (t-t_0)] & t > t_0 \end{cases}$$

However, while the rise time of the first flare, $\tau_{\rm rise} \approx 45$ min was 3 times longer that the decay time, $\tau_{\rm decay} \approx 15$ min, the rise time of the second flare, $\tau_{\rm rise} \approx 30$ min, was almost an order of magnitude shorter than the decay time $\tau_{\rm decay} \approx 4.5$h.

For a deep understanding of acceleration and radiation processes in jets, it is crucial to search for *spectral* variability on timescales comparable to the characteristic dynamical times of flares. The strongest flares of Mkn 421 in 2001 provide us with such unique data. In particular, the HEGRA IACT system was able to perform spectral studies on hour timescales, when the flux was larger than 1-2 Crab units. This is seen in Fig. 2.18 where in the lower panels the hardness ratios of these two flares are shown as a function of time. During the rising stage, the spectra in both cases show a significant hardening, correlated with the absolute flux increase, and a softening at the end of the flare.

This behaviour is clearly seen in Fig.2.19, where the energy spectra are shown for two different time intervals that cover the pre-flare and flare stages (indicated by vertical lines in Fig. 2.18). For both flares, the pre-flare energy spectra are described by a power-law with a photon index of ≈ 3, with a significant hardening by $\Delta\alpha = 0.75 \pm 0.25$ during the flare.

The analysis of other flares detected by HEGRA show that the flux variations are not always accompanied by spectral changes (Aharonian et al., 2002c). For example, one flare has been observed with very high flux, but with power-law photon index of ≈ 2.9. On the other hand, very strong

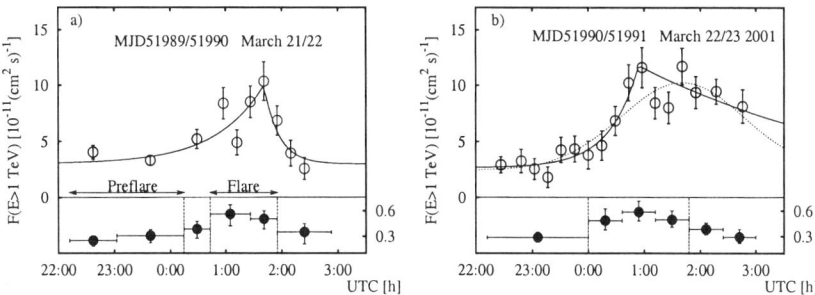

Fig. 2.18 The time profiles of two Two flares of Mkn 421 during the night March 21/22, 2002 and March 21/22, 2002 (right panel). The solid lines indicate a fit of the function $F(t)$ (see the text). The lower panels show the time evolution of the spectral hardness ratios of these flares (from Aharonian et al., 2002c).

flares of a comparable strength sometimes may show significantly different energy spectra. Such an "irregular" relation between the absolute flux and the energy spectrum indicates that the spectral variations are result of dramatic nonlinear changes of conditions and principal parameters in the particle acceleration and γ-ray production regions, rather than a consequence of quantitative changes of the power that initiates and drives these flares. For a proper understanding of complex processes in the blazar jets, the TeV data should be treated in the context of multiwavelength observations, first of all together with X-rays. Remarkably, TeV observations of Mkn 421 in 2001 have been perfectly coordinated with X-ray observations. Many strong flares of this exceptionally high state of Mkn 421 have been continuously observed by the RXTE detectors together with the HEGRA and Whipple telescopes.

As mentioned above, simultaneous X-ray (RXTE) and TeV (CAT,

HEGRA, Whipple) data of high quality have been obtained also during the high state of 1ES 1959+650 in 2002. These data, together with the multiwavelength observations of Mkn 501 in 1997 and Mkn 421 in 2001 constitute solid observational material for further phenomenological and theoretical studies. At the same time, it would be naive to expect that

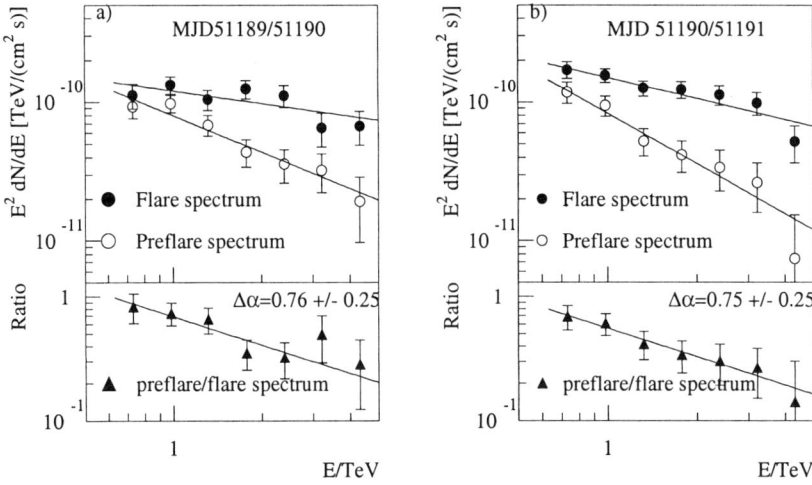

Fig. 2.19 The evolution of the energy spectra of two flares detected during the nights March 21/22 and March 22/23. Both flares show a significant spectral hardening while the flux increases (from Aharonian et al., 2002c).

these result will immediately bring harmony to the current paradigm of the TeV blazars. There is little doubt that these data will require significant modification, if not a dramatic revision of the existing models of TeV emission, including the currently popular synchrotron-self-Compton (SSC) or external Compton models (see Chapter 10).

A real problem for the existing models could come also from the TeV blazar 1ES 1426+428 detected recently in TeV γ-rays by the Whipple (Horan et al., 2002, Petry et al., 2002), HEGRA (Aharonian et al., 2002a) and CAT (Djannati-Ataï et al., 2002) groups. Although the statistical significances of reported signals in each case were at the level of 5σ, the three independent measurements do not leave a doubt in the discovery of the 4th TeV blazar.

Motivated by this result, the HEGRA CT-system has been used for very deep (more than 200 hours) observations of the source in 2002 (Aharonian et al., 2003b). Although during this period, the flux was by a factor of ≈ 2.5

lower than during the previous observations in 1999/2000 (Fig. 2.20), the source was again detected at 5.3 σ level, thus confirming this object as a TeV source.

Because of the relatively large redshift of 1ES 1426+428 ($z = 0.129$), the TeV radiation from this extreme BL Lac object suffers severe intergalactic absorption. Therefore the discovery of TeV γ-rays was, to a certain extent, a big surprise. Despite large uncertainties in the *Cosmic-Infrared-Background* (CIB), which is a part of the overall diffuse extragalactic background radiation (DEBRA) at optical and IR wavelengths, there is little doubt that the TeV γ-rays from this source arrive after significant absorp-

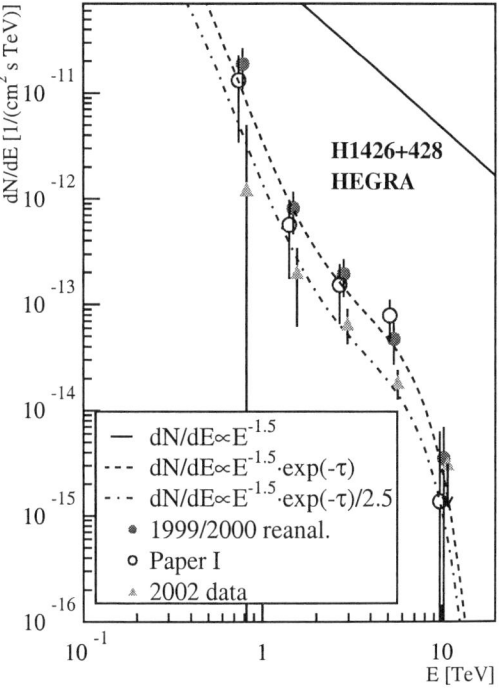

Fig. 2.20 The energy spectra of 1ES 1426+428 derived by the HEGRA CT-system for the different data sets. The reanalysed data set taken in the years 1999 and 2000 reproduces the initially published spectrum (Aharonian *et al.*, 2002a). The solid curve represents a possible intrinsic (source) γ-ray spectrum, and the dashed and dot-dashed curves represent the observed spectrum after intergalactic absorption (adopting the CIB model of Primack *et al.* (2001); see Chapter 10) for the 1999/2000 and 2002 measurements, respectively. The data set taken in 2002 is compatible with the 1999/2000 data with a flux level reduced by a factor of 2.5. (From Aharonian *et al.*, 2003b).

tion. For any reasonable CIB model, the absorption effect is at least factor of 10 or, more likely, factor of 100 (see Fig.10.10). This implies not only significant deformation of the primary spectrum, but also very large TeV luminosity of the source. The spectrum of γ-rays detected by HEGRA, CAT and Whipple has a quite unusual shape – it is steep below 1 TeV, but becomes flat above 1.5 TeV (see Fig. 2.21). However, the source spectra (i.e. the spectra after correction for the intergalactic absorption), look significantly different compared to the observed spectrum. It is demonstrated in Fig. 2.21 for three different models of the diffuse extragalactic background radiation. While for the CIB model suggested by Primack *et al.* (2001) the intrinsic γ-ray spectrum has a surprisingly "decent" power-law form with a photon index close to 1.5 (Fig. 2.21a), for two other CIB models the reconstructed source spectra have unusual forms with very sharp positive slopes ($E^2 \, dN/dE \propto E^3$) below 1 TeV (Fig. 2.21b; for an extreme CIB model discussed in Aharonian *et al.*, 2002a) or above 1 TeV (Fig. 2.21c; for the CIB spectrum suggested by Malkan and Stecker, 2001). Such a strong dependence of the γ-ray source spectra on the CIB model, which is explained by very large optical depth at all energies of detected γ-rays, implies that gamma-ray astronomers are quite close to resolve the diffuse infrared background between 1 an 10 μm (see Chapter 10).

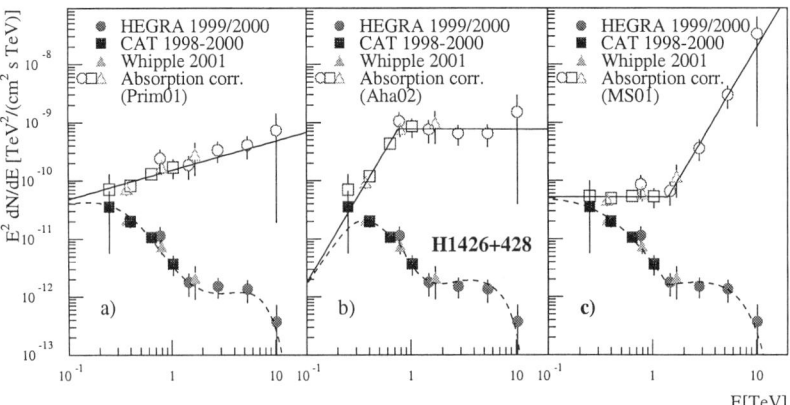

Fig. 2.21 Energy spectra of 1ES 1426+428. The filled symbols indicate the measured fluxes reported by the Whipple, HEGRA and CAT groups, the open symbols – intrinsic fluxes after correction for the intergalactic absorption based on three models of CIB.

This would require, however, significant experimental efforts. In particular, one should be careful when drawing conclusions about the energy spectrum based on a compilation of fluxes obtained by different groups at different observation epochs. Derivation of the energy spectrum from 100 GeV to 10 TeV by a single instrument during a relatively short observation period is a crucial condition for unbiased conclusions concerning the spectral shape of radiation relevant to the well defined states of the source. The low fluxes and the significant variability make this a rather difficult task, which however can be successfully performed with the next generation of IACT arrays. The detection of at least one more object at similar redshift, $z \sim 0.1 - 0.15$, is highly desirable for conclusions concerning the impact of intergalactic absorption on deformation of the intrinsic γ-ray spectra. And finally, the reduction of the energy threshold down to 100 GeV at which the absorption effect becomes rather small (see Fig.10.10), is another important issue. Observations of 1ES 1426+428 by the MAGIC and VERITAS telescopes and PKS 2155-304 by the CANGAROO-III and H.E.S.S. telescope arrays, promise cosmological information of extraordinary significance about the diffuse extragalactic background at optical and infrared wavelengths.

In any case, this source seems to be a representative of a new class of γ-ray blazars that can be different from both the GeV blazars and the "standard" TeV BL Lac objects. It should be noted, however, that the classification of TeV blazars based on a few detected objects is foolhardy. Therefore, in addition to detailed studies of spectral and temporal characteristics of selected individual objects, the discovery of a large number of new TeV blazars is the second key issue for further development of the field. In this respect, with the arrival of the new generation of IACT arrays, crucial results are expected in the near future. Both the significant improvement of flux sensitivities at TeV energies, and reduction of the energy threshold to 50-100 GeV, should result in a dramatic increase in the statistics of TeV blazars.

2.3.2.7 Other extragalactic objects

Although the TeV blazars were not initially predicted as TeV emitters, now it is clear that the discovery of large TeV fluxes from these distant objects was possible because of the relativistic bulk motion of γ-ray production regions towards the observer, leading to the Doppler boosting of fluxes by several orders of magnitude. The detection of isotropically radiating sources of similar intrinsic power can hardly be realized in the foreseeable future.

In order to produce TeV γ-ray flux at the level of 0.1 Crab (approximately several time 10^{-12} erg/cm^{2}s), the γ-ray luminosity of an "isotropic" point source at a distance d should exceed $L \geq 3 \times 10^{42}$ $(d/100$ Mpc$)^2$ erg/c. Thus, one may hope to detect isotropically emitted γ-rays only from the most powerful extragalactic objects, like radiogalaxies or clusters of galaxies. On the other hand, the requirement for the γ-ray luminosity of sources within 10 Mpc is quite modest, so one may expect, on pure phenomenological grounds, detectable γ-ray fluxes from some nearby prominent galaxies. The radiogalaxies Centaurus A and M 87, as well the starburst galaxies M 82 and NGC 253 are obvious TeV source candidates, basically because of their overall power, a significant part of which is released in nonthermal forms. The first evidence of TeV radiation from an extragalactic object, the radiogalaxy Centaurus A, was obtained already in the 1970s (Grindlay et al., 1977). Recently, the CANGAROO and HEGRA collaborations reported TeV signals from two other nearby galaxies – NGC 253 (Itoh et al., 2002) and M 87 (Aharonian et al., 2003c).

Although the flux of γ-rays above 730 GeV from M 87 has been obtained using the reliable technique of shower reconstruction with the HEGRA stereoscopic system of telescopes, the statistical significance of about 4σ is marginal, and therefore the result needs an independent confirmation. Unfortunately, the tiny signal detected at ≈ 0.03 Crab level is not achievable by other current instruments. But the next generation of IACT arrays in both hemispheres should be able to verify this discovery rather quickly, and, in the case of confirmation, to provide detailed studies of the energy spectrum over two decades, from 100 GeV to 10 TeV, localise the position and measure the possible extension of the TeV source with an accuracy of about 1-2 arcminutes. For the distance to M 87 of about 16 Mpc, this angular scale corresponds to 5-10 kpc. If the extension of the TeV source exceeds several arcminutes, this would be an indication that the TeV emission is produced within the giant elliptical galaxy M 87. Both the inverse Compton scattering of relativistic electrons on 2.7 K CMBR and interactions of cosmic ray protons with the ambient interstellar gas can be responsible for the observed TeV emission. In particular, approximately $10^{57}(n/0.1$ cm$^{-3})^{-1}$ erg total energy in ≥ 10 TeV protons is required to be accumulated in the elliptical galaxy (n is the average number density of the interstellar gas) in order to explain the observed γ-ray flux by p-p interactions. Another possible site of TeV emission is the famous kpc scale jet with several bright knots detected at radio, optical and X-rays. The latter is believed to have a synchrotron origin and be produced by electrons with energies up to 100

TeV. If so, within reasonable model parameters, detectable fluxes of inverse Compton TeV γ-rays also can be expected.

The detection of a TeV signal from the starburst galaxy NGC 253 has been reported by the CANGAROO group at the 11σ level (Itoh et al., 2002). Despite the very high statistical significance, this detection perhaps should be considered yet as tentative, because of possible systematic effects which are less controllable in the case of a single telescope, especially for the large angular size of the excess region $\sim 0.25°$ claimed by the authors. For the distance to the source 2.5 Mpc, this corresponds to the liner size of about 10 kpc, just the size of the radio halo discovered earlier around NGC 253. This prompted the authors to propose that the TeV radiation originates from the same radio halo region due to the inverse Compton scattering of electrons (Itoh et al., 2003a), which proceeds with very high efficiency in this region. Because of severe radiative losses, it is unlikely that the TeV electrons are produced in the galactic disk and then transported to the halo region. Thus, one has to assume that the electron accelerator(s) are located immediately in the radio halo. If so, the particle acceleration by strong termination shocks of the galactic wind (see e.g. Jokipii and Morfil, 1985) could be a possible source of TeV electrons in the halo.

The measured TeV spectrum is very steep (see Fig. 2.22), with a slop described by a photon index 3.74 ± 0.27 (Itoh et al., 2003b). Such a steep spectrum can be explained, most naturally, assuming an exponential cutoff in the acceleration spectrum at energies around several TeV. On the other hand in order to avoid the conflict with the flux upper limits set by EGRET at lower energies, it is necessary to assume a rather flat electron spectrum with spectral index p close to 2. Note that these values characterise the cooled (steady-state) electron spectrum, and since the radiative (synchrotron and Compton) cooling time of ≥ 100 GeV electrons, which are responsible for ≥ 100 MeV γ-rays, does not exceed 10^6 years, one has to assume unusually hard acceleration spectrum with differential spectral index $p - 1 \sim 1$. Also, the magnetic field in the halo required by this model is somewhat lower (by a factor of 2 or 3) than the field $B \sim 6$ μG derived from the polarization measurements (Beck, 1994).

For different assumptions about the steady-state spectrum of electrons, Itoh et al. (2003b) estimated the total energy of electrons between 10^{55} and 10^{56} erg (Itoh et al., 2003). It is interesting to compare this estimate with the energy in cosmic ray protons, assuming that the bulk of TeV γ-rays is produced in interactions of cosmic rays with the ambient gas. A proton differential spectrum with a power-law index slightly less than 2 and with a break or exponential cutoff around 10 TeV can readily describe the TeV spectrum observed by CANGAROO, as well as accommodate the EGRET upper limits at GeV energies. For such a spectrum, the energy required in TeV protons is estimated $W_\mathrm{p} \simeq L_\gamma \, t_{\pi^0} \simeq 5 \times 10^{57} (n/0.01)^{-1}$ erg, where $L_\gamma \approx 10^{40}$ erg/s is the detected TeV γ-ray luminosity, t_{π^0} is the characteristic time of π^0-production (see Chapter 3), and n is the number density of the ambient gas. Even for a very low gas density in the halo, the total energy in protons seems not unreasonably high. Moreover, in the galactic disk, where the average gas density can be as large as $n \sim 10 \, \mathrm{cm}^{-3}$, the required energy in cosmic rays can reduced to a rather modest level, $\leq 10^{55}$ erg. The model of γ-ray production in the galactic disk does not agree, however, with the angular size of the TeV source reported by the CANGAROO collaboration. On the other hand, any strong assumption concerning the origin of the TeV emission seems a bit premature. Both the energy spectrum and the angular size of the source (and, perhaps, even the very fact of the TeV signal) should be verified by independent measurements. Such measurements with the CANGAROO-III and H.E.S.S. telescope arrays will be available within the next 1 or 2 years.

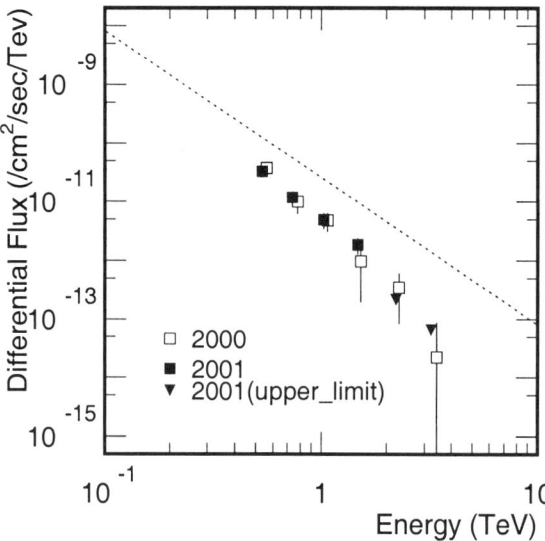

Fig. 2.22 Differential fluxes of γ-rays from the starburst galaxy NGC 253 based on the data obtained by the CANGAROO group in 2000 and 2001 (from Itoh *et al.*, 2003b). By the dotted line the Crab flux is shown for a reference.

2.3.3 Next generation of IACT arrays

The success of ground-based gamma-ray astronomy in the 1990s not only led to remarkable astrophysical discoveries, but also elucidated the promising detection techniques to be developed in the context of next generation instruments. Among a variety of competing designs, the stereoscopic arrays consisting of 10m diameter or larger aperture telescopes is identified as the most convincing approach that can facilitate a qualitative improvement in performance at an affordable cost, and, at the same time, promises a fast scientific return.

2.3.3.1 Atmospheric Cherenkov radiation

Observations of cosmic γ-rays from the ground is possible by detecting the secondary products of interactions of the primary γ-ray photon with the atmosphere, either directly or through their electromagnetic radiation. The atmospheric Cherenkov telescopes detect the Cherenkov light emitted in the atmosphere by secondary electrons which are produced in the cascades initiated by primary cosmic rays – protons, nuclei, electrons, γ-rays. Generally, the ground-based technique based on the *direct* detection of secondary cascade products – electrons, photons, muons, hadrons – effectively works in the ≥ 1 TeV primary energy region, even for particle detectors installed at very high mountain altitudes, closer to the shower maximum. At smaller primary energies the cascades die out in the upper atmosphere. But, fortunately, the atmospheric Cherenkov radiation potentially makes accessible primary γ-rays down to energies of about several GeV.

In the Cherenkov light, a γ-ray induced shower looks like an object of (approximately) elliptical shape that starts at an altitude of some 20-25 km (the point of the first interaction) and extends down to several km. The relativistic electrons move along the shower axis without significant lateral displacement, thus the light is emitted predominantly at the characteristic Cherenkov angle, which at the the maximum of the shower development of about 10 km above sea level (a.s.l.) is close to 1° (see e.g. Hillas, 1996). This results in a pool of Cherenkov light on the ground with a radius ≈ 120 m at average mountain altitude of about 2 km a.s.l. (see Fig. 2.23).

Because of absorption and scattering in the atmosphere, the Cherenkov light arrives with a spectrum that peaks at wavelengths around 300-350 nm. This light is very faint, with a density ranging from 100 to several 100 photons/m^2 (depending on the altitude) for a 1 TeV γ-ray photon, and very brief - it lasts only several nanoseconds. This implies that the

84 Very High Energy Cosmic Gamma Radiation

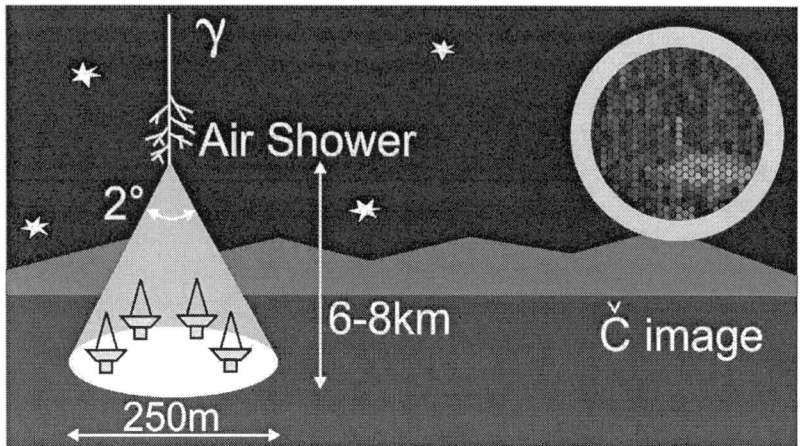

Fig. 2.23 Illustration of detection of the Cherenkov light from a shower initiated by a primary very high energy γ-ray photon.

Cherenkov telescopes must have large ($\gg 1 \text{ m}^2$) optical reflectors to image the Cherenkov light onto a multi-pixel camera. The latter should be sensitive to the visible (closer to blue) light and sufficiently fast with a typical "time gate" of about 10 nanoseconds. A characteristic image of a γ-ray induced shower obtained with such a camera is shown in Fig. 2.23. The total intensity of the image is a measure of the primary energy, the orientation of the image in the camera correlates with the arrival direction of the primary particle, and the shape of the image contains information about the the origin of the air shower (induced by a cosmic ray proton/nucleus or by a γ-ray photon). These three features comprise the basis of the Imaging Atmospheric Cherenkov Telescope (IACT) technique.

2.3.3.2 *Stereoscopic detection of Cherenkov images*

The huge detection area of showers determined effectively by the radius of the Cherenkov light pool, the effective separation of electromagnetic and hadronic showers and the good accuracy of reconstruction of shower parameters are three remarkable attributes of the IACT technique. The critical advantage of air-shower imaging is the discrimination it provides against a potentially overwhelming rate of the cosmic-ray induced shower events (Weekes and Turver, 1977, Stepanian *et al.,* 1983, Plyasheshnikov and Bignami, 1985, Hillas, 1985). Two independent image criteria are roughly equal in CR rejection power by a single telescope – the intersection of the shower track with the source location and the apparent shower width. The first criterion is valid for point sources, but does not depend on the origin (hadronic or electromagnetic) of showers. The second one is based on the inherently larger transfer momentum of secondaries produced in hadronic interactions. With single imaging telescopes, the Whipple, CAT and HEGRA groups have achieved approximately 300:1 rejection of cosmic ray showers with a factor of 2 loss in gamma-ray events.

Viewing the shower from several vantage points provides additional degrees of freedom, and thus allows further significant improvement of the γ-ray registration technique. The concept of *stereo imaging* is based on the simultaneous detection of a single air shower in different projections by at least two telescopes separated at a distance comparable with the "effective radius" of the Cherenkov light pool. The stereoscopic approach allows (i) unambiguous and precise reconstruction of shower parameters on an *event-by-event* basis, (ii) superior rejection of hadronic showers, and (iii) effective suppression of the background light from different sources - the night sky background (NSB), local muons, *etc.*.

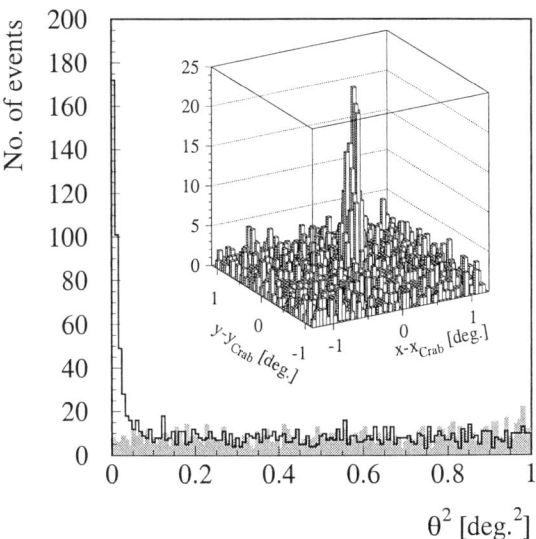

Fig. 2.24 Performance of the HEGRA stereoscopic system of telescopes for reconstruction of the arrival direction of primary γ-rays. Distribution of reconstructed directions of showers detected in the 2-telescope coincidence mode for 12 h ON-source observations of the Crab Nebula (from Daum *et al.*, 1997).

Compared with single ("stand alone") telescopes, which can adequately measure the shower inclination only in the direction perpendicular to the plane containing the telescope axis, the stereoscopic approach allows full reconstruction of the arrival direction of an individual primary photon with accuracy of about 0.1° or less. Apart from the good directional information, the stereoscopic systems make use of the fact that the Cherenkov images of a shower detected in different projections are only partially correlated. Therefore, the stereoscopic measurements increase the efficiency of rejection of hadronic (background) showers at both the hardware (trigger) and software levels, and consequently improve significantly the flux sensitivity compared to the sensitivity of single telescopes. Finally, the precise detection of shower parameters, in particular the shower maximum and the position of the shower core, allows energy determination of primary γ-rays with resolution as good as 10 per cent.

The only disadvantage of the stereoscopic approach is a non-negligible loss in the detection rate because of the overlap of the shower detection areas of individual telescopes. However, this loss of statistics is largely compensated for by a significant reduction of the energy threshold when the telescopes operate in the coincidence mode.

The first attempt of stereoscopic observations of air showers using the so-called "double-beam" technique, which also contained certain elements of the modern imaging Cherenkov technique, led to the tentative detection of a γ-ray signal from the radiogalaxy Centaurus A (Grindlay et al., 1975). Both the energy threshold and the flux sensitivity ($\sim 10^{-11}$ ph/cm^{2}s above 300 GeV) achieved by the two 7-m aperture optical reflectors were quite impressive even in the standards of the current ground-based instruments. However, the full potential of the stereoscopic approach has been convincingly demonstrated two decades later by the HEGRA system of 5 relatively small, 3 m diameter telescopes equipped with medium resolution (0.25° pixel size) cameras. The observations of the standard TeV candle, the Crab Nebula (Daum et al., 1997; Konopelko et al., 1999) thoroughly confirmed the early predictions (Aharonian, 1993) for the performance of this instrument. Fig. 2.24 shows the potential of the HEGRA IACT array for the reconstruction of the arrival direction of γ-rays. The angular resolution of about 0.1° coupled with the effective rejection of hadronic showers allowed observations of point γ-ray sources with a sensitivity characterised by a robust 10σ detection of a TeV signal from a Crab-like source for just 1 hour observation time. This is a factor of 1.5-2 better than the sensitivity of the best single telescopes, CAT and Whipple, despite the fact that they by intrinsic fluctuations in cascade development – the further improvement of the flux sensitivities for a given energy threshold can be achieved by an increase of the shower collection area, i.e. construction of an array consisting of several stereoscopic systems - *cells*. A possible design of a homogeneous multi-cell array of 100 GeV-threshold class telescopes is shown in Fig. 2.27a. The detection area of a telescope array consisting of a large number of 100m×100m 4-telescope cells, $n_0 \gg 10$, is almost energy-independent, $A \approx n_0 A_0$ where $A_0 \approx 10^4$ m^2 is the area of a single cell. Thus, approximately $n_0 = 100$ cells are required in order to approach to a highly desired detection area of about 1 km^2. This is, however, an expensive and hardly affordable approach. Also, it has another major disadvantage – significant loss in the collection area per cell above the energy threshold 100 GeV, and, therefore, an inadequate (for the total number of telescopes) sensitivity at TeV energies.

Fig. 2.26 The first H.E.S.S. telescope equipped with 960-pixel camera (from the H.E.S.S project web page, http://www.mpi-hd.mpg.de/hfm/HESS).

From this point of view, an array consisting of several cells located at large distances from each other (significantly larger than the Cherenkov pool radius), seems a more effective, and economically better justified, approach. A possible design for such an array is shown in Fig.2.27b. At energies close to the energy threshold, the cells operate independently, thus we should expect the same sensitivity as in the case of a "homogeneous" array consisting of the same number of cells. However, at energies significantly above the energy threshold there should a significant gain in the collection area. At very high energies, a "stereoscopic trigger" applied to some of the telescopes from different cells should provide also a significant gain of shower detection in the so-called large-zenith-angle mode.

It is important to note that the so-called 100 GeV energy threshold arrays provide, in fact, the best energy flux sensitivity at TeV energies. For optimisation of the γ-ray detection at energies around 100 GeV, we must reduce the energy threshold of individual cells to 10 GeV or so, by using larger, ≥ 20 m aperture class reflectors and/or using higher quantum efficiency optical detectors. On the other hand, reduction of the detection threshold to such low energies is a big issue itself. It will help not only to improve significantly the flux sensitivities at 100 GeV, but also will open an exciting scientific research area in the intermediate for the ground- and space-based gamma-ray astronomies interval between 10 and 100 GeV.

2.3.3.4 Sub-10 GeV ground based detectors?

The forthcoming stereoscopic systems of large imaging telescopes, with their superior energy-flux sensitivity between $10^{-13} - 10^{-12}$ erg/cm^{2}s, perfectly suit the energy range from approximately 10 TeV down to 100 GeV (or perhaps even 30 GeV) – a spectral domain in its own right from the point of view of both the main scientific motivations and the specific astrophysical source populations that emit most effectively in this energy band.

It is expected that the next generation major satellite γ-ray mission GLAST with a large pair-conversion telescope (Fig. 2.28) will have similar energy-flux sensitivity in the 30 MeV to 10 GeV energy range, and extent the exploration of the gamma-ray sky with still reasonable sensitivity to 100 GeV, or even beyond (see Fig. 1.2). Thus the gap between space-based and ground-based instruments will finally disappear. This is an important condition for cross-calibration of two different detection techniques.

At the same time, the astrophysical significance of the overlap of detection domains of the satellite-borne and ground-based instruments should

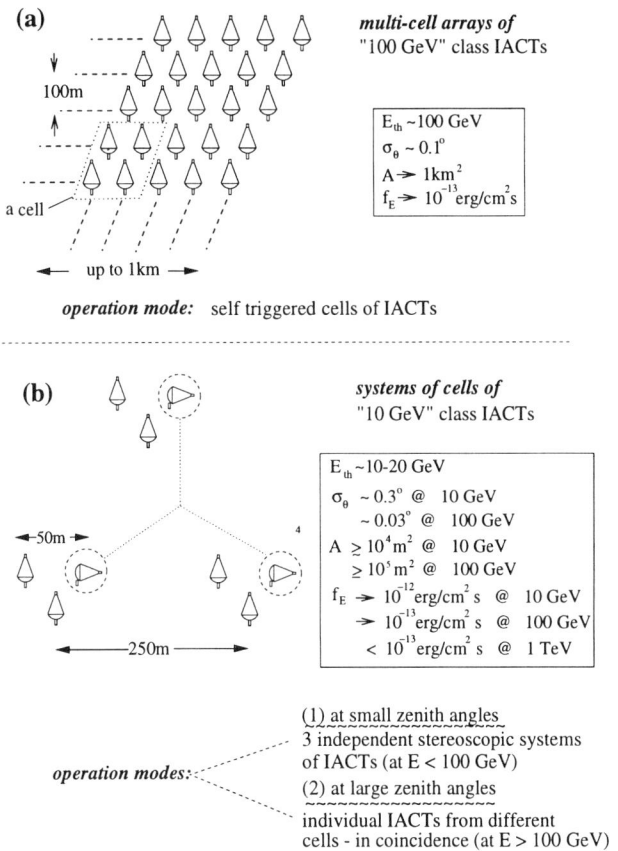

Fig. 2.27 Two possible designs of future IACT arrays (from Aharonian, 1997)

not be overestimated. Although at GeV energies GLAST will improve the EGRET sensitivity by almost two orders of magnitude, the capability of GLAST and, in fact, of any post-GLAST space project at energies well beyond 10 GeV will be quite limited because of the limited detection area, even if the Moon would be used in (far) future as a possible platform for installation of very large, e.g. 100 m^2 area pair-conversion detectors.

The impressive sensitivity of GLAST shown in Fig. 1.2 can be achieved from an approximately one year all-sky survey. For the persistent γ-ray sources this is an adequate estimate, taking into account that a huge num-

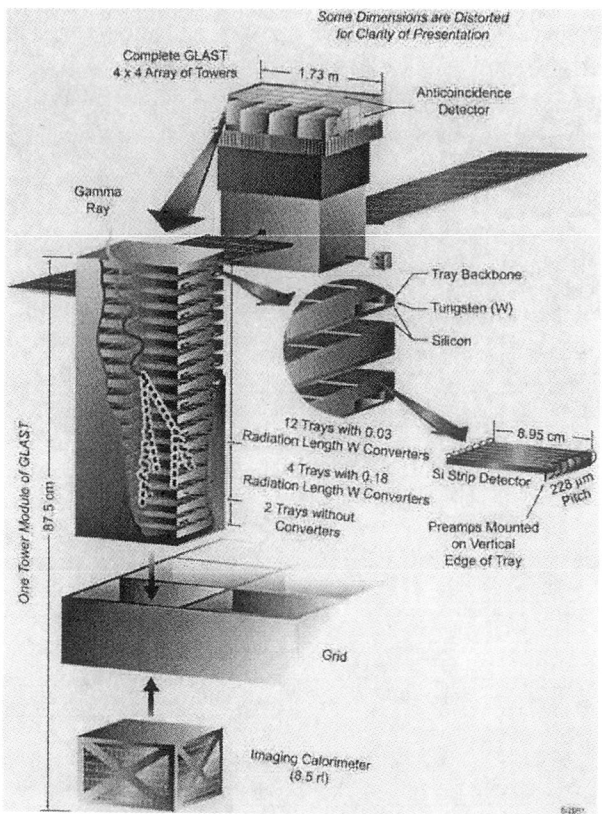

Fig. 2.28 The GLAST γ-ray telescope (from Morselli, 2002). The main element of the telescope is the *tracker* consisting of four-by-four array of tower modules. Each module consists of layers of silicon-strip detectors and thin large-Z material sheets converting γ-rays into electron positron pairs. The silicon-strip detectors track with high precision the secondary electrons. The segmented CsI(Tl) *calorimeter* measures the energy of the absorbed electron-positron pair, and thus gives information about the the energy of the primary γ-ray photon. The active *anti-coincidence shield* consisting of segmented plastic scintillator tiles, provides effective rejection of the charged particle background from primary cosmic rays and from the Earth albedo secondary products.

ber of sources will be simultaneously covered by the large, almost $\sim 2\pi$ steradian homogeneous FoV of GLAST. However, the small, approximately $1\,\mathrm{m}^2$, detection area limits the potential of this instrument for detailed studies of the temporal and spectral characteristics of highly variable sources like blazars or solitary events like GRBs at multi-GeV energies. In this regard, GLAST can hardly match the performance of current X-ray detectors that have similar detection areas but operate in a regime of photon fluxes that exceeds the fluxes of MeV/GeV γ-rays by many orders of magnitude.

The need in a powerful instrument to study transient phenomena with adequate high energy γ-ray photon statistics, has motivated the idea/concept of extension of the domain of the imaging atmospheric Cherenkov technique, with its huge shower collection area $\geq 10^4$ m^2, down to energies of about 5 GeV. This requires high elevations of the order of 5 km. That is why the concept is called 5@5 (Aharonian et al., 2001).

Very high altitudes allow significant gain in the density of the Cherenkov light. Two other key requirements to achieve sub-10 GeV energy threshold are the stereoscopic approach and very large optical reflectors. Monte Carlo studies show that the operation of a stereoscopic system consisting of ≥ 3 telescopes, each of which is a 25 to 30 m diameter optical reflector equipped with a conventional PMT-based multichannel camera with pixel size less than 0.1°, at an altitude of about 5 km above sea level, should indeed allow reduction of the effective detection threshold down to several GeV.

The successful realization of a several GeV energy threshold telescope array would largely depend on the availability of exceptional sites with a dry and transparent atmosphere at an altitude as high as 5 km. Nature does provide us with such an extraordinary site - the Liano de Chajnantor in the Atacama desert in Northern Chile. This site with its very arid atmosphere has been chosen for the installation of one of the most powerful future astronomical instruments – the Atacama Large Millimeter Array (ALMA), a US-European project which will consists of sixty four 12m aperture radio antennas. The very large flat area of this unique site could certainly accommodate a Cherenkov telescope array as well. Another attractive feature of this site seems is an adequate infrastructure which will be built up for ALMA. A potential site for installation of a similar array in the Northern Hemisphere would be a site close to the Yangbajing Laboratory (Tibet, China) at 4.3 km a.s.l. where the low-threshold air-shower array ARGO-YBJ is under construction, and other detectors have operated for a number of years.

An array of such large telescopes located at a relatively modest, but more comfortable altitude around 2 km a.s.l. would have somewhat larger, by a factor of 2 or 3, energy threshold. The lower Cherenkov light density at these elevation can be compensated for by more sensitive cameras based on novel (unfortunately, yet not available) fast (nanosecond) detectors of optical radiation with quantum efficiency exceeding 50%. Because of the unavoidable increase of the night sky background. such cameras cannot, however, provide proportional reduction of the energy threshold. Finally we note that combination of several such (sub)arrays with spacing of approximately 200 to 400 m, as shown in Fig. 2.27b, can provide an excellent flux sensitivity around 100 GeV.

Reduction of the energy threshold down to several GeV is a critical issue for a number of astrophysical and cosmological problems, e.g. for study of γ-radiation from pulsars and cosmologically distant objects like quasars and gamma-ray bursts. Therefore, preference should be given to very high elevations. Moreover, with sub-10 GeV threshold arrays one may significantly gain in sensitivity because of the effect related to the so-called rigidity cutoff. At such low energies the cosmic ray background is dominated by electrons. However, for the geomagnetic latitudes of both the ALMA and Yangbajing sites, the geomagnetic field effectively prevents the electrons and protons with energies less than 10 to 15 GeV from entering the Earth's atmosphere (see e.g. Lipari, 2002). This may have an impact on the background caused by electrons, and consequently would improve the flux sensitivity at energies below the rigidity cutoff.

The range of sensitivities that can be achieved by a 5@5 type instrument after approximately 100 hours observation of a point source is shown in Fig. 1.2. Comparing the sensitivity curves in Fig. 1.2 one may conclude that 5@5 and GLAST are competitors. However, they are, in fact, highly complementary. Note that the sensitivity curves for GLAST and 5@5 shown in Fig. 1.2 correspond to very different times (100 h for 5@5 versus 1 year for GLAST). While GLAST with its almost 2π FoV can provide very effective simultaneous monitoring of a large number (hundreds or even thousands) of sources, and also enables studies of the galactic and extragalactic components of the diffuse γ-ray background, 5@5 has an obvious advantage for the search and study of highly variable or transient γ-ray sources. On the other hand, 5@5 is a detector with a small field of view, therefore it requires special strategies for the search and study of multi-GeV emitters. Because of overlap of the energy intervals covered by these two instruments, GLAST may serve as a perfect "guide" for 5@5. Generally, all sources that will be detected by GLAST can be potential targets for observations with 5@5.

The concept of 5@5 is not only motivated by the possibility of coverage of the as yet unexplored region of multi-GeV γ-rays. In fact, 5@5 combines two advantages of the current ground-based and satellite-borne γ-ray domains - large photon fluxes at GeV energies (typically, 10^{-8} to 10^{-6} ph/cm^{2}s above 1 GeV versus 10^{-12} to 10^{-10} ph/cm^{2}s above 1 TeV), and enormous detection areas in the TeV domain (up to 10^5 m^2 at TeV energies to ≈ 1 m^2 at GeV energies). This makes 5@5 a unique "Gamma-ray timing explorer" (with a sensitivity to detect EGRET sources with hard spectra extending to 10 GeV, in exposure times of a few seconds to several minutes) for the study of transient non-thermal γ-ray phenomena like rapid variability of Blazars, synchrotron flares in Microquasars, the high energy (GeV) counterparts of Gamma Ray Bursts, *etc.*

The capability of 5@5 will not be limited to highly variable source studies. This instrument, in fact, has significantly broader objectives related to detailed γ-ray spectrometry of emission in the energy interval from 10 to 100 GeV from γ-ray sources like SNRs, Plerions, Pulsars, the Galactic Disk as a whole, large (kpc to Mpc) scale jets in radiogalaxies, rich galaxy clusters, *etc.*. Therefore, 5@5 will be complementary in its capabilities to the forthcoming 10m class telescope arrays (spectrometry typically above 100 GeV) and GLAST (spectrometry below 10 GeV). The good energy resolution in the energy interval between 10 and 100 GeV, supported by an adequate γ-ray photon statistics, is crucial also for cosmological studies through (i) probing the cosmic optical and near infrared background radiation, (ii) detecting γ-rays from large scale cosmological structures, and (iii) searching for characteristic line emission from the non-baryonic Dark Matter Halos.

The successful realization of the 5@5 concept in the form of an operating high-altitude Cherenkov telescope array during the lifetime of GLAST was identified as a major observational goal and classified as one of the highest priority objectives in the field of gamma-ray astronomy (Buckley *et al.*, 2002).

2.3.3.5 *Large field-of-view detectors*

The imaging atmospheric Cherenkov telescopes are designed for observations of γ-rays from objects with well determined positions. However, the high sensitivity of stereoscopic arrays coupled with relatively large (4 degree or more) field-of-view homogeneous imaging cameras, may allow quite effective sky surveys as well, as has been demonstrated by the HEGRA CT-system. The next generation telescope arrays with an energy threshold around 100 GeV and with significantly improved sensitivity at TeV energies, will provide deeper surveys. In particular, it is expected that all point TeV sources with fluxes at the 0.1 Crab level can be be revealed within one steradian of the sky during a one-year survey. A 5@5 type instrument will have a similar potential at GeV energies.

Nevertheless, the development of a ground-based technique allowing *simultaneous* coverage of a significant (1 steradian or so) fraction of the sky is recognised as a high priority issue. This would be important, for example, for monitoring of γ-ray activity of a large number of highly variable blazars and, for independent detection of solitary VHE γ-ray events, e.g. TeV counterparts of GRBs. The general requirements to the performance of the future wide FoV γ-ray ground-based detectors is dictated by the parameters of the upcoming imaging Cherenkov telescope arrays. Namely, the energy threshold of these instruments should be close to 100 GeV and their sensitivity for a one year survey should be at least 0.1 Crab in order to be compatible to the sensitivity of the Cherenkov telescope arrays achieved for several hour (1 night) observation time. Two possible approaches have been proposed in this regard - (i) very large FoV imaging air Cherenkov telescope technique based on refractive optics (Kifune, 2001) and (ii) dense air shower particle arrays or large water Cherenkov detectors installed at very high, 4 km or higher altitudes (for a review see Hoffman *et al.*, 1999).

The first technique requires several technological innovations – very large UV transparent Fresnel lenses, mega-pixel/nanosecond/high-quantum-efficiency detectors, a huge data handling capability, etc. This approach may receive a significant support from the cosmic ray community which recently proposed to build a similar, but space-based downward looking atmospheric fluorescence detectors for registration of extremely high energy cosmic rays (the OWL and EUSO projects).

The second approach does not face serious technological challenges. The feasibility of both the high altitude air shower array and water Cherenkov techniques have been convincingly demonstrated by the Tibet and Milagro collaborations (for a review, see e.g. Buckley *et al.*, 2002). Moreover, ARGO, a new air shower detector under construction by the Chinese-Italian collaboration in Tibet, is expected to have performance rather close to the above requirements (Bacci *et al.*, 2002). An ambitious idea of constructing a Milagoro type instrument with 10 times more physical area and located at 4 km a.s.l. (G. Sinnis, private communication), promises further significant improvement of sensitivity of the 100 GeV threshold, large field-of-view ground-based technique.

2.3.3.6 *IACT arrays for probing PeV γ-rays*

The current trend to reduce the energy threshold of ground-based γ-ray detection technique concerns both the atmospheric Cherenkov telescopes and the air-shower detectors. As a result, presently there is only little instrumental activity in the energy domain above 10 TeV. On the other hand, this energy region is of high astrophysical interest. For example, the detection of 100 TeV γ-rays from shell type SNRs would be a key proof that the shocks in SNRs accelerate protons up to energies 10^3 TeV. Another interesting issue is related to 10-100 TeV γ-rays from nearby extragalactic sources, for example from radiogalaxies M 87 or Cen A – potential sites of acceleration of the observed highest energy cosmic rays.

The fluxes of γ-rays typically decrease very rapidly with energy. This limits the capability of the traditional air-shower arrays because of both limited proton/gamma separation power and limited detection area, and consequently low γ-ray photon statistics. Any meaningful study of cosmic γ-rays beyond 10 TeV requires a detection area of about 1 km^2.

At large zenith angles of about 60°, the collection area of the atmospheric Cherenkov detectors increases rapidly. Thus the use of IACT systems at such large angles can improve the γ-ray statistics in the multi-TeV region. For many astrophysical objects, on the other hand, the observation time within a single night is very short in this mode, because the γ-ray source sets rapidly below the horizon. Besides, even small variations of the atmospheric transparency add non-negligible uncertainties in the derivation of the shower parameters obtained at large zenith angles. Also, an exploitation of the large-zenith angle technique requires very small pixel size, and therefore quite expensive multi-channel cameras.

An effective and straightforward approach seems the use of an array of imaging telescopes optimised for detection of γ-rays in the 10 to 100 TeV region. Such an array can consists of telescopes of rather modest, approximately 10 m^2 mirror size, and separated from each other by a distance essentially larger than the conventional 100 m. Depending on the tasks, and the configuration of the imagers (multi-pixel cameras), the optimum distance varies between 300 to 500 m. The requirement to the pixel size of the imagers is also quite modest, from 0.25° to 0.5°, however they have to cover very large field-of-view, 6 to 8 degree, in order to collect showers from distances up to 500 m or more. Monte Carlo simulations show (Plyasheshnikov et al., 2000) that an array consisting of 9 or more such telescopes can provide an extraordinary collection area exceeding 1 km^2, a reasonable efficiency for suppression of hadronic showers, good angular resolution of several arcmin and reasonable energy resolution. Such an array can serve also as an effective tool for the study of the energy spectrum and mass composition of cosmic rays up to the so-called "knee" around 10^3 TeV.

Chapter 3

Gamma Ray Production and Absorption Mechanisms

Any interpretation of an astronomical observation requires, by definition, unambiguous identification of the relevant radiation mechanism(s). The adequate knowledge of the features of the principal radiation processes is another key issue. Therefore the physics of radiation and absorption mechanisms is one of the central subjects of astronomy. With some exceptions, all basic radiation processes relevant to astronomy have been studied in great detail using the methods and tools of the modern experimental and theoretical physics, in particular the atomic and molecular physics, nuclear physics, high energy (particle) physics. Generally, each wavelength band in astronomy is characterised by one or two basic radiation mechanisms. However, this is not the case of gamma-ray astronomy. Here we deal with a large number of competing processes which makes the theoretical studies very interesting, but also challenging. Very often, we face a dilemma when the same experimental result can be equally well explained by more than one radiation mechanism. For example, for explanation of the TeV γ-radiation from blazars at least four different possible channels associated with the inverse Compton scattering of electrons, as well as with interaction of protons with magnetic field (synchrotron radiation), photon fields (photo-meson processes) and matter (inelastic p-p collisions) have been proposed, and, at this stage, none of them can be firmly discarded. Even in simpler scenarios, like γ-ray production in supernova remnants, a basic question of the dominant radiation mechanism remains unsolved. Moreover, sometimes two or more radiation processes contribute comparably to the same energy interval of the observed γ-ray flux. This is the case, for example, of diffuse radiation of the Galactic Disk at MeV and GeV energies.

The unambiguous identification of γ-ray production mechanisms in celestial objects requires not only comprehensive experimental data containing information about the spectral, temporal and spatial properties of radiation[1], but also clear understanding of the features of relevant elementary processes and their relationships to each other, especially in the context of multiwavelength studies.

Apart from the emission by thermal relativistic plasmas, γ-ray production in astronomical environments implies interactions of accelerated particles (electrons, protons and nuclei) with ambient targets - thermal gas, low-frequency photons, magnetic fields. There are several possible ways of grouping different radiative processes by character of interactions (leptonic or hadronic, absorption or radiation) or by type of the target. Below we adopt the latter classification which allows a rather convenient description of relative contributions of different processes for a given astronomical environment dominated by matter, radiation or magnetic field. The cross-sections of all electromagnetic processes are calculated within the framework of Quantum Electrodynamics. These results, obtained with an extremely high accuracy, are summarised in many famous monographs (e.g. Heitler, 1954; Jauch and Rohrlich, 1955; Akhiezer and Berestetskii, 1965). On the other hand, our knowledge about the hadronic cross-sections are based essentially on accelerator data. In the energy region of interest, up to 10^{15} eV or even higher, all relevant cross-sections are measured with an adequate accuracy for astrophysical applications.

Below we will discuss the properties of the most important absorption and radiation processes in the γ-ray domain with an emphasis on the mechanisms that operate effectively in the high and very high energy regimes, while the processes relevant to low energy domain will be covered briefly.

3.1 Interactions with Matter

Two γ-ray production mechanisms related to electron bremsstrahlung and the decay of neutral π-mesons produced at nucleon-nucleon interactions are among the most effective processes resulting in high and very high energy γ-ray emission. The annihilation of positrons, as well as the nuclear processes associated with neutron-capture reactions and de-excitation of nuclei, dominate at lower, MeV energies.

---

[1]Polarization is another inherent characteristic of radiation, but unfortunately practical realization of polarization measurements in the high energy γ-ray regime is an extremely difficult task.

3.1.1 *Electron bremsstrahlung and pair-production*

Comprehensive analysis of features of the cross-sections of these processes can be found in Heitler (1954) and Akhiezer and Berestetskii (1965). Although the two processes have many similarities, generally electron bremsstrahlung has broader astrophysical applications than pair production. The latter becomes important only in the context of cascade development in optically thick environments.

The integral cross-sections of the bremsstrahlung and pair production processes in hydrogen gas are shown in Fig. 3.1. The energies of electrons and γ-rays are expressed in units of $m_e c^2$. The cross-sections are normalised to the asymptotic value of the pair production cross-section at $\varepsilon_0 \to \infty$:

$$\sigma_0 = 7/9 \times 4\alpha_\mathrm{f} r_\mathrm{e}^2 Z(Z+1) \frac{\ln(183 Z^{-1/3})}{1 + 0.12(Z/82)^2} \quad (3.1)$$

where Z is the charge of the target nucleus, and r_e is the classical electron

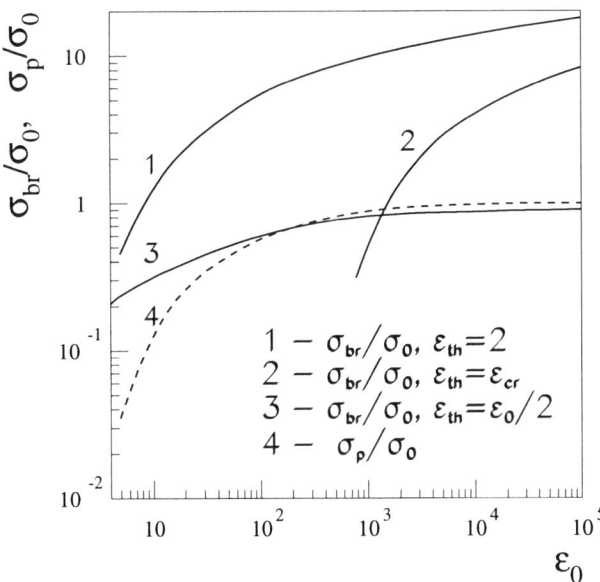

Fig. 3.1 Total cross-sections of the bremsstrahlung (σ_br) and pair production (σ_p) processes in hydrogen gas normalised to the asymptotic value (σ_0) of the pair production cross-section at $\varepsilon_0 \to \infty$. The bremsstrahlung cross-sections are calculated for secondary γ-rays produced with energies exceeding (1) the pair-production threshold, $\varepsilon_\mathrm{th} = 2$; (2) the critical energy, $\varepsilon_\mathrm{th} \simeq 700$; (3) half of the energy of the primary electron, $\varepsilon_\mathrm{th} = \varepsilon_0/2$.

radius. This actually implies introduction of the so-called radiation length

$$X_0^{(m)} = 7/9(n\sigma_0)^{-1} , \qquad (3.2)$$

the physical meaning of which is the average distance over which the ultra-relativistic electron loses all but $1/e$ of its energy due to bremsstrahlung. The same parameter also implies the mean free path of γ-rays. This convenient parameter is widely used to describe cascade development in optically thick sources. The cascade effectively develops at depths exceeding the radiation length. Usually the radiation length is expressed in units of g/cm^2. For hydrogen gas $X_0^{(m)} \simeq 60 \text{ g/cm}^2$. The second important parameter that characterises the cascade development is the so-called critical energy below which ionization energy losses dominate over bremsstrahlung losses. In hydrogen gas, $\varepsilon_{cr} \simeq 700$. Effective multiplication of particles in a cascade is possible only at energies $\varepsilon \geq \varepsilon_{cr}$. At lower energies electrons dissipate their energy by ionization rather than producing more high energy γ-rays which would support further development of the electron-photon shower.

In Fig. 3.1 the bremsstrahlung total cross-sections is calculated for 3 different values of minimum energy of emitted photons: $\varepsilon_{th} = 2$, ε_{cr} and $\varepsilon_e/2$. The first value corresponds to the cross-section of production of all γ-rays capable of producing electron-positron pairs. The second value corresponds to the cross-section for γ-rays produced above the critical energy, and thus capable of supporting the cascade. And finally, the third value corresponds to the cross-section of production of the "most important" γ-rays which play the major role in the cascade development. It is seen from Fig. 3.1 that while for $\varepsilon_{th} = 2$ the pair-production cross-section is an order of magnitude smaller compared to the bremsstrahlung cross-section, for $\varepsilon_{th} = \varepsilon_e/2$ the cross-sections of two processes become almost identical at $\varepsilon \geq 100$.

The differential cross-sections of bremsstrahlung and pair production are presented in Fig. 3.2 The pair-production cross-section obviously is a symmetric function around the point $x = \varepsilon_e/\varepsilon_0 = 0.5$. The bremsstrahlung differential cross-section has a $1/\varepsilon_\gamma$ type singularity at $\varepsilon_\gamma \to 0$, but because of the hard spectrum of bremsstrahlung photons the energy losses of electrons contribute mainly to emission of high energy γ-rays. Thus bremsstrahlung should be treated as an essentially catastrophic process. Nevertheless, it is convenient to introduce the so-called average energy loss-rate,

$$-\left(\frac{d\varepsilon_e}{dt}\right)_{br} = \left(\frac{cm_p n}{X_0}\right)\varepsilon_e . \qquad (3.3)$$

Correspondingly, the lifetime of electrons due to the bremsstrahlung losses is

$$t_{\rm br} = \frac{\varepsilon_e}{-d\varepsilon_e/dt} \simeq 4 \times 10^7 (n/1 \text{ cm}^{-3})^{-1} \text{ yr}, \qquad (3.4)$$

where n is the number density of the ambient gas.

Note that the electron energy loss rate given by Eq.(3.3) is proportional to the electron energy, and, correspondingly, the lifetime given by Eq.(3.4) is energy independent. This implies that for a initial (acceleration) power-law spectrum $Q(\varepsilon_e)$, bremsstrahlung losses do not change the original electron

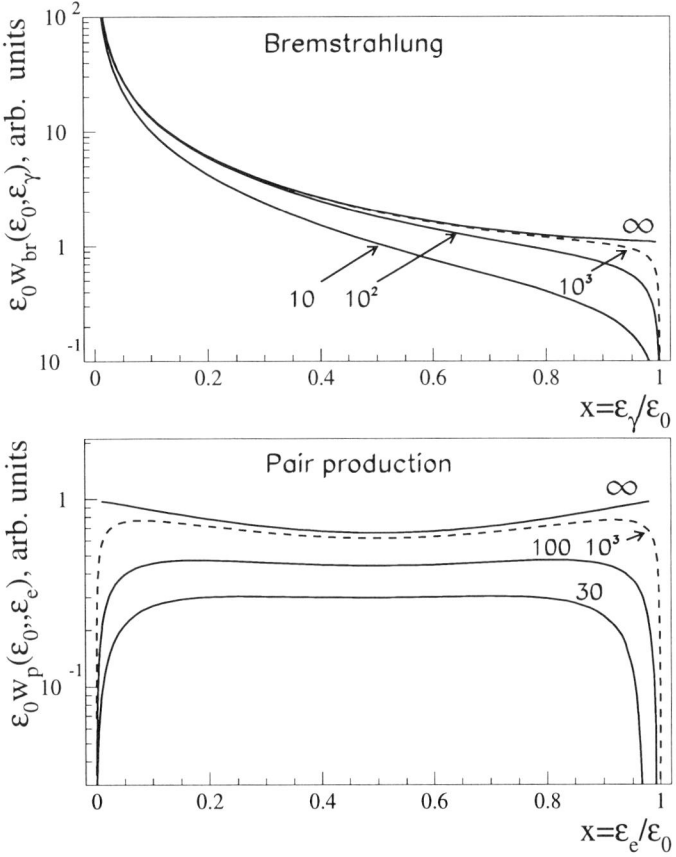

Fig. 3.2 Differential cross-sections of the bremsstrahlung (upper panel) and pair production (bottom panel) processes in hydrogen. The cross-sections are normalised to one radiation length. The energies of primary electrons and γ-rays ε_0 (in units of $m_e c^2$) are indicated at the curves.

spectrum, taking into account that the cooled steady-state spectrum $N(\varepsilon_e)$ is determined as (see e.g. Blumenthal and Gould, 1970)

$$N(\varepsilon_e) \propto (d\varepsilon_e/dt)^{-1} \int_{\varepsilon_e} Q(E_e) d\varepsilon_e \ . \qquad (3.5)$$

Interestingly, in the case of power-law spectrum of electrons $N(\varepsilon_e)$, the spectrum of bremsstrahlung γ-rays is also power low with the same power-law index (this is a result of $1/\varepsilon_\gamma$ dependence of the differential cross-section). Thus, the bremsstrahlung γ-ray spectrum simply repeats the shape of the electron acceleration spectrum $Q(\varepsilon_e) \propto \varepsilon_e^{-\Gamma}$.

This is true, however, only when the energy losses are dominated by bremsstrahlung. In hydrogen gas, at energies below $\sim 700 m_e c^2 \simeq 350$ MeV ionization dominates over the bremsstrahlung. Because both the ionization and bremsstrahlung loss rates are proportional to n, this condition does not depend on the ambient gas density. On the other hand, in the relativistic regime, the ionization loss rate does not depend on the electron energy. Thus, in accordance with Eq.(3.5), the steady-state electron spectrum becomes flatter, $N(\varepsilon_e) \propto \varepsilon_e^{-\Gamma+1}$, and correspondingly at energies below several hundred MeV we should expect a very hard bremsstrahlung γ-ray spectrum with power-law photon index $\Gamma - 1$.

In many astrophysical scenarios the inverse Compton and synchrotron losses may well dominate, especially at very high energies, over bremsstrahlung, depending on the ratio of the energy density of the radiation and magnetic fields to the number density of the ambient gas. The synchrotron and inverse Compton energy loss rates are proportional to the electron energy. This makes the steady-state electron spectrum steeper, $N(\varepsilon_e) \propto \varepsilon_e^{-\Gamma-1}$, and correspondingly the bremsstrahlung γ-rays emerge with photon index $\Gamma + 1$.

Eq.(3.3) and (3.4) are derived for neutral gas. In a fully ionized gas, i.e. in the absence of the screening effect, the corresponding equations are somewhat different (see, e.g., Akhiezer and Berestetskii, 1965). In particular, an additional term proportional to $\ln \varepsilon_e$ appears in the energy loss-rate. However these corrections do not essentially change the above conclusions.

Due to ionization losses, the efficiency of bremsstrahlung in the energy range below the critical energy ε_{cr} decreases approximately proportional to the electron energy, and becomes very low at sub-relativistic energies. In the nonrelativistic regime, proton bremsstrahlung (often called "inverse" bremsstrahlung) becomes identical to electron bremsstrahlung, when the proton and electron kinetic energies are express in their rest mass units (see e.g. Hayakawa, 1969). Although in some specific cases like in Solar flares, the non-relativistic bremsstrahlung of electrons and protons may play a major role in production of nonthermal X-rays, generally this is a very inefficient radiation mechanism. Because of severe ionization losses only a tiny fraction, $\leq 10^{-5}$, of the kinetic energy of particles is released in X-rays. Finally we note that since the hard X-rays of energy ε_x are produced by protons with energy $(m_e/m_p)\varepsilon_x \simeq 10(\varepsilon_x/5\,\text{keV})$ MeV, this process is tightly connected with the prompt γ-ray emission due to excitation of the nuclei of the ambient matter by the same ≥ 10 MeV protons (see Sec.3.1.3.2).

3.1.2 *Electron-positron annihilation*

The astrophysical significance of this process, $e^+e^- \to 2\gamma$, generally is attributed to the annihilation line at energy $m_e c^2 = 0.511$ MeV, as well as to the 3-photon positronium continuum, produced by annihilation of thermalized positrons with relatively cold thermal electrons of the ambient gas/plasma. However, if the positrons are injected into the production region with relativistic energies, a significant fraction (from 10 to 20 per cent, depending on the ionization state of the ambient plasma) of positrons annihilate in flight before they cool down to the temperature of the thermal background gas (Aharonian and Atoyan, 1981a).

The differential spectrum of the γ-rays produced in the annihilation of fast positrons with Lorentz-factors $\varepsilon_+ = E_+/m_e c^2$ on the ambient electrons with density n_e is described by a simple analytical expression (Aharonian and Atoyan 1981b)

$$q_{\text{ann}}(\varepsilon_\gamma) = \frac{3\sigma_T^2 c n_e}{8\varepsilon_+ p_+} \left[\left(\frac{\varepsilon_\gamma}{\varepsilon_+ + 1 - \varepsilon_\gamma} + \frac{\varepsilon_+ + 1 - \varepsilon_\gamma}{\varepsilon_\gamma} \right) + 2\left(\frac{1}{\varepsilon_\gamma} + \frac{1}{\varepsilon_+ + 1 - \varepsilon_\gamma}\right) - \left(\frac{1}{\varepsilon_\gamma} + \frac{1}{\varepsilon_+ + 1 - \varepsilon_\gamma}\right)^2 \right] \quad (3.6)$$

where the photon energy $\varepsilon_\gamma = E/m_e c^2$ varies in the limits

$$\varepsilon_+ + 1 - p_+ \leq 2\varepsilon_\gamma \leq \varepsilon_+ + 1 + p_+ \,. \quad (3.7)$$

Here $p_+ = \sqrt{\varepsilon_+^2 - 1}$ is the dimensionless momentum of the positron.

For a power-law steady-state spectrum of positrons, $N_+ \propto \varepsilon_+^{-\Gamma+1}$ (here Γ can be interpreted as the primary positron spectral index, assuming that ionization losses make the positron spectrum harder, $\Gamma \to \Gamma - 1$), the spectrum of annihilation radiation at $\varepsilon_\gamma \gg 1$ has a power-law form

$$J_{\text{ann}}(\varepsilon_\gamma) \propto \varepsilon_\gamma^{-(\Gamma)} \left[\ln(2\varepsilon_\gamma) - 1\right]. \qquad (3.8)$$

Thus, the spectrum of annihilation radiation is steeper than the (steady-state) spectrum of the parent positrons, but almost repeats the shape of the *primary* spectrum of positrons. At lower energies the spectrum has a more complicated form with a broad maximum around 1 MeV.

The total cross-section of annihilation of a relativistic positron of energy ε_+ is given by $\sigma_{\text{ann}} = \frac{3}{8}\sigma_T^2 \varepsilon_+^{-1}[\ln(2\varepsilon_+) - 1]$. Correspondingly the annihilation time is

$$t_{\text{ann}} = \frac{8}{3\sigma_T^2 cn} \frac{\varepsilon_+}{\ln(2\varepsilon_+) - 1} \simeq 4 \times 10^6 \frac{\varepsilon_+}{\ln(2\varepsilon_+) - 1} (n/1 \text{ cm}^{-3})^{-1} \text{ yr}. \quad (3.9)$$

Comparing Eq.(3.9) with Eq.(3.4) one finds that for positron energies $\varepsilon \leq 30$, the annihilation times becomes shorter than the bremsstrahlung cooling time. This implies that at energies less than approximately 15 MeV the annihilation continuum starts to dominate over the bremsstrahlung spectrum, taking into account that in both processes the leading photon receives a substantial part of the positron kinetic energy.

3.1.3 Gamma rays produced by relativistic protons

3.1.3.1 π^0-decay gamma rays

This process provides a unique channel of information about the hadronic component of cosmic rays. The important role of this process in gamma-ray astronomy was recognised by the pioneers of the field (see e.g. Ginzburg and Syrovatskii, 1964) long ago.

Relativistic protons and nuclei produce high energy γ-rays in inelastic collisions with ambient gas due to the production and decay of secondary pions, kaons and hyperons. The neutral π^0-mesons provide the main channel of conversion of the kinetic energy of protons to high energy γ-rays. For the production of π^0-mesons the kinetic energy of protons should exceed $E_{\text{th}} = 2m_\pi c^2(1 + m_\pi/4m_p) \approx 280$ MeV, where $m_\pi = 134.97$ MeV is the mass of the π^0-meson. This particle immediately decays to two γ-rays. The mean lifetime of π^0-decay, $t_{\pi^0} = 8.4 \times 10^{-17}$ s, is significantly shorter than the lifetime of charged π-mesons ($\approx 2.6 \times 10^{-8}$ s). At high energies, all three types of pions are produced with comparable probabilities. The spectral form of π-mesons is generally determined by a few (one or two) leading particles (that carry a significant fraction of the nucleon energy) rather than by the large number of low-energy secondaries.

The decays of charged pions lead to ν_e and ν_μ neutrinos with spectra quite similar to the spectrum of the accompanying π^0-decay γ-rays. However, this symmetry can be violated in environments with high gas or radiation densities. In certain conditions, the characteristic time for inelastic interactions of charged pions with nucleons or photons could be shorter than the decay time, so the energy of pions degrades before they decay. At very high energies this would result in significantly smaller fluxes of neutrinos compared to γ-rays.

The distinct feature of the spectrum of π^0-decay γ-rays is the maximum at $E_\gamma = m_\pi c^2/2 \simeq 67.5$ MeV, independent of the energy distribution of π^0 mesons, and consequently of the parent protons. The appearance of such a bump in the γ-ray spectrum is a result of the $\pi^0 \to 2\gamma$ decay kinematics. It is easy to show (see e.g. Stecker, 1971; Ozernoy et al., 1973) that the spectrum of γ-rays from decays of monoenergetic pions of energy E_π and velocity v_π is constant $f(E_\pi) = c/(v_\pi E_\pi)$ within the interval between $E_1 = 0.5 E_\pi(1 - v_\pi/c)$ and $E_2 = 0.5 E_\pi(1 + v_\pi/c)$. The spectrum of γ-rays for an arbitrary distribution of π^0-mesons $\Pi(E_\pi)$ can be presented as superposition of rectangles for which only one point at $m_\pi c^2/2$ is always presented. Obviously this should result in a spectral maximum independent of the distribution of parent pions.

The spectral features of γ-rays through the channel $pp \to \pi^0 \to 2\gamma$ has been extensively studied by many authors (e.g. Stecker, 1971; Dermer, 1986; Berezinsky et al., 1993; Mori, 1997, etc.). Although precise calculations of γ-ray spectra require quite heavy integrations over differential cross-sections obtained experimentally at particle accelerators, the emissivity of γ-rays for an arbitrary *broad* energy distribution of protons can be derived within a simple formalism which nevertheless provides surprisingly good accuracy over a broad γ-ray energy range.

The γ-ray emissivity $q_\gamma(E_\gamma)$ is directly defined by $q_\pi(E_\pi)$ as

$$q_\gamma(E_\gamma) = 2 \int_{E_{\min}}^{\infty} \frac{q_\pi(E_\pi)}{\sqrt{E_\pi^2 - m_\pi^2 c^4}} dE_\pi , \qquad (3.10)$$

where $E_{\min} = E_\gamma + m_\pi^2 c^4/4 E_\gamma$. The emissivity of secondary pions q_π from inelastic proton-proton interactions can be calculated with high accuracy using accelerator measurements of the inclusive cross-sections $\sigma(E_i, E_p)$ (see e.g. Gaisser, 1990). The emissivity of π^0-mesons calculated in the δ-function approximation for the cross-section $\sigma(E_\pi, E_p)$ then becomes

$$q_\pi(E_\pi) = c\, n_{\rm H} \int \delta(E_\pi - \kappa_\pi E_{\rm kin}) \sigma_{\rm pp}(E_{\rm p}) n_{\rm p}(E_{\rm p}) {\rm d}E_{\rm p}$$

$$= \frac{c\, n_{\rm H}}{\kappa_\pi} \sigma_{\rm pp}\!\left(m_{\rm p} c^2 + \frac{E_\pi}{\kappa_\pi}\right) n_{\rm p}\!\left(m_{\rm p} c^2 + \frac{E_\pi}{\kappa_\pi}\right) \quad (3.11)$$

where $\sigma_{\rm pp}(E_{\rm p})$ is the total cross section of inelastic pp collisions, and κ_π is the mean fraction of the kinetic energy $E_{\rm kin} = E_{\rm p} - m_{\rm p} c^2$ of the proton transferred to the secondary π^0-meson per collision; $n_{\rm p}(E_{\rm p})$ is the energy distribution of the protons.

In a broad region from GeV to TeV energies $\kappa_\pi \approx 0.17$ which includes a $\sim 6\%$ contribution from η-meson production (Gaisser 1990). From the threshold at $E_{\rm kin} \simeq 0.3\,{\rm GeV}$, $\sigma_{\rm pp}$ rises rapidly to about 30 mb. But after $E_{\rm kin} \sim 2\,{\rm GeV}$, $\sigma_{\rm pp}$ increases only logarithmically. In the GeV to TeV energy region, the total cross section can be approximated by

$$\sigma_{\rm pp}(E_{\rm p}) \approx 30\,[0.95 + 0.06\,\ln(E_{\rm kin}/1\,{\rm GeV})]\ {\rm mb} \quad (3.12)$$

for $E_{\rm kin} \geq 1\,{\rm GeV}$, with the assumption $\sigma_{\rm pp} = 0$ at lower energies. More accurate approximations of the cross-section below 1 GeV do not noticeably change the fluxes of γ-rays even at very low energies provided that the broad power-law spectrum of protons extends beyond 10 GeV, and thus the overall flux is contributed to by protons with energies above a few GeV.

The rather good accuracy of this simple approach (Aharonian and Ayoyan, 2000) is demonstrated in Fig. 3.3 where the emissivity of π^0-decay γ-rays calculated on the basis of Eq.(3.10) – (3.12) is compared with the results of Monte-Carlo calculations by Mori (1997) based on a detailed treatment of the cross-sections of secondary pion production.

The dashed curve corresponds to a similar calculation but for the local CR proton flux given in the form of Eq. (4.30) in Chapter 4, which has been used for the detailed γ-ray emissivity calculations by Dermer (1986) (the original spectrum from Dermer (1986) is not shown in order not to overload the figure with almost coinciding curves). The two other curves in Fig. 3.3 correspond to the emissivities calculated for power law proton spectra with spectral indices $\Gamma = 2.5$ (dot–dashed curve) and $\Gamma = 2$ (3-dot–dashed curve), normalised to the same energy density of CR protons $w_{\rm p} = \int n_{\rm p}(E_{\rm p}) E_{\rm p} {\rm d}E_{\rm p} \approx 1.2\,{\rm eV/cm}^3$ derived from the CR proton flux given by Eq. (4.30). For CR proton spectra with $\Gamma \geq 2.4$ the emissivities, in terms of $E_\gamma^2 q_\gamma$, reach their maximum at $E \simeq 1\,{\rm GeV}$, and then at lower energies the spectra sharply decline.

The characteristic cooling time of relativistic protons due to inelastic p-p interactions in the hydrogen medium with number density n_0 is almost independent of energy. Assuming an average cross-section at very high energies of about 40 mb (see Eq.(3.12)), and taking into account that on average the proton loses about half of its energy per interaction (for the coefficient of inelasticity $f \approx 0.5$), we find

$$t_{\rm pp} = (n_0 \sigma_{\rm pp} f c)^{-1} \simeq 5.3 \times 10^7 (n/1\ {\rm cm}^{-3})^{-1}\ {\rm yr}\ . \qquad (3.13)$$

Since the $t_{\rm pp}$ cooling time is almost energy-independent in the energy region above 1 GeV, where the nuclear losses well dominate over ionization losses, the initial (acceleration) spectrum of protons remains unchanged. On the other hand, the γ-ray spectrum essentially repeats the spectrum of the parent protons. This implies that at high energies γ-rays carry direct information about the acceleration spectrum of protons. To a certain extent, this is similar to the bremsstrahlung γ-rays of relativistic electrons.

Assuming that the electrons and protons are accelerated with the same

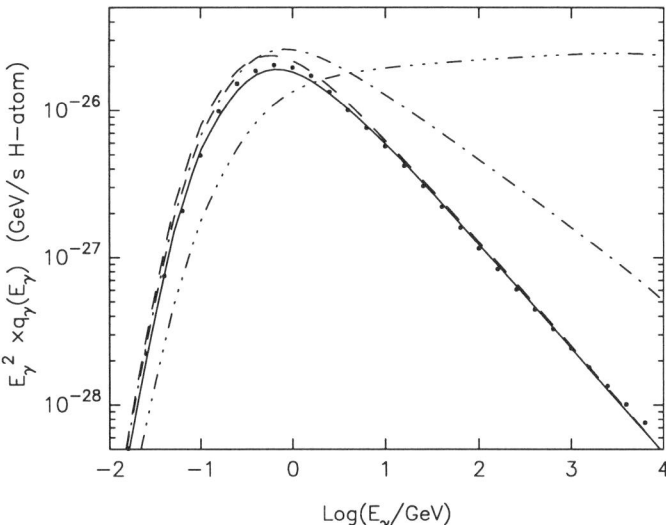

Fig. 3.3 The emissivities, per 1 H-atom, of π^0-decay γ-rays calculated using an approximate method given by Eq. (3.10)-(3.12) for the spectra of CR protons corresponding to the 'median' proton flux (solid line) of Mori (1997), and the flux given by Eq. (4.30) (dashed line), as compared with the results of detailed calculations by Mori (1997) shown by full dots. The dot-dashed and 3-dot–dashed curves correspond to the emissivities calculated for the single power-law spectra of protons with indices $\Gamma_{\rm p} = 2.5$ and $\Gamma_{\rm p} = 2$.

power-law spectrum, the ratio of γ-ray emissivities of these two processes can be estimated as $q_\gamma^{br}/q_\gamma^{\pi^0} \simeq r^{-1}\, 3 t_{\rm pp}/t_{\rm br} \simeq 4r^{-1}$, where r is the electron to proton ratio. Thus for $r \geq 10$ (which is the case for the galactic cosmic rays), the π^0-production at high energies dominates over the bremsstrahlung γ-rays. Moreover, at very high energies, due to inverse Compton and synchrotron losses, the contribution from electron bremsstrahlung is further suppressed.

3.1.3.2 Nuclear gamma-ray line emission

Although below 300 MeV the protons cannot produce pions, they still play an important role in the production of γ-rays through nuclear excitation of the ambient medium (Ramaty et al., 1979). De-excitation of the target nuclei leads to γ-ray lines in the energy region between several hundred keV to several MeV. The excitation cross-sections for the isotopes most relevant for astrophysical applications have been recently comprehensively reviewed by Kozlovsky et al. (2002). The energy region of nonthermal particles responsible for the prompt γ-ray line emission, extends approximately from several MeV to 100 MeV per nucleon. The γ-ray emission is formed by contributions from both target and projectile nuclei. The lines from both components are broad. In the first case the typical line width in the cold, low density medium is of the order of several tens of keV; it is determined essentially by the recoil momentum of the target nucleus. The γ-ray lines from projectile nuclei are broader due to their high velocities, $0.1c$ or more. Apart from these broad line components, one would expect also very narrow γ-ray lines, if the target nuclei are contained in cosmic grains. In this case, because of ionization losses, the excited nuclei come to the rest before they decay.

The spectrum of de-excitation γ-ray line emission depends on the abundance of elements in cosmic rays (very broad lines) and in the ambient medium both in the form of the gaseous component (broad lines) and grains (narrow lines). The most distinct features in the overall nuclear γ-ray spectrum appear around 4.4 MeV (from ^{12}C*), 6.1 MeV (from ^{16}O*), 0.85 MeV (^{56}Fe*), 0.45 MeV (from α-α reactions), etc.

Generally, the formation of nuclear γ-ray emission through excitation of nuclei by low energy (sub-relativistic) cosmic rays is quite an inefficient mechanism in the sense that only a very small fraction, 10^{-5} to 10^{-6} of the kinetic energy of fast particles is released in the form γ-ray lines (Ozernoy and Aharonian, 1979; Skibo et al., 1996). The rest goes to the heating and ionization of the ambient gas. The problem of ionization or, in the fully ionized plasma, Coulomb losses can be somewhat relaxed, assuming that γ-ray production takes place in a hot two temperature plasma with $kT_e \leq 10^{10}$K and $kT_i \geq 10^{10}$K. However, the nuclear de-excitation line production efficiency in such environment remains rather small. Indeed, in a plasma with an ion temperature high enough for nuclear excitation, inelastic spallation reactions proceed at a similar or higher rate, leading to nuclear destruction during the time equal or smaller to the stationary high-temperature plasma formation time. As a result, the nuclear line luminosity cannot exceed 10^{-4} of the total luminosity radiated away by the electron component (Aharonian and Sunyaev, 1984; Bildsten et al., 1992).

An interesting outcome of the spallation reactions in compact objects is the formation of a proton-neutron plasma without a noticeable content of nuclei, but, due to electro-neutrality, with electrons and, possibly, positrons. In addition, if the plasma density is sufficiently high, $n \geq 10^{16}$ cm^{-3} (thus the neutrons are effectively captured by protons before they decay), deuterium nuclei co-exist in equilibrium with protons and neutrons. Because of rapid radiative cooling of electrons through the thermal bremsstrahlung and/or Comptonization, a two temperature plasma can be formed around a black hole in which the nucleons lose their energy mainly through p-e Coulomb exchange. Besides, the nucleons radiatively cool due to the capture of neutrons by protons and proton-neutron bremsstrahlung (Aharonian and Sunyaev, 1984). Although the energy spectra of these two radiation components are different, the peak luminosities of both processes appear in the γ-ray domain, $E_\gamma \sim kT_i$. At temperatures $kT_i \gg 10$ MeV, the pion production becomes an additional γ-ray production channel (Dahlbacka et al., 1974, Kolikhalov and Sunyaev, 1979; Mahadevan et al., 1997; see however Aharonian and Atoyan (1983) concerning the low efficiency of this process, at least in spherical accretion flow).

Observations of γ-rays with characteristic spectra in the 1 to 100 MeV energy range can provide direct evidence of the formation of two temperature plasmas under conditions when the heavy nuclei are destroyed, and the emission in γ-ray lines is suppressed. Another interesting effect connected with such a scenario is the escape ("evaporation") of neutrons from the accretion flow, and their capture by the atmosphere of the companion star (Aharonian and Sunyaev, 1984; Guessoum and Kazanas, 1990; Guessoum and Jean, 202). In dense cold regions, the neutrons are quickly thermalized and captured by protons resulting in narrow 2.22 MeV γ-ray line emission. Moreover, it has been shown (Belyanin and Derishev, 2001) that for a broad range of accretion parameters, neutrons may effectively decouple from protons leading to the formation of a self-sustained halo. This implies that new neutrons in the halo are supplied mainly by the destruction of helium nuclei by the existing neutrons. Thus, once formed, such a halo can exist even if the proton temperature is much lower than the energy threshold of helium dissociation.

Finally we note that γ-ray line emission is expected also from radioactive isotopes synthesised in stellar interiors or during supernova explosions (Clayton, 1982). Since nucleosynthesis can be effective only in very dense environments, of the large number of γ-ray lines only a few can survive and be observed. These are γ-rays related to the abundant isotopes with long lifetimes. The best candidates are lines from ^{26}Al and ^{60}Fe for the production of diffuse galactic emission, as well as γ-ray lines from ^{7}Be, ^{44}Ti and ^{56}Ni that can observed during solitary transient phenomena (see, for a review, e.g. Diehl and Timmes, 1998; Tatischeff, 2002).

3.2 Interactions with Photon Fields

The interaction of relativistic electrons with radiation fields through inverse Compton scattering provides one of the principal γ-ray production processes in astrophysics. It works effectively almost everywhere, from compact objects like pulsars and AGN to extended sources like supernova remnants and clusters of galaxies. Because of the universal presence of the 2.7 K CMBR, as well as low gas densities and low magnetic fields, inverse Compton scattering proceeds with very high efficiency in the intergalactic medium over the entire γ-ray domain. Since the Compton cooling time decreases linearly with energy, the process becomes especially effective at very high energies.

The electron-positron pair production in photon-photon collisions is tightly coupled with inverse Compton scattering. First of all, it is an absorption process that prevents the escape of energetic γ-rays from compact objects, and determines the "γ-ray horizon" of the Universe. At the same time, in an environment where the radiation pressure dominates over the magnetic field pressure, the photon-photon pair production and the inverse Compton scattering "work" together supporting the effective transport of high energy radiation via electromagnetic "Klein-Nishina" cascades.

Although the inverse Compton scattering of protons is suppressed by a factor of $(m_e/m_p)^4$, very high energy protons effectively interact with the ambient photon fields through electron-positron pair-production and photomeson processes. While in the first process γ-rays are produced indirectly, via inverse Compton scattering of the secondary electrons, photomeson reactions result in the direct production of π^0-mesons and their subsequent decay to γ-rays. Typically, at extremely high energies these interactions proceed effectively both in compact objects and large scale structures.

3.2.1 *Inverse Compton scattering*

The derivation of the cross-section for Compton scattering with given four-vector momenta of the electron and photon can be found in Akhiezer and Berestetskii (1965). The basic expressions of the inverse Compton scattering (i.e. when the energy of the electron significantly exceeds the energy of the target photon) have been comprehensively analysed by Jones (1968), Blumenthal and Gould (1970) and Coppi and Blandford (1990) for the case of isotropically distributed photons and electrons. The anisotropic case has been studied by Aharonian and Atoyan (1981c), Nagirner and Putanen (1993), Brunetti (2000), and Sazonov and Sunyaev (2000).

The angle-averaged total cross-section of inverse Compton scattering depends only on the product of the energies of the interacting electron ε and photon ω_0, $\kappa_0 = \omega_0 \varepsilon_e$ (where all energies are in units of $m_e c^2$). In the nonrelativistic regime ($\kappa_0 \ll 1$) it approaches the classical (Thomson) cross-section $\sigma_{\rm IC} \approx \sigma_{\rm T}(1 - 2\kappa_0)$, while in the ultrarelativistic regime ($\kappa_0 \gg 1$) it decreases with κ_0 as $\sigma_{\rm IC} \approx (3/8)\sigma_{\rm T}\kappa_0^{-1}\ln(4\kappa_0)$. With an accuracy of better than 10 per cent in a very broad range of κ_0, the cross-section can be represented in the following simple form (Coppi and Blandford, 1990)

$$\sigma_{\rm IC} = \frac{3\sigma_{\rm T}}{8\kappa_0}\left[\left(1 - \frac{2}{\kappa_0} - \frac{2}{\kappa_0^2}\right)\ln(1 + 2\kappa_0) + \frac{1}{2} + \frac{4}{\kappa_0} - \frac{1}{2(1 + 2\kappa_0)^2}\right] \quad (3.14)$$

The total cross-section of Compton scattering as a function of κ_0 is shown in Fig. 3.4.

The energy distribution of up-scattered γ-rays is determined by the differential cross-section of the process. Assuming that a monoenergetic beam of low energy photons ω_0 penetrates an isotropic and homogeneous region filled with relativistic electrons of energy ε_e, the spectrum of radiation scattered at the angle θ relative to the initial photon beam is written as (Aharonian and Atoyan, 1981c)

$$\frac{d^2 N(\theta, \varepsilon_\gamma)}{d\varepsilon_\gamma d\Omega} = \frac{3\sigma_T}{16\pi\omega_0\varepsilon_e^2}\left[1 + \frac{z^2}{2(1-z)} - \frac{2z}{b_\theta(1-z)} + \frac{2z^2}{b_\theta^2(1-z)^2}\right], \quad (3.15)$$

where $b_\theta = 2(1-\cos\theta)\omega_0\varepsilon_e$, $z = \varepsilon_\gamma/\varepsilon_e$. The energy of the high energy γ-ray photon ε_γ varies in the limits $\omega_0 \ll \varepsilon_\gamma \ll \varepsilon_{\gamma,\max}$, where $\varepsilon_{\gamma,\max} = \varepsilon_e b/(1+b)$, $b = 4\kappa_0$.

In the case of isotropically distributed electrons and photons, the integration of Eq.(3.15) over the angle θ gives (Jones, 1968, Blumenthal and Gould, 1970)

$$\frac{dN(\varepsilon_\gamma)}{d\varepsilon_\gamma} = \frac{3\sigma_T}{4\omega_0\varepsilon_e^2}\left[1 + \frac{z^2}{2(1-z)} + \frac{z}{b(1-z)} - \frac{2z^2}{b^2(1-z)^2} + \frac{z^3}{2b(1-z)^2} - \frac{2z}{b(1-z)}\ln\frac{b(z-1)}{z}\right]. \quad (3.16)$$

The differential energy spectra of γ-rays for several fixed values of κ_0 are shown in Fig. 3.5. In the deep Klein-Nishina regime ($\kappa_0 \gg 1$) the spectrum

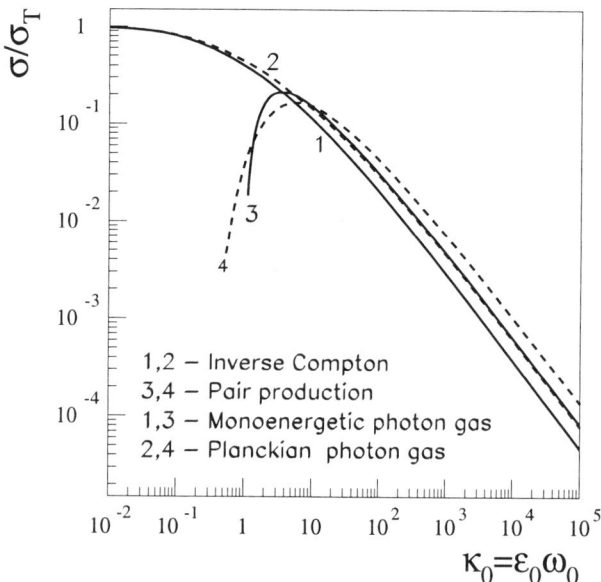

Fig. 3.4 Total cross-sections of inverse Compton scattering and photon-photon pair production in isotropic radiation fields. Two spectral distributions for the ambient photon gas are assumed: (i) monoenergetic with energy ω_0 (curves 1 and 3), and (ii) Planckian with the same mean photon energy $\omega_0 \simeq 3kT/m_e c^2$ (curves 2 and 4).

grows sharply towards the maximum at $\varepsilon_{\gamma,\mathrm{max}}$. This implies that in this regime just one interaction is sufficient to transfer a substantial fraction of the electron energy to the upscattered photon (see also Table 3.1). In the Thomson regime ($\kappa_0 \ll 1$) the average energy of the upscattered photon is $\varepsilon_\gamma \approx \omega_0 \varepsilon_e^2$, thus only a fraction $\varepsilon_\gamma/\varepsilon_e \sim \kappa_0 \ll 1$ of the primary electron energy is released in the upscattered photon.

For a power-law distribution of electrons, $\mathrm{d}N_e/\mathrm{d}\varepsilon_e \propto \varepsilon_e^{-\Gamma}$, the resulting γ-ray spectrum in the nonrelativistic regime ($a = 4\omega_0 \varepsilon_\gamma \ll 1$) has a power-

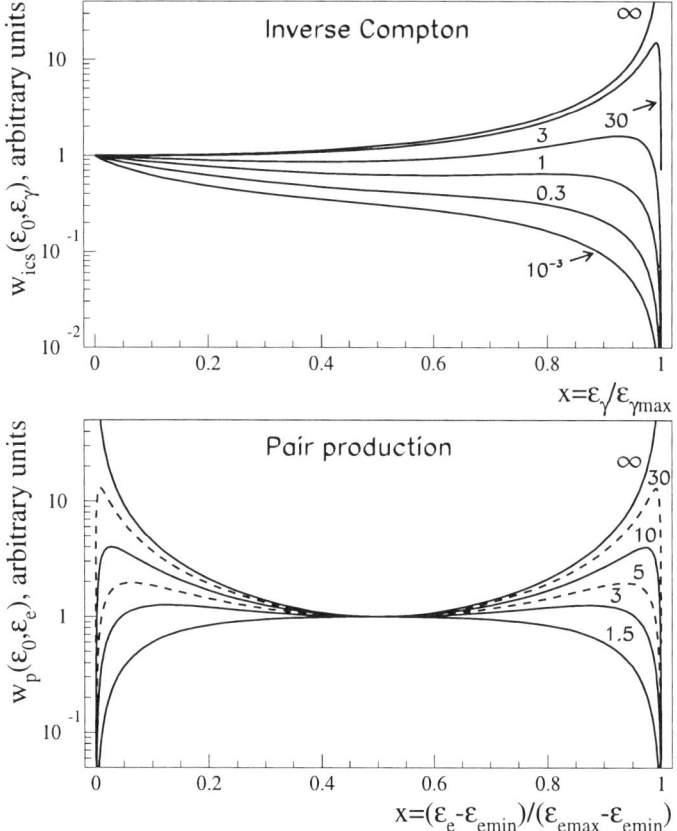

Fig. 3.5 Differential spectra of γ-rays from inverse Compton scattering (upper panel) and electrons from photon-photon pair production (bottom panel) in an isotropic and mono-energetic photon field. The parameters $\varepsilon_{\gamma,\mathrm{max}}$, $\varepsilon_{e,\mathrm{min}}$ and $\varepsilon_{e,\mathrm{max}}$ are defined as $\varepsilon_{\gamma\mathrm{max}} = 4\varepsilon_0(\kappa_0/1 + 4\kappa_0)$ and $\varepsilon_{\mathrm{emin,emax}} = 0.5\varepsilon_0(1 \mp \sqrt{1 - 1/\kappa_0})$. The same values of the parameters $\kappa_0 = \varepsilon_e \omega_0$ and $s_0 = \varepsilon_\gamma \omega_0$ are indicated by the curves.

law form with photon index $\alpha = (\Gamma+1)/2$ (Ginzburg and Syrovatskii, 1964). In the ultrarelativistic ($a \gg 1$) regime the γ-ray spectrum is noticeably steeper, $\propto \varepsilon_\gamma^{-\alpha}(\ln a + \mathrm{const})$ with $\alpha = (\Gamma+1)$ (Blumenthal and Gould, 1970). Several useful analytical approximations for γ-ray spectra over a broad energy interval, including these two regimes and the Klein-Nishina transition region ($a \sim 1$), can be found in Aharonian and Atoyan (1981c) and Coppi and Blandford (1990). The spectral features of the inverse Compton radiation in power-law target photon fields have been studied by Zdziarski (1988)

Table 3.1 The mean fraction of primary energy ($\bar\varepsilon_\gamma/\varepsilon_0$) transferred to the secondary photon in the inverse Compton scattering (ics) and the synchrotron radiation (syn) processes for different values of the parameters $\kappa_0 = \varepsilon_0\omega_0$ and $\chi_0 = \varepsilon_0 H/H_{\mathrm{cr}}$, respectively.

κ_0, χ_0	0.01	0.1	1	10^2	10^4	10^6
$(\bar\varepsilon_\gamma/\varepsilon_0)_{\mathrm{ics}}$	0.014	0.099	0.358	0.760	0.867	0.910
$(\bar\varepsilon_\gamma/\varepsilon_0)_{\mathrm{syn}}$	$0.44\cdot 10^{-2}$	0.033	0.118	0.241	0.250	0.250

The energy-loss rate of relativistic electrons in a monoenergetic field of photons with energy ω_0 and number density n_{ph} is given by the following equation (Aharonian and Atoyan, 1981c)

$$\frac{\mathrm{d}\varepsilon_e}{\mathrm{d}t} = \frac{3\sigma_\mathrm{T} c n_{\mathrm{ph}}}{4\omega_0 b}\left[\left(6+\frac{b}{2}+\frac{6}{b}\right)\ln(1+b)-\ln^2(1+b)-\right.$$
$$\left. 2\mathrm{Li}\left(\frac{1}{1+b}\right)-\frac{(11/12)b^3+8b^2+13b+6}{(1+b)^2}\right]. \quad (3.17)$$

where $\mathrm{Li}(x) = \int_x^1 (1-y)^{-1}\ln(y)\mathrm{d}y$.

In the Thomson and Klein-Nishina regimes Eq.(3.17) reduces to the well known expressions (e.g. Blumenthal and Gould, 1970)

$$\frac{\mathrm{d}\varepsilon_e}{\mathrm{d}t} = \frac{4}{3}\sigma_\mathrm{T} c\omega_0 n_{\mathrm{ph}}\varepsilon_e^2 \qquad \text{at}\quad b \ll 1, \quad (3.18)$$

and

$$\frac{\mathrm{d}\varepsilon_e}{\mathrm{d}t} = \frac{3}{8}\frac{\sigma_\mathrm{T} c n_{\mathrm{ph}}}{\omega_0}(\ln b - 11/6) \qquad \text{at}\quad b \gg 1. \quad (3.19)$$

The energy losses in these two regimes have quite a different dependence on the electron energy. While in the Thomson regime the loss rate is proportional to ε_e^2, in the Klein-Nishina regime it is almost energy independent. This implies that in the first case the steady-state electron spectrum becomes steeper, whereas the Compton losses in the Klein-Nishina regime make the electron spectrum harder.

For calculations of electron energy losses in radiation fields with more realistic distributions one should integrate Eqs.(3.17)-(3.19) over ω_0. It is interesting to note that in the Thomson regime the energy-loss rate does not depend on the spectral distribution of target photons, but depends only on the total energy density of radiation u_r. Correspondingly, the cooling time of electrons due to Thomson scattering is given by

$$t_{\rm IC} \approx 3 \times 10^8 (u_{\rm r}/1{\rm eV/cm}^3)^{-1}(E_{\rm e}/1{\rm GeV})^{-1} \; {\rm yr} \;, \qquad (3.20)$$

where the radiation energy density $u_{\rm r}$ and the electron energy E_e are expressed in units of eV/cm^3 and GeV, respectively. Note that the same equation describes the synchrotron energy losses if we replace u_r by the magnetic field energy density $B^2/8\pi$.

The comparison of Eq.(3.20) with the bremsstrahlung cooling time given by Eq.(3.4) shows that Compton and synchrotron losses dominate over the bremsstrahlung cooling, if

$$E_{\rm e} \geq 10 \left(\frac{1{\rm eV/cm}^3}{u_{\rm r} + B^2/8\pi} \right) \left(\frac{n}{1{\rm cm}^{-3}} \right) \; {\rm GeV}. \qquad (3.21)$$

3.2.2 *Photon-photon pair production*

Photon-photon pair production is the inverse process to pair annihilation. Therefore the differential cross-section is identical to the pair annihilation cross-section, except for a different phase-space volume. In the relativistic regime this process is quite similar also to inverse Compton scattering. However, unlike the pair annihilation and Compton scattering, the photon-photon pair production has a strict kinematic threshold given by

$$\varepsilon_{\gamma 1} \varepsilon_{\gamma 2} (1 - \cos \theta) \geq 2 \;, \qquad (3.22)$$

where $\varepsilon_{\gamma 1}$ and $\varepsilon_{\gamma 2}$ are the energies of two photons in units of $m_e c^2$ colliding at an angle θ (in the laboratory frame).

The large cross-section makes the photon-photon pair production one of the most relevant elementary processes in high energy astrophysics. The role of this process in the context of intergalactic absorption of γ-rays was first pointed out by Nikishov (1962). Bonometto and Rees (1971) were first who emphasised the importance of this process in dense radiation fields of compact objects.

Several convenient approximations for the total cross-section of this process in the isotropic radiation field have been proposed by Gould and Schrèder (1967), Aharonian et al. (1983a) and Coppi and Blandford (1990). With an accuracy of better than 3 per cent, the total cross-section in the monoenergetic isotropic photon field can be represented in the following analytical form

$$\sigma_{\gamma\gamma} = \frac{3\sigma_{\rm T}}{2s_0^2} \left[\left(s_0 + \frac{1}{2}\ln s_0 - \frac{1}{6} + \frac{1}{2s_0} \right) \ln(\sqrt{s_0} + \sqrt{s_0-1}) - \left(s_0 + \frac{4}{9} - \frac{1}{9s_0} \right) \sqrt{1 - \frac{1}{s_0}} \right]. \quad (3.23)$$

The total cross-sections of inverse Compton scattering and pair production in an isotropic monoenergetic photon field of energy ω_0 are shown in Fig. 3.4 (curves 1 and 3, respectively). Both cross sections depend only on the product of the primary (ε_e or ε_γ) and target photon (ω_0) energies, $\kappa_0 = \varepsilon_e \omega_0$ and $s_0 = \varepsilon_\gamma \omega_0$. While as $\kappa_0 \to 0$, the inverse Compton cross-section approaches the Thomson cross-section, $\sigma_{\rm IC} \approx \sigma_{\rm T}(1-2\kappa_0)$, as $s_0 \to 1$ the pair production cross-section approaches zero, $\sigma_{\gamma\gamma} \approx (1/2)\sigma_{\rm T}(s_0-1)^{3/2}$. For $\kappa_0, s_0 \gg 1$ the two cross-sections are quite similar and decrease with κ_0 and s_0: $\sigma_{\rm IC} \approx (3/8)\sigma_{\rm T}\kappa_0^{-1}\ln(\kappa_0)$, $\sigma_{\gamma\gamma} \approx (2/3)\sigma_{\rm T}s_0^{-1}\ln(s_0)$. The pair-production cross-section has a maximum at the level of $\sigma_{\gamma\gamma} \approx 0.2\sigma_{\rm T}$ achieved at $s_0 \approx 3.5 - 4$.

The parameter that characterises γ-ray absorption at photon-photon interactions in a source of size R is the so-called optical depth

$$\tau(\varepsilon_\gamma) = \int_0^R \int_{\omega_1}^{\omega_2} \sigma(\varepsilon_\gamma, \omega) n_{\rm ph}(\omega, r) {\rm d}\omega {\rm d}r , \quad (3.24)$$

where $n_{\rm ph}(\omega, r)$ describes the spectral and spatial distribution of the target photon field in the source. For a homogeneous source with a narrow spectral distribution of photons, for order of magnitude estimates one can use the approximation $\tau(\varepsilon_\gamma) = R\sigma(\varepsilon_\gamma, \bar{\omega}) n(\bar{\omega})$, where $\bar{\omega}$ is the average target photon energy. However, generally one has to be careful with this type of estimate, especially at low energies; while this approximation implies a completely transparent source (i.e. $\tau = 0$) at $\varepsilon_\gamma < 1/\bar{\omega}$, in fact non-negligible absorption can take place at low energies. For example, for a Planckian distribution of target photons, the optical depth τ cannot be disregarded at energies $\varepsilon_\gamma \leq 1/\bar{\omega} \simeq m_e c^2/3kT$, because of interactions with photons from the Wien tail region (see Fig. 3.4). Very useful approximations for the optical depth of γ-rays in a Planckian photon gas can be found in Gould and Schrèder (1967).

Because of narrowness of the pair-production cross-section, for a large class of broad band target photon energy distributions $n_{\rm ph}(\omega)$, the optical depth at given γ-ray energy ε_γ is essentially determined by a relatively narrow band of target photons with energy centered on $\omega_* = 4/\epsilon_\gamma$ (Herterich, 1974). Therefore, the optical depth can be written in the form $\tau(\varepsilon_\gamma) = \eta(\sigma_{\rm T}/4)\omega_* n_{\rm ph}(\omega_*) R$, where the normalization factor η depends on the spectral shape of the background radiation. For a power-law target photon spectrum, $n_{\rm ph}(\omega) = n_0 \omega^{-\alpha}$, the parameter η is calculated analytically, $\eta = (7/6) 4^\alpha \alpha^{-5/3} (1+\alpha)^{-1}$ (Svensson, 1987).

The energy spectrum of electrons produced at photon-photon pair production has been studied by Aharonian et al. (1983a), Zdziarski and Lightman (1985), Coppi and Blandford (1990) and Böttcher and Schlickeiser (1997). For a low-energy monoenergetic photon field ($\omega_0 \ll 1$), and correspondingly $\varepsilon_\gamma \gg 1$, the spectrum of electron-positron pairs can be represented, with an accuracy of better than a few per cent, in the following analytical form (Aharonian et al., 1983a):

$$\frac{{\rm d}N(\varepsilon_{\rm e})}{{\rm d}\varepsilon_{\rm e}} = \frac{3\sigma_{\rm T}}{32\omega_0^2 \varepsilon^3} \left[\frac{4\varepsilon_\gamma^2}{(\varepsilon_\gamma - \varepsilon_{\rm e})\varepsilon_{\rm e}} \ln \frac{4\omega_0 (\varepsilon_\gamma - \varepsilon_{\rm e})\varepsilon_{\rm e}}{\varepsilon_\gamma} - 8\omega_0 \varepsilon_\gamma + \right.$$

$$\left. \frac{2(2\omega_0 \varepsilon_\gamma - 1)\varepsilon_\gamma^2}{(\varepsilon_\gamma - \varepsilon_{\rm e})\varepsilon_{\rm e}} - \left(1 - \frac{1}{\omega_0 \varepsilon_\gamma} \right) \frac{\varepsilon_\gamma^4}{(\varepsilon_\gamma - \varepsilon_{\rm e})^2 \varepsilon_{\rm e}^2} \right] . \quad (3.25)$$

The kinematic range of variation of $\varepsilon_{\rm e}$ is

$$\frac{\varepsilon_\gamma}{2} \left(1 - \sqrt{1 - \frac{1}{\omega_0 \varepsilon_\gamma}} \right) \leq \varepsilon_{\rm e} \leq \frac{\varepsilon_\gamma}{2} \left(1 + \sqrt{1 - \frac{1}{\omega_0 \varepsilon_\gamma}} \right) . \quad (3.26)$$

The differential energy spectra of γ-rays for several fixed values of the parameters $s_0 = \omega_0 \varepsilon_\gamma$ are shown in Fig. 3.5. The spectra are symmetric around the point $x = \varepsilon_{\rm e}/\varepsilon_\gamma$. Although the average energy of the secondary electrons is $\varepsilon_\gamma/2$, for very large s_0 the interaction has a catastrophic character - the major fraction of the energy of the primary γ-ray photon is transferred to the leading electron. This fraction exceeds 0.5 and asymptotically approaches 1 (see Table 3.2).

For calculation of electron spectra produced in radiation fields by γ-rays with more realistic spectral distributions, $N(\varepsilon_\gamma)$, one should integrate the product of spectrum given by Eq.(3.25) and $N(\varepsilon_\gamma)$ over a broad primary γ-ray energy interval. For a power-law spectrum of γ-rays, $N(\varepsilon_\gamma) \propto E^{-\Gamma}$, the spectrum of secondary pairs can be approximated, with an accuracy of better than 20 per cent, in a simple form (Aharonian and Atoyan 1991b):

$$q_\pm(E)\,dE = f(\Gamma)\frac{\exp\left[-(1/(x-1))\right]}{E_* x(1+0.07\,x^\Gamma/\ln x)}\,dE, \qquad (3.27)$$

where $x = \varepsilon_e/\varepsilon_* \geq 1$, $\varepsilon_* = 1/4\omega_0$, $f(\Gamma) = (1.11 - 1.60\Gamma + 1.17\Gamma^2)$, and $\int q_\pm(E)\,dE = 2$ (two electrons per interaction). Starting from the minimum (allowed by kinematics) energy at $\varepsilon_e = \varepsilon_*$, the electron spectrum sharply rises achieving its maximum at $\approx 2.4\varepsilon_*$, and then at $\varepsilon_e \gg \varepsilon_*$, it behaves as $q_\pm \propto \varepsilon_e^{-(\Gamma+1)} \ln \varepsilon_e$.

Table 3.2 The mean fraction of energy of the primary γ-ray photon transferred to the leading secondary electron at electron-positron pair production in a mono-energetic radiation field (rad) and in the magnetic field (B) for different values of the parameters $\kappa_0 = \varepsilon_0 \omega_0$ and $\chi_0 = \varepsilon_0 H/H_{\rm cr}$, respectively.

κ_0, χ_0	1	3	10	10^2	10^4	10^6
$(\bar\varepsilon_e/\varepsilon_0)_{\rm rad}$	0.500	0.701	0.797	0.891	0.948	0.966
$(\bar\varepsilon_e/\varepsilon_0)_{\rm B}$	0.634	0.693	0.746	0.782	0.824	0.825

Two pairs of coupled processes – inverse Compton scattering and photon-photon pair production – determine the basic features of interactions of electrons and γ-rays in the radiation dominated environments. At extremely high energies higher order QED processes may compete with these basic channels. Namely, when the product of the energies of colliding cascade particles (electrons or photons) E and the background photons ω significantly exceed $10^5 m_e^2 c^4$, the processes $\gamma\gamma \to e^+e^-e^+e^-$ (Brown et al., 1973) and $e\gamma \to e\gamma e^+ e^-$ (Mastichiadis, 1991; Dermer and Schlickeiser, 1991) dominate over single (e^+, e^-) pair production and Compton scattering, respectively. For example, in the 2.7 K CMBR the first process stops the linear increase of the mean free path of the highest energy γ-rays around 10^{21} eV, and puts a robust limit on the mean free path of γ-rays of about 100 Mpc. Analogously, above 10^{20} eV the second process becomes more important than the conventional inverse Compton scattering. Because the $\gamma\gamma \to 2e^+2e^-$ and $e\gamma \to e\gamma e^+e^-$ channels result in production of 2 additional electrons, they substantially change the character of interactions. Note, however, that an effective realization of these processes is possible only under very specific conditions with an extremely low magnetic field and narrow energy distribution of the background photons.

3.2.3 Interactions of hadrons with radiation fields

In astrophysical environments the radiation density often exceeds the density of gas component. In these conditions the interactions of high energy hadrons with radiation can dominate over interactions with matter, albeit the relevant cross-sections are relatively small. The main processes of hadron-photon interactions include (i) inverse Compton scattering: $p+\gamma \to p+\gamma'$, (ii) electron-positron pair production: $p+\gamma \to p\,e^+e^-$, (iii) photodisintegration of nuclei: $A + \gamma \to A' + kN$, (iv) photomeson production : $N+\gamma \to N+k\pi$. In extremely dense radiation fields the secondary $\pi^\pm$-mesons may effectively interact with photons before they decay.

Except for the inverse Compton scattering, all other processes take place only above certain kinematic thresholds: ~ 1 MeV, 10 MeV, and 140 MeV (in the rest frame of projectile particles) for the pair production, photodisintegration, and pion production, respectively.

The process of inverse Compton scattering of protons is identical to the inverse Compton scattering of electrons, but the energy loss rate of protons is suppressed, for the fixed energy of both particles, by a factor of $(m_e/m_p)^4 \approx 10^{-13}$. Generally, this process does not have noticeable astrophysical applications. At energies above the pair production threshold, the inverse Compton energy loss rate is significantly (by a factor of $\alpha(m_p/m_e)^2 \sim 10^4$) slower compared to the losses caused by pair-production.

In certain conditions the pair-production may result in significant spectral distortions of highest energy protons propagating through dense photon fields. The cross-section of this process is quite large (the same Bethe-Heitler cross-section in the rest frame of proton), but in each interaction only a small fraction of the proton energy is transferred to the secondary electrons (Blumenthal, 1990). Therefore the energy loss rate of protons remains relatively slow. Moreover, the energy region where this process dominates is quite narrow. It is limited by the energy interval of protons $\sim (1-100) \times 10^{15}(\omega_0/1 \text{ eV})^{-1}$ eV, (ω_0 is the average energy of target photons). When the proton energy exceeds the pion production threshold, the hadronic photomeson interactions well dominate over the pair production (see e.g. Berezinsky and Grigoreva, 1988; Geddes et al., 1996)

The photodisintegration of nuclei may have a strong impact on the formation of the chemical composition of very high energy cosmic rays in compact astrophysical objects (Karakula and Tkaczyk, 1993) as well as in the intergalactic medium (e.g. Stecker, 1969). However, this process does not lead to significant production of high energy γ-rays.

Photomeson production is the most important channel for transformation of the kinetic energy of protons into high energy γ-rays, electrons and neutrinos. Close to the energy threshold, the process proceeds through single-pion production, $p + \gamma \to p + \pi^0$, and $p + \gamma \to n + \pi^+$. At higher energies, multi-pion production channels begin to dominate.

Cross-sections of these processes are basically well known from particle accelerator experiments. For astrophysical applications the data obtained with γ-ray beams at energies from 140 MeV to 10 GeV are quite sufficient, if one takes into account the fact that for typical broad-band target photon spectra the hadron-photon interactions are contributed mainly from the region not far from the energy threshold, i.e. $E_\gamma \leq 1$ GeV.

The photomeson processes in radiation fields have been studied by Mücke et al. (1999) using the Monte-Carlo code SOPHIA. Recently Atoyan and Dermer (2003) suggested a simple approach for approximation of the pion production cross-sections by the sum of two step-functions $\sigma_1(E_\gamma)$ and $\sigma_1(E_\gamma)$ for the single-pion and multi-pion channels respectively, with $\sigma_1 = 340$ μb for 200 MeV $\leq E_\gamma \leq 500$ MeV and $\sigma_2 = 120$ μb for $E_\gamma \geq 500$ MeV. The inelasticities in these two energy intervals are approximated by $f_{p\gamma} = 0.2$ and 0.6, respectively. Finally, applying the δ-function approximation to calculations of the spectra of secondary particles (assuming $E_\gamma \approx 0.1 E_p$ with 2 photons per π^0-decay, and $E_{\gamma,\nu} \approx 0.05 E_p$, with 1 electron and 3 neutrinos produced in every charged pion decay), this simple approach appears quite accurate for an adequate treatment of production rates and energy spectra of secondary products.

The cross-sections of interactions of secondary electrons and γ-rays with the ambient photons exceed by three orders of magnitude the photomeson cross-sections. Therefore the electrons and γ-rays cannot leave the active region of pion production, but rather initiate electromagnetic cascades in the surrounding photon and magnetic fields. The standard spectra of the low-energy cascade γ-rays that eventually escape the source are not sensitive to the initial spectral distributions, and thus contain information only about the total hadronic power of the source. On the other hand the secondary neutrinos freely escape the production region, and thus carry direct information about the energy spectra of accelerated protons.

Another interesting feature of the mixed hadronic/electromagnetic cascades in radiation dominated environments is the effective transport of primary nonthermal energy released in accelerated protons further away from a central engine through production and escape of secondary neutrons. This important channel of energy transport, pointed out first by Eichler and Wiita (1978), has been explored by many authors (Kirk and Mastichiadis, 1989; Sikora *et al.*, 1989; Giovanoni and Kazanas, 1990, Atoyan and Dermer, 2003, *etc.*). The presence of dense photon fields in the compact particle accelerators may have an even more fundamental impact. It has been recently recognised (Derishev *et al.*, 2003) that in relativistic flows, e.g. in GRBs or AGN jets, the multiple conversions of relativistic particle from charged to neutral state (proton→neutron→proton..., e → γ → e...) may allow a strong (up to the bulk Lorenz factor squared) energy gain in each cycle, whereas in the standard relativistic shock acceleration scenario the energy gain $\sim \Gamma^2$ occurs only in the first circle (Achterberg *et al.*, 2001). This novel acceleration mechanism, which is capable of boosting protons in GRBs and AGN jets to maximum available energies (limited by the condition of confinement in the magnetic field), could be a key to the solution of the problem of the highest energy, $E \geq 10^{20}$ eV, particles observed in cosmic rays.

3.3 Interactions with Magnetic Fields

3.3.1 *Synchrotron radiation and pair-production*

Many important results of the theory of synchrotron radiation are obtained within the framework of classical electrodynamics (see e.g. Jackson, 1975). The classical treatment of the synchrotron radiation is limited by the condition

$$\frac{E_e}{m_e c^2} \frac{B}{B_{\rm cr}} \ll 1 , \qquad (3.28)$$

where $B_{\rm cr} = m_e^2 c^3/e\hbar \approx 4.4 \times 10^{13}$ G is the so-called critical value of the magnetic field relevant to quantum effects.

Generally, the energy of synchrotron photons are much less than the energy of parent electrons. However, in specific astrophysical environments, e.g. in pulsar magnetospheres or in magnetised accretion disks, the synchrotron radiation could be close to the "quantum threshold" of about $EB \simeq 10^7$ TeV Gauss. The production of electron-positron pairs in a magnetic field by high energy γ-rays is, by definition, a quantum process. In the quantum regime these two processes are tightly coupled, and lead to an effective cascade development.

For interactions of electrons and photons with magnetic fields it is convenient to introduce interaction probabilities instead of standard total cross-sections (Anguelov and Vankov, 1999). But in the literature this parameter is still formally called a cross-section. These probabilities, normalised to the strength of the magnetic field, are shown in Fig. 3.6.

The probabilities of both synchrotron and pair production processes depend on a single parameter – $\chi_0 = \varepsilon_0 B/B_{\rm cr}$, where B is the component of the magnetic field perpendicular to the vector of the particle speed. This parameter is an analog of the parameters κ_0 and s_0 in the photon field. While the probability of synchrotron radiation at $\chi_0 \ll 1$ is constant, the probability of pair production below $\chi_0 = 1$ drops dramatically (proportional to $\exp\left[-8/(3\chi_0)\right]$). After achieving its maximum at $\chi_0 \simeq 10$, the probability of pair production decreases with χ as $\propto \chi_0^{-1/3}$ (see Erber, 1966). At large χ_0, the probability of synchrotron radiation has a similar behaviour, but its absolute value exceeds by a factor of 3 the probability of pair production.

The differential spectra of γ-rays due to synchrotron radiation and the electrons due to the magnetic pair-production are shown in Fig. 3.7. The cross-sections for synchrotron radiation and pair production are from

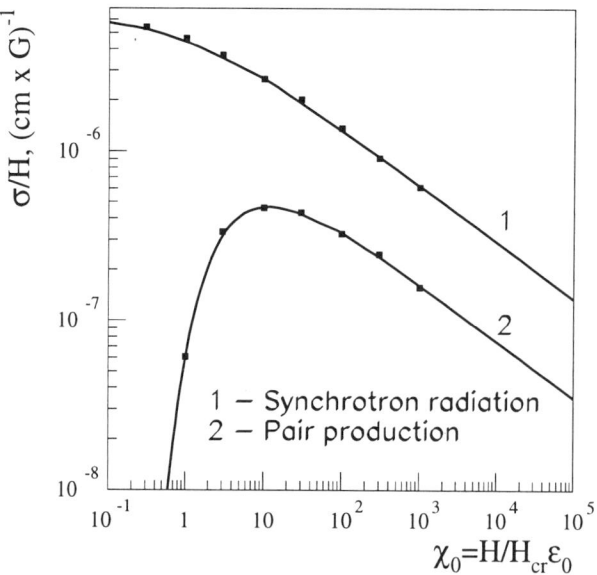

Fig. 3.6 Total cross-sections (interaction probabilities) of the synchrotron radiation and magnetic pair production. Solid curves from Aharonian and Plyasheshnikov (2003), points – Monte Carlo calculations of Anguelov and Vankov (1999).

Akhiezer et al. (1994). At $\chi_0 \ll 1$, the synchrotron γ-ray spectra are very steep, but at large values, $\chi_0 \geq 1$, γ-rays are characterised by a flat distribution. This implies a rather catastrophic character of interaction like in the photon gas. Note that in the photon gas at $\kappa_0 \gg 1$ the fraction of the parent electron energy that is transferred to γ-rays exceeds 0.5 and asymptotically approaches 1. In the magnetic field the energy transfer is smaller. At $\chi_0 \sim 1$, it is approximately 0.1, and asymptotically approaches to $16/63 \simeq 0.25$ at extremely large χ_0 (see Table 3.1). The energy distribution of pair-produced electrons is obviously a symmetric function.

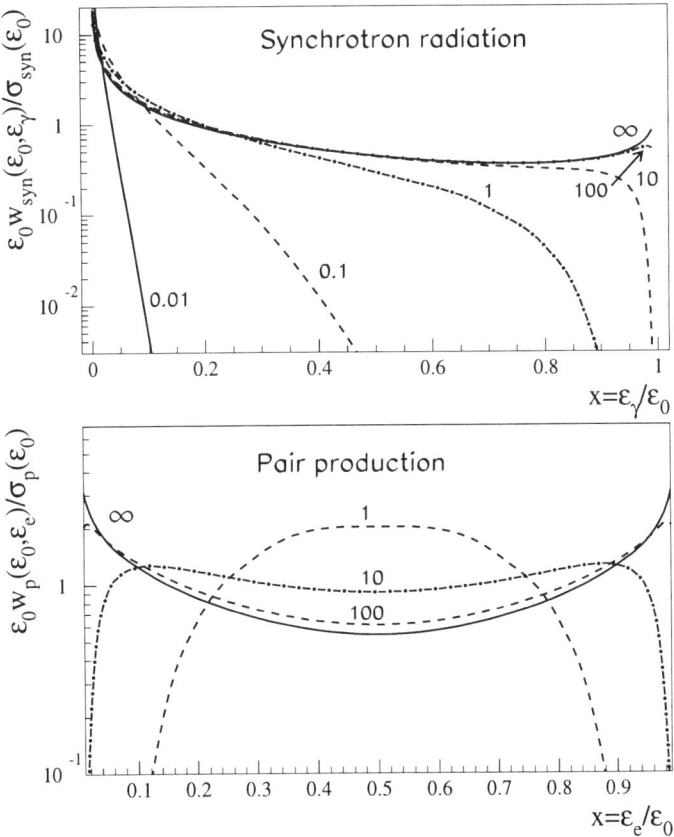

Fig. 3.7 Differential cross-sections of synchrotron radiation (upper panel) and the magnetic pair production (bottom panel) normalised to the total cross-sections of these processes. The values of the parameter $\chi_0 = B/B_{cr}\varepsilon_0$ are indicated by the curves.

Although at large χ_0 the electron spectrum increases when $\varepsilon_e \to \varepsilon_0$, the pair production spectra in a magnetic field are flatter than in radiation fields, and correspondingly the energy of the primary γ-ray in the photon gas is transferred to the leading electron with somewhat higher efficiency (see Table 3.2).

3.3.2 Synchrotron radiation of protons

Generally, proton synchrotron radiation is treated as an inefficient process. However, under certain conditions the synchrotron cooling time of protons can be comparable or even shorter than other timescales that characterise the acceleration and confinement regions of ultrarelativistic protons. Moreover, in compact accelerators of $\sim 10^{20}$ eV protons, very high energy synchrotron or curvature γ-radiation of protons always accompanies acceleration of the highest energy particles (Levinson, 2000; Aharonian et al., 2002; Derishev et al., 2003). The formalism of proton synchrotron radiation is quite simple and identical to the theory of electron synchrotron radiation. Nevertheless, it is worth discussing some basic features of this process relevant to the magnetic-field dominated environment in which the protons are accelerated at the theoretically highest possible rate.

The comprehensively developed theory of electron synchrotron radiation (see e.g. Ginzburg and Syrovatskii, 1965) can be readily applied to the proton-synchrotron radiation by re-scaling the Larmor frequency $\nu_L = eB/2\pi mc$ by the factor $m_p/m_e \simeq 1836$. For the same energy of electrons and protons, $E_p = E_e = E$, the energy loss rate of protons $(dE/dt)_{sy}$ appears $(m_p/m_e)^4 \simeq 10^{13}$ times slower than the energy loss rate of electrons. Also, the characteristic frequency of the synchrotron radiation $\nu_c = 3/2\, \nu_L\, (E/mc^2)^2$ emitted by a proton is $(m_p/m_e)^3 \simeq 6 \times 10^9$ times smaller than the characteristic frequency of synchrotron photons emitted by an electron of the same energy. The synchrotron cooling time of the proton, $t_{sy} = E/(dE/dt)_{sy}$, and the characteristic energy of the synchrotron photon $\epsilon_c = h\nu_c$ are then

$$t_{sy} = \frac{6\pi m_p^4 c^3}{\sigma_T\, m_e^2\, E\, B^2} = 4.5 \times 10^4\, B_{100}^{-2}\, E_{19}^{-1}\, \text{s}, \qquad (3.29)$$

and

$$\epsilon_c = h\nu_c = \sqrt{\frac{3}{2}}\, \frac{heBE^2}{2\pi m_p^3 c^5} \simeq 87\, B_{100}\, E_{19}^2\, \text{GeV}, \qquad (3.30)$$

where $B_{100} = B/100\,\text{G}$ and $E_{19} = E/10^{19}\,\text{eV}$. Hereafter it is assumed that the magnetic field is distributed isotropically, i.e. $B_\perp = \sin\psi\, B$ with $\sin\psi = \sqrt{2/3}$.

The average energy of synchrotron photons produced by a particle of energy E is equal to $\epsilon_m \simeq 0.29\,\epsilon_c$ (Ginzburg and Syrovatskii, 1965). Correspondingly, the characteristic time of radiation of a synchrotron photon of energy ϵ by a proton in a magnetic field B is

$$t_{\text{sy}}(\epsilon) \simeq 2.2 \times 10^5\, B_{100}^{-3/2}\, (\epsilon/1\,\text{GeV})^{-1/2}\,\text{s}\,. \tag{3.31}$$

For comparison, the time needed for radiation of a synchrotron γ-ray photon by an electron is shorter by a factor of $(m_p/m_e)^{5/2} \simeq 1.5 \times 10^8$.

The spectral distribution of synchrotron radiation is given by

$$P(E,\epsilon) = \frac{\sqrt{2}}{h}\, \frac{e^3 B}{mc^2}\, F(x)\,, \tag{3.32}$$

where $x = \epsilon/\epsilon_c$, and $F(x) = x\int_x^\infty \mathrm{d}x K_{5/3}(x)$; $K_{5/3}(x)$ is the modified Bessel function of $5/3$ order. The function $F(x)$ can be presented in a simple analytical form $F(x) = C\, x^{1/3}\exp(-x)$ (e.g. Melrose, 1980). Numerical calculations show that with $C \approx 1.85$ this approximation provides very good, less than 1 per cent error, accuracy in the region of the maximum at $x \sim 0.3$, and still reasonable (less than several per cent error) accuracy in the broad dynamical region $0.1 \leq x \leq 10$.

In Fig. 3.8a four different examples of possible proton spectra are presented. Curve 1 corresponds to the most "standard" assumption for the spectrum of accelerated particles - power-law with an exponential cutoff at energy E_0: $N_p(E) = N_0 E^{-\alpha_p}\exp(-E/E_0)$. Curve 2 corresponds to a less realistic, truncated power-law spectrum, i.e. $N_p(E) \propto E^{-\alpha_p}$ at $E \leq E_0$, and $N_p(E) = 0$ at $E \geq E_0$.

While the cutoff energy E_0 in the spectrum of accelerated particles could be estimated quite confidently from the balance between the particle acceleration and energy loss rates, the shape of the resulting spectrum in the cutoff region depends on several circumstances - the specific mechanisms of acceleration and energy dissipation, the diffusion coefficient, etc. For example, it has been argued that in the shock acceleration scheme one may expect not only spectral cutoffs, but perhaps also pile-ups preceding the cutoffs (Melrose and Crouch 1997; Protheroe and Stanev 1999; Drury et al., 1999). Two such spectra are shown in Fig. 3.8a. Curve 3 represents the extreme class of spectra containing a sharp (with an amplitude of factor of 10) spike at the very edge of the spectrum. The Curve 4 corresponds to a smoother spectrum with a modest pile-up (or "bump") and super-exponential (but not abrupt) cutoff. The corresponding spectral en-

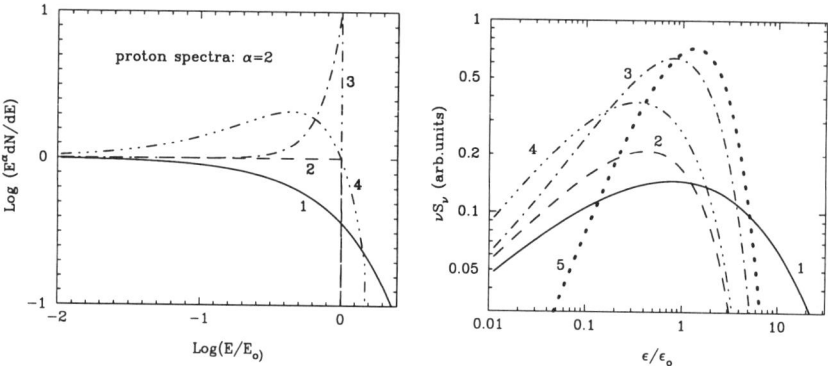

Fig. 3.8 (a) Possible spectra of accelerated protons (left panel), and (b) the corresponding Spectral Energy Distributions of their synchrotron radiation (right panel). At energies $E \ll E_0$ all proton spectra have power-law behaviour with $\alpha_p = 2$, but in the "cutoff" region around E_0 they have very different shapes. Curve 1 corresponds to the proton spectrum described by a power-law with exponential cutoff; curve 2 corresponds to the truncated proton spectrum; curve 3 corresponds to the proton spectrum with a sharp pile-up and an abrupt cutoff at E_0; curve 4 corresponds to the proton spectrum with a smooth pile-up and a super-exponential cutoff. For comparison, in the right panel the spectrum of the synchrotron radiation of mono-energetic protons, $xF(x) \propto x^{4/3} \exp(-x)$, is also shown (curve 5).

ergy distributions ($\nu S_\nu = \epsilon^2 J(\epsilon)$) of synchrotron radiation are shown in Fig. 3.8b. In the high energy range, $\epsilon \geq \epsilon_0$, where ϵ_0 is defined by Eq.(3.30) as $\epsilon_0 = \epsilon_c(E_0)$, the radiation spectrum from the proton distribution with sharp pile-up and abrupt cut-off is quite similar to the synchrotron spectrum from mono-energetic protons, $xF(x) = x^{4/3} e^{-x}$; $x = \epsilon/\epsilon_0$.

All synchrotron spectra shown in Fig. 3.8b exhibit, despite their essentially different shapes, spectral cutoffs at approximately $x \sim 1$, if one defines the cutoff as *the energy at which the differential spectrum drops to $1/e$ of its extrapolated (from low energies) power-law value.* Therefore the energy $\epsilon_0 = \epsilon_c(E_0)$ could be treated as an appropriate parameter representing the synchrotron cutoff for a quite broad class of proton distributions. In the case of mono-energetic protons, the cutoff energy coincides exactly with ϵ_0. This is true also for the power-law proton spectrum with exponential cutoff for which the SED of the synchrotron radiation has a shape close to

$\nu S_\nu \propto \epsilon^{1/2} \exp\left[-(\epsilon/\epsilon_0)^{1/2}\right].$

The maximum energy of the synchrotron radiation depends on the spectrum of accelerated protons. The high energy cutoff in the spectrum of protons is determined by the balance between the particle acceleration and cooling times. It is convenient to present the acceleration time of particles $t_{\rm acc}$ in the following general form

$$t_{\rm acc} = \eta(E)\, r_{\rm g}/c = 1.36 \times 10^4\, E_{19}\, B_{100}^{-1}\, \eta(E)\, {\rm s}\,, \qquad (3.33)$$

where $r_{\rm g} = E/(eB_\perp)$. The so-called gyro-factor $\eta(E) \geq 1$ characterises the energy-dependent rate of acceleration. For almost all proposed models η remains a rather uncertain model parameter. On the other hand, any postulation of acceleration of EHE protons in compact objects like in small scale AGN jets or in transient objects like GRBs, requires η to be close to 1 (Aharonian et al., 2002b).

If the energy losses of protons are dominated by synchrotron radiation, the maximum energy of accelerated particles is determined by the condition $t_{\rm sy} = t_{\rm acc}$:

$$E_0 = (3/2)^{3/4}\,\sqrt{\frac{1}{e^3 B}}\, m_{\rm p}^2 c^4 \simeq 1.8 \times 10^{19}\, B_{100}^{-1/2}\, \eta^{-1/2}\,{\rm eV}\,. \qquad (3.34)$$

The relevant cutoff in the electron spectrum appears at a lower energy, $E_{e,0} = (m_{\rm e}/m_{\rm p})^2 E_0 \simeq 5.3 \times 10^{12}\, B_{100}^{-1/2}\, \eta^{-1/2}\,{\rm eV}$.

Substituting Eq.(3.34) into Eq.(3.30) we find that the position of the cutoff in the synchrotron spectrum is determined by only two fundamental physical constants, the mass of the emitting particle and the fine-structure constant $\alpha_{\rm f} = 1/137$:

$$\epsilon_0 = \frac{9}{4}\alpha_{\rm f}^{-1} mc^2\, \eta^{-1}\,. \qquad (3.35)$$

Thus, for $\eta = 1$ the self-regulated cutoffs in the spectra of synchrotron radiation by electrons and protons appear at energies $\simeq 160$ MeV and $\simeq 300$ GeV, respectively. If γ-rays are produced in a relativistically moving source with Doppler factor $\delta_{\rm j} > 1$, the electron and proton spectral cutoffs are shifted towards the GeV and TeV domains, respectively.

In relatively small magnetic fields, synchrotron radiation is the only channel of proton interactions. However, at sufficiently large values of the product $E_{\rm p} \times B$, protons start to produce secondary particles – electron-positron pairs, pions, etc. The energy threshold of production of a particle of mass $m_{\rm x}$ is estimated as $E_{\rm p} \sim m_{\rm p}(m_{\rm x}/m)^2(B_{\rm crit}/B)$ (e.g. Ozernoy et al., 1973). For example, in pulsars with magnetic fields as strong as 10^{12} G pair production starts at proton energies of about 100 GeV. The production of π-mesons requires ≈ 5 orders of magnitude larger energies.

3.4 Relativistic Electron-Photon Cascades

Relativistic electrons – directly accelerated, or secondary products of various hadronic processes – may result in copious γ-ray production caused by interactions with ambient targets in forms of *gas (plasma), radiation* and *magnetic fields*. In different astrophysical environments γ-ray production may proceed with high efficiency through *bremsstrahlung, inverse Compton scattering* and *synchrotron (and/or curvature) radiation*, respectively.

Generally, γ-ray production in a given process is effective when the relevant radiative cooling time does not significantly exceed *(i)* the source age, *(ii)* the time of non-radiative losses caused by adiabatic expansion or by particle escape, and *(iii)* the cooling time of competing radiation mechanisms resulting in low-energy photons *outside* the γ-ray domain. As long as the charged particles are effectively confined to the γ-ray production region, in some circumstances these conditions could be fulfilled even in environments with relatively low gas and photon densities or a weak magnetic field. More specifically, the γ-ray production efficiency could be close to 1 even when $t_{\rm rad} \gg R/c$ (R is the characteristic linear size of the production region, c is the speed of light). In such cases the secondary γ-rays escape the source without significant internal absorption.

Each of the above mentioned gamma-ray *production* mechanisms has its major "counterpart" – a gamma-ray *absorption* mechanism of the same electromagnetic origin – resulting in electron-positron pair production in matter (the counterpart of bremsstrahlung), in photon gas (the counterpart of inverse Compton scattering), and in a magnetic field (the counterpart of synchrotron radiation). As discussed above, the γ-ray production mechanisms and their absorption counterparts have similar cross-sections, therefore the condition for radiation $t_{\rm rad} \geq R/c$ generally implies small optical depths for the corresponding γ-ray absorption process, $\tau_{\rm abs} \leq 1$.

But in many astrophysical scenarios, in particular in compact galactic and extragalactic objects with favourable conditions for particle acceleration, the radiation processes proceed so fast that $t_{\rm rad} \leq R/c$. At these conditions the internal γ-ray absorption becomes unavoidable. If the γ-ray production and absorption processes occur in relativistic regimes, namely when (i) $E_{\gamma,e} \geq 10^3 m_e c^2$ in hydrogen gas, (ii) $E_{\gamma,e}\omega_0 \gg m_e^2 c^4$ in photon gas (Klein-Nishina regime), or (iii) $(E_{\gamma,e}/m_e c^2)(B/B_{\rm crit}) \gg 1$ in the magnetic field (quantum regime), the problem cannot be reduced to a simple absorption effect. In this regime, the secondary electrons produce a new generation of high energy γ-rays, and these photons again produce electron-positron pairs, so an electromagnetic cascade develops.

The characteristics of electromagnetic cascades in matter have been comprehensively studied in the context of interactions of cosmic rays with the Earth's atmosphere (see e.g. Rossi and Greisen, 1941; Nishimura, 1967; Ivanenko, 1968), as well as for calculations of the performance of detectors of high energy particles (e.g. Nelson et al., 1985). The theory of electromagnetic cascades in matter can be applied to some sources of high energy cosmic radiation, in particular to the "hidden source" scenarios like massive black holes in centers of AGN or young pulsars inside the dense shells of recent supernovae explosions (see e.g. Berezinsky et al., 1990). Also, within the so-called "beam dump" models (see e.g. Halzen and Hooper, 2002) applied to X-ray binaries, protons accelerated by the compact object (a neutron star or a black hole), hit the atmosphere of the normal companion star (Berezinsky, 1976; Eichler and Vestrand, 1984) or the accretion disk (Cheng and Ruderman, 1989; Anchordoqui et al., 2003) and thus result in the production of high energy neutrinos and γ-rays. In such objects, the thickness of the surrounding gas can significantly exceed 100 g/cm^2, thus the protons produced in the central source would initiate (through the production of high energy γ-rays and electrons) electromagnetic showers. These sources perhaps represent the "best hope" of neutrino astronomy, but they are generally considered as less attractive targets for gamma-ray astronomy. However, the γ-ray emission in these objects is not fully suppressed. The recycled radiation with spectral features determined by the thickness ("grammage") of the gas shell, should be seen in γ-rays in any case, unless the synchrotron radiation of secondary electrons dominates over the bremsstrahlung losses and channels the main fraction of the nonthermal energy into the sub-gamma-ray domain.

The development of electromagnetic cascades in photon gas and magnetic fields is a more common phenomenon in astrophysics. In photon fields such cascades can be created on almost *all* astronomical scales, from compact objects like accreting black holes, fireballs in gamma-ray bursts, and sub-pc jets of blazars, to large-scale (up to ≥ 100 kpc) AGN jets and ≥ 1 Mpc clusters of galaxies. Very high energy γ-rays emitted by astronomical objects and interacting with diffuse extragalactic photon fields initiate electromagnetic cascades in the entire Universe (Hayakawa, 1966; Prilutsky and Rozental, 1970; Berezinsky and Smirnov, 1975; Aharonian *et al.*, 1992; Protheroe and Stanev, 1996; Coppi and Aharonian, 1997; Sigl *et al.*, 1997). The superposition of contributions of γ-rays from these cascades should constitute a significant fraction of the observed diffuse extragalactic background.

Bonometto and Rees (1971) were the first who realized the astrophysical importance of electron-photon cascades supported by γ-γ pair-production and inverse Compton scattering in dense photon fields. When the so-called compactness parameter (Guilbert *et al.*, 1983) $l = L\sigma_T/Rm_ec^3$ (L is the luminosity and R is the radius of the source) is less than 10, then the cascade develops in the linear regime, i.e. when the soft radiation produced by cascade electrons does not have a significant feedback effect on the cascade development. In many cases, including the cascade development in compact objects, this approximation works quite well.

The properties of linear cascades in photon fields have been quantitatively studied using the method of Monte Carlo simulations (Aharonian *et al.*, 1985; Protheroe, 1986; Protheroe and Stanev, 1993; Mastichiadis *et al.*, 1994, Müke *et al.*, 1999) or by solving the cascade equations (Aharonian *et al.*, 1990; Coppi, 1992; Ivanenko and Lagutin, 1991; Kalashev *et al.*, 2001, Aharonian and Plyasheshnikov, 2003). Generally, the kinetic equations that describe the cascade development can be solved only numerically. However, with some simplifications it is possible to derive useful analytical approximations (Svensson, 1987; Zdziarski, 1988; Coppi and Blandford, 1990) which help to understand the features of the steady-state solutions for cascades in photon fields.

Cascade development in a magnetic field is a key element for understanding of the physics of pulsar magnetospheres (Sturrok, 1971; Baring and Harding, 2001). Therefore it is generally treated as a process associated with very strong magnetic fields. However, such cascades could be triggered in some other (at first glance unusual) sites like the Earth's geomagnetic field (Anguelov and Vankov, 1999; Plyasheshnikov and Aharonian, 2002), accretion disks of massive black holes (Bednarek, 1997), *etc.* In general, the pair cascades in magnetic fields are effective when the product of the particle (photon or electron) energy and the strength of the B-field becomes close to the "quantum threshold" of about $B_{\rm crit}m_ec^2 \simeq 2 \times 10^7 C$ TeV Gauss, unless we assume a specific, regular field configuration. An approximate method, similar to the so-called approximation A of cascade development in matter (e.g. Rossi and Greisen, 1941), has been applied by Akhiezer *et al.* (1994). Although this theory quite satisfactorily describes the basic features of photon-electron showers, it does not provide adequate accuracy for a quantitative description of the cascade characteristics (Anguelov and Vankov, 1999).

As long as we are interested in the one-dimensional cascade development (which seems to be quite sufficient for many astrophysical purposes), all 3 types of cascades can be described by the same integro-differential equations as the ones derived by Landau and Rumer (1938), but in each case specifying the cross-sections of the relevant interaction processes. The solution of these equations over a broad range of energies is, however, not a trivial task. Such a study based on the numerical solutions of the so-called adjoint cascade equations (Uchaykin and Rizhov, 1998) has been recently conducted by Aharonian and Plyasheshnikov (2003). The results of this investigation shows that the electron-photon cascade curves in photon gas and a magnetic field have features quite different from the cascade development in matter. The energy spectra of cascade particles are also considerably different from the conventional cascade spectra in matter. The spectra of cascade particles in the magnetic field have properties intermediate between those for cascade spectra in matter and in radiation fields. Although for certain astrophysical scenarios the development of cascades in "pure" environments can be considered as an appropriate and fair approximation, in some conditions the interference of the processes associated with interactions of cascade electrons and γ-rays with both the ambient photon gas and magnetic field (or matter) can significantly change the character of cascade development, and consequently the spectra of observed γ-rays. The impact of such interference is very complex and quite sensitive to the choice of the principal parameters. Therefore each practical case is a subject to independent studies.

Chapter 4

Gamma Rays and Origin of Galactic Cosmic Rays

4.1 Origin of Galactic Cosmic Rays: General Remarks

Since the discovery of Cosmic Rays (CRs) by Victor Hess in 1912, the origin of this radiation has remained a mystery. Despite extensive efforts we still do not have a coherent theory which can explain a great variety of the features of CRs.

4.1.1 *What do we know about Cosmic Rays?*

Actually, we know a lot. In particular, we know that CRs consist mainly of *primary* protons, nuclei and electrons, i.e. particles directly accelerated to relativistic energies by powerful objects, which plausibly are different from ordinary stars. At the same time, a major fraction of some species of CRs, in particular the nuclei of the (Li,Be,B) group, as well as the anti-particles (positrons and antiprotons) have a *secondary* origin. They are produced by *primary* CRs interacting with the ambient interstellar gas, and partly with the thermal plasma (and, perhaps, also with low-frequency photon fields) *inside* the accelerators. Anti-particles can be produced also in some exotic processes like evaporation of primordial black holes or annihilation of dark matter. However the current data do not show convincing evidence of a significant fraction of "exotic" positrons and antiprotons in CRs. The secondary particles carry important information about the history of CRs during their passage through the galactic magnetic fields. In particular they tell us about the average time spent by CRs in the disk before they escape the Galaxy, $t_{\rm esc} \sim 10^7$ yr. We know quite well the flux and the energy spectrum of CRs, and we know that the energy spectrum of CRs extends to extremely high energies, $E \sim 10^{20}$ eV and even beyond (see

Fig. 4.1). The energy spectrum of electrons is measured up to $\simeq 2$ TeV. The proton-to-electron ratio at GeV energies is about 100; at 1 TeV the content of electrons does not exceed 10^{-3} (see Fig. 4.2).

The CR spectrum has two distinct features - the so-called *knee* and *ankle* around 10^{15} eV and 10^{18} eV, respectively (see Fig. 4.1). It is believed that all particles below the knee are of galactic origin, and that the Extremely High Energy Cosmic Rays (EHECRs) above the *ankle* are produced/accelerated outside of the Galactic Disk - in the Halo of our Galaxy, or in powerful extragalactic objects like AGN, Radiogalaxies and Clusters of Galaxies.

The acceleration, accumulation and effective mixture of nonthermal particles, through their diffusion and convection in galactic magnetic fields, produce the so-called "sea" of Galactic Cosmic Rays (GCRs). The average density of GCRs throughout the Galactic Disk is determined by operation of all galactic sources over a relatively long time period, comparable with the escape time of CRs of about $\sim 10^7$ yr. Assuming that the level of the "sea" of GCRs is not far from the directly measured fluxes of CRs, we can estimate the average energy density of CRs in the Galactic Disk, $w_{\rm CR} \approx 1$ eV/cm^3. More than 90 percent of this density is contributed by particles with energy ≤ 100 GeV. Within the homogeneous disk model, we then can derive the luminosity (acceleration power) of the Galaxy in CRs: $\dot{W}_{\rm CR} = V w_{\rm CR}/t_{\rm esc} \sim 3 \times 10^{40}$ erg/s, where $V \sim 10^{67}$ cm^3 is the volume of the disk where CRs are effectively confined. Although this estimate cannot guarantee an accuracy of better than a factor of a few, it is quite independent of model parameters. For the fixed mass of the diffuse gas $M_{\rm g} = m_{\rm p} n V$, the ratio $V/\tau_{\rm esc}$ is reduced to $M_{\rm g} c/\bar{x}$, where $\bar{x} = \tau_{\rm esc} m_{\rm p} n c$ is the mean amount of matter ("grammage") traversed by CRs. This parameter is determined by the content of secondary nuclei in CRs, and for the bulk of the observed CRs is about several g/cm^2. Thus, the CR production rate in the Galaxy can be estimated solely on the basis of CR measurements, namely from the total flux and the secondary-to-primary ratio of CRs, being rather independent of details characterising their confinement region (density, volume, *etc.*). In particular, both the disk and halo confinement models give approximately the same CR production rates (see e.g. Berezinsky *et al.*, 1990).

We know, from the energy-dependence of the secondary-to primary ratio of CRs, that the acceleration spectra of individual CR sources are significantly harder than the spectrum of the locally measured CRs, and therefore (supposedly) the spectrum of the "sea" of GCRs. The average source (acceleration) spectrum of CRs is believed to be described by a differential power-law index Γ close to 2.1 (see e.g. Swordy, 2001), although a steeper source spectrum with Γ up to 2.4 cannot be excluded, if CRs are additionally re-accelerated in the interstellar medium (e.g. Seo and Ptuskin, 1994).

And finally, we know that the pressure of CRs is comparable with the pressure of galactic magnetic fields, as well as with the turbulent and thermal pressure of the interstellar gas. This implies that GCRs play an important role in the dynamical balance of our Galaxy, and perhaps have also a non-negligible impact on *interstellar chemistry* through the heating and ionization of the interstellar medium (see e.g. Wolfendale, 1993).

4.1.2 What we do not know about Cosmic Rays?

The irony of the discipline called *Astrophysics of Cosmic Rays* is that, in spite of the considerable experimental material and extensive theoretical efforts, we still do not have a definite opinion about the origin of these relativistic particles.

We do not know what part of the observed CR spectrum is in fact

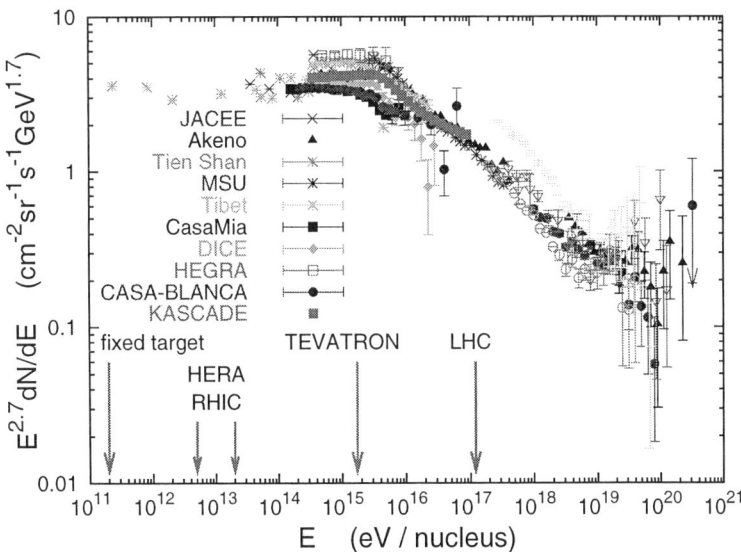

Fig. 4.1 Summary of measurements of the broad-band spectrum of high energy cosmic rays (from Gaisser, 2001).

of Galactic origin - below the knee around 10^{15} eV or does it extend to the *ankle* at 10^{18} eV? There is more confidence in the assumption that particles from the Extremely High Energy (EHE) domain above 10^{18} eV are produced outside the Galaxy (see e.g. Cronin, 1999; Nagano and Watson, 2000). These particles cannot be produced in the Galactic Disk, otherwise significant anisotropies would then be expected, in contrast to observations. However, the association of these particles with the Halo of our Galaxy cannot be ruled out, in particular within the so-called "top-down" scenarios, in which the observed particles are not result of classical acceleration (the "bottom-up" scenario), but may originate from decays of relic topological defects (Berezinsky *et al.*, 1998) or super-massive particles (Berezinsky *et al.*, 1997a; Birkel and Sarkar *et al.*, 1998) clustered in the Galactic Halo.

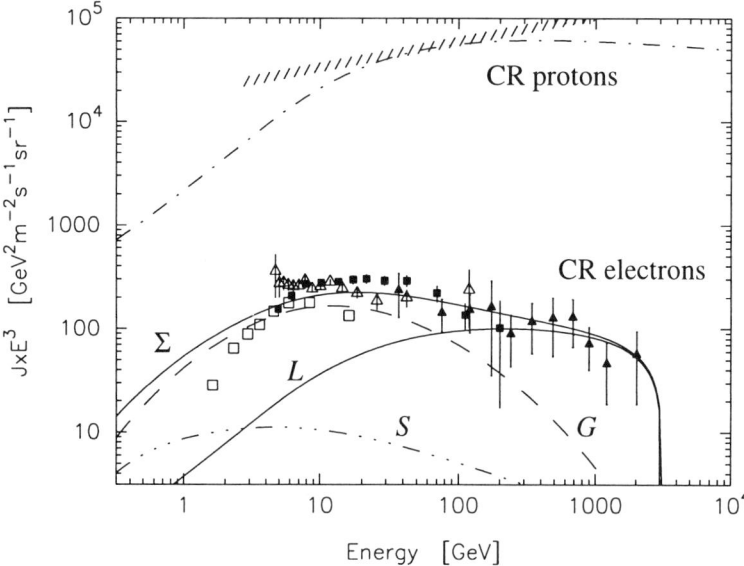

Fig. 4.2 Two-component approach to the observed CR electron flux (from Aharonian *et al.*, 1995). The thin solid line represents the Local ("L") component of electrons that originates from a single, $t = 10^5$ yr old burst-like source at $r = 100$ pc and $t = 10^5$ yr. The dashed line represents the Galactic ("G") component assuming a homogeneous distribution of CR sources in the Galactic Disk. The 3dot-dashed line corresponds to secondary electrons and positrons produced by galactic cosmic rays. The calculations are normalised to the observed flux at 10 GeV. The required energy release in electrons with power-law index $\Gamma = 2.2$ is $W_e = 1.1 \times 10^{48}$ erg. The spectrum of protons from the same *local* source assuming $W_p = 3 \cdot 10^{50}$ erg is also shown (dot-dashed line). The range of measured CR proton fluxes is indicated by the hatched region.

We do not know what powers the CR accelerators, and how they work. We do not know how many and which type of sources are responsible for the observed CR fluxes. Moreover, we are not fully confident that the bulk of directly observed CRs are contributed by sources distributed in the entire Galactic Disk. Paradoxically, we cannot exclude the scenario in which these particles may have a *local* origin, being contributed by a few sources or even a single nearby object. This statement is true at least for very high energy CR electrons. Because of severe radiative losses, the source(s) of the observed TeV electrons cannot be located well beyond a few 100 pc (Nishimuara *et al.*, 1980), and therefore the sheer fact of extension of the observed electron spectrum to TeV energies is an unambiguous indicator of existence of a nearby cosmic Tevatron(s) (Aharonian et al. 1995, Nishimuara *et al.*, 1997). This means that the assumption of a continuous distribution of sources of CR electrons would have to be valid down to scales of ~ 100 pc. Otherwise, the correct approach to the interpretation of the observed CR electron flux requires a separate treatment for two different components: (1) the contribution from one or a few nearby local sources (L-component), and (2) the contribution from sources at large distances, typically beyond 1 kpc, which may still be treated in the framework of the traditional assumption of a uniform and continuous (in space and time) source distribution in the Galactic Disk (G-component). Fig. 4.2 shows that the two-component approach describes reasonably well the observed flux of CR electrons from GeV to TeV energies.

Formally, a single local source can explain the entire CR population up to the knee around 10^{15} eV. Such a possibility is demonstrated in Fig. 4.2. Although the assumed total energy budget $W_\mathrm{p} = 3 \cdot 10^{50}$ erg seems somewhat high, it can be reduced by a significant factor assuming a somewhat smaller distance to the source. The contribution from several local sources is another possible option. Also, better agreement with the experimental data, especially at low energies, can be readily achieved assuming a specific energy dependence for the diffusion coefficient. Should we take such an attractive (or rather provocative) possibility too seriously? In recent years Erlykin and Wolfendale (1997, 2000) have claimed empirical evidence for a complex structure in the CR spectrum in the region of the knee attributing it to the effect of explosion of a single, recent, nearby supernova. Although this claim has not received a supportive response from other experts in the field of air-shower physics (see e.g. Schatz, 2002), the idea of a single CR source is a hypothesis in its own right and should not be necessarily related to the existence or lack of a specific structure in the knee.

The "local single CR source" hypothesis implies a dramatic revision of the current belief that the bulk of cosmic rays we detect are part of the "sea" of GCRs. Therefore it needs thorough inspection both on theoretical and experimental grounds. At the same time, the fact that we cannot firmly rule out such a possibility reflects the poor status of the field. A pessimist may even argue that CR measurements alone are *a priori* not sufficient to solve the problem. However, the *Cosmic Ray Community* does not share such a pessimistic view.

4.1.3 Common beliefs and "nasty" problems

It is widely believed that Supernova Remnants (SNRs) are the major source population in our Galaxy responsible for the observed CRs. The main *phenomenological* argument, recognised at very beginning of astrophysical interest in the problem (see e.g. Ginzburg and Syrovatski, 1964), is based on the fact that the power to maintain the galactic population of CRs is estimated to be a few percent of the total mechanical energy released by SNe explosions in our Galaxy. It is notable that as early as 1933 W.Baade and F.Zwickey realized the possible association of cosmic rays with supernovae, based on the comparable energies characterising these two phenomena. This is, of course, an important, but not decisive, argument, given that other potential source populations like pulsars, young stars with powerful mechanical winds, microquasars, gamma-ray bursts *etc.* can also meet, at least formally, this energy requirement. In this regard, note, for example, that the mechanical power of the jets of the galactic microquasar SS 433 is comparable with the total production rate of CRs in the Galaxy.

The second, equally important argument in favour of SNRs comes from theory. Actually, the only model of particle acceleration developed at a level which allows quantitative calculations is diffusive shock acceleration applied to the strong shocks in SNRs (see e.g. Drury, 1983; Blandford and Eichler, 1987; Berezhko and Krymsky, 1988; Jones and Ellison, 1991). Over last 20 years the basic properties of this model have been comprehensively checked by many theorists using different mathematical approaches. Recently, there have been important developments in the field based on the non-linear treatment of the problem (for review see Malkov and Drury, 2001; Drury *et al.*, 2001). In particular, it has been clearly understood that nonlinear reaction effects on the shock structure are unavoidable, if the process is to operate with high efficiency. On the other hand, the efficiency of acceleration of particles, i.e. the fraction of the mechanical energy of the shock transfered to non-thermal particles, should be very high, 10 per cent or more, in order to explain the observed CR flux. Therefore, nonlinear shock acceleration seems to be a key element in the SNR paradigm of GCRs. This allows conclusive observational predictions given the *inflexibility* (in a good sense) of the nonlinear shock acceleration theory. The distinct feature of this model is the very hard, power-law type (although not precisely power-law) energy distribution with differential spectral index Γ close to 2.

The high efficiency coupled with hard acceleration spectra extending well beyond 10 TeV, should lead to detectable γ-ray fluxes of hadronic origin. Thus, the best way to check the hypothesis is to search for π^0-decay γ-ray signals, especially at TeV energies, from 10^3-10^4 yr old shell type SNRs (Drury *et al.*, 1994; Naito and Takahara, 1994). TeV γ-rays have indeed been reported from three famous SNRs - SN 1006, RX J1713.7-3946 and Cas A (see Chapter 2). These are, however, objects where ultrarelativistic electrons are at least equally plausible as parent particles. On the other hand, other SNRs, like γ Cygni and IC 433, where γ-rays of hadronic origin are expected to dominate , have not shown TeV emission. Currently, this fact is interpreted by many as a failure of SNRs in general, and diffusive shock acceleration in particular, to produce the bulk of GCRs. However, given the limited sensitivity of current detectors, as well as large uncertainties in key model parameters, these conclusions in many cases are poorly justified and, in fact, misleading. Driven by an ultimate desire for dramatic revisions of the current concepts, the claims about the difficulties associated with γ-ray observations are premature and, to a large extent, exaggerated. At the same time, GLAST and the new generation of IACT arrays like H.E.S.S., VERITAS and CANGAROO-III, will be able to probe the SNR visibility in π^0-decay γ-rays at a level which must provide, even under the most pessimistic model assumptions, a decisive test for the SNR origin of galactic cosmic rays (Aharonian, 1999).

Another test of diffusive shock acceleration may come from studies of the secondary component of CRs produced by primary particles interacting with the interstellar medium (ISM). The CR data for the B/C ratio detected up to ~ 100 GeV/amu, derived assuming a simple propagation model, favour a quite strong energy dependence of the escape time of CRs from the Galactic Disk, $\tau_{\rm esc} \propto E^{-\delta}$ with $\delta \simeq 0.6 - 0.7$. Applied to protons, this requires a source spectral index of $\Gamma \simeq 2.0 - 2.1$ which agrees with the index anticipated by nonlinear shock acceleration. Note that the source spectral index derived in this way is distinctly harder than the values of

$\Gamma \sim 2.3 - 2.4$ favoured by re-acceleration models of CR propagation (Drury et al., 2001). Thus, any independent evidence of significant re-acceleration of CRs in the interstellar medium would work against the nonlinear shock acceleration.

On theoretical grounds, the diffusive shock acceleration model faces several challenges or "nasty problems" (Drury et al., 2001) like the "injection problem" and the "maximum energy problem", recently critically reviewed by Kirk and Dendy (2001), Drury (2001) and Malkov and Drury (2001). Diffusive shock acceleration requires particles with energy at least several times larger than the thermal energy of the plasma, and it is not yet clear how to get particles from the thermal pool accelerated to supra-thermal energies. Recent theoretical progress in this direction (e.g. Malkov and Völk, 1995; Dieckmann et al., 2000) provides optimism that eventually the injection problem will be resolved, most likely through extensive numerical simulations (Kirk and Dendy, 2001).

The problem of the maximum achievable energy problem is an old one and has a vital implication for the SNR paradigm of GCRs. In diffusive shock acceleration theory, the maximum energy of particles is achieved during the so-called free-expansion phase which, however, does not last long enough to allow acceleration of particles up to the *highly desired point*, the knee around 10^{15} eV. Therefore, violation of the so-called "upper limit" of Lagage and Cesarsky (1983), which, for the standard SNR parameters, the shock speed, duration of the free-expansion phase, and the ambient magnetic field, cannot significantly exceed 10^{14} eV, remains as one of the highest priorities of current theoretical studies.

A promising way has recently been suggested by Lucek and Bell (2000). They showed that cosmic ray streaming drives large-amplitude Alfvènic waves which may amplify the magnetic field non-linearly to many times the pre-shock value. Thus, the cosmic rays themselves provide the field necessary for their effective acceleration! The increased magnetic field reduces the acceleration time, and correspondingly increases the maximum particle energies to 10^{15} eV and even beyond. Needless to say that this effect, if confirmed by independent theoretical investigations, would be the solution of the 20-year old "maximum energy" problem. Ideally speaking, the most elegant version of the SNR paradigm of galactic cosmic rays should allow shock acceleration of particles up to the "ankle" around 10^{18} eV. The smooth transition of the spectrum from the "sub-knee" to the "above the knee" region of the spectrum, which over 3 decades up to 10^{18} eV continues as steep power-law with $\Gamma \sim 3$, not only indicates a possible galactic origin for this part of the spectrum, but also favours the same acceleration mechanism (and sources) responsible for the CR spectrum from 1 GeV to 10^{18} eV. Although the model of Lucek and Bell (2000) allows acceleration of protons up to 10^{17} eV, in their second paper Bell and Lucek (2001) argued that expansion into a pre-existing stellar wind may increase the maximum cosmic ray energy by an additional factor of 10. Acceleration of particles well beyond the knee is possible also by shocks in so-called Superbubbles - a "multiple supernova remnant" powered by SN explosions and winds of luminous stars in OB associations (e.g. Parizot, 2000; Bykov, 2001). Particle acceleration by multiple shocks (Bykov and Toptygin, 2001) in such systems has features similar to the standard picture in isolated SNRs, but because of larger dimensions particles may achieve energies up to 10^{18} eV.

Finally, this brief overview of the status of origin of galactic cosmic rays would be biased if we concluded without remarking that in the future we may need to invoke, despite all the pleasing features and advantages of the SNR paradigm of GCRs, other source populations, pulsars, X-ray binaries with relativistic jets, or something else, and develop new acceleration theories to explain the phenomenon called Galactic Cosmic Rays.

4.1.4 *Searching for sites of production of GCRs*

As discussed above, after several decades of intensive experimental and theoretical studies, our knowledge about the accelerators of galactic cosmic rays continues to be quite limited and inconclusive. The main obstacle to revealing the production sites and acceleration mechanisms of GCRs is the effective diffusion of charged particles in interstellar magnetic fields, which results in the confusion of individual contributors to the "sea" of GCRs, and significantly modifies the original (source) spectra of accelerated particles. Therefore, it is believed that the resolution of these long-standing questions will be provided by gamma-ray astronomy, i.e. through *indirect* but (almost) model-independent measurements of secondary γ-rays . The basic idea of this approach is straightforward and concerns both the acceleration and propagation aspects of the problem. While the localised sources of γ-rays pinpoint the *sites* of particle acceleration, the angular and spectral distributions of the diffuse γ-ray emission of the Galactic Disk contain information about the character of *propagation* of both the *electronic* and *nucleonic* components of CRs in the galactic magnetic fields. The realization of this seminal prediction recognised by pioneers of the field in the 1950s and 1960s is still considered as one of the major goals of γ-ray astronomy. The non-thermal synchrotron radiation contains additional and complementary information at radio and possibly also at X-ray wavelengths, but it concerns only the *electronic* component of CRs in two extreme energy bands, typically below 1 GeV and above 100 TeV, respectively. TeV neutrinos carry adequate information about the *nucleonic* component of CRs as well, but the sensitivities of the current high energy neutrino projects do not adequately match, even under extreme model assumptions, the neutrino fluxes expected from interactions of GCRs.

The study of the diffuse γ-radiation by the SAS-II and COS B satellite experiments, and especially by the EGRET instrument aboard the Compton GRO, have already made a significant contribution to the current knowledge of spatial distribution of relatively low energy, 1 to 100 GeV, CRs in the Galactic Disk. Furthermore, many famous galactic objects representing different classes of potential accelerators of GCRs like pulsars (e.g. Crab, Vela, and Geminga), shell-type supernova remnants (e.g. IC 433 and γ Cygni), giant molecular clouds (GMCs) and associated star formation regions (e.g. Orion and ρ Ophiuchus complexes), are identified by EGRET as sources of 100 MeV radiation (Hartman et al., 1999). At the same time, most of the EGRET sources, especially the objects at low- and mid-galactic latitudes, still do not have clear counterparts at other wavelengths (see Chapter 2). The next generation space-based γ-ray detector, GLAST, with its superior flux sensitivity and good angular resolution should be able to reveal the nature of these γ-ray hot spots. For bright sources with flat γ-ray spectra, GLAST can provide spectral coverage up to energies of $\sim$100 GeV. This ensures the great role of GLAST for future studies of GCRs of intermediate energies, typically between 1 GeV and 1 TeV.

Even so, the energy coverage of GLAST will not tell us much about the sources responsible for the formation of the most energetic part of the spectrum of GCRs which extends to the knee around $E \sim 10^3$ TeV. This is the domain of ground-based γ-ray detectors. The range of flux sensitivities that could be achieved by future γ-ray detectors is shown in Fig. 1.2. It is seen that GLAST and the forthcoming IACT arrays can probe γ-ray point sources in a very broad energy region from 0.1 GeV to 10 TeV, at the level of energy fluxes between 10^{-13} and 10^{-12} erg/cm^{2}s. Thus, all point galactic sources with luminosities down to $L_\gamma(E) = 4\pi d^2 E^2 J(E) \simeq 10^{31} \, (d/1 \, \text{kpc})^2$ erg/cm^{2}s can be detected by GLAST and/or IACT arrays. Note that for both GLAST and stereoscopic IACT systems with angular resolution of about $0.1° - 0.2°$, the "point" source implies an angular size less than a few arcminutes. Thus, the detection of extended sources with angular size of about 1° would require an order of magnitude higher γ-ray luminosities. The GeV luminosities of the EGRET sources detected at low galactic latitudes exceed $10^{34} \, (d/1\,\mathrm{kpc})^2$ erg/s. Therefore, most of the EGRET sources should be seen in TeV γ-rays, provided that the γ-ray spectra extend unbroken to the TeV region with differential photon index $\simeq 3$ for point sources and $\simeq 2.7$ for extended "1°" sources.

Since the spectra of particle acceleration (e.g. by SNR shocks) generally are expected to be significantly harder than the locally observed spectrum of CRs, the failure to detect TeV γ-rays from EGRET sources (see Chapter 2) can be interpreted as a result of "early" cutoffs in the source spectra below 1 TeV. If so, the EGRET sources cannot (at first glance) be considered as important contributors, at least at high energies, to the "sea" of GCRs. However, the lack of TeV γ-rays from GeV sources can be result of propagation effects, namely energy-dependent escape of accelerated particles from the source. Generally the confinement time of particles in the source decreases with energy (the leakage of particles becomes easier at high energies), therefore the quasi-stationary spectrum of particles established in the source could be significantly steeper than the acceleration spectrum. Correspondingly, even for a hard, e.g. E^{-2} type, particle acceleration spectrum, the secondary γ-rays produced by interactions of relativistic particles inside the sources, could have quite a steep spectrum – just opposite to the common belief in which the hardest γ-ray spectra are expected from CR accelerators themselves.

This effect is illustrated in Fig. 4.3. It is assumed that high energy protons are injected into a dense region of size $R = 3$ pc, gas density $n = 100$ cm^{-3}, and magnetic field $B = 100$ μG. These parameters are typical for the so-called giant molecular clouds – possible sites of particle acceleration and gamma-ray production. The acceleration spectrum of protons is assumed to be a power-law with an index $\alpha = 2.1$, and exponential cutoff at 10^{15} eV. The time history of acceleration is assumed as $L = L_0(1+t/\tau_0)^{-2}$, with $L_0 = 10^{38}$ erg/s and $\tau_0 = 10^3$ yr. This assumption implies that the acceleration rate was essentially constant over the first 10^3 years, but has later decreased with time as t^{-2}. Finally, the confinement time of particles was approximated in the form $t_{\rm esc} = R^2/2D(E) \approx 4\times 10^4 \kappa^{-1}(E/100\,\mathrm{TeV})^{-1}$ yr, where $\kappa = 1$ corresponds to the slowest possible escape in the Bohm diffusion regime. One can see that for the chosen parameters of the ambient medium and acceleration rate, the proton escape results in a significant suppression of TeV γ-rays, especially at observation epochs $t \geq 10^4$ yr, even if the particle escape proceeds in the regime close to the Bohm diffusion.

Thus, the study of a cosmic accelerator by detecting γ-rays from the central source cannot be complete because it contains information only about relatively low-energy particles effectively confined in the source. In many cases the detection of γ-rays from regions surrounding the accelerator could add much to our knowledge about the highest energy particles which

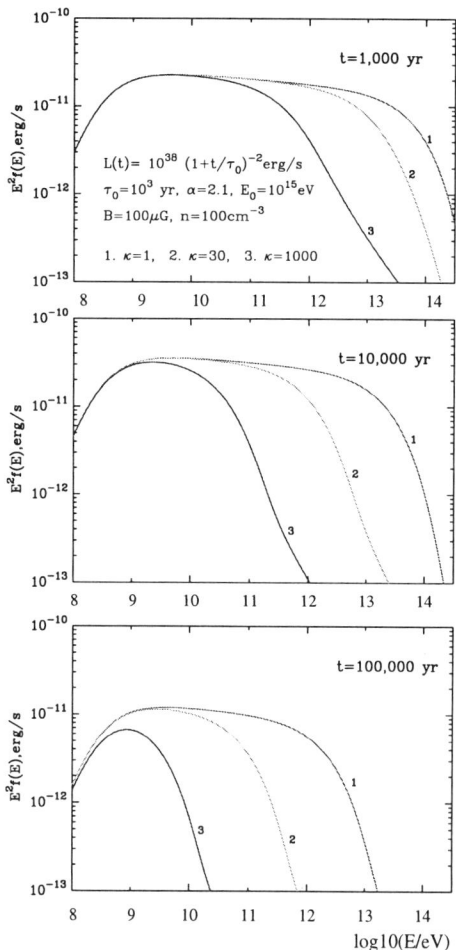

Fig. 4.3 Expected γ-ray spectra at different observation epochs – $t = 10^3, 10^4$ and 10^5 years after the start of operation of the proton accelerator, for three different assumptions concerning the escape time of particles from the γ-ray production region: $\kappa = 1, 30, 1000$.

quickly escape from the source and thus do not contribute to the γ-ray production inside the source.

For "typical" CR accelerators, e.g. SNR with a total energy release in protons of less than $W \leq 10^{50}$ erg, the extension of these regions cannot significantly exceed several tens of parsecs, because at such large distances from the source, the density of relativistic particles becomes negligible compared to the level of the "sea" of GCRs (see below). Also, the detection of extended sources is quite difficult due to the backgrounds caused by diffuse galactic γ-rays (for GLAST) and local CRs (for ground-based detectors).

The existence of a powerful particle *accelerator* by itself is not yet sufficient for effective γ-ray production. Clearly, an additional component – a dense gas target – is required. Giant Molecular Clouds (GMCs) are perfect objects to play that role in our Galaxy. They are physically connected with star formation regions which are believed to be the most probable sites for the production of galactic cosmic rays. Fig. 1.5 illustrates the γ-ray production by molecular clouds located in the vicinity of a particle accelerator where the energy density of CRs can significantly exceed the level of the "sea" of GCRs of about 1 eV/cm^3.

The low-energy ambient photon fields play a similar role for tracing VHE (multi-TeV) electrons through their inverse Compton radiation. With some exceptions (e.g. Crab Nebula, Cas A) the 2.7 K CMBR significantly dominates over other photon fields for the production of IC γ-ray emission. Therefore, the detection and identification of the inverse Compton radiation component associated with the 2.7 K CMBR should allow us to find the sources of ultra-relativistic electrons, and study the propagation features of these electrons in the regions not far from their birth sites.

Because of its universal character and the precisely measured energy spectrum and density, the 2.7 K CMBR may serve as an ideal target also for the search and study of extremely high energy, $E \geq 10^{19}$ eV, protons in the vicinity of their extragalactic accelerators like AGN, radiogalaxies, clusters of galaxies, *etc.*, through the radiation components initiated by the secondary products of interactions of protons with the 2.7 K CMBR.

4.2 Giant Molecular Clouds as Tracers of Cosmic Ray

The first studies of the diffuse galactic γ-radiation by the SAS-2 and COS B satellite missions revealed a relatively small, by a factor of ≤ 2, spatial gradient of CRs in the Galactic Disk (see e.g. Bloemen, 1989). EGRET measurements generally confirm this conclusion (Hunter et al., 1997a). A small variation of the CR density on large galactic scales can be naturally explained by effective mixture of contributions from individual sources during the CR propagation in the Galactic Disk on timescales $\tau \geq 10^7$ yr. Meanwhile, the gradient of cosmic rays, both in the absolute flux and spectral shape, can be much stronger on smaller (sub 100-pc) spatial scales, especially in the vicinity of young CR accelerators. Consequently, we may expect significantly higher γ-ray emissivity in these regions relative to the average γ-ray production rate in the Galactic Disk.

Below, the emissivity of π^0-decay γ-rays in the $R \leq 100$ pc region of a proton accelerator will be discussed for both *impulsive* (i.e. burst-like) and *continuous* scenarios of injection of relativistic protons into the ISM. Though in this section the sources of cosmic rays are not specified, it is assumed that the total CR energy per "average" particle accelerator does not exceed $W_p \sim 10^{50}$ erg, which is the typical amount of energy believed to be produced in the form of relativistic particles during their diffusive acceleration by SNR shocks (e.g. Drury, 1983). On the other hand, approximately this amount of nonthermal energy per supernova is needed to explain the current fluxes of galactic cosmic rays (e.g. Ginzburg and Syrovatskii, 1964). Pulsars comprise the second important class of potential suppliers of galactic cosmic rays. An upper limit on the total energy release in cosmic rays by a pulsar may be obtained assuming that during $t \leq 10^5$ years a pulsar effectively accelerates particles with a rate L_p close to the rotational energy loss rate $L_0 \sim 10^{37}$ erg/s, i.e. $W_p \sim L_0 \times t \sim 3 \cdot 10^{49}$ erg. Assuming now that the accelerated particles during their propagation in the ISM reach a radius $R(t)$ at time t, the mean energy density of particles in the occupied region is estimated as $w_p = W_p/(4/3)\pi R^3 \approx 0.55 \, (W_p/10^{50} \, \text{erg})(R/100 \, \text{pc})^{-3} \, \text{eV/cm}^3$.

Thus, in the regions up to several tens of parsecs around CR accelerators with $W_p \sim 10^{49} - 10^{50}$ erg, the fluxes of relativistic particles at certain stages, depending on the time history of injection and the character of their propagation in the ISM, may exceed the average level of the "sea" of GCRs, $w_0 \approx 1 \, \text{eV/cm}^3$. Correspondingly, at these stages we should expect higher γ-ray fluxes. Moreover, in the case of energy dependent diffusive propagation of particles, the spectral shape of the expected γ-ray flux may differ significantly from the spectrum of γ-rays produced by the "sea" of galactic cosmic rays. This circumstance, coupled with the possible location of high density regions near the particle accelerators, would result in enhanced γ-radiation. Also, the hard proton spectra which appear at some stages of their propagation significantly increase the probability for detection of TeV γ-rays (Aharonian, 1995; Aharonian and Atoyan, 1996a), and thus allow us to probe CR sources to very high energies. The star formation regions, which are believed to be potential settings of acceleration of CRs by supernova shocks, strong stellar winds, pulsars, *etc.*, are of special interest (Montmerle, 1979; Casse and Paul, 1980; Paul 2001).

4.2.1 *Proton fluxes in the ISM near the accelerator*

Let assume that relativistic particles accelerated by a single source escape the source and enter the ISM. The energy spectrum of particles at a given time and distance from the source depends on (1) the *time history* and the *injection spectrum*, (2) the *energy loss rate*, and (3) the *character of propagation* of cosmic rays. In the standard diffusion approximation, the spherically symmetric propagation of cosmic rays produced by an *impulsive* source is described by Eq.(A.1); the general solution for an arbitrary energy source of relativistic particles injected into the ISM $Q(E, R, t)$, energy loss rate $P(E)$, and the diffusion coefficient $D(E)$, is given by Eq.(A.2). The energy losses of the protons in the gas are due to ionization and nuclear interactions, $P(E) = P_{\text{ion}} + P_{\text{nucl}}$. In the hydrogen medium, at kinetic energies of protons above 1 GeV the nuclear energy losses dominate over ionization losses.

4.2.1.1 *Impulsive source*

In the case of a power-law injection spectrum of an impulsive source, i.e. $Q(E, t) = N_0 \, E^{-\Gamma} \, \delta(t)$, and power-law diffusion coefficient, $D(E) \propto E^{\delta}$, the general solution reduces to

$$f(E, R, t) \approx \frac{N_0 \, E^{-\Gamma}}{\pi^{3/2} \, R_{\text{dif}}^3} \, \exp\left(-\frac{(\Gamma - 1)t}{\tau_{\text{pp}}} - \frac{R^2}{R_{\text{dif}}^2}\right), \qquad (4.1)$$

where

$$R_{\text{dif}} \equiv R_{\text{dif}}(E, t) = 2\sqrt{D(E) \, t \, \frac{\exp(t\delta/\tau_{\text{pp}}) - 1}{t\delta/\tau_{\text{pp}}}} \qquad (4.2)$$

is the diffusion radius. It corresponds to the radius of the sphere up to which particles of energy E effectively propagate during the time t after their injection into the interstellar medium.

Generally, the spectrum of cosmic rays at the given time t and distance R from the accelerator may noticeably differ from the source (acceleration) spectrum. While the energy-independent diffusion leads to variation of the cosmic ray flux only in time, and does not change the form of the primary spectrum, the energy-dependent diffusion results in significant spectral changes. The modification of the particle spectrum is defined mainly by the parameter $g(E, R, t) = \zeta^3 \exp(-\zeta^2)$, where $\zeta \equiv \zeta(E, R, t) = R/R_{\rm dif}$. For timescales less than the energy loss time, $t \ll \tau_{\rm pp}$, the effective diffusion radius is reduced to the form $R_{\rm dif} = 2\sqrt{D(E)\, t}$.

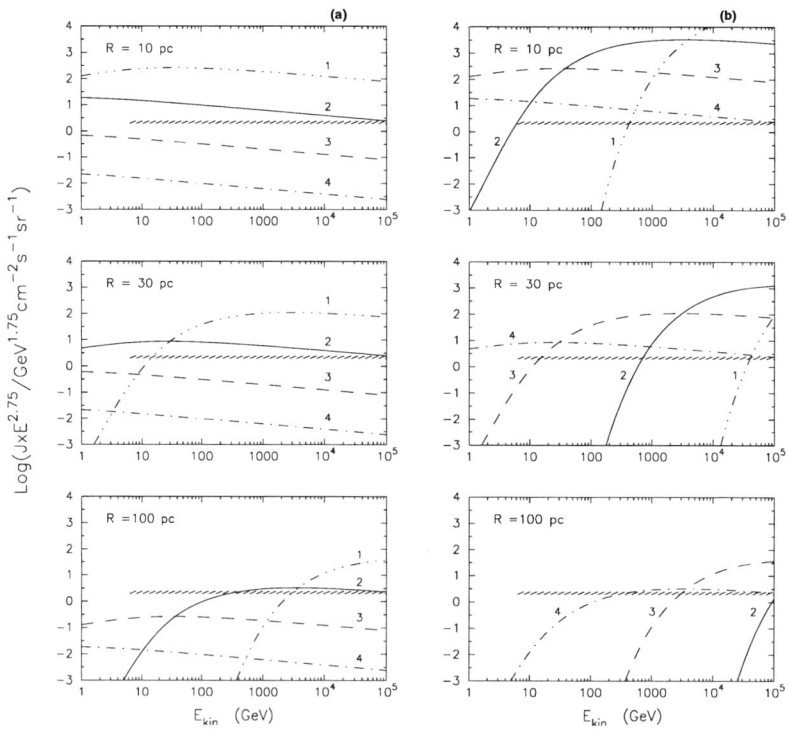

Fig. 4.4 The temporal evolution of the energy spectrum of relativistic protons in the vicinity of an *impulsive* accelerator during their energy-dependent propagation (from Atoyan and Aharonian, 1996a). The power-law source spectrum with exponent $\Gamma = 2.2$ and total energy $W_{\rm p} = 10^{50}$ erg are assumed. The differential fluxes of protons at different distances R and times t are shown. The curves plotted by fancy (1), solid (2), dashed (3), and dot-dashed (4) lines correspond to the age of the source $t = 10^3$ yr, 10^4 yr, 10^5 yr, and 10^6 yr, respectively. The energy-dependent diffusion coefficient $D(E)$ with power-law index $\delta = 0.5$ is assumed. (a) and (b) are for $D_* = 10^{28}$ cm^2/s and $D_* = 10^{26}$ cm^2/s, respectively. The hatched curve shows the fluxes of cosmic ray protons observed near the Earth.

For a given energy, the maximum cosmic ray flux at a fixed distance from the source is reached when $\zeta = \sqrt{1.5}$, i.e. at

$$t = t_{\max}(E, R) = R^2/6D(E). \quad (4.3)$$

At $t \ll t_{\max}(E)$ the particles have not yet reached the point R, while at $t \gg t_{\max}(E)$ the cosmic ray flux decreases due to the spherical expansion as $R_{\text{dif}}^{-3} \propto t^{-3/2}$. At sufficiently high energies, for which the time of the maximum flux has already passed, the modification factor at a distance R from an *impulsive* source is proportional to $E^{-(3/2)\delta}$. Therefore the particles are distributed in a power-law form with

$$\Gamma' = \Gamma + (3/2)\delta. \quad (4.4)$$

At lower energies, for which the maximum flux has not yet been reached, the primary spectrum is exponentially suppressed. Note that for $D(E) = $ const the differential flux of cosmic rays just repeats the shape of the injection spectrum with a time-dependent amplitude proportional to $g(t)$.

The evolution of the differential fluxes of protons,

$$J(E, R, t) = (c/4\pi)f, \quad (4.5)$$

during their energy-dependent propagation in the interstellar medium from a single *impulsive* source with $W_p = 10^{50}$ erg is shown in Figs. 4.4. The power-law diffusion coefficient in the form of

$$D(E) = D_*(E/10\,\text{GeV})^\delta \quad (4.6)$$

with $\delta = 0.5$ is assumed, where $D_* = D(10\,\text{GeV})$ is the value of the diffusion coefficient at $E = 10\,\text{GeV}$. The commonly used diffusion coefficient at 10 GeV is of about $D_* \sim 10^{28}\,\text{cm}^2/\text{s}$ (see e.g. Berezinsky *et al.*, 1990). However, much smaller values, e.g. $D_* \sim 10^{26}\,\text{cm}^2/\text{s}$, in particular in dense regions of the interstellar gas, cannot be excluded (Ormes *et al.*, 1988).

4.2.1.2 *Continuous source*

The *impulsive* source of particles corresponds to a scenario in which the time interval for acceleration of the bulk of relativistic protons, Δt, is significantly less than the age of the accelerator, t. Therefore, for relatively short timescales, $t \leq 10^3 - 10^4$ yr, the assumption of "impulsive particle acceleration" should be considered rather as an *idealised working hypothesis*. In particular, for SNRs the assumption of *impulsive* source would be actually valid only for timescales $t > 10^4 - 10^5$ yr. Though the possibility of the effective production of CRs at early stages of supernova evolution cannot be excluded, a more realistic model of diffusive shock acceleration in SNRs predicts particle acceleration in the so-called Sedov phase with a typical duration $\Delta t \sim 10^3 - 10^4$ yr. Thus, SNRs may be treated rather as *continuous* accelerators. On the other hand, the case of *impulsive* source could be applied to the classical "continuous accelerators" like pulsars, if one assumes that the bulk of particles is produced at the early stages, e.g. before the braking of the pulsar due to the magnetic dipole radiation.

In the case of continuous acceleration of particles by a single source with time dependent evolution, $Q(E,t) = Q_0 E^{-\Gamma} q(t)$, Eq.(4.1) should be convolved with the function $q(t-t')$ in the time interval $0 \leq t' \leq t$. Although the acceleration of particles may have rather complicated time history, to make the interpretation of the results easier, in Figs. 4.5 are shown the spatial and temporal evolution of CRs calculated for the *continuous* source with a constant acceleration rate.

For the assumed luminosity of $L_p = 10^{37}$ erg/s, the total energy released in relativistic protons during 10^5 yr by a stationary accelerator is about 3×10^{49} erg, which is only 3 times less than the total energy of accelerated protons assumed in Fig. 4.4 for an *impulsive* source. From comparison of Fig. 4.4 and Fig. 4.5 it is seen, however, that the character of the evolution in time of the fluxes from *impulsive* and *continuous* accelerators is qualitatively different. To understand this difference, note that for the time intervals less than the energy loss time, the energy distribution function of particles injected from a *continuous* source is given by the expression:

$$f(E,R,t) = \frac{Q_0 E^{-\Gamma}}{4\pi D(E) R} \operatorname{erfc}\left(\frac{R}{R_{\text{dif}}(E,t)}\right), \tag{4.7}$$

where erfc(z) is the error-function (see e.g. Atoyan et al., 1995). At a given distance R, the flux of protons with given energy E at initial stages, $t \leq r^2/4D(E)$ (i.e. when $R_{\text{dif}} \leq R$), increases exponentially with time t, and when $R_{\text{dif}}(E,t) \gg R$ it gradually saturates at the level

$$J(E,R) = \frac{c Q_0 E^{-\Gamma}}{(4\pi)^2 D(E) R}. \tag{4.8}$$

This is the maximum flux at given distance R from a *continuous* source, which is relevant to the situation when the decrease of the cosmic ray density due to spherical expansion of the volume occupied by the cosmic rays injected earlier is compensated for by the arrival of new particles. For a power-law diffusion coefficient, the spectrum of accelerated particles from a single *continuous* source observed at distance R is described by the power-law exponent

$$\Gamma' = \Gamma + \delta, \tag{4.9}$$

i.e. by a factor of $\delta/2$ smaller than in the case of an *impulsive* accelerator (see Eq. 4.4)). For $\Gamma = 2.2$ and $\delta = 0.5$, the index $\Gamma' = 2.7$ is close the spectral index of the cosmic rays observed locally near the Earth. For these

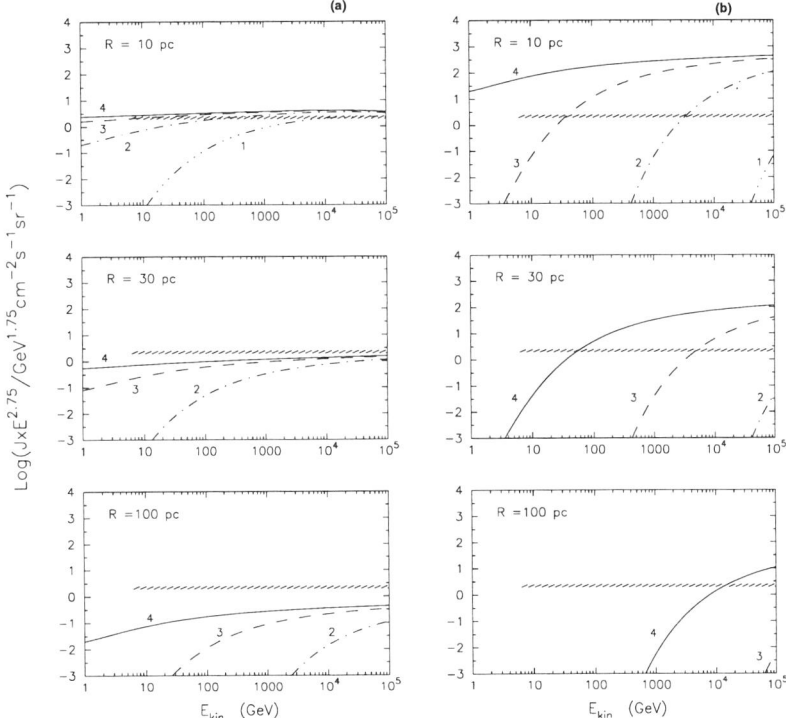

Fig. 4.5 The temporal evolution of the energy spectrum of relativistic protons in the vicinity of a *continuous* accelerator during their energy-dependent propagation (from Atoyan and Aharonian, 1996a). The power-law primary spectrum with the exponent $\Gamma = 2.2$ and the source luminosity $L_p = 10^{37}$ erg/s in the protons are assumed. The differential fluxes of protons at different distances R and times t are shown. The curves plotted by fancy (1), dot-dashed (2), dashed (3), and solid (4) lines correspond to the age of the source $t = 10^2$ yr, 10^3 yr, 10^4 yr, and 10^5 yr, respectively. The energy-dependent diffusion coefficient $D(E)$ with power-law index $\delta = 0.5$ is assumed. **(a)** and **(b)** are for $D_* = 10^{28}$ cm^2/s and $D_* = 10^{26}$ cm^2/s, respectively. The hatched curve shows the fluxes of cosmic ray protons observed near the Earth.

parameters Eq. (4.8) is reduced to

$$J(E,R) = 2.5 \, L_{37} \, D_{28}(R/10\,\text{pc})^{-1} \, E_{\text{GeV}}^{-2.7} \, \text{cm}^{-2}\text{s}^{-1}\text{sr}^{-1}\text{GeV}^{-1} \quad (4.10)$$

where $L_{37} = L/10^{37}\,\text{erg/s}$, and $D_{28} = D_*/10^{28}\,\text{cm}^2/\text{s}$. From comparison of Eqs.(4.10) with the locally observed CR spectrum given by Eq.(4.30), we find that for a large diffusion coefficient, $D_* = 10^{28}\,\text{cm}^2/\text{s}$, the flux of protons from a *continuous* accelerator with $L_p = 10^{37}\,\text{erg/s}$ could exceed the level of the "sea" of GCRs only in the near vicinity of the source. However, for a smaller diffusion coefficient, $D_{28} = 0.01$, much larger regions around the source, up to distances $R \sim 100\,\text{pc}$, could be significantly enhanced by cosmic rays (see Fig. 4.5).

4.2.1.3 *The case of dense gas regions*

The results presented in Fig. 4.4 and Fig. 4.5 correspond to the propagation of particles in low-density medium, $n \sim 1\,\text{cm}^{-3}$, where the energy losses of protons during the timescales of interest, $t \leq 10^6\,\text{yr}$, are negligible. The effect of energy losses of relativistic protons in a dense medium with $n = 10^3\,\text{cm}^{-3}$ and $D_* = 10^{26}\,\text{cm}^2/\text{s}$ is demonstrated in Fig. 4.6 for **(a)** *impulsive* and **(b)** *continuous* sources, respectively.

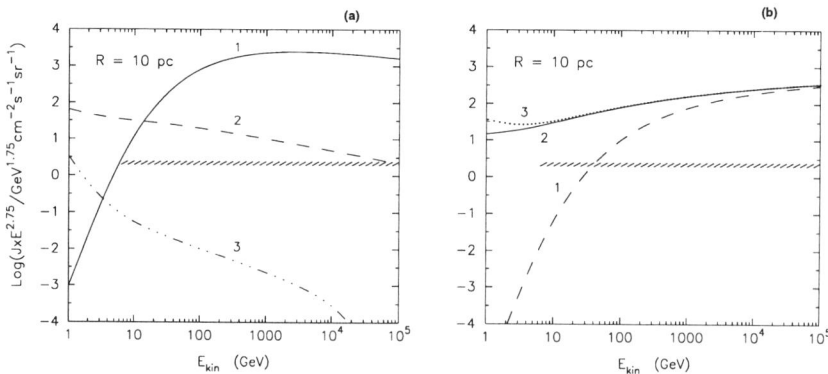

Fig. 4.6 The temporal evolution of the proton fluxes in dense medium with $n = 10^3\,\text{cm}^{-3}$ at $R = 10\,\text{pc}$ from the accelerator (from Aharonian and Atoyan, 1996a). The diffusion coefficient with $D_* \sim 10^{26}\,\text{cm}^2/\text{s}$ and $\delta = 0.5$ are assumed. The curves marked as 1, 2 and 3 correspond to the fluxes at times $t = 10^4\,\text{yr}$, $10^5\,\text{yr}$ and $3 \times 10^5\,\text{yr}$, respectively. (**a**) - *impulsive* accelerator with $W_p = 10^{50}\,\text{erg}$. (**b**) - *continuous* accelerator with $L_p = 10^{37}\,\text{erg/s}$. The hatched curve indicates the level of the locally observed cosmic ray proton flux.

For time periods up to $t = 10^4$ yr the proton fluxes in the high and low density media coincide, because the cooling time of the protons t_{pp} in a medium with $n \sim 10^3 \, \text{cm}^{-3}$ still is significantly larger than 10^4 yr (see Eq. 3.13). However, for $t \geq 10^5$ yr the fluxes of protons propagating in a dense medium are strongly affected by interactions with the ambient gas. In particular, in the case of an *impulsive* injection of CRs in the dense medium, the fluxes of protons at $t = 3 \times 10^5$ yr are exponentially suppressed compared with the relevant fluxes in a low-density medium. For a *continuous* accelerator, the effect caused by energy losses of protons is less profound; it is substantially masked by the flux of recently injected particles.

4.2.2 Gamma rays from a cloud near the accelerator

One of the principal parameters determining the γ-ray visibility of GMCs is M_5/d_{kpc}^2, where $M_5 = M_{\text{cl}}/10^5 M_\odot$ is the diffuse mass of a GMC in units of 10^5 solar masses, and $d_{\text{kpc}} = d/1$ kpc. Assuming that γ-rays are produced by interactions of CRs with the ambient gas,

$$F_\gamma(\geq E_\gamma) \simeq 10^{-7} \left(\frac{M_5}{d_{\text{kpc}}^2}\right) q_{-25}(\geq E_\gamma) \, \text{cm}^{-2}\text{s}^{-1}, \quad (4.11)$$

where $q_{-25}(\geq E_\gamma) = q_\gamma(\geq E_\gamma)/10^{-25} \, (\text{H} - \text{atom})^{-1}\text{s}^{-1}$ is the γ-ray emissivity. In a "passive" GMC, i.e. in a cloud submerged in the "sea" of GCRs, assuming that the level of the latter is the same as the proton flux measured at the Earth, given by Eq.(4.30), the γ-ray emissivity above 100 MeV is equal to $q_{-25}(\geq 100 \, \text{MeV}) = 1.53 \times 10^{-25}\eta_A \, (\text{H} - \text{atom})^{-1}\text{s}^{-1}$, where the parameter $\eta_A \simeq 1.5$ takes into account the contribution of nuclei both in CRs and in the interstellar medium (Dermer, 1986; Mori, 1997). A "passive" cloud could be detectable by GLAST, if $M_5/d_{\text{kpc}}^2 \geq 0.1$, taking into account the large angular size (typically 1 degree or more) of relatively close clouds. At energies $E \gg 1 \, \text{GeV}$, $J(\geq E) \simeq 1.5 \times 10^{-13} \, (E/1 \, \text{TeV})^{-1.75}(M_5/d_{\text{kpc}}^2) \, \text{ph/cm}^2\text{s}$, thus even for $M_5/d_{\text{kpc}}^2 \sim 10$ (there are only several clouds in the Galaxy with such a large value of M_5/d_{kpc}^2), detection of TeV γ-rays from passive clouds is extremely difficult. Nevertheless, searches for VHE γ-rays from GMCs are of a great interest because of the possible existence of high-energy cosmic ray accelerators nearby or inside GMCs (Aharonian, 1991).

Indeed, since the density of CRs in the regions up to 100 pc around the accelerators at some stages may exceed the average level of the "sea" of GCRs, $w_{\rm GCR} \leq 1\,{\rm eV/cm^3}$, we may expect significantly enhanced γ-ray emissivities in these regions. This is demonstrated in Fig. 4.7, where the π^0-decay γ-ray emissivities, in terms of $E_\gamma^2 q_\gamma(E_\gamma)$, are presented for an impulsive proton accelerator at two different instants t and two different distances R from the accelerator. For comparison, the γ-ray emissivity of the "sea" of galactic cosmic ray protons is also shown.

The expanding "bubble" of cosmic rays penetrating the nearby molecular clouds initiates intense γ-ray emission. The right-hand side axis in Fig. 4.7 shows the differential γ-ray fluxes expected from a cloud with $M_5/d_{\rm kpc}^2 = 1$. The temporal evolution of the integral γ-ray fluxes in the energy intervals between 0.3-3 GeV and 1-10 TeV from a cloud with $M_5/d_{\rm kpc}^2 = 1$, located at 10 pc, 30 pc, and 100 pc distances from an *impulsive* proton accelerator with $W_{\rm p} = 10^{50}$ erg, are shown in Fig. 4.8. For the diffusion coefficient, the values of $D_* = 10^{27}\,{\rm cm^2/s}$ and $\delta = 0.5$ are

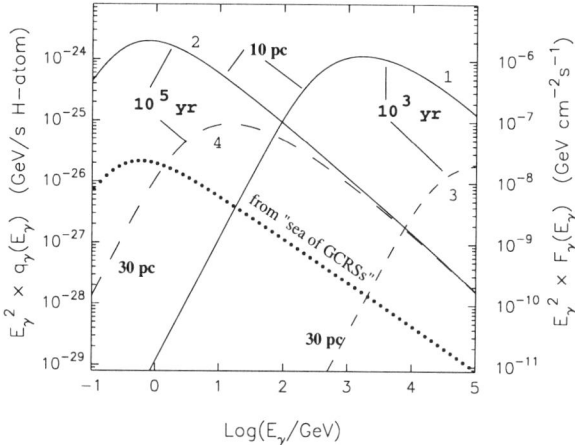

Fig. 4.7 The differential emissivities and fluxes of γ-rays at different times t and different distances R from an impulsive proton accelerator with $W_{\rm p} = 10^{50}$ erg. The left-hand side axis shows the γ-ray emissivities, in terms of $E_\gamma^2 q_\gamma(E_\gamma)$. The right-hand side axis shows the γ-ray fluxes, $E_\gamma^2 F_\gamma(E_\gamma)$, which are expected from a cloud with parameter $M_5/d_{\rm kpc}^2 = 1$ for the given emissivity q_γ. The curves 1 and 3 correspond to times $t = 10^3$ yr, and the curves 2 and 4 to $t = 10^5$ yr. The emissivities at the distance $R = 10\,{\rm pc}$ are plotted by solid lines (curves 1 and 2), and the emissivities at $R = 30\,{\rm pc}$ are shown by dashed lines (curves 3 and 4). The primary proton spectrum and the diffusion coefficients are the same as in Fig. 4.4 with $D_{10} = 10^{26}\,{\rm cm^2/s}$. The curve shown by the full dots corresponds to the γ-ray emissivity for the locally observed proton flux.

assumed. The emissivities and fluxes of γ-rays from the "sea" of galactic cosmic rays interacting with the cloud, are also shown.

The comparison of the expected γ-ray fluxes with the sensitivities of GLAST (for approximately 1 year observation time), and the future IACT

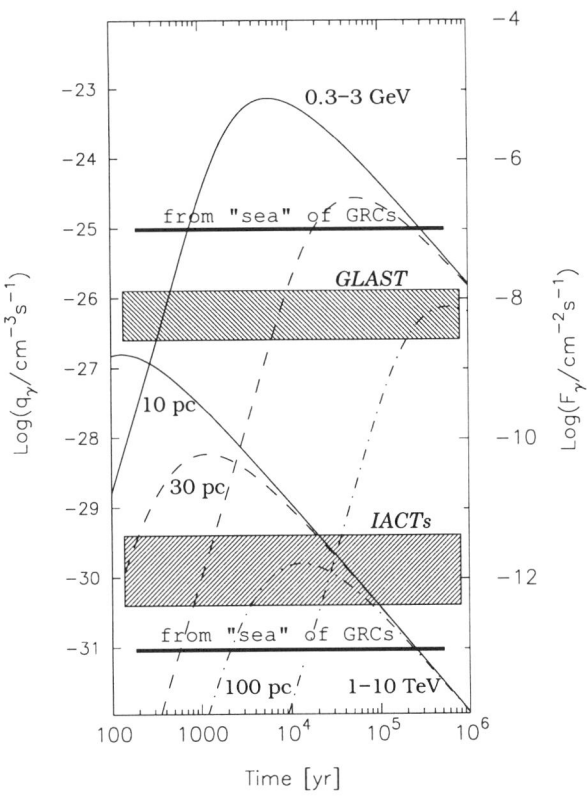

Fig. 4.8 Time dependence of the π^0-decay γ-ray emissivities (the left-hand side ordinate axes) and the fluxes (the right-hand side ordinate axes) from a cloud in the energy intervals 0.3-3 GeV and 1-10 TeV at three different distances from an *impulsive* accelerator: 10 pc, 30 pc, and 100 pc. The fluxes are calculated for a GMC with $M_5/d^2_{\rm kpc} = 1$ and a proton accelerator with $W_{\rm p} = 10^{50}$ erg. The horizontal lines indicate the corresponding emissivities and fluxes of γ-rays expected from the "sea" of GCRs. The expected sensitivities of GLAST and the next generation of IACT arrays in the same energy intervals are shown for the range of the γ-ray source size between 0.1° and 1°.

arrays (for ≈ 10 h of observation time), show that at certain (not necessarily same) epochs the cloud could be visible at GeV and/or TeV energies, up to distances to the accelerator $R \sim 30$ pc even for the source size of about 1°, provided that $W_{50} M_5/d_{\rm kpc}^2 \geq 0.01$ ($W_{50} = W_{\rm p}/10^{50}$ erg).

Typical estimates of distances to most of the unidentified EGRET sources, based on their spatial distribution in the galactic plane, is between 1.2 and 6 kpc. The corresponding γ-ray luminosities $\geq 10^{35}$ erg/s can be explained by the CR irradiation of GMCs with reasonable values of the parameter $M_5/d_{\rm kpc}^2$ between 0.1 and 1, provided that powerful and relatively young particle accelerators with $W_{\rm p} \sim 10^{49} - 10^{50}$ erg, operate in the vicinity of these clouds.

The remarkable feature of γ-radiation of GMCs is the strong evolution in time of both the absolute fluxes and the spectra of γ-rays. The character of the evolution is different for *impulsive* and *continuous* injection of cosmic rays into the ISM, and essentially depends on the diffusion coefficient $D(E)$ and distance R between the target and accelerator (see Fig. 4.7 and Fig. 4.8). Depending on the combination of the diffusion coefficient $D(E)$, distance R, as well as the age of the accelerator t, one should expect quite different γ-ray spectra from source to source. Namely, in the case of a cloud near a relatively young accelerator the differential γ-ray spectrum is expected to be much harder than the primary spectrum of the accelerated particles, i.e. $\alpha_\gamma < 2$. Meanwhile, the γ-ray spectra from clouds located near old accelerators would be soft, with a spectral index $\alpha_\gamma \geq 2.7$. Depending on the energy of γ-rays under consideration, the accelerator may be classified as "young" for a cloud at the given distance R, if $R/R_{\rm diff} \geq 1$, i.e. during the times until $t \leq R^2/4D(10E_\gamma)$. Note that in the case of energy-independent propagation of cosmic rays the spectra of γ-rays are the same at all epochs, and at $E_\gamma \geq 1$ GeV they repeat the spectrum of the parent protons, though the evolution of the absolute fluxes is strong.

Thus, the detection of γ-rays from different clouds located at different distances from the accelerator may provide unique information about the diffusion coefficient $D(E)$ as well as about the age of the accelerator. Similar information may be obtained detecting γ-rays from the same cloud, but in different energy domains, e.g. at GeV and TeV energies. However, in the case of energy-dependent propagation of cosmic rays the chance of simultaneous detection of a cloud in GeV and TeV γ-rays could be not very high, because the maximum fluxes at these energies are reached at different epochs. Since the higher energy particles propagate faster and therefore reach the cloud earlier, the maximum of GeV γ-radiation appears at an epoch when the maximum of the TeV γ-ray flux has already passed. In the case of energy-independent propagation (e.g. due to strong convection) the ratio of fluxes $F_\gamma(\geq 100\,\mathrm{MeV})/F_\gamma(\geq 1\,\mathrm{TeV})$ is independent of time, therefore the clouds which are visible at GeV energies would be detectable also at TeV energies.

4.2.3 *Accelerator inside the cloud*

The second interesting possibility for enhanced cosmic ray density in GMCs is the existence of an accelerator inside the cloud (see e.g Ginzburg and Ptuskin, 1984; Morfill *et al.*, 1984; Ormes *et al.*, 1988; Aharonian and Atoyan, 1996a). For a constant gas density and power-law CR spectrum, the γ-ray flux at energies above 1 GeV reduces to

$$F_\gamma(\geq E_\gamma) = 10^{-6}\, g \left(\frac{W_{50}}{d_{\mathrm{kpc}}^2}\right) n_2\, E_{\mathrm{GeV}}^{-\Gamma+1}\, \eta_\mathrm{A}\;\mathrm{cm}^{-2}\mathrm{s}^{-1}, \qquad (4.12)$$

where W_{50} is the total kinetic energy of CR protons in the cloud at instant t of observation in units of 10^{50} erg, $n_2 \equiv n/10^2\,\mathrm{cm}^{-3}$, and $g \simeq 0.7 - 0.95$ for Γ between 2.0 and 2.8.

This equation allows us to understand qualitatively the energy budget of cosmic rays which is required to make a cloud visible in γ-rays. However, for quantitative calculations one needs to take into account the effects of the spectral modification due to energy-dependent propagation (end escape) of relativistic particles in the cloud. The γ-ray fluxes expected from a cloud with an *impulsive* and *continuous* accelerator in its center are shown in Fig 4.9a,b. For the chosen radius of the cloud, $R_{\mathrm{cl}} = 20\,\mathrm{pc}$, a mean gas density $n = 130\,\mathrm{cm}^{-3}$ is assumed, which corresponds to the mass of the cloud $M_{\mathrm{cl}} = 10^5 M_\odot$. The distance to the cloud is taken to be $d = 1\,\mathrm{kpc}$. At initial epochs, $t \leq 10^3$ yr, when all relativistic protons remain inside the cloud, the γ-ray fluxes repeat the hard power-law spectrum of protons with an index $\Gamma = 2.2$. At $t \geq 10^4$ yr the γ-ray spectrum gradually steepens due to the escape of high energy particles from the cloud. For the chosen gas density of $n \sim 100\,\mathrm{cm}^{-3}$, the cloud with the *continuous* accelerator inside becomes visible at GeV energies, i.e. $F_\gamma(\geq 1\,\mathrm{GeV}) \sim 10^{-8}\,\mathrm{cm}^{-2}\mathrm{s}^{-1}$, only at late epochs, $t \sim 10^4$ yr, when the total energy output in accelerated protons becomes of order of 3×10^{48} erg. On the other hand, because of the hard source spectrum of accelerated protons, the cloud may already be visible in TeV γ-rays at epochs $t \geq 10^3$ yr when $F_\gamma(\geq 1\,\mathrm{TeV}) \geq 10^{-12}\,\mathrm{cm}^{-2}\mathrm{s}^{-1}$.

In the case of an *impulsive* accelerator the expected GeV-TeV correlations look quite different. In this scenario γ-ray fluxes monotonically decrease with time because of proton losses, especially due to the escape of particles from the cloud. The effect is especially strong at TeV energies. While, for the chosen radius, density and the diffusion coefficient, the flux of γ-rays at 1 GeV remains almost stable until $t \sim 10^5$ yr, the flux of TeV γ-rays at $t \sim 10^5$ yr drops to a very low (although still detectable) level. Remarkably, this feature is quite different from the one which is expected if the accelerator is outside the cloud. In this case the low energy particles need more time to reach the cloud, therefore the cloud becomes visible at GeV γ-rays later than at TeV γ-rays (compare Figs. 4.7 and 4.9a).

At $t \gg 10^5$ yr γ-ray fluxes are suppressed not only due to the escape of the particles from the cloud, but also due to their energy losses. At stages when the energy losses of protons become important, i.e. $t \geq \tau_{\rm pp}$, a significant fraction of the total energy of accelerated protons is deposited in the secondary electrons from the $\pi^\pm$-mesons, which results in additional γ-ray production due to the bremsstrahlung of these electrons. This component of radiation, together with the bremsstrahlung associated with the primary (i.e. directly accelerated) electrons, may have a significant impact on the

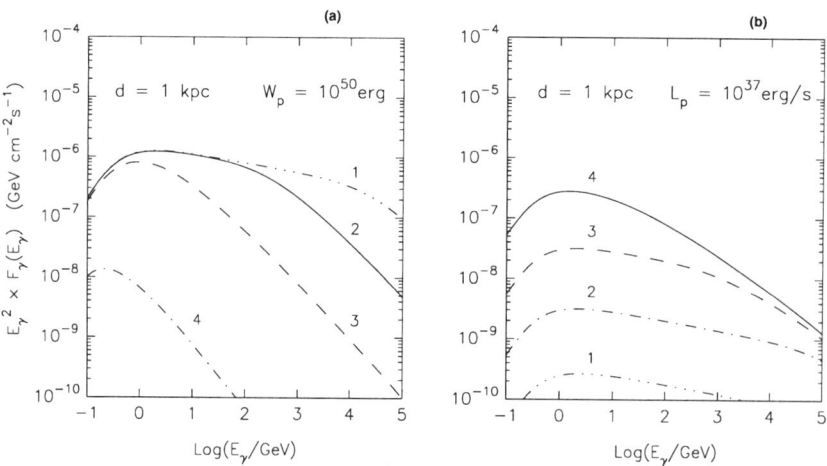

Fig. 4.9 Gamma-ray fluxes expected from a cloud with an accelerator in its center. The parameters of the cloud and the diffusion coefficient are: $R_{\rm cl} = 20\,{\rm pc}$, $n = 130\,{\rm cm}^{-3}$, $d = 1\,{\rm kpc}$, $D_{10} = 10^{26}\,{\rm cm}^2/{\rm s}$, $\delta = 0.5$. (a) - *impulsive* accelerator; the curves 1, 2, 3, 4 correspond to the ages $t = 10^3$, 10^4, 10^5, 10^6 yr, respectively. (b) - *continuous* accelerator; the curves 1, 2, 3, 4 correspond to the ages $t = 10^2$, 10^3, 10^4, 10^5 yr, respectively.

γ-ray spectrum below 1 GeV.

4.2.4 *On the level of the "sea" of galactic cosmic rays*

It is generally believed that the local CR flux (hereafter LCRs), i.e. that directly measured at the Earth, gives a correct estimate for the level of the "sea" galactic cosmic rays. However, strictly speaking, this is an *ad hoc* assumption. In fact, it is not obvious that LCRs should be taken as being representative of the whole galactic population of relativistic particles. In other words, we cannot exclude the possibility that the flux of LCRs could be dominated by a single or few local sources, especially given the fact that the Solar system is located in a rather extraordinary region - inside active star formation complexes which constitute the so-called Gould Belt. This statement is certainly true for the observed $\geq 1\,\mathrm{TeV}$ electrons which suffer severe synchrotron and inverse Compton losses, and thus reach us, for any reasonable diffusion coefficient, from regions no farther than a few hundred parsecs (see Sec. 4.3).

Because of possible contamination of the "sea" of GCRs by nearby sources, we may expect a non-negligible deviation of both the spectrum and the energy density of LCRs from the spectrum and density of GCRs. Therefore, the "GCRs≡LCRs" hypothesis needs a reliable *observational* confirmation. This can be achieved by observations of high energy γ-rays from "passive" GMCs, i.e. from clouds located in environments free of strong CR accelerators. If the flux of LCRs does reflect the level of the "sea" of GCRs, the shape and the absolute flux of γ-rays from "passive" clouds can be predicted with reasonable accuracy. The standard spectral shape of γ-radiation would be an important indicator of quietness of γ-ray emitting clouds, although the absolute γ-ray fluxes from individual clouds can be quite different because of different values of the parameter M_5/d_{kpc}^2. Therefore this method requires detailed spectroscopic measurements for an ensemble of GMCs with known distances d_{kpc} and masses M_5. Also, detection of enhanced γ-ray fluxes, compared with the "standard" γ-ray flux calculated for LCRs, would imply a presence of nearby CR sources. Although these clouds cannot be used for precise determination of the level of the "sea" of GCRs, a reliable detection of γ-radiation from even a single *under-luminous* GMC, i.e. a cloud emitting γ-rays below the expected "standard" flux, would imply that the CR flux we observe at the Earth is our local *"fog"* which obscures the genuine flux of GCRs. Below we argue that the γ-observations of the Orion complex contain evidence for such a drastic conclusion with important astrophysical implications for the origin of cosmic rays.

The extensive study of the high energy diffuse emission towards Orion by EGRET allowed an accurate measurement of the emissivity in the Orion region above 100 MeV, $q_\gamma/4\pi = (1.65 \pm 0.11) \times 10^{-26}\,\text{s}^{-1}\text{sr}^{-1}$, and, more importantly, a derivation of the differential emissivity in a broad energy region from 30 MeV to 10 GeV (Digel et al., 1999). These measurements, together with γ-ray emissivities calculated under different assumptions about the CR spectrum and flux, are shown in Figs. 4.10a,b and Fig. 4.11 (Aharonian, 2001a). The emissivity of π^0-decay γ-rays that corresponds to the local proton flux represented by Eq.(4.30), assuming an additional 50 per cent contribution from α-particles and other nuclei, is shown in Fig. 4.10a (curve 1). Although the predicted integral emissivity above 100 MeV, $q_\gamma/4\pi \simeq 1.8 \times 10^{-26}\,\text{s}^{-1}\text{sr}^{-1}$, almost coincides with the measured emissivity, it is difficult to fit the derived γ-ray spectrum without assuming an additional, low energy component, e.g. a power-law with a photon index $\Gamma = 2.4$ (dotted line). This component could be naturally attributed to the bremsstrahlung of primary (i.e. directly accelerated) and partly also

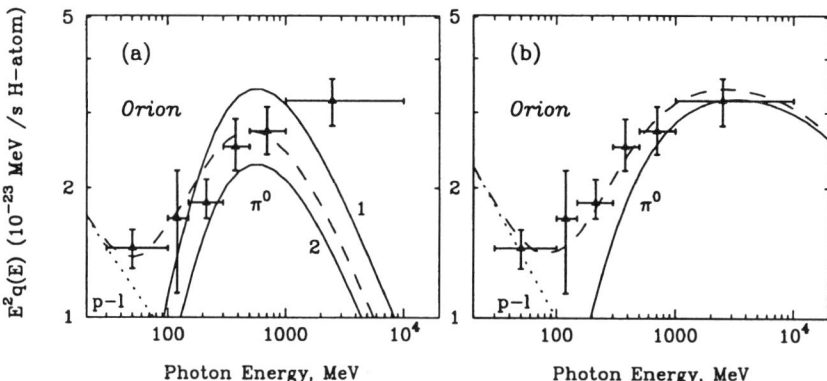

Fig. 4.10 Gamma-ray emissivity in the Orion region. The spectra of π^0-decay γ-rays are shown by solid lines. The power-law (p-l) components at low energies, with the photon index $\Gamma = 2.4$, are shown by dotted lines. The dashed lines correspond to the total, 'π^0'+'p-l' spectra. (a) π^0-decay γ-ray emissivity is calculated for the flux of LCRs (curve 1), and a flux lower by a factor of 1.5 (curve 2). The dashed curve corresponds to the sum of the latter and the low energy p-l component; (b) π^0-decay γ-ray emissivity (solid curve) is calculated for a proton spectrum with power-low index index $\Gamma = 2.1$ and energy density $w = 0.7\,\text{eV}/\text{cm}^3$. The dashed curve corresponds to the superposition of the π^0-decay and low energy p-l components. (From Aharonian, 2001a).

secondary (i.e. $\pi^{\pm}$-decay) electrons. It is seen from Fig. 4.10a that the superposition of the 'power-law' and 'π^0-decay' contributions (dashed line) satisfactorily fits the derived emissivity, if one ignores the last point measured by EGRET between 1 and 10 GeV. Note that such a fit is achieved assuming a somewhat reduced, by a factor of 1.5, CR flux in Orion compared with the local CR flux (curve 2).

One should probably not overemphasise this difference, taking into account possible systematic errors in measurements of the CR flux, as well as uncertainties in the CO measurements used for derivation of the column density of the molecular hydrogen. However, the discrepancy becomes quite significant if we include in the fit the measured highest energy point at 1-10 GeV. This point apparently requires a harder CR spectrum. In particular, Fig. 4.10b shows that a good fit to the EGRET data could be achieved assuming a flat spectrum of protons with power-law index $\Gamma = 2.1$, and energy density $w = 0.7\,\text{eV}/\text{cm}^3$ at $E \leq 1$ TeV, which is by a factor of 1.7 less than the energy density of LCRs. The lack of γ-ray measurements above

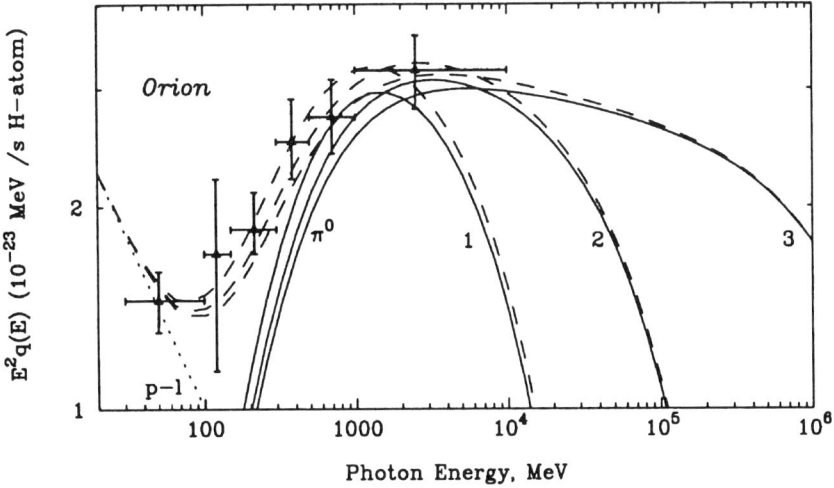

Fig. 4.11 Differential γ-ray emissivities in the Orion region. The curves 1, 2 and 3 correspond to emissivities calculated for CR spectra with the same power-law index $\Gamma = 2.1$, but with exponential cutoff at 3 different energies, $E_0 = 100$ GeV, 1 TeV, and 10 TeV, respectively. The corresponding energy densities of CR protons are: $w_\text{p} = 0.55$, 0.7, and 0.85 eV/cm^3. At low energies an additional power-law component of radiation ($\propto E^{-2.4}$) is also assumed (dotted curve). The superpositions of π^0-decay and power-law components are shown by dashed curves. (From Aharonian et al., 2001a).

10 GeV do not allow a robust constraint on the power-low index, but, most probably, it cannot exceed 2.4. This almost excludes the possibility that Orion is a passive γ-ray source (i.e. its γ-radiation is merely contributed by the "sea" of GCRs interacting with massive molecular clouds). A more plausible interpretation of this result would be a scenario when the bulk of the observed γ-ray flux is produced by particles accelerated within the Orion complex, with a less (although unavoidable) contribution from GCRs homogeneously distributed throughout the Galactic Disk, and freely passing through the molecular clouds. Indeed, such an extra contribution by GCRs around 1 GeV would lead to an overproduction of γ-rays, unless we assume that the level of the "sea" of GCRs at energies $\sim 1 - 10\,\mathrm{GeV}$ does not exceed $\approx 1/3$ of the directly measured flux of LCRs.

This rather dramatic conclusion formally could be avoided by speculating that the low-energy CRs cannot freely penetrate the dense clouds (see, however, Cesarsky and Völk, 1978). For example, assuming that the coefficient of reflectivity of CRs from a cloud decreases with energy, e.g. approximately as $E^{-\kappa}$, with $\kappa \sim 0.6 - 0.7$, one can obtain the required hard spectrum of particles in Orion without invoking additional nearby accelerators . Both these possibilities do not agree, to a certain extent, with the current concepts of origin and propagation of GCRs. Therefore, more careful analysis is needed before claiming any deviation from the conventional models of GCRs. Note that the conclusion about the hard CR spectrum in Orion is essentially based on the reported γ-ray flux above 1 GeV. This indicates the importance of future accurate spectrometric measurements of γ-radiation of Orion, as well as of other individual γ-ray emitting clouds detected by EGRET towards Cepheus, Monoceros, and Ophiuchus (Digel et al., 1999). These measurements should allow an effective separation of the contributions from the "sea" of GCRs and from the local CR accelerators. It is especially important to extend the measurements to higher energies. If the "excess" flux at 1-10 GeV is indeed due to the accelerator(s) nearby or inside the Orion complex, we may expect continuation of this hard spectrum well beyond 10 GeV, depending on the high energy cutoff in the CR spectrum (see Fig. 4.11). On the other hand, the hypothesis of the energy-dependent reflection of CRs from dense clouds (assuming that GCRs≡LCRs), predicts an essentially different spectrum of γ-rays - a flat part below a few GeV, with a quite steep power-law tail with $\Gamma \simeq 2.75$ at higher energies. This part of the spectrum is produced by particles of the "sea" of GCRs which freely enter the cloud. Such standard high energy γ-ray tails should be observed from other local clouds as well.

Large-scale CO surveys of molecular clouds in the Milky Way by Dame *et al.* (1987) revealed two dozen local GMCs within 1 kpc. The study of high energy γ-rays from these clouds with a broad-distribution of distances from $\sim 100 - 200$ pc (like the Taurus dark clouds or Aquila Rift) to 800 pc (like Cyg OB7 and Cyg Rift), which should be visible for GLAST, is of great interest for the derivation of the spatial and spectral distribution of CRs in our local environment.

4.3 Probing the Sources of VHE CR Electrons

The standard interpretation of the energy spectrum of CRs usually assumes a uniform and continuous distribution of sources in the Galaxy both in space and time. Whereas for the nucleonic component of CRs this approximation can be considered as a reasonable (although not undisputable) working hypothesis, the validity of this assumption for the electrons is questionable, at least for the high energy part of the measured spectrum which extends up to TeV energies.

In Fig. 4.12 the energy spectrum of CR electrons calculated assuming a uniform and continuous distribution of the sources in the Galactic Disk (solid curve) is decomposed to show the contributions from sources located at distances $r \geq r_0$ for different r_0. It is seen that even at energies ~ 10 GeV the total flux of the observed electrons is dominated by particles produced and injected into the ISM at distances $r \leq 1$ kpc from the Sun. At TeV energies, the sources beyond 1 kpc contribute less than 0.1% of the total electron flux. Note that the calculations are performed assuming rather fast diffusion, with a diffusion coefficient given by Eq.(4.6) with $D_* = 10^{28}$ cm/s and $\delta = 0.6$ (the power-law acceleration spectrum of electrons is assumed to have $\Gamma = 2.4$). For a smaller diffusion coefficient the contributions from large distances would be even smaller. Since a homogeneous distribution of CR sources (in space and in time) on spatial scales $\ll 1$ kpc seems unlikely, both the spectrum and the absolute flux of TeV electrons may significantly vary from site to site in the Galactic Disk (Aharonian *et al.*, 1995, Atoyan *et al.*, 1995).

The extended IC γ-radiation of electrons produced in the vicinity of their accelerators, provides a unique tool to search for the electron *Tevatrons* in our Galaxy.

4.3.1 Distributions of VHE electrons

The energy-loss rate of electrons in Eq.(A.1) for the standard regions of ISM can be presented in the form $P(\gamma) = p_0 + p_1 \gamma + p_2 \gamma^2$ (in units of s^{-1}). Here $p_0 \simeq 6 \cdot 10^{-13} n\, s^{-1}$ is for the ionization losses of electrons in the neutral interstellar gas with number density n (in units of cm^{-3}). The second term with $p_1 \simeq 10^{-15} n\, s^{-1}$ corresponds to the bremsstrahlung energy losses and the last one, with $p_2 \simeq 5.2 \cdot 10^{-20}\, w_0 s^{-1}$ and $w_0 = w_B + w_{2.7K} + w_{opt}$ (in units of eV/cm^3), represents synchrotron and inverse Compton losses ($w_{2.7K} = 0.25\, eV/cm^3$ is the 2.7 K CMBR energy density, $w_{opt} \simeq 0.5\, eV/cm^3$ is the energy density of optical-IR radiation in interstellar space; the energy density of the magnetic field is $w_B = 0.6\, eV/cm^3$ for $B = 5\,\mu G$). Note that a more accurate expression for $P(\gamma)$ should take into account that the Compton scattering of electrons with energies $\gamma \geq 10^5$ off the optical photons corresponds to the Klein-

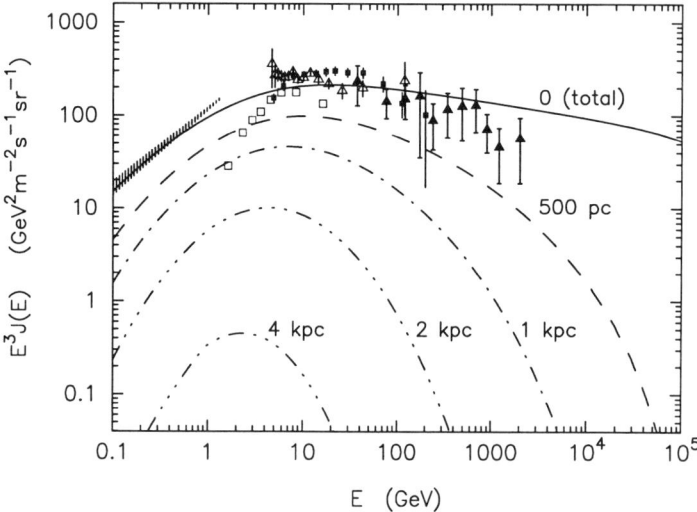

Fig. 4.12 The flux of CR electrons. The hatched region corresponds to the estimate of the mean flux of low energy electrons derived from radio data in the direction of galactic poles. The calculations demonstrate the relative contributions from different regions of the Galactic Disk to the overall flux of electrons assuming that the sources are continuously and uniformly distributed in the Galactic Disk. The overall flux (solid curve, $r_0 = 0\,pc$) is decomposed in order to show the contributions from the sources located at distances $r \geq r_0$ for different r_0 indicated near the curves. (From Aharonian et al., 1995).

Nishina rather than to the Thomson limit. However, as long as the energy density of the magnetic field and 2.7 K CMBR is comparable with the density of optical radiation, this approximation provides quite accurate results. Above several GeV, Compton and synchrotron losses dominate over bremsstrahlung and ionization losses. Moreover, for timescales $t \leq 10^7$ yr, only Compton and synchrotron losses are important, thus $P(\gamma) \approx p_2 \gamma^2$. This allows a simple analytical description of the density of electrons at a distance R and at time t after their injection from an *impulsive* source (Atoyan et al., 1995):

$$f(R,t,E) = \frac{Q_0 E^{-\Gamma}}{\pi^{3/2} R^3} (1 - p_2 t E)^{\Gamma-2} \left(\frac{E}{E_{\text{dif}}}\right)^3 e^{-(R/R_{\text{dif}})^2} , \qquad (4.13)$$

where $E < E_{\text{cut}} \equiv E_{\text{cut}}(t) = (p_2 t)^{-1} m_e c^2$ (otherwise $f=0$), and

$$R_{\text{dif}}(E,t) \simeq 2 \sqrt{D(E) t \, \frac{1 - (1 - E/E_{\text{cut}})^{1-\delta}}{(1-\delta) E/E_{\text{cut}}}} \qquad (4.14)$$

Fig. 4.13 demonstrates the temporal and spectral features of the electron

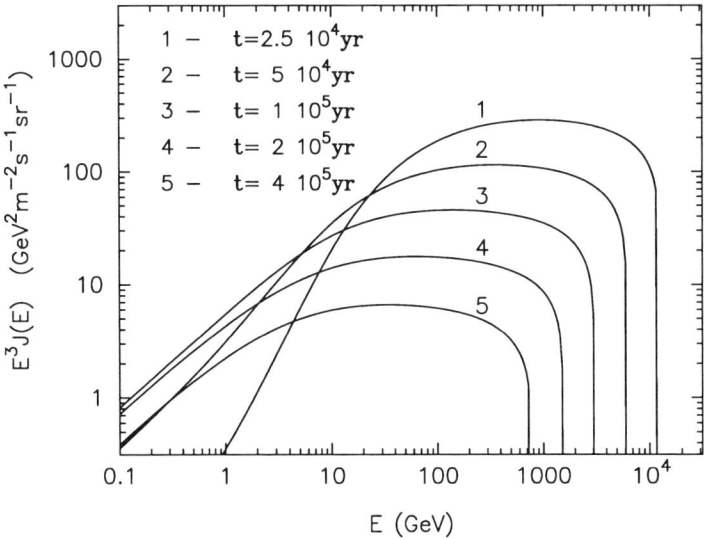

Fig. 4.13 The energy spectra of electrons at different epochs t after their injection into the ISM at 100 pc from an impulsive accelerator. The calculations correspond to the total energy output $W_e = 10^{48}$ erg in power-law electrons with $\Gamma = 2.2$ extending up to $E_{\text{max}} = 500$ TeV. The diffusion coefficient is the same as in Fig. 4.12.

flux at 100 pc from an impulsive accelerator.

In the case of continuous injection of electrons from a point source the energy spectrum of electrons is described by

$$f(R,t,E) = \frac{Q_0 \gamma^{-\Gamma}}{4\pi D(E) R} \, erfc\left(\frac{R}{2\sqrt{D(E) t_E}}\right), \qquad (4.15)$$

where $t_E = \min(t, m_e c^2/p_2 E)$. At low energies Eq.(4.15) has an energy dependence similar to Eq.(4.13) for burst-like injection. However, at higher energies it is qualitatively different. In particular, the steepening of this spectrum at high energies corresponds to a power-law index $\Gamma' \approx \Gamma + \delta$. Also, the cutoff of the energy spectrum above $\gamma_{\rm cut}$ expected in the case of burst-like injection, now disappears.

4.3.2 Extended regions of IC gamma radiation

Electrons ejected by a single accelerator can be traced by their inverse Compton radiation outside the accelerator. The characteristics of this radiation, produced in relatively compact environments of accelerators (e.g. in the pulsar-driven nebulae in the case of pulsars or in the shells in the case of SNRs) are described in the next Chapters. Here we discuss the characteristics of radiation of these electrons beyond the (possible) compact nebulae, assuming that the cloud of electrons expands in a conventional region of the ISM with magnetic field $B_{\rm ISM} \leq 5\mu$G and a diffuse optical/infrared photon density of $w_r \leq 1$ eV/cm^3. The synchrotron radiation of multi-TeV electrons peaks at optical or UV frequencies, and because of the diffuse galactic background cannot be detected. At the same time, the X-ray tail of this radiation produced by ≥ 100 TeV electrons may significantly contribute to the diffuse X-ray background of the Galactic Disk (see below).

The cloud of electrons expanding in the ISM radiates high energy γ-rays through inverse Compton scattering on the 2.7 K CMBR and on the diffuse interstellar infrared/optical (IR/O) radiation. The high production rate of IC γ-rays is provided predominantly by 2.7 K CMBR photons, although the energy density of IR/O is a factor of 2 to 4 higher than the density of 2.7 K CMBR, $w_{2.7K} \simeq 0.25$ eV/cm^3. Indeed, the emissivity of IC γ-rays in the case of a blackbody distribution of ambient photons in the Thomson regime is proportional to

$$q(E_\gamma) \propto \zeta T_r^{(p+5)/2} E_\gamma^{-(p+1)/2} \qquad (4.16)$$

where T_r and ζ are the temperature and dilution factor of the radiation, and p is the current power-law index of electrons (Blumenthal and Gould, 1970). Thus, for the parameters characterising the 2.7 K CMBR ($T_r = 2.7K$, $\zeta = 1$) and the interstellar IR/O starlight ($T_r \approx 5000K$, $\zeta \approx 1.5 \cdot 10^{-13}$), the contribution from the 2.7 K CMBR dominates as long as $p \leq 2.8$. In addition, the efficiency of IC scattering on IR/O photons at very high energies drops significantly due to the Klein-Nishina effect. The dominance of the IC radiation due to the up-scattering of the 2.7 K CMBR is a nice feature which allows derivation of robust information about the electrons independent of (poorly known) details, e.g. concerning the spatial variations of the diffuse IR/O radiation in the Galactic Disk.

The IC luminosities of the expanding cloud of electrons continuously ejected by an accelerator (e.g. by an isolated pulsar) into the ISM, are shown in Fig. 4.14. Due to the assumed hard spectrum of the electrons, IC scattering on the 2.7 K CMBR dominates over other γ-radiation chan-

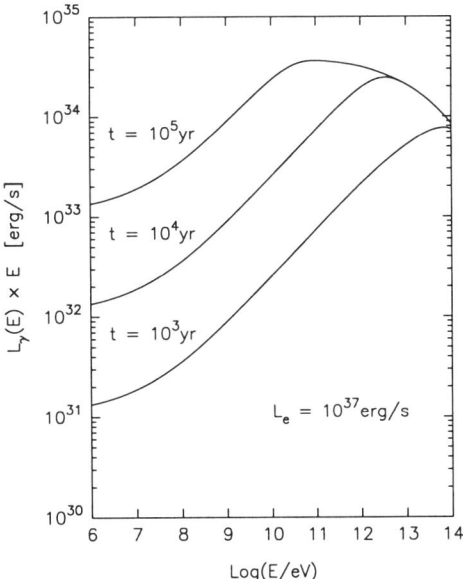

Fig. 4.14 The γ-ray luminosity of the expanding "cloud" of relativistic electrons at different epochs. A constant electron injection rate of $L_e = 10^{37}$ erg/s with a power-law acceleration spectrum ($\Gamma = 2$) extending up to 10^{15} eV is assumed. The magnetic field is $B_{\rm ISM} = 3\,\mu{\rm G}$, the ambient gas density is $n = 1\,{\rm cm}^{-3}$, and energy density of O/IR starlight radiation is $w_{\rm IR} = 0.5\,{\rm eV/cm}^3$.

nels. At lower energies, $E \leq 1\,\text{GeV}$, the contribution of the bremsstrahlung photons becomes noticeable, especially for steep electron spectra.

The results presented in Fig. 4.14 are obtained by integrating the fluxes over all angles from the central source. Therefore they depend *only* on the initial spectrum of electrons, but not on the diffusion coefficient. However, the character of diffusion of electrons in the ISM has a crucial impact on the angular distribution of radiation, and thus on the visibility of γ-ray fluxes. This is illustrated in Figs 4.15a,b where the γ-ray fluxes expected within different angles in the direction of the source of electrons are shown. The fluxes are calculated assuming that the relativistic electrons are continuously injected into the ISM with a constant rate $L_e = 3.5 \cdot 10^{36}\,\text{erg/s}$ during the last 10^4 years from a source at a distance of 0.5 kpc.

Fig. 4.15a corresponds to the "worst" combination of the model parameters in the sense of the detectability of γ-rays. The fast expansion of the cloud of electrons due to the large diffusion coefficient and the relatively soft ($\Gamma = 2.4$) initial spectrum of electrons (which during propagation becomes even steeper due to the energy-dependent diffusion) results in radiation of

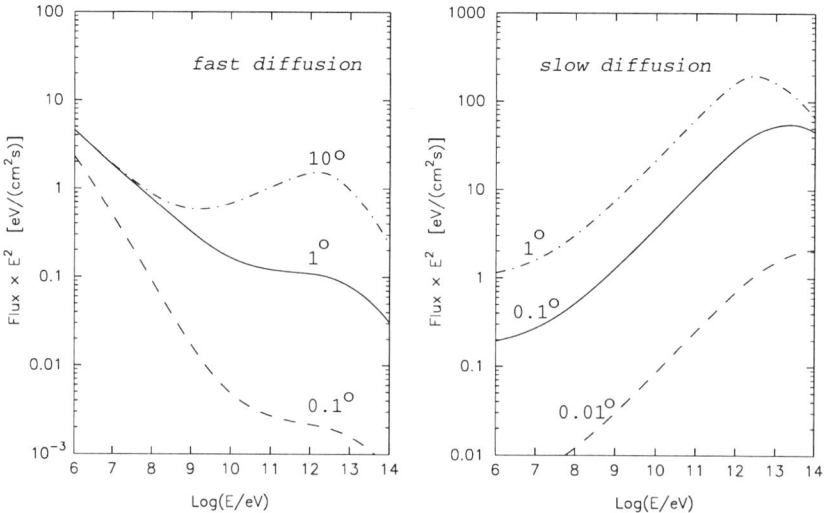

Fig. 4.15 IC γ-ray fluxes expected within different detection angles from an electron accelerator located at a distance of 0.5 kpc. The injection rate of electrons is $Le = 3.5 \cdot 10^{36}\,\text{erg/s}$. **(a)** (left) fast diffusion – power-law diffusion coefficient $D(E) = D_*(E/10\text{GeV})^{-\delta}$ with $D_* = 10^{27}\,\text{cm}^2/\text{s}$ and $\delta = 0.5$; steep spectrum of accelerated electrons with $\Gamma = 2.4$. **(b)** (right) slow diffusion – constant diffusion coefficient with $D_* = 10^{27}\,\text{cm}^2/\text{s}$ and $\delta = 0$; hard spectrum of accelerated electrons with $\Gamma = 2$.

the bulk of the photons within large angles and at low energies. Correspondingly, the γ-ray fluxes in both GeV and TeV energies turn out to be, for the assumed injection rate of electrons, well below the sensitivities of current γ-ray telescopes. The situation could be quite different if electrons with a hard initial spectrum propagate more slowly in the ISM (Fig. 4.15b). For the small diffusion coefficient the bulk of γ-rays are produced with a very hard spectrum in rather compact regions surrounding the source. As a result of this optimistic scenario, the γ-ray fluxes reach a level detectable by GLAST and future IACT arrays

In Fig. 4.16 γ-ray fluxes within different angular distances from an impulsive source are shown. It is assumed that $W_e = 10^{49}$ erg, released in the form of relativistic electrons, was injected into ISM at early stages of "operation" of a 10^4 year old source. It is interesting to note that $W_e = 10^{49}$ erg corresponds to the total energy required to explain the (directly) measured fluxes of high energy cosmic ray electrons above $E \geq 100\,\mathrm{GeV}$ solely by a *nearby single source* like the Geminga pulsar (Aharonian et al., 1995). Thus, the next generation of IACT systems with extended ($\sim 1°$) flux sensitivity better than $E^2 J(E) = 10$ eV/cm³ should be able to discover similar objects in our Galaxy by detecting characteristic TeV γ-radiation from isolated pulsars located at distances beyond 1 kpc.

The sharp cutoffs in γ-ray spectra shown in Fig. 4.16 are due to radiative loses of electrons produced 10^4 years ago and not supported by freshly accelerated particles. Therefore such objects could be detected most effectively at $E \sim 1$ TeV. Older objects obviously require observations at lower energies.

The γ-ray fluxes in Figs. 4.15a,b and 4.16 are calculated for certain sets of model parameters. Note that for the fixed diffusion coefficient and the spectral index of accelerated electrons, the absolute flux and angular size of the IC radiation shown are proportional to L_e/d^2 (or W_e/d^2) and $1/d$, respectively. Therefore, the angular and spectral characteristics of IC radiation of other sources with different combinations of L_e (or W_e) and d, can be calculated by renormalization of the curves presented in Figs. 4.15a,b and 4.16.

Finally, it is worth noting that the predicted fluxes of GeV γ-rays produced during the propagation of the electrons in the conventional regions of the ISM can be marginally detected by GLAST. However, in the case of expansion of the cloud of electrons in dense ($n \gg 1\,\mathrm{cm}^{-3}$) regions surrounding an electron accelerator, the fluxes of the bremsstrahlung photons could increase significantly to observable levels.

The injection of VHE electrons into the ISM by SNRs and pulsars (or by any other possible sources of high energy electrons) may result in rather high diffuse radiation from some regions of the Galactic Disk. In fact, due to the short propagation paths of VHE electrons, the IC component of the diffuse radiation of the Galactic Disk at high energies should be structured with an increase of the flux towards the discrete electron accelerators. The separation of the high energy IC component from the overall diffuse background of the Galactic Disk, e.g. by searching for specific spatial and spectral variations on angular scales $\leq 10°$, would allow identification of the high energy electron accelerators in our Galaxy.

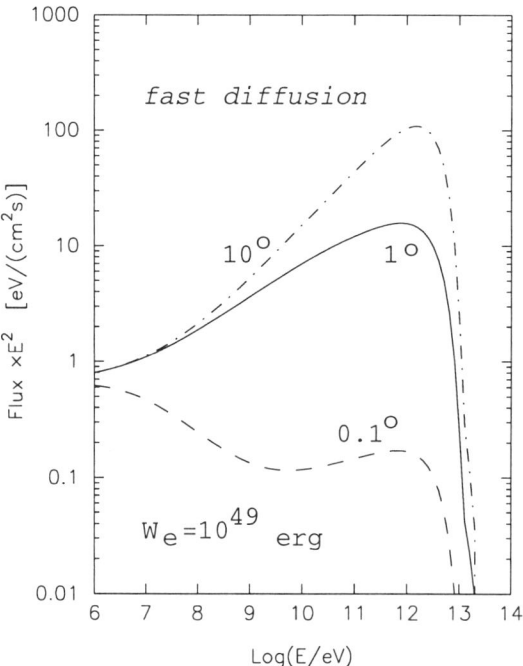

Fig. 4.16 IC γ-ray fluxes expected within different detection angles from an impulsive electron accelerator located at 2 kpc. It is assumed that $We = 10^{49}$ erg energy of relativistic electrons has been injected into ISM at early stages of 10^4 year old accelerator. The power-law electron spectrum and the diffusion coefficient, as well as the parameters characterising ISM are the same as those used in Fig.4.14.

4.4 Diffuse Radiation from the Galactic Disk

The study of diffuse galactic γ-ray emission, i.e. the radiation components produced in interactions of electronic and nucleonic components of CRs with the interstellar gas and photon fields, provides information about the density and energy spectra of CRs in different parts of the Galactic Disk, and thus provides a key insight into the character of propagation of CRs in the Galaxy on kpc scales. A proper understanding of these processes is a necessary condition for accurate estimates of the production rates of relativistic electrons and protons/nuclei in our Galaxy. Presently, our knowledge of GCRs is based on conclusions derived from the concept that interprets the *secondary* CR nuclei and anti-particles (positrons, antiprotons) as result of interactions of *primary* CRs with interstellar gas throughout the entire Galaxy. It is not obvious, however, that the *locally* observed CRs can be taken as undisputed representatives of the whole galactic population of relativistic particles (see Sec. 4.2.4). Therefore, the diffuse galactic γ-radiation is believed to be the most informative and model-independent channel telling us about the energy and spatial distributions of CRs in the Galactic Disk. The synchrotron radiation of the ISM at radio and possibly also at X-ray wavelengths provides additional and complementary information, but it concerns only the *electronic* component of CRs in two extreme energy bands below 1 GeV and above 100 TeV, respectively.

The identification and separation of the truly diffuse γ-ray emission is not an easy task because of a non-negligible contamination due to contributions from weak but numerous unresolved discrete sources. Before the launch of the Compton Gamma Ray Observatory the observations of the diffuse galactic γ-ray background were limited to the energy range between 100 MeV and few GeV explored by the SAS-2 and COS-B γ-ray satellites. The results of these important missions revealed noticeable correlations between the high energy γ-ray fluxes and the column density of the interstellar hydrogen, and thus demonstrated the existence of a truly diffuse galactic γ-radiation (for a review see Bloemen, 1989).

The highly successful observations conducted in the 1990s by detectors aboard the Compton Gamma Ray Observatory resulted in good quality data from the Galactic Disk over five decades of energy, from 1 MeV to 10 GeV (see Hunter *et al.*, 1997b, and references therein), and initiated new theoretical studies of diffuse high-energy γ-rays (e.g. Bertsch et al., 1993; Gralewicz *et al.*, 1997, Porter and Protheroe, 1997; Pohl and Esposito, 1998; Moskalenko and Strong, 1998; Strong *et al.*, 2000; Aharonian and Atoyan, 2000, *etc.*). Because of several competing processes, the problem of identification of specific radiation mechanisms is rather complicated and confused. Nevertheless the existing data do allow definite conclusions concerning the relative contributions of different processes in each specific γ-ray energy band. The discussion of the problem in the following sections will be limited to the inner parts of the Galactic Disk within $\geq 315°$ and $l \leq 45°$, which is not only the best experimentally studied region, but is also of prime interest because the CR sources are believed to be concentrated in that region of the Galaxy.

4.4.1 CR spectra in the inner Galaxy

In this section a simple model (Aharonian and Atoyan, 2000) of propagation of relativistic particles in the Galactic Disk will be described, which will be used for calculations of the resulting γ-radiation from the disk at galactic latitudes $|b| \leq 5°$. Although the halo of galactic CRs may extend up to heights of a few kpc (e.g. Bloemen *et al.*, 1993), one needs to know the mean spectrum of CRs only in a region close to the galactic plane. Below this region is approximated as a disk with a half-thickness $h \simeq 1\,\mathrm{kpc}$ and radius $R_\mathrm{G} \sim 15\,\mathrm{kpc}$.

The diffusion equation for the energy distribution $f = f(\mathbf{r}, E, t)$ of relativistic particles can be written in a general form as (Ginzburg and Syrovatskii, 1964)

$$\frac{\partial f}{\partial t} = \mathrm{div}_\mathbf{r}(D\,\mathrm{grad}_\mathbf{r} f) - \mathrm{div}_\mathbf{r}(\mathbf{u} f) + \frac{\partial}{\partial E}(P f) + A[f]\,, \qquad (4.17)$$

where $\mathbf{u} \equiv \mathbf{u}(\mathbf{r})$ is the fluid velocity of the gas containing relativistic particles, and $P \equiv P(\mathbf{r}, E) = -\mathrm{d}E/\mathrm{d}t$ describes their total energy losses, including the adiabatic energy loss term $P_\mathrm{adb} = \mathrm{div}\,\mathbf{u}\,E/3$ (e.g. Owens and Jokipii, 1977; Lerche and Schlickeiser, 1980). $D \equiv D(\mathbf{r}, E)$ is the spatial diffusion coefficient, and $A[f]$ is a functional representing acceleration terms of relativistic particles.

The integration of Eq.(4.17) over the volume $V = 4\pi h R_\mathrm{G}^2$ results in a convenient equation for the total energy distribution function of particles $N(E, t) = \int f\,\mathrm{d}^3 r$ in the Galactic Disk at $|z| \leq h$:

$$\frac{\partial N}{\partial t} = \frac{\partial(\overline{P} N)}{\partial E} - \frac{N}{\tau_\mathrm{esc}} + Q\,, \qquad (4.18)$$

where τ_{esc} is the "diffusive + convective" escape time of particles:

$$\tau_{esc}(E) = \left[\frac{1}{\tau_{dif}(E)} + \frac{1}{\tau_{conv}}\right]^{-1} . \qquad (4.19)$$

The parameter τ_{conv} describes the convective escape of particles from the disk through its surface due to the galactic wind driven by the pressure of CRs and the thermal gas (Jokipii, 1976; Berezinsky et al., 1990; Breitschwerdt et al., 1991; Ptuskin et al., 1997),

$$\tau_{conv} \simeq ah/u , \qquad (4.20)$$

where u is the wind velocity on the surface of the Galactic Disk, which could reach $\sim 50\,\mathrm{km/s}$ (Zirakashvili et al. 1996) at the height $h = 1\,\mathrm{kpc}$, and the (generally energy-dependent) parameter $a \geq 1$ is the ratio of the mean density of particles in the disk to their density at the disk surface (Aharonian and Atoyan, 2000). Correspondingly, the mean convective escape time of CRs from the Galactic Disk is estimated to be $\tau_{conv} \simeq 2 \times 10^7 a\,\mathrm{yr}$.

The diffusive escape time is very often presented in a convenient power-law form

$$\tau_{dif}(E) = \tau_{10}\,(E/10\,\mathrm{GeV})^{-\delta} + \tau_{min} , \qquad (4.21)$$

where τ_{10} corresponds to the particle escape time at energy $E = 10\,\mathrm{GeV}$. Because the diffusive escape time cannot be less than the light travel time, $\tau_{min} = h/c$, the energy dependence of τ_{esc} above a certain energy disappears. On the other hand, since τ_{dif} increases with decreasing E, τ_{esc} given by Eq.(4.19) becomes energy-independent below some E_* defined from the condition $\tau_{dif}(E_*) = \tau_{conv}$. Neglecting at these energies τ_{min} in Eq.(4.21), from Eq.(4.19) it then follows that the overall escape time can be presented in the form

$$\tau_{esc} \approx \tau_{conv}/[1 + (E/E_*)^{\delta}] . \qquad (4.22)$$

The implications of the Leaky-Box type Eq.(4.18) and the limits of its validity are discussed by Aharonian and Atoyan (2000). Here we briefly mention that the power-law index δ in Eq.(4.21) for the diffusive escape time in the Leaky-Box type Eq.(4.18) does not coincide with the power-law index δ_1 of the diffusion coefficient $D(E) \propto E^{\delta_1}$ in the general Eq.(4.17), but is effectively confined within the limits $\delta_1/2 < \delta < \delta_1$. Generally, CR propagation requires a more detailed treatment of diffusion and convection effects (see Strong et al. (2000) for numerical 3D-simulations of CR diffusion, and Breitschwerdt *et al.* (2002) for a self-consistent analytical approach to the problem), as well as certain assumptions about the distribution of CR sources in the Galactic Disk (e.g. Pohl and Esposito, 1998). Also, the disk-halo transition at distances ~ 1 kpc is an important part of the picture of CR transport by the galactic wind, and should be carefully taken into account in detailed calculations of the diffuse γ-radiation, in particular at high galactic latitudes.

Such treatments, however, contain a number of *ad hoc* assumptions, thus for the study of diffuse galactic γ-radiation the accuracy of sophisticated theoretical models should not be overrated, given the uncertainties concerning both the parameters of the interstellar medium (especially in distant, confused parts of the Galaxy) and the origin of sources of GCRs. Therefore, at this stage the simple phenomenological approach based on the approximate description of CR spectra by the Leaky-Box type equations is reasonably justified, at least for the narrow region of the galactic plane, and allows unambiguous conclusions from the existing γ-ray data.

For a time-dependent injection $Q(E,t)$, the solution to Eq.(4.18), in terms of the spatial density functions $n = N/V$ and $q = Q/V$, is

$$n(E,t) = \frac{1}{P(E)} \int_0^t P(\zeta_t) Q(\zeta_t, t_1) \times \exp\left(-\int_{t_1}^t \frac{\mathrm{d}x}{\tau_{\mathrm{esc}}(\zeta_x)}\right) \mathrm{d}t_1 \ . \quad (4.23)$$

Here the variable ζ_t corresponds to the energy of a particle at an instant $t_1 \leq t$ which has energy E at the time t, and is determined from the equation

$$t - t_1 = \int_{E_e}^{\zeta_t} \frac{\mathrm{d}E_1}{P(E_1)} \ . \quad (4.24)$$

For a quasi-stationary injection of electrons into the ISM on time-scales exceeding the escape time $\tau_{\mathrm{esc}}(E)$, the energy distribution of particles becomes time-independent. For both CR protons and electrons injected into the ISM we assume a stationary (continuous) source function (per unit volume) in a 'standard' power-law form with an index Γ_0 and an exponential cutoff at E_0. The flux of diffuse radiation with energy E_γ in a given direction is defined by the unit volume emissivity $q_\gamma(\mathbf{r}, E_\gamma)$ integrated along the line of sight:

$$J(E_\gamma) = \int \frac{q_\gamma(\mathbf{r}, E_\gamma)}{4\pi} \mathrm{d}l = \frac{\bar{q}_\gamma(E_\gamma) l_\mathrm{d}}{4\pi} \ , \quad (4.25)$$

where $\bar{q}_\gamma(E_\gamma)$ is the mean emissivity, and l_d is the characteristic line-of-sight depth of the emission region.

4.4.2 Diffuse radiation associated with cosmic ray electrons

There are four principal processes for the production of non-thermal hard X-rays and γ-rays in the interstellar medium by CR electrons: inverse Compton (IC) scattering, bremsstrahlung, annihilation of positrons, and synchrotron radiation (provided that the electrons are accelerated up to energies exceeding 100 TeV). The first three mechanisms are discussed below; the synchrotron radiation of hard X-rays will be discussed in Sec. 4.4.4 in the context of IC radiation from the highest energy electrons.

4.4.2.1 IC gamma rays

Calculations of the diffuse IC γ-rays require knowledge of the low-frequency target photons fields. The photon fields which are important for production of IC γ-rays are the 2.7 K CMBR and the diffuse galactic radiation – the starlight and dust photons at near and far infrared frequencies, respectively. While the density of the 2.7 K CMBR is universal, with $w_{2.7\mathrm{K}} \approx 0.25\,\mathrm{eV/cm^3}$, the densities of diffuse galactic photon fields vary from site to site. The studies of Chi and Wolfendale (1991) show that the starlight energy density increases from the local value $w_{\mathrm{NIR}} \simeq 0.5\,\mathrm{eV/cm^3}$ (Mathis et al. 1983) up to $\simeq 2.5\,\mathrm{eV/cm^3}$ in the central 1 kpc region of the Galaxy. For the calculations below the mean value, $w_{\mathrm{NIR}} \simeq 1.5\,\mathrm{eV/cm^3}$ is used. The energy density of dust emission in the galactic plane is rather uncertain; it is estimated from $w_{\mathrm{FIR}} \simeq 0.05-0.1\,\mathrm{eV/cm^3}$ (e.g. Mathis et al. 1983) to $w_{\mathrm{FIR}} \simeq 0.2-0.3\,\mathrm{eV/cm^3}$ (Chi & Wolfendale 1991). For the calculations below $w_{\mathrm{FIR}} \simeq 0.2\,\mathrm{eV/cm^3}$ is used. Fortunately, large uncertainties in w_{FIR} do not appear to have a strong impact on the calculations because at all γ-ray energies the contribution from IC up-scattering of 2.7 K CMBR photons significantly exceeds the IC fluxes produced on FIR photons (see below).

Another source of uncertainties in calculations of the diffuse IC fluxes is the lack of independent information about the high energy electrons. While radio measurements allow definite conclusions about the average electron flux below a few GeV, at higher energies the electron fluxes in the Galaxy are rather uncertain. The standard interpretation of the energy spectrum of CRs usually assumes a uniform and continuous distribution of sources in the Galaxy both in space and time. The validity of this assumption is questionable, at least for TeV electrons. Because of severe radiative losses, the source(s) of observed TeV electrons should be younger than $\approx 10^5$ yr, and cannot be located much beyond a few 100 pc. Therefore, the measured electron spectrum might not be applicable for calculations of γ-radiation from distant parts of the Galactic Disk.

Both the spectrum and the absolute flux of high energy electrons at TeV and higher energies may show significant variations from site to site in the Galactic Disk. In particular, one could expect a significant enhancement of the electron flux in the central region of the Galaxy due to the suspected concentration of CR sources there. Therefore one may allow deviations of the predicted electron distribution in the inner Galaxy from the observed fluxes, except perhaps for the region below a few GeV where the radio observations do provide information about the average spectrum of galactic electrons along the line of sight.

In Fig. 4.17 the average spectrum of electrons calculated for the inner part of the Galaxy is shown. The calculations are performed assuming a power-law injection spectrum with $\Gamma_{e,0} = 2.15$. For the assumed energy density of resulting electrons, $w_e = 4\pi/c \int E\, J(E)\, dE = 0.05\,\mathrm{eV/cm^3}$, the calculated spectra $J(E)$ below 1 GeV match the electron fluxes derived from radio observations. This normalisation requires an acceleration rate

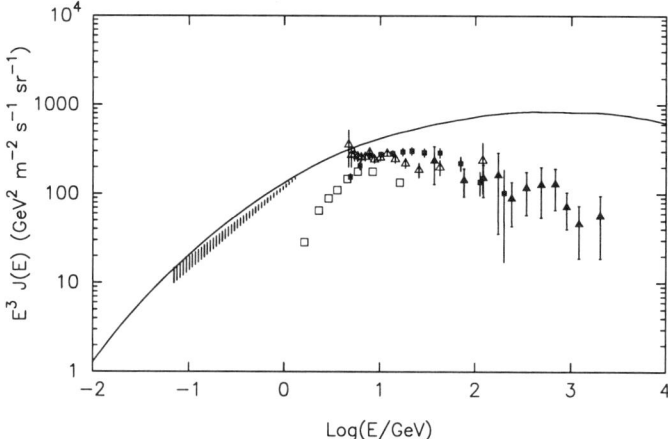

Fig. 4.17 The mean flux of electrons assuming a quasi-stationary production of electrons with a power-law acceleration index $\Gamma_{e,0} = 2.15$ and exponential cutoff at $E_0 = 100\,\mathrm{TeV}$. For the CR propagation the following escape times are assumed: $\tau_{\mathrm{conv}} = 2 \times 10^7\,\mathrm{yr}$, $\tau_{\mathrm{dif}} = 10^7 (E/10\,\mathrm{yr})^{-0.6} + \tau_{\mathrm{min}}$; $\tau_{\mathrm{min}} = 3 \times 10^3\,\mathrm{yr}$. For the ISM the following parameters are assumed: $B = 6\,\mu\mathrm{G}$, $\bar{n}_\mathrm{H} = 0.15\,\mathrm{H-atom/cm^3}$, and $w_{\mathrm{FIR}} = 0.2\,\mathrm{eV/cm^3}$, $w_{\mathrm{NIR}} = 1.5\,\mathrm{eV/cm^3}$. The electron injection rate is normalised so that the energy density in the resulting spectrum of electrons is $w_e = 0.05\,\mathrm{eV/cm^3}$. The hatched region and experimental points are the same as in Fig. 4.12.

of electrons in the Galaxy per 1 km^3

$$L_e \simeq 1.6 \times 10^{37} \, \text{erg/kpc}^3 \, \text{s} \,. \tag{4.26}$$

This implies that for the inner part of the Galactic Disk, with $R \leq 8.5$ kpc and a half thickness 1 kpc, the overall acceleration power should be about 7×10^{39} erg/s. Between 100 MeV and 1 GeV the dissipation of electrons is dominated by adiabatic and bremsstrahlung losses, the rates of which are proportional to energy, $dE/dt \propto E$ (or $t = E/(dE/dt) = const$), and thus do not change the original (acceleration) spectrum. Therefore the spectral index of the observed synchrotron radio emission contains direct information about the acceleration spectrum of electrons in this energy region. Below 100 MeV, the electron spectrum suffers significant deformation (flattening) because of energy-independent ionization losses, while above 1 GeV the spectrum steepens because of both the escape and radiative (synchrotron and inverse Compton) losses.

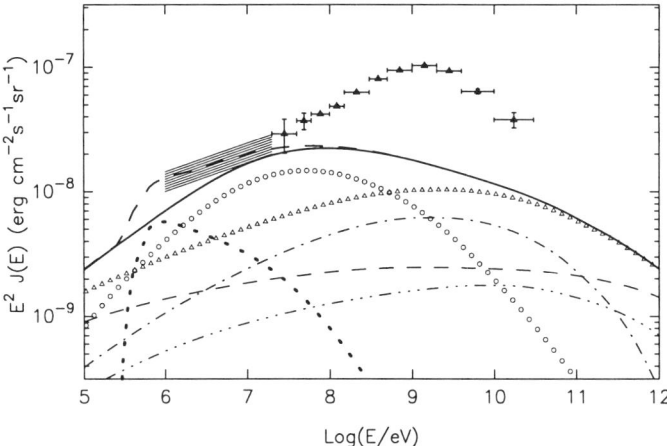

Fig. 4.18 The flux of diffuse γ-rays produced by CR electrons. The open dots show the bremsstrahlung flux, and the open triangles show the overall flux of the IC radiation components due to the 2.7 K CMBR (thin dashed line), diffuse NIR/optical radiation (dot–dashed line), diffuse FIR radiation (3-dot–dashed line). The heavy dotted line shows the flux of γ-rays due to annihilation of relativistic positrons in flight. The sum of the bremsstrahlung and IC fluxes is shown by the solid line. The heavy dashed line corresponds to the overall γ-ray flux including also the annihilation radiation. The data points show the mean flux of diffuse high energy γ-rays observed by EGRET, and the hatched region shows the range of average diffuse γ-ray fluxes detected by COMPTEL from the direction of the inner Galaxy at low galactic latitudes.

The fluxes of diffuse γ-rays produced by electrons are shown in Fig. 4.18. The spectrum of IC γ-rays below the highest energy observed by EGRET, $E \leq 30\,\mathrm{GeV}$, is not very sensitive to the exact value of the cutoff energy, E_0, in the injection spectrum of electrons, provided that E_0 exceeds several TeV. For the energy density of the diffuse interstellar NIR/optical radiation $w_\mathrm{NIR} = 1.5\,\mathrm{eV/cm^3}$ is assumed, for which the IC radiation component produced on galactic starlight photons (dot-dashed line) somewhat exceeds, at energies between 10 MeV amd 30 GeV, the IC flux from the up-scattered 2.7 K CMBR (dashed line). The 'FIR' component of IC radiation (3-dot–dashed line), calculated for $w_\mathrm{FIR} = 0.2\,\mathrm{eV/cm^3}$, in any energy band contributes less than 25% of the total IC flux.

In its turn, the overall IC γ-ray flux can account, for the chosen infrared photon field densities, for only $\leq 20\,\%$ of the γ-ray fluxes observed both at MeV ("COMPTEL") and GeV ("EGRET") energies (see Fig. 4.18). Formally, the IC fluxes could be increased assuming larger depths l_d of the emission region. However, the value $l_\mathrm{d} = 15\,\mathrm{kpc}$ assumed in Fig. 4.18 is already large, and could not realistically be increased further.

Another way to increase the flux of IC γ-rays would be to assume that the electron density in the inner Galaxy is significantly larger than $w_\mathrm{e} = 0.05\,\mathrm{eV/cm^3}$. The freedom here is however restricted by radio observations. Below a few GeV the spectral index of electrons is well determined, $\Gamma_\mathrm{e} = 2\,\alpha_\mathrm{r} - 1 \simeq 2.15$ where $\alpha_\mathrm{r} = 1.57 \pm 0.03$ is the photon index of the observed radio emission. Normalising the GeV electrons to the radio flux at $\nu = 10\,\mathrm{MHz}$, F_10MHz, we find a direct relation between the IC γ-ray and synchrotron radio fluxes:

$$F_\mathrm{IC}^\mathrm{(NIR)}(E) \simeq 2.6 \left(\frac{w_\mathrm{NIR}}{1\,\mathrm{eV/cm^3}}\right) \left(\frac{B}{6\,\mu\mathrm{G}}\right)^{-\frac{1+\Gamma_\mathrm{e}}{2}} \times$$

$$\left(\frac{\epsilon_0}{1\,\mathrm{eV}}\right)^{\frac{\Gamma_\mathrm{e}-3}{2}} \left(\frac{E}{5\,\mathrm{MeV}}\right)^{\frac{3-\Gamma_\mathrm{e}}{2}} F_\mathrm{10\,MHz}. \qquad (4.27)$$

For the radio flux at $\nu = 10\,\mathrm{MHz}$, $F_\nu \leq 4 \times 10^{-10}\,\mathrm{erg/cm^2\,s\,ster}$, a decrease of the magnetic field by a factor of two would lead to an increase of the electron flux shown in Fig. 4.17 by a factor of $2^{\alpha_\mathrm{r}} \sim 3$, and correspondingly to an increase of the IC γ-ray fluxes by the same factor. Thus an increase of the IC flux would also imply very high radio fluxes, exceeding by more than one order of magnitude the flux detected from the direction of the galactic poles, unless we assume unrealistically low magnetic field, $B \sim 1\,\mu\mathrm{G}$ in the galactic plane. Besides, this would also automatically increase the flux of the bremsstrahlung γ-rays, resulting in an overproduction of diffuse radiation in the 10 − 30 MeV region.

4.4.2.2 Electron bremsstrahlung

Below 1 GeV, bremsstrahlung γ-rays are produced by the same electrons responsible for the galactic synchrotron radio emission. Therefore the differential flux $J(E)$ of this radiation in the region from 30 MeV to ~ 1 GeV should have a characteristic power-law slope with an index coinciding with the spectral index of the radio electrons $\Gamma_e \simeq 2.1 - 2.2$. The results of numerical calculations are shown in Fig. 4.18 by open dots assuming a standard composition of the interstellar gas.

The comparison of the bremsstrahlung gamma- and radio synchrotron fluxes produced by the same 70 MeV $\leq E_e \leq 1$ GeV electrons gives:

$$F_{\rm brem}(30\,{\rm MeV}) \simeq 28 \frac{N_{\rm H}}{10^{22}{\rm cm}^{-2}} \left(\frac{l_{\rm d}}{10{\rm kpc}}\right)^{-1} \left(\frac{B}{6\mu{\rm G}}\right)^{-1.57} F_{10{\rm MHz}}, \quad (4.28)$$

where $F_{10{\rm MHz}}$ is the radio flux from the galactic plane. Since the latter cannot be less (and, presumably, is even several times higher) than the flux of $10^{-10}\,{\rm erg\,cm}^{-2}\,{\rm s}^{-1}\,{\rm sr}^{-1}$ detected from the direction of galactic poles (Webber et al. 1980), the contribution of bremsstrahlung to the overall flux of $E \sim 10$ MeV γ-rays should be significant, unless we assume either an unrealistically high magnetic field ($B \gg 10\,\mu{\rm G}$) and/or unrealistically low gas column density ($N_{\rm H} \ll 10^{22}\,{\rm cm}^{-2}$).

It is interesting to compare the bremsstrahlung flux in the energy region $E \sim 30$ MeV with the 'NIR' component of the IC flux in the region $E \sim 10$ MeV. Both radiation components are due the same electrons, although contributed by two different, low-energy and high-energy, parts of the power-law distribution of radio electrons, respectively:

$$\frac{F_{\rm brem}(30\,{\rm MeV})}{F_{\rm IC}^{\rm (NIR)}(10\,{\rm MeV})} \simeq 7.6\,\frac{N_{\rm H}}{10^{22}\,{\rm cm}^{-2}} \left(\frac{l_{\rm d}}{10\,{\rm kpc}}\right)^{-1} \left(\frac{w_{\rm NIR}}{1\,{\rm eV/cm^3}}\right)^{-1}. \quad (4.29)$$

In the region $E \leq 30$ MeV the bremsstrahlung flux is contributed by electrons with $E_e < 70$ MeV (i.e. outside the domain of the radio emitting electrons) where the ionization losses cause significant flattening of the electron spectra. This results in a drop of $F_{\rm brem}$ at 10 MeV by a factor of 1.5 compared to $F_{\rm brem}(30\,{\rm MeV})$. For $w_{\rm NIR} \sim 1\,{\rm eV/cm^3}$, the overall IC flux at 10 MeV is comparably contributed by both NIR and 2.7 K CMBR target photons. Taking all these effects into account, we may conclude that around $E \sim 10\,\mathrm{MeV}$ bremsstrahlung should dominate the overall diffuse emission of the galactic plane. This is confirmed by the accurate numerical calculations presented in Fig. 4.18.

4.4.2.3 Annihilation of CR positrons in flight

A non-negligible fraction of the diffuse γ-radiation below 10 MeV could be due to annihilation of relativistic positrons in flight (Aharonian and Atoyan, 1981a; Aharonian et al., 1983b). The differential spectrum of γ-rays produced in the annihilation of a fast positron on electrons of the ambient thermal gas with density n_e is described by Eq.(3.6). For a power-law distribution of positrons, the spectrum of annihilation radiation at energies $E \gg m_e c^2$ has an almost power-law behaviour given by Eq.(3.8). The annihilation to bremsstrahlung flux ratio $J_{\mathrm{ann}}/J_{\mathrm{brem}}$ does not depend on the ambient gas density and, for given content of positrons, $\kappa_+ = e^+/(e^+ - e^-)$, depends only on the energy distribution of electrons.

The flux of annihilation radiation calculated for the electron spectrum from Fig. 4.17 assuming $\kappa_+ = 0.5$, is shown in Fig. 4.18. It is seen that the contribution of the annihilation radiation at 1 MeV exceeds the contributions of other radiation processes. Thus, depending on the (unknown) content of low-energy ($\leq 10\,\mathrm{MeV}$) positrons in CRs, this process may result in a significant enhancement of the diffuse radiation at MeV energies. In this regard it is interesting to note that at $E_e < 1\,\mathrm{GeV}$ the fraction of positrons in the local (directly measured) component of CR electrons gradually increases, reaching $\kappa_+ \geq 0.3$ at $E_e \sim 100\,\mathrm{MeV}$ (Fanslow et al., 1969).

Generally, the assumption that the annihilation of supra-thermal positrons in flight may significantly contribute to the diffuse low-energy γ-radiation of the inner Galaxy would imply also a high flux of 0.511 MeV annihilation line radiation. The OSSE measurements (Purcell et al., 1993) indicate that the positron annihilation in the ISM proceeds predominantly ($\simeq 97\%$, Kinzer et al., 1996) through the positronium state, the annihilation radiation of which consists of two components - the narrow 0.511 annihilation line and the γ-ray continuum below 0.5 MeV. At energies $E \sim (0.2 - 0.5)\,\mathrm{MeV}$ the diffuse γ-ray emission of the inner Galaxy is contributed mainly by the fluxes of these two components of the annihilation radiation (not shown in Fig. 4.18). The total flux of these photons is at the level $\simeq 10^{-3}\,\mathrm{cm}^{-2}\mathrm{s}^{-1}$ (see Kinzer et al. 1996), which is equivalent to $\simeq 2.4 \times 10^{-2}\,\mathrm{cm}^{-2}\,\mathrm{s}^{-1}\mathrm{sr}^{-1}$ for the $3.8° \times 11.4°$ field of view of the OSSE instrument.

Approximately 20 per cent of relativistic positrons annihilate in flight in the same medium where they cool due to Coulomb and bremsstrahlung energy losses (Aharonian and Atoyan, 1981a). The integrated flux of the annihilation radiation by relativistic electrons shown in Fig. 4.18 (dotted curve), is $\simeq 6.7 \times 10^{-3}\,\mathrm{cm}^{-2}\,\mathrm{s}^{-1}\,\mathrm{sr}^{-1}$. Therefore one could expect that the photon flux of annihilation radiation of thermalized positrons should be as high as $3 \times 10^{-2}\,\mathrm{cm}^{-2}\,\mathrm{s}^{-1}\,\mathrm{sr}^{-1}$, i.e. quite comparable with the OSEE observations. And, *vice versa*, if the observed diffuse annihilation line at 0.511 MeV is due to positrons injected in the ISM with relatively high, $E \gg 1$ MeV, energies (e.g. by pulsars), we should expect collateral continuous MeV radiation from positrons (before their thermalization) at a level comparable with the fluxes shown in Fig. 4.18. If the positrons are produced with low energies, $E \leq 1$ MeV, e.g. they arise from β-decay of radioactive nuclei, the continuous MeV radiation obviously cannot have an annihilation origin.

The overall fluxes of diffuse γ-rays of electronic origin without and with the annihilation radiation are shown in Fig. 4.18 by heavy solid and dashed curves, respectively. It is seen that the γ-radiation produced by CR electrons is significantly below the measured fluxes in the entire energy range from 100 MeV to 30 GeV. Although, due to the lack of independent information on the spectrum of high energy electrons, the predicted γ-ray fluxes contain significant uncertainties, the gap by a factor of 5 to 7 around 1 GeV is not easy to fill by any reasonable set of model parameters. Moreover, the flat shape of the overall flux produced by CR electrons cannot explain the "GeV bump" without violation of the fluxes observed at lower energies. More naturally, this bump can be related to interactions of CR protons and nuclei with the interstellar gas.

4.4.3 *Gamma rays of nucleonic origin*

The inelastic collisions of CR protons and nuclei with the interstellar gas result in γ-ray emission through production and subsequent decay of secondary π^0-mesons. For a power-law energy distribution of protons, the differential production spectrum of this radiation $J(E)$ has a distinct maximum at 100 MeV, which in the $\nu F_\nu = E^2 J(E)$ presentation is shifted to ≈ 1 GeV, thus shaping the so-called "GeV bump" in the spectral energy distribution (SED), provided that the spectrum of protons is steeper than E^{-2}.

In Fig. 4.19a the SED of π^0-decay γ-rays calculated for power-law CR spectra are shown, assuming for the product of the CR energy density and the hydrogen column density $<w_\mathrm{p}\, N_\mathrm{H}> = 2.5 \times 10^{22}\,\mathrm{eV/cm^5}$. The fluxes produced by protons are multiplied by the factor $\eta_A = 1.5$ which takes into account the overall contribution from nuclei both in CRs and the ISM.

The dashed curve in Fig. 4.19a is the diffuse γ-ray flux calculated for the local CR proton flux which is described by a power-law energy distribution (e.g. Simpson, 1983) with spectral index $\Gamma = 2.75$:

$$J_\odot(E) = 2.2\, E_\mathrm{GeV}^{-2.75}\,\mathrm{cm^{-2}s^{-1}sr^{-1}GeV^{-1}}\,. \qquad (4.30)$$

Above 1 GeV, the calculated γ-ray spectrum fails to explain the diffuse flux observed γ-ray flux by a factor of 1.5-2. This deficit cannot be overcome by increasing w_p or N_H, otherwise the flux at sub-GeV energies would be over-predicted by the same factor. The assumption of a hard power law index for CRs in the Galactic Disk, $\Gamma_\mathrm{p} = 2.3$ (dot-dashed line), explains the data at ~ 1 GeV, but over-predicts the flux at higher energies. And finally, a moderately steep spectrum of protons with $\Gamma_\mathrm{p} \simeq 2.5$ (solid line) satisfactorily explains the spectral shape of the observed excess around several GeV, except for the point at 2 GeV which appears $\approx 20\%$ higher.

There is another, perhaps more attractive, possibility providing a somewhat better fit for the GeV γ-ray spectrum, namely assuming the following

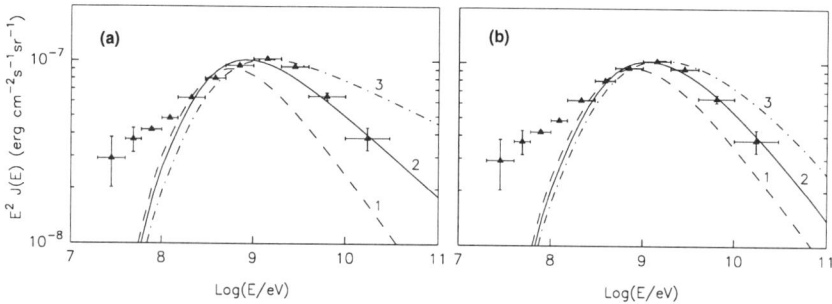

Fig. 4.19 Spectra of π^0-decay γ-rays. The CR proton energy distributions are given: (a) (left) by a single power-law with 3 different indices Γ_p: 2.75 (curve 1), 2.5 (curve 2), and 2.3 (curve 3); and (b) (right figure) in the form of Eq.(4.31) with $\Gamma_0 = 2.1$, $\delta = 0.65$ and three different values for E_*: 3 GeV (curve 1), 20 GeV (curve 2), and 100 GeV (curve 3). The fluxes are calculated assuming $\eta_A = 1.5$, and $<w_\mathrm{p}\, N_\mathrm{H}> = 2.5 \times 10^{22}\,\mathrm{eV/cm^5}$ (a), and $<w_\mathrm{p}\, N_\mathrm{H}> = 2.1 \times 10^{22}\,\mathrm{eV/cm^5}$ (b). The data points are the fluxes detected by EGRET from the inner Galaxy.

energy distribution of protons in the inner Galaxy:

$$n_{\rm p}(E_{\rm p}) \propto E_{\rm p}^{-\Gamma_0}\left[1+\left(\frac{E_{\rm p}}{E_*}\right)^{\delta}\right]^{-1}. \qquad (4.31)$$

This equation implies a power-law spectrum of protons with $\Gamma_{\rm p} \approx \Gamma_0$ below some E_*, but gradual steepens to $\Gamma_{\rm p} \approx \Gamma_0+\delta$ at higher energies, $E \gg E_*$. Actually, a proton energy distribution of this kind can be naturally formed in the ISM. Indeed, the energy distribution of the protons, which do not suffer significant energy losses, can be approximated by $n_{\rm p}(E_{\rm p}) \simeq q_{\rm p}(E_{\rm p}) \cdot \tau_{\rm esc}^{-1}(E_{\rm p})$. For a power-law acceleration spectrum with Γ_0, this is easily reduced to the form of Eq.(4.31) if the diffusive escape time of CRs becomes comparable with the convective escape time at $E_{\rm p} = E_*$ (see Eq. 4.22). The results of numerical calculations presented in Fig. 4.19b show that the GeV bump in the observed γ-ray spectrum can be nicely explained by a hard spectrum of accelerated protons with $\Gamma_0 \simeq 2.1$, additionally assuming an energy-dependent ($\delta = 0.65$) diffusive escape of particles from the Galactic Disk which leads to the steepening of the CR spectrum above $E_* = 20\,\mathrm{GeV}$.

4.4.4 Overall gamma ray fluxes

The radiation mechanisms discussed above significantly contribute to one or more energy intervals of the broad-band diffuse γ-ray emission of the Galactic Disk. Therefore, any attempt to interpret the observed diffuse γ-radiation, even in a specifically limited spectral band, cannot be treated separately, but rather should be conducted within a multi-wavelength approach to the problem.

Fig. 4.20 demonstrates that the γ-ray data from $\sim 1\,\mathrm{MeV}$ to $30\,\mathrm{GeV}$ can be adequately explained with a set of quite reasonable parameters characterising both the cosmic rays and the interstellar medium. For the calculations a mean line-of-sight depth for the inner Galactic Disk of $l_{\rm d} = 15\,\mathrm{kpc}$ and a column density of $N_{\rm H} = 1.5 \times 10^{22}\,\mathrm{cm}^{-3}$ are assumed. This corresponds to a mean gas density along the line of sight at low galactic latitudes of about $n_{\rm H} = N_{\rm H}/l_{\rm d} \simeq 0.33\,\mathrm{cm}^{-3}$. The energy densities of CR protons and electrons in the inner Galaxy are normalised to $w_{\rm p} = 1\,\mathrm{eV/cm^3}$ and $w_{\rm e} = 0.075\,\mathrm{eV/cm^3}$, respectively. Interestingly, the latter is larger by a factor of 1.5 than the energy density of local CR electrons, while the required energy density of protons is $\approx 20\,\%$ less than the observed one. It should be noticed, however, that these estimates are inversely proportional to the hydrogen column density $N_{\rm H}$, the uncertainty of which exceeds the above deviations, and therefore does not allow the derivation of the the average density of CRs in the Galactic Disk to an accuracy of better than $\approx 50\%$. At the same time, the required spectral shape of protons in the inner Galaxy at low energies deviates noticeably from the local proton spectrum given by Eq.(4.30).

In Fig. 4.20, injection spectra of electrons and protons are assumed with power-law index $\Gamma = 2.1$. For the adopted escape times of $\tau_{\rm diff} = 1.4 \times 10^7 (E/10\,{\rm GeV})^{-0.65}$ yr and $\tau_{\rm conv} = 2 \times 10^7$ yr, the diffusive escape time becomes equal to the time of convective escape at $E = 5.8\,{\rm GeV}$. Note that the decrease of E_*, as compared with the 'best fit' value $E_* = 20\,{\rm GeV}$ in Fig. 4.19b, is necessitated by the significant contribution of the IC component to the overall flux (especially at high energies). This demonstrates the importance of the multi-wavelength approach to the modelling of the

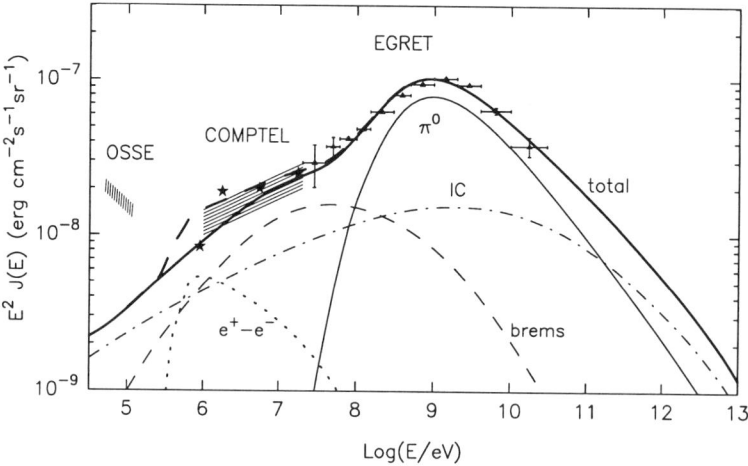

Fig. 4.20 The fluxes of diffuse radiation produced by electronic and nucleonic components of cosmic rays in the inner Galaxy. The calculations are performed for hard power-law injection spectra of electrons and protons with $\Gamma = 2.1$, and for the following parameters characterising the CR propagation: $\tau_{\rm diff} = 1.4 \times 10^7 (E/10\,{\rm GeV})^{-0.65}$ yr, $\tau_{\rm conv} = 2 \times 10^7$ yr. Other model parameters are: $w_{\rm e} = 0.075\,{\rm eV/cm^3}$, $w_{\rm p} = 1\,{\rm eV/cm^3}$, $N_{\rm H} = 1.5 \times 10^{22}\,{\rm cm^{-2}}$, $l_{\rm d} = 15\,{\rm kpc}$, $B = 6\,\mu{\rm G}$. The contributions from π^0-decay (thin solid line), bremsstrahlung (dashed), the total inverse Compton (dot-dashed) and positron annihilation in flight for $\kappa_+ = 0.5$ (dotted line) are shown. The heavy dashed and solid lines represent the overall fluxes with and without the contribution from the positron annihilation radiation, respectively.

diffuse γ-radiation even for specific, relatively narrow, energy bands.

Below 100 MeV the observed γ-ray fluxes are due mainly to the IC and bremsstrahlung components. At energies below several MeV the annihilation radiation of relativistic electrons contributes significantly to the overall flux, and may even exceed the individual fluxes of both bremsstrahlung and IC γ-rays, provided that at energies below 100 MeV the positron content in the electronic component of the galactic cosmic rays is significant. The heavy dashed curve in Fig. 4.20 shows the overall ("IC+bremsstrahlung+annihilation") flux. For comparison, the COMP-TEL data points (Hunter et al., 1997b) corrected for contamination caused by annihilation of thermalised positrons (Purcell et al., 1993), are shown by stars. One can see that the annihilation of positrons "in flight", on top of the IC and bremsstrahlung fluxes, fits the COMPTEL measurements rather well.

Among the principal radiation mechanisms of diffuse γ-rays in the region $E \leq 10$ MeV, the prompt γ-ray line emission produced by sub-relativistic cosmic rays (SRCRs) via nuclear de-excitation should also be mentioned. The emissivity of the total (unresolved) γ-ray line emission in the energy range between several hundred keV and several MeV, normalised to the energy density of SRCRs $w_{\rm scr} = 1\,{\rm eV/cm^3}$, and calculated for the standard cosmic compositions of CRs and the interstellar gas, is about 2×10^{-25} ph/s H-atom (Ramaty et al., 1979). This implies that for $N_{\rm H} = 1.5 \cdot 10^{22}\,{\rm cm}^{-2}$ and $w_{\rm scr} \leq 1\,{\rm eV/cm^3}$, the energy flux of this component of gamma radiation cannot exceed 10^{-9} erg/cm^{2}s sr. Thus, this radiation mechanism cannot be responsible for more than several per cent of the observed γ-ray flux at MeV energies, unless the energy density of SRCRs in the ISM exceeds by 1.5-2 orders of magnitude the "nominal" energy density of CRs in the relativistic regime of about 1 eV/cm^3. Although quite speculative, such high densities of SRCRs cannot be ruled out. Moreover, recently Boldt (1999) and Dogiel et al. (2002) argued that the hard X-ray emission from the galactic ridge, which likely has non-thermal origin, is produced by bremsstrahlung of sub-relativistic protons or electrons. The bremsstrahlung of sub-relativistic particles, both of protons or electrons, in the cold medium is a rather inefficient mechanism of radiation. Namely, because of ionization losses only $\leq 10^{-5}$ part of the kinetic energy is released in the form of non-thermal X-rays (Aharonian et al., 1979; Skibo et al., 1996; Dogiel et al., 2002). Therefore the "sub-relativistic bremsstrahlung" models require continuous injection of low-energy electrons and/or protons, e.g. by SNRs, into the ISM with uncomfortably large rates if one tries to explain the fluxes of nonthermal X-rays detected by the GINGA, RXTE and ASCA satellites. Assuming that bremsstrahlung X-rays are produced directly in the regions of particle acceleration with plasma temperatures of about several hundred eV, one may somewhat reduce the requirement to the energy output of quasi-thermal electrons, which however still remains uncomfortably high, at the level of 10^{41} erg/s (Dogiel et al., 2002).

The production of X-rays through proton bremsstrahlung is tightly connected with the prompt γ-ray line emission due to the excitation of the nuclei (primarily Fe, C, O, etc.) of the of ambient gas by the same subrelativistic protons (Aharonian et al., 1979). This allows robust upper limits on the contribution of proton bremsstrahlung to X-rays from the fluxes of diffuse γ-rays observed at MeV energies. The studies by Pohl (1998) and Valinia et al. (2000), based on the comparison of the observed keV and MeV fluxes, lead to the conclusion that indeed the proton bremsstrahlung alone could not be responsible for the bulk of the diffuse galactic X-ray emission. And finally, the hard X-ray fluxes extending to 200 keV cannot be explained by proton bremsstrahlung, because the energy of protons producing 200 keV X-rays, $E \geq (m_p/m_e)E_X \simeq 400$ MeV would over-produce π^0-decay γ-rays.

Synchrotron radiation of ultra-relativistic electrons is an alternative mechanism to explain the diffuse X-ray emission of the Galactic Disk (Porter and Protheroe, 1997; Aharonian and Atoyan, 2000). An obvious advantage of this mechanism, compared with the bremsstrahlung of subrelativistic particles, is its almost 100% efficiency of transformation of the particle kinetic energy into X-rays. On the other hand, this mechanism requires, for any reasonable ambient magnetic field, more than 100 TeV electrons in the ISM. Since the rates of acceleration by SNR shocks are not sufficient to compensate for the severe synchrotron losses, these electrons can hardly be produced in SNRs. Pulsar-driven nebulae (plerions) seem more probable sites for the acceleration of such energetic electrons through the wind termination shocks. Interestingly, since the same electrons also produce, through IC scattering on the 2.7 K CMBR, ultra-high energy γ-rays, the magnetic field in the regions of production of hard synchrotron X-rays should be quite large, $B \geq 20$ μG (Aharonian and Atoyan, 2000), otherwise the IC fluxes would exceed the flux upper limits at energies $E \geq 100$ to 1000 TeV reported by the by CASA-MIA (Borione et al., 1998) and KASCADE (G. Schatz, private communication) collaborations (see Fig. 4.21).

The life-time of $E_e \gg 100$ TeV electrons does not exceed several hundred years, therefore they cannot propagate more than a few tens of parsecs from their acceleration sites. This means that the diffuse X-ray background, as well as the accompanying IC radiation should have a "cell - structure", i.e. they consist of *superpositions* of contributions from a large number of unresolved *extended* sources along the line of sight (Aharonian, 1995; Pohl and Esposito, 1998).

The broad-band diffuse radiation flux shown in Fig. 4.21 is calculated in the framework of a model which assumes that besides the main population of CRs (presumably accelerated by the SNR shocks), there is also a second electron population beyond 100 TeV (accelerated presumably by the pulsar wind termination shocks). The acceleration power in the second electron component, which is needed to explain the hard X-ray background, is about 6×10^{36} erg/s per kpc^3, or $L_e^{(II)} \simeq 3 \times 10^{39}$ erg/s in the entire inner Galactic Disk. If these electrons are associated with pulsar winds, for $\approx 10^4$ sources this would imply a rather modest mean acceleration power per "old pulsar" of about 3×10^{35} erg/s, i.e. three orders of magnitude less than the power of the relativistic electron-positron wind of the Crab pulsar. The kick velocities of pulsars can be of order from a few 100 to ~ 1000 km/s, thus the 10^6 yr old pulsars would be able to propagate to distances $\leq (0.3-1)$ kpc, contributing therefore to the emission at galactic latitudes of up to several degrees. If so, we may expect an interesting effect of a *gradual spectral hardening* of radiation arriving from higher galactic latitudes. The detection of this flat IC component, which at high latitudes may dominate, at least at TeV energies, over the significantly suppressed π^0-decay radiation, should be possible by GLAST and forthcoming IACT arrays.

4.4.5 *Probing the diffuse γ-ray background on small scales*

The diffuse galactic gamma-ray background consists of the truly diffuse emission that is produced in the interaction of CRs with the interstellar gas and photon fields, and of contributions from unresolved discrete sources. Actually, if the "sea" of GCRs is described by the steep, $E^{-2.75}$ type spectrum of local CRs (curve 1 Fig. 4.19a), a new component of radiation would be required to explain the so-called "GeV excess" detected by EGRET[1]. Pohl *et al.* (1997) have noticed that pulsars can account for up to 20 per cent of the diffuse emission above 1 GeV in selected regions of the ISM, albeit that they cannot be responsible for all the GeV excess.

---

[1] The "GeV excess" should be understood as a discrepancy between the observed flux and model predictions for the spectrum of local CRs observed in the solar vicinity.

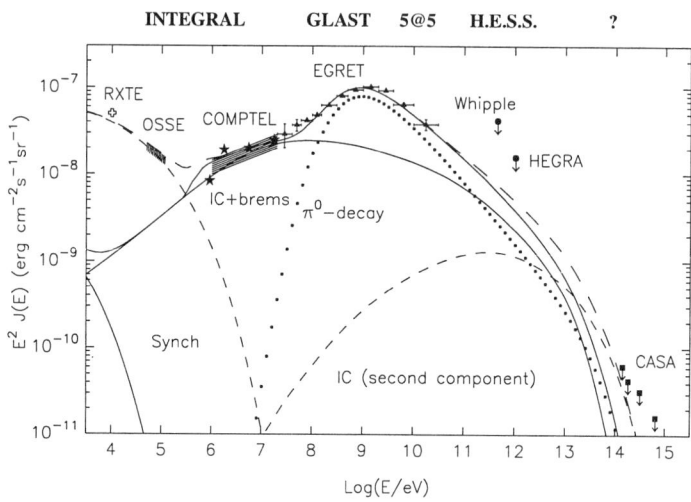

Fig. 4.21 The broad-band diffuse background radiation from the Galactic plane in terms of the two-component model of relativistic electrons. The heavy solid line corresponds to the flux produced by the electrons of the first (main) CR population, with the same model parameters as in Fig. 4.20, but for $\delta = 0.7$ and $\kappa_+ = 0.3$. The heavy dashed line shows the overall flux including the contribution from the second population of electrons accelerated to energies $E_0 = 250\,\mathrm{TeV}$. The local mean magnetic field of the region where the second electron population is confined is assumed $B = 25\,\mu\mathrm{G}$. The fluxes of galactic diffuse radiation at γ-rays detected by COMPTEL and EGRET, and at X-rays detected by RXTE (Valinia and Marshall, 1998) and OSSE (Kinzer et al., 1997), as well as the upper flux limits at very high energies reported by the Whipple (Le Bohec et al., 2000), HEGRA (Aharonian et al., 2001) and CASA-MIA (Borione et al., 1998) collaborations, are also shown.

Berezhko and Völk (2000) argued that a significant fraction of the diffuse radiation from inner Galaxy can be contributed by unresolved SNRs, assuming that approximately 10 SNRs of an age younger than 10^5 yr can on average lie within $1°$ of the galactic center. This model predicts a hard γ-ray spectrum up to 1 TeV energies with an absolute flux around 1 TeV of about $10^{-8}\,\mathrm{erg/cm^2 s\,sr}$, i.e. a factor of 2 higher than the fluxes of the truly diffuse radiation shown in Fig. 4.20 and 4.21. This flux from the direction of the galactic center can be readily detected by the H.E.S.S. IACT array. Important support for this model would be the resolution of individual contributors and their identification with known SNRs. Since this radiation is supposed to be a superposition of contributions from ≤ 10 SNRs within a $1°$ field of view, the fluxes of individual contributors are expected at the level of $10^{-12}\,\mathrm{erg/cm^2 s}$ at 1 TeV and a factor of 3 higher at several GeV.

In principle, these sources can be resolved both at GeV and TeV energies by GLAST and H.E.S.S., respectively.

A large contribution to the diffuse background at low galactic latitudes may come from "active" molecular clouds. The propagation of CRs in the Galactic Disk implies an effective mixture of contributions from individual sources/accelerators of CRs on a $\sim 10^7$ yr timescale. Therefore one cannot expect a strong gradient of CR density on large, kpc scales. However, strong variations are possible on smaller scales, in particular in the $l \leq 100$ pc regions around the CR accelerators, where the CR density may significantly exceed the average density of the "sea" of GCRs. If these regions also host massive gas clouds, we may expect enhanced γ-ray emission within $l/d \sim 0.5°(d/10 \text{ kpc})^{-1}$. Speculating now that a significant fraction of the diffuse γ-ray background is produced selectively, being a result of radiation from regions containing both particle accelerators and massive gas clouds, one may explain in a quite natural way the hard spectra of radiation observed by EGRET above 1 GeV. Indeed, if γ-rays are produced at interactions of clouds with relatively fresh (recently accelerated) particles with spectra that have not yet suffered strong modulation (steepening) due to propagation effects, the resulting γ-ray spectra should be significantly harder than the typical γ-ray spectrum produced by the "sea" of GCRs in regions of the ISM far from the cosmic accelerators. Such an assumption agrees with the correlation observed between γ-ray intensity and the hydrogen column density (Hunter et al., 1997) which in the galactic plane is predominantly due to molecular clouds.

If a significant fraction of the diffuse γ-ray background is indeed contributed by regions of enhanced CR density surrounding the particle accelerators, then we should expect non-negligible fluctuations, both in the spectral shape and in the absolute flux on scales less than 1°. In Fig. 4.22a the diffuse γ-ray fluxes detected by EGRET within a 10^{-3} sr solid angle, corresponding to regions on sky with angular radius $\approx 1°$, are shown. The largest flux, marked as "GD(C)", corresponds to the radiation that arrives from direction of the Galactic Center, namely to the average flux detected from the region with galactic coordinates $l \sim 0°$ and $2° \leq b \leq 6°$ (Hunter et al., 1997). It is approximated as

$$\left(\frac{dN}{dE}\right)_{GC} \approx 1.7 \times 10^{-7} E^{-2.5} (\Delta\Omega/10^{-3} \text{sr}) \text{ ph/cm}^2\text{s GeV} . \quad (4.32)$$

The curve marked as "GD(A)" in Fig. 4.22a corresponds to the flux from the Anti-center region, namely $l \sim 180°$ and $6° \leq b \leq 10°$ (Hunter et al., 1997). It is approximated as

$$\left(\frac{dN}{dE}\right)_{GA} \approx 6.1 \times 10^{-9} E^{-2.33} (\Delta\Omega/10^{-3} \mathrm{sr}) \; \mathrm{ph/cm^2 s \; GeV} \; . \qquad (4.33)$$

Note that the EGRET measurements of the diffuse galactic radiation do not extend beyond 30 GeV, but in Fig. 4.22a we extrapolate the fluxes up to 300 GeV. Interestingly, the highest energy γ-rays detected by EGRET around 300 GeV belong to the isotropic (extragalactic) diffuse background radiation (Sreekumar et al., 1998). This component, marked in Fig. 4.22 as "EGB", is described by a hard power-law spectrum,

$$\left(\frac{dN}{dE}\right)_{EGB} \approx 1.1 \times 10^{-9} E^{-2.15} (\Delta\Omega/10^{-3} \mathrm{sr}) \; \mathrm{ph/cm^2 s \; GeV} \; . \qquad (4.34)$$

For comparison, in Fig. 4.22a π^0-decay γ-ray fluxes of individual supernova remnants and "passive" molecular clouds are also shown. For the differential flux of a "typical" γ-ray emitting SNR, the following approximation is used (see Sec. 5.1)

$$\left(\frac{dN}{dE}\right)_{SNR} \approx 10^{-8} E^{-2} \, A_{SNR} \; \mathrm{ph/cm^2 s \; GeV} \; . \qquad (4.35)$$

where $A_{SNR} = W_{50} n_0 / d_{kpc}^2$; $W_{50} = W_p/10^{50}$ erg is the total energy of accelerated CRs contained in the remnant, $n_0 = n/1 \; \mathrm{cm}^{-3}$ is the gas density in the remnant, and $d_{kpc} = d/1$ kpc is the distance to the source.

The differential flux of γ-rays from a "passive" GMC (see Sec. 4.2) at energies above 1 GeV is approximated as

$$\left(\frac{dN}{dE}\right)_{GMC} \approx 4.7 \times 10^{-8} E^{-2.75} \, A_{GMC} \; \mathrm{ph/cm^2 s \; GeV} \; . \qquad (4.36)$$

where $A_{GMC} = M_5/d_{kpc}^2$; $M_5 = M/10^5 M_\odot$ is the total mass of the cloud.

And finally, in Fig. 4.22a we present the spectrum of CR ray electrons observed in the solar vicinity. Although the intensity of these particles exceeds by two orders of magnitude or more the flux of diffuse γ-rays, the active anti-coincidence shields of the satellite based detectors effectively protect the γ-ray telescopes from charged particles. At the same time the showers produced by CR electrons constitute the main background for imaging Cherenkov telescope arrays operating in the sub-100 GeV regime.

The minimum size of angular cells chosen for the the study of the spatial variation of the spectrum of diffuse γ-rays is determined by the angular resolution of detectors and the accumulated photon statistics. Because of the small detection area and limited angular resolution, the spatially resolved spectral analysis of EGRET has been performed using 10° × 4° bins with ≤ 40 per cent statistical errors and a highest interval of 1–30 GeV (Hunter, 2001). GLAST, with significantly improved angular resolution and larger detection area, should be able to derive γ-ray spectra for smaller bins and upper energies of 100 GeV. Fig. 4.23 shows the number of γ-rays above given energy E per degree^2 expected for 1 year of scanning mode operation by GLAST (Hunter, 2001). It is seen that GLAST has the potential to derive spectra up to 80 GeV and 15 GeV for 4 deg^2 bins from the directions of the galactic center and anti-center, respectively, if one requires at least 10 detected photons. At the same time, GLAST should be able to image the γ-ray emission towards the galactic center above 1 GeV on much smaller scales, ≤ 1 deg^2 (down to 0.3° × 0.3° limited by angular resolution at 1 GeV), albeit without adequate spectral information.

The extension of spectral studies of the diffuse background beyond 10

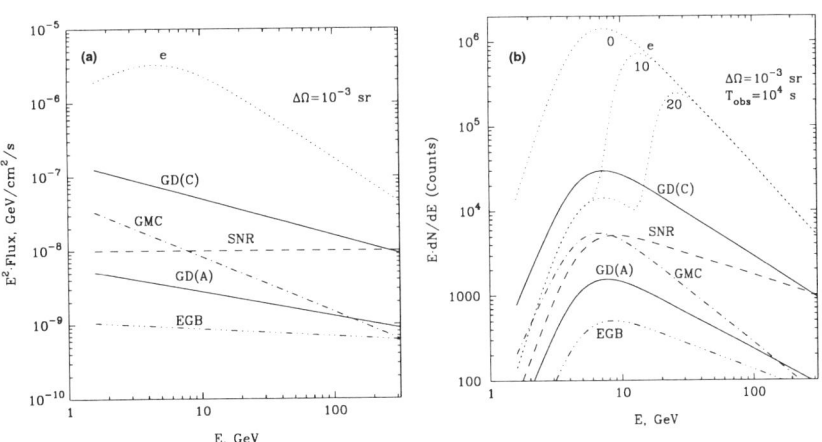

Fig. 4.22 (a) Fluxes of diffuse galactic γ-ray emission within a 10^{-3} sr solid angle detected by EGRET towards the galactic center, GD(C), and anti-center, GD(A). For comparison, the flux of the (isotropic) extragalactic background radiation (EGB) detected by EGRET up to 300 GeV is also shown. The curves marked "SNR" and "GMC" represent the fluxes of a "standard" SNR with the scaling-factor $A_{\rm SNR} = 1$ (see Eq.(4.35)) and from a "standard" passive molecular cloud with the scaling-factor $A_{\rm GMC} = 1$ (see Eq.(4.36)), respectively. The flux of CR electrons observed in the solar vicinity is also shown. (b) The differential counts of radiation components presented in figure (a) expected after a $T_{\rm obs} = 10^4$ s observation with the high altitude IACT array 5@5. The detection rates of CR electrons are shown for 3 different values of the geomagnetic cutoff: 0, 10, and 20 GeV. (From Aharonian and Bogovalov, 2003).

GeV with $\leq 1/3°$ resolution is of great interest, because these measurements would allow us to trace propagation of CRs on scales less than 100 pc even for very distant ($d \geq 10$ kpc) regions of the Galactic Disk. Because of the limited detection area, the potential of GLAST is still limited, especially for spectroscopic measurements from regions outside of the inner Galaxy and at large galactic latitudes. Planned imaging atmospheric Cherenkov detectors like CANGAROO-III, H.E.S.S. and VERITAS will have very good angular resolutions of about $0.1°$ and detection areas $10^4 - 10^5$ m^2. Even so, above the energy threshold of these instruments around 100 GeV, the γ-ray fluxes drop significantly, thus spectroscopic studies in this energy regime will be limited by the low photon fluxes. In the energy region from several GeV to several 100 GeV an adequate potential for studies of the diffuse galactic background may require future low-threshold IACT arrays like 5@5 installed at very high mountain altitudes (see Sec. 2.3.3). A challenge for such instruments would be the spectroscopy of the diffuse γ-radiation on $1/3°$ or perhaps even smaller angular scales. The number of γ-rays detected by 5@5 at given energy E with $\Delta E = E$ (i.e. $E(\mathrm{d}N/\mathrm{d}E)$) within 10^{-3} sr solid angle and for an exposure time 10^4 s, are shown in Fig. 4.22b (Aharonian and Bogovalov, 2003). It is seen that the number of detected γ-rays in each $0.3° \times 0.3°$ bin ($\Delta\Omega \simeq 10^{-4}$ sr) toward the galactic center could be as large as 10^3 at 30 GeV and 250 at 100 GeV. Although the detection rate of CR electrons which constitute the main source of background for 5@5, is an order of magnitude higher, the signal-to-noise ratio, $S/\sqrt{B}$ is sufficiently high for the detection of statistically significant signals for just

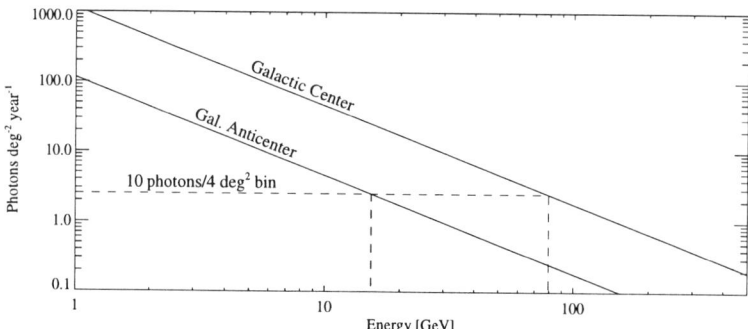

Fig. 4.23 The number of γ-rays above an energy E expected per 1 deg^2 from the directions of the galactic center and anti-center, after 1 year of operation of GLAST in the scanning mode (from Hunter, 2001).

several hours observations by this instrument. For the effective field of view of about 3 degree (in diameter), these observations may allow a *simultaneous* probe of 25 "$0.3° \times 0.3°$" bins. This should be sufficient to resolve point-like sources with luminosities several times $10^{33}(d/10 \text{ kpc})^2$ erg/s, and thus to separate the flux of truly diffuse γ-ray emission from contributions of discrete sources. Note that the $1/3°$ angular scale corresponds to $50(d/10 \text{ kpc})^2$ pc linear size, thus these observations may reveal the sites of CR production in the Galaxy even when the accelerators themselves are not visible in γ-rays. Remarkably, this could be a quite widespread case, for example due to the lack of adequate targets in accelerators and/or fast escape of accelerated particles from their production sites. In any case, the study of ≤ 100 pc proximity of CR accelerators should allow correct estimates of the overall energy released in accelerated relativistic particles and an important study of CR propagation effects (e.g. the diffusion coefficient) in the (presumably) very active and turbulent regions surrounding the CR accelerators.

4.4.6 *Concluding remarks*

The diffuse galactic γ-ray emission carries unique information, a proper understanding of which should eventually result in a quantitative theory for the origin of galactic cosmic rays. The problem is complicated and confused by the operation of several competing γ-ray production mechanisms. Although the data obtained by the COMPTEL and EGRET detectors aboard the Compton GRO allow rather definite conclusions concerning the relative contributions of different γ-ray production mechanisms in the energy region from 1 MeV to 30 GeV, many details concerning both the spatial distribution of CRs and their energy spectra in different parts of the Galactic Disk remain unresolved. The next generation of space- and ground-based γ-ray detectors with significantly improved performance should be able to provide a framework for a coherent understanding of many aspects of the origin and propagation of galactic cosmic rays.

Gamma-rays below 100 MeV. In this energy region the diffuse γ-radiation has an electronic origin, namely the observed γ-ray fluxes from 10 MeV to 100 MeV can be well explained by superposition of the bremsstrahlung and IC components of radiation . At lower energies, a non-negligible flux can be contributed by the annihilation of relativistic positrons with the ambient thermal electrons, and perhaps also by superposition of prompt γ-ray lines from interaction of sub-relativistic protons and nuclei with the ambient interstellar gas. The detailed spectroscopic studies in the region between several 100 keV and 10 MeV by detectors of the INTEGRAL mission could reveal the spectral features associated with these radiation channels. The detection of prompt γ-ray lines would allow determination of presently highly uncertain flux of sub-relativistic CRs in the ISM, and thus should give a definite estimate of the proton bremsstrahlung contribution to the nonthermal hard X-ray emission of the galactic ridge. The direct measurements of hard X-rays, in particular the mapping and spectroscopy of radiation from 20 keV to several 100 keV by the detectors of the INTEGRAL mission may distinguish between two models relating this radiation to the bremsstrahlung of sub-relativistic particles (electrons and/or protons) and to the synchrotron radiation of extremely high energy electrons. The second model predicts also IC γ-radiation extending well beyond 1 TeV at the flux level marginally detectable by the next generation of IACT arrays. The detailed spectroscopic studies of diffuse radiation by GLAST, in particular the identification of spectral features in the "electron-dominated" to the "nucleon-dominated" transition region around 100 MeV, should allow determination one of the crucial parameters characterising the production and propagation of low-energy CRs, the ratio of the CR electron density to the proton density in the Galactic Disk.

The "GeV bump" and beyond. Above 100 MeV, π^0-decay γ-rays dominate the diffuse radiation at low galactic latitudes for any reasonable set of parameters characterising the electron-to-proton ratio in CRs and the ambient photon and gas densities in the Galactic Disk. Moreover, the excess emission detected by EGRET at energies of several GeV can be naturally explained by γ-rays of nucleonic origin, assuming that the spectrum of CR protons in the inner Galaxy is harder than the spectrum of directly observed CRs. In particular, the results presented in Figs. 4.19a and 4.20 show that the proton spectrum given by Eq.(4.31) with $E_* \sim 5 - 10\,\mathrm{GeV}$, $\Gamma_0 = 2.1$ and $\delta \simeq 0.6$, which could be formed due to a certain combination of diffusive and convective escape time-scales, is able to explain reasonably well the observed excess GeV emission. In the total "π^0+IC" spectrum the contribution of π^0-decay γ-ray component gradually decreases, however, this decline is compensated for by the hard IC component, which at energies above 100 GeV becomes the dominant contributor to the overall γ-ray flux (see Figs. 4.20 and 4.21). The "GeV bump" can be explained also by a single power-law spectrum of CR protons with $\Gamma_p \sim 2.5$ (Fig. 4.19a), but then we should assume strong suppression of the IC contribution to the overall γ-ray flux. These two scenarios predict essentially different origins of γ-rays in the VHE domain. While in the case of the proton spectrum given by Eq.(4.31), the γ-ray emission at $E \geq 100\,\text{GeV}$ is contributed mainly by IC scattering of multi-TeV electrons on the 2.7 K CMBR, in the case of a single power-law spectrum of protons the VHE γ-ray flux is dominated by the π^0-decay component. The upper limits for the diffuse γ-ray fluxes at TeV energies shown in Fig. 4.21 are significantly higher than the predicted fluxes. Moreover, the upper limits reported by the Whipple (LeBohec *et al.*, 2000) and HEGRA (Aharonian *et al.*, 2001) collaborations are derived for galactic longitudes around 40°, and thus are irrelevant to the radiation of the inner Galaxy. This part of the Galactic Disk can be much better surveyed by IACT arrays located in the Southern Hemisphere. These instruments should be able to detect the VHE diffuse γ-ray emission of the Galactic Disk, and thus to provide crucial information about the character of propagation of ≥ 110 TeV cosmic rays in the Galactic Disk.

Resolving small-scale features of the diffuse background. The removal of contributions from unresolved γ-ray sources is a key condition for detailed study of truly diffuse radiation, in particular its spectral and spatial variations which carry direct information about the character of propagation of GCRs. If the "unresolved source component" is contributed by objects with fluxes of more than 10 per cent of the minimum detectable flux by EGRET, GLAST will be able to resolve the truly diffuse component from the overall diffuse γ-ray background of the Galactic Disk, and probe its spatial distribution on an angular scale $\approx 0.3°$. This corresponds to linear scales of less than 100 pc even in remote galactic regions. This implies that GLAST can reveal the sites of enhanced CR density expected within 100 pc around strong cosmic ray accelerators. The potential of GLAST for spectroscopic measurements is, however, limited by photon statistics, especially for energies beyond 10 GeV. In this energy region a unique spectroscopic performance on the same angular scales of about $\approx 0.3°$ may demonstrate the future sub-10 GeV IACT arrays like 5@5. Before then, the forthcoming IACT arrays should be able to detect the diffuse galactic background at TeV energies. In particular, a deep survey of the inner Galaxy by H.E.S.S. may provide the first observational probe of the average density of TeV cosmic rays and, hopefully, allow the study of their variations in the inner Galaxy on 1° angular scales.

Chapter 5

Gamma Ray Visibility of Supernova Remnants

For more than 40 years, the galactic SNRs have been believed to be the sites of production of the bulk of the observed CR flux. The strong shocks in SNRs may provide - through the diffusive shock acceleration mechanism – very effective, up to 10 to 30 per cent, conversion of the total SN explosion into relativistic protons and nuclei, as well as explaining the hard, $E^{-(2.0-2.1)}$ type, source spectra, as it follows from the CR propagation studies in the galactic disk (see Chapter 4). Although quite plausible, these arguments still remain as a theoretical conjecture, and thus cannot supersede direct evidence. The detection and identification of π^0-decay γ-rays from SNRs, primarily at TeV energies, would be the first straightforward proof of the acceleration of protons by SNR shocks. On the other hand, the failure to detect π^0-decay γ-ray signals from several selected SNRs would impose strong constraints on the energy in accelerated protons, $W_{\rm CR} \leq 10^{49} - 10^{50}$ erg. This would indicate an inability of the ensemble of galactic SNRs to explain the observed CR fluxes.

5.1 Gamma Rays as a Diagnostic Tool

For a standard "power-law with exponential cutoff" energy distribution of protons,

$$dN/dE \propto E^{-\Gamma} \exp\left(-E/E_0\right), \tag{5.1}$$

the flux of γ-rays produced in the interactions of CR protons with the ambient gas is basically determined by the scaling parameter

$$A = \left(\frac{W_{\rm CR}}{10^{50}\,\rm erg}\right)\left(\frac{d}{1\,\rm kpc}\right)^{-2}\left(\frac{n}{1\,\rm cm^{-3}}\right), \tag{5.2}$$

where W_{CR} is the total energy in accelerated protons, n is the ambient gas density, and d is the distance to the source. The γ-ray spectrum at high energies repeats the spectral shape of parent protons – a power-law with approximately the same power-law index Γ and a cutoff at $E \sim 0.1 E_0$. For

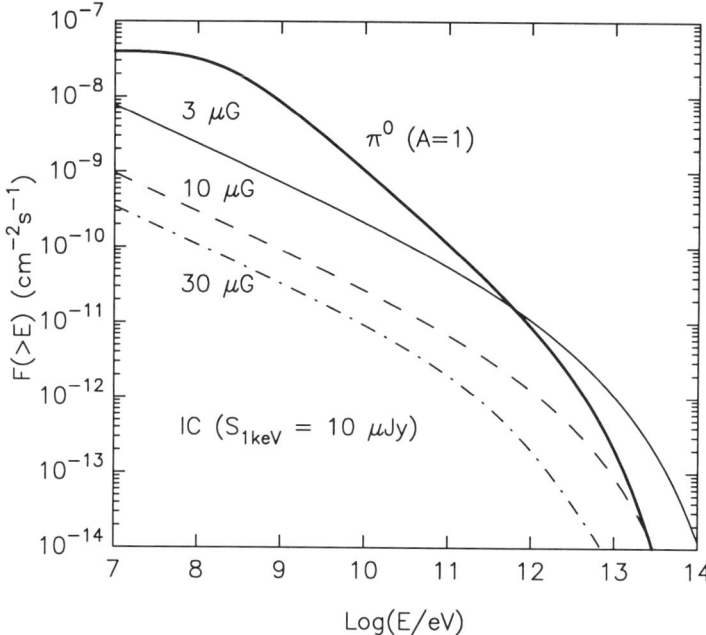

Fig. 5.1 Fluxes of π^0-decay (heavy solid line) and IC (thin lines) γ-rays from a 10^3 year old SNR. The π^0 decay γ-rays are calculated for the scaling parameter $A = 1$. The IC γ-ray fluxes are calculated for 3 different values of the magnetic field $B = 3$ (solid curve), 10 (dashed curve), and 30 μG (dot-dashed curve), assuming that the electrons produce the same flux of synchrotron radiation $S_\nu = 10\,\mu$Jy. For the protons and electrons we assume the same acceleration spectrum with $\Gamma = 2$ and $E_0 = 100\,\mathrm{TeV}$.

a given energy density of accelerated protons, the integral flux of γ-rays above 300 MeV is almost independent of the proton spectral index,

$$J_\gamma (\geq 300\,\mathrm{MeV}) \approx 3 \cdot 10^{-8} A \ \mathrm{cm}^{-2}\mathrm{s}^{-1} \ . \tag{5.3}$$

At energies $1\,\mathrm{GeV} \leq \mathrm{E} \leq 0.1 \mathrm{E}_0$, and for the standard chemical composition of CRs and the ambient gas,

$$J_\gamma (\geq E) = 10^{-11} E_{\mathrm{TeV}}^{-\Gamma+1} A f_\Gamma \ \mathrm{cm}^{-2}\mathrm{s}^{-1}, \tag{5.4}$$

where $f_\Gamma \approx 1$ and 0.2 for the the proton spectral indices $\Gamma = 2$ and 2.3, respectively. The integral fluxes of π^0-decay γ-rays from a 10^3 yr old SNR for the scaling parameter $A = 1$ are shown in Fig. 5.1. The γ-ray fluxes are calculated for a proton spectrum with $\Gamma = 2$ and $E_0 = 100$ TeV.

Within the diffusive shock acceleration model, the amount of relativistic particles increases with time as the remnant passes through its free expansion phase, and reaches the maximum when the SNR enters the so-called Sedov phase – the phase when the mass of the swept-up matter becomes comparable with the mass of the ejecta. Correspondingly the peak luminosity of γ-rays appears at the early Sedov phase (Drury et al., 1994, Berezhko and Völk, 1997), typically $10^3 - 10^4$ years after the SN explosion. At this stage the radius of the shell exceeds several parsecs which implies an angular size of about $1°$ for relatively close ($d \leq 1$ kpc) SNRs. The conflict between the angular size ($\phi \propto 1/d$) and the γ-ray flux ($J_\gamma \propto 1/d^2$) significantly limits the number of SNRs which could be detected in γ-rays by the current generation of IACTs. The best candidates would be young SNR at distances of a few kpc or less, which have already entered their Sedov phase, show nonthermal synchrotron radiation, and are expanding into regions of enhanced gas density. An additional criterion would be the possible association of these SNRs with γ-ray sources detected by EGRET. The Whipple collaboration has chosen for observations 6 SNRs which more or less satisfy these condition. No positive signal from any of these SNR has been detected (Buckley et al., 1998). The corresponding upper limits on the fluxes above 300 GeV are shown in Fig. 2.8 (Chapter 2) along with the EGRET integral fluxes or upper limits (Esposito et al., 1996).

These points are compared with the theoretical expectations normalised to the EGRET data points as well as with a range of fluxes based on conservative estimates of the scaling parameter A. Although being very important and meaningful, these upper limits cannot be used as an evidence against the SNRs as sites of acceleration of galactic cosmic rays, especially if one takes into account large, typically up to a factor of 10, uncertainties in the scaling parameter A. For example, even for the most stringent constraint on the TeV flux set so far, the upper limit obtained by the HEGRA collaboration for the Tycho SNR at the 33 mCrab level (Aharonian et al., 2001e) is still above the prediction of the "nominal theory" . The model predictions in Fig. 2.9 assume that Tycho has progressed well into the Sedov phase. In fact this young supernova remnant could still be in the pre-Sedov phase. If so, the fraction of the mechanical energy converted into relativistic particles could be below 10 percent ($\theta = 0.1$), and correspondingly lower γ-ray fluxes would be expected.

The TeV upper limits become uncomfortable only if we assume that the MeV/GeV fluxes detected by EGRET are dominated by interactions of accelerated protons with the ambient nebular gas. However, due to the poor angular resolution of EGRET, the association of most of the "excess GeV" regions with SNRs is questionable and needs further observational evidence. For example, in the case of γ Cygni, the GeV data are not consistent with the spatial extent of the remnant, and can be associated with a weak X-ray source, RX J2020.2+4026 (Brazier et al., 1996). The theoretical predictions do not unconditionally support the associations of GeV sources with SNRs either. Drury et al.(1994) argued that EGRET could not detect GeV γ-rays from standard SNRs, because even for the best candidates the parameter A is less than 1, with the flux level determined by $A = 1$ being only marginally compatible with the EGRET sensitivity. Only special configurations like dense molecular clouds overtaken by supernova shells may provide detectable γ-radiation at both GeV and TeV energies at the level of sensitivities of EGRET and current ground-based instruments (Aharonian et al., 1994a).

If we nevertheless accept that the EGRET detections have *physical* links to SNRs, i.e. assume that the GeV radiation is produced at interactions of accelerated particles with ambient gas, there are at least 3 possible reasons that could explain the lack of TeV γ-rays from the same SNRs.

Large contributions of the electron bremsstrahlung. Assuming that only 20 percent (or less) of the observed fluxes below 1 GeV from IC 433, γ Cygni and W44 is contributed by protons, the flux of π^0-decay γ-rays at TeV energies would be correspondingly reduced by a factor of 5, i.e. to the level which does not contradict the TeV upper limits shown in Fig. 2.8. If so, the major fraction of GeV radiation should be attributed to electrons, most likely of bremsstrahlung origin. This would require the electron-to-proton ratio of accelerated particles close to 1, i.e. strong, up to a factor of 100, enhancement of electrons in the total energy budget of accelerated particles compared to their relative content in the locally observed CRs. If true, this would imply that SNRs cannot be responsible for the bulk of observed cosmic ray protons and nuclei.

Cutoffs in the proton spectrum below 1 TeV. This assumption can be tightly connected with the requirement of dense ambient gas. In order to provide the observed absolute fluxes of GeV γ-rays by interactions of accelerated electrons and/or protons we should assume high density environments, otherwise it is difficult to keep the total energy budget in accelerated particles within reasonable limits. Indeed, the EGRET fluxes above 300 MeV shown in Fig. 2.8 are close to 10^{-7} ph/cm^{2}s, which implies that the parameter A in Eq.(5.2) should be as large as 3. For the maximum available content of CRs in a SNR, $W_{\rm CR} \simeq 10^{50}$ erg, the density of the ambient hydrogen gas should exceed 10 cm^{-3}, given the ≥ 2 kpc distances to the SNRs in the Whipple selection list. If particle acceleration takes place in the same region, a variety of new effects, in particular the role of wave damping on the maximum energy to which particles can be accelerated (Drury et al., 1996), should be included in theoretical treatments. For example, Baring et al. (1999) have argued, based on their study of nonlinear shock acceleration, that SNRs expanding into high density regions cannot effectively accelerate particles beyond 1 TeV. Another mechanism for "early cutoffs" in the SNR spectra, associated with a feedback effect in the highly turbulent plasma, has been suggested by Malkov et al. (2002). Although the cutoffs below 1 TeV in the spectra of accelerated protons could be the simplest solution to the problem of lack of γ-rays from the "EGRET SNRs", it leaves unanswered the question of whether it should be treated as an argument against the SNR paradigm of galactic cosmic rays in general, or it simply implies that another type of SNRs should be invoked for explanation of the spectrum of GCRs extending to 10^{15} eV.

Steep acceleration spectra. For a fixed scaling parameter A, a proton spectrum with $\Gamma = 2.3$ results in a factor of 5 lower γ-ray flux above 1 TeV than one with a harder spectrum of $\Gamma \sim 2$. Thus, a steeper acceleration spectrum of protons would be another simple way to avoid TeV γ-rays at the flux levels exceeding the Whipple and HEGRA upper limits, even if the observed fluxes around 1 GeV are entirely due to the π^0-decay component of radiation. In this regard we note that the presently favoured source spectrum $\propto E^{-\Gamma}$ with $\Gamma \leq 2.1$, derived from the observed energy dependence of the ratio of secondary to primary cosmic rays (e.g. Swordy, 2001), implies negligible reacceleration of CRs in the ISM. Otherwise softer source spectra, as steep as $E^{-2.4}$, should be expected. This actually agrees with the so-called Kolmogorov type spectrum for the interstellar turbulence which predicts CR propagation with a diffusion coefficient $\propto E^{-1/3}$. Although CR source spectra with $\Gamma \simeq 2.3 - 2.4$ seem to be in conflict with models of the nonlinear diffusive shock acceleration, such relatively steep source spectra do not rule out SNRs as major contributors to the galactic cosmic rays.

An example of a good fit to the γ-ray data from the supernova remnant IC 443, based on the above assumptions to avoid high TeV fluxes, is shown in Fig. 5.2. In accordance with this phenomenological study (Gaisser et al., 1998), for a specific set of parameters it is possible to accommodate both the EGRET fluxes and TeV γ-ray upper limits.

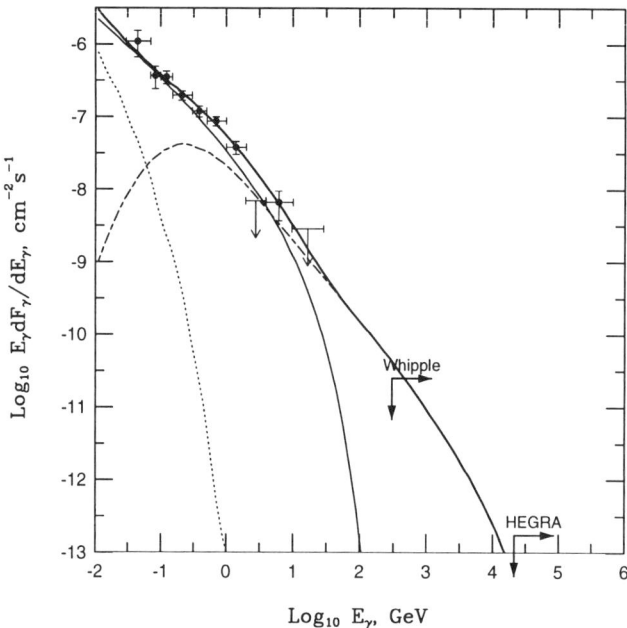

Fig. 5.2 Interpretation of γ-ray fluxes from IC 433 (from Gaisser et al., 1998). The solid, dashed and dotted lines correspond to the contributions from bremsstrahlung, π^0-decay, and inverse Compton processes, calculated for the following model parameters: $\Gamma_p = \Gamma_e = 2.32$, $E_0 = 80$ TeV, e/p $= 0.16$, $B = 54\mu G$, $n = 10^3$ cm^{-3}.

In all γ-ray production scenarios in SNRs the inverse Compton component does not noticeably contribute to the overall flux of low energy ($E \leq 1 \text{GeV}$) γ-rays (see e.g. Gaisser et al., 1998). Therefore, the absolute EGRET fluxes require quite a large density of ambient gas which in turn could be the reason for the large power-law indices of $\Gamma \sim 2.3$-2.4 or "early cutoffs" ($E_0 \leq 1$ TeV) in the proton spectra, if the particle acceleration and γ-ray production regions coincide. In this context one may conclude that the SNRs in *low density* and homogeneous environments may appear as more effective TeV emitters. The scaling factors A of such SNRs are, however, small, $A \leq 0.1$, thus the integral γ-ray fluxes of hadronic origin at 1 TeV could be below 10^{-12} ph/cm²s at 1 TeV. In addition to the low fluxes, the detection of TeV radiation of hadronic origin from such objects is not a easy task because of the extended character of radiation (1° or so) and the significant "contamination" induced by inverse Compton γ-rays of directly accelerated electrons.

5.2 Inverse Compton Versus π^0-Decay Gamma Rays

When deriving information about the accelerated protons one has to subtract a possibly non-negligible "contamination" caused by directly accelerated electrons that upscatter photons of the 2.7 K CMBR (which is the dominant target photon field in most of SNRs) to γ-ray energies. The same multi-TeV electrons responsible for IC γ-rays of TeV energies produce also synchrotron UV/X-ray radiation. The typical energies E_γ and ε_x of the IC and synchrotron photons produced by an electron are related by

$$E_\gamma \simeq 2(\varepsilon_x/0.1\,\text{keV})\,(B/10\,\mu\text{G})^{-1}\,\text{TeV}. \qquad (5.5)$$

This relation neglects the Klein-Nishina effect which however becomes important at energies $E_\gamma \geq 10\,\text{TeV}$. The ratio of the synchrotron and IC fluxes $f_E \equiv E^2 F(E) = \nu S_\nu$ at these energies does not practically depend on the shape of the spectrum of parent relativistic electrons, but strongly depends on the magnetic field:

$$\frac{f_{\text{IC}}(E_\gamma)}{f_{\text{sy}}(\varepsilon_x)} \simeq 0.1\,(B/10\,\mu\text{G})^{-2} \qquad (5.6)$$

For a flat X-ray spectrum with photon index $\simeq 2$, the energy flux f_X is almost energy-independent. Therefore f_X at a typical energy of 1 keV could serve as a good indicator of the IC γ-ray fluxes expected at TeV energies, although for magnetic fields $B \leq 100\,\mu\text{G}$ the energy of synchrotron photons (produced by the same parent electrons) relevant to $\sim 1\,\text{TeV}$ γ-rays is in the soft X-ray domain.

The contribution of π^0-decay γ-rays dominates over the contribution of the IC component when

$$A \geq 0.1\,(S_{1\text{keV}}/10\mu\text{Jy})\,(B/10\,\mu\text{G})^{-2}, \qquad (5.7)$$

where $S_{1\text{keV}}$ is the flux of nonthermal synchrotron radiation at 1 keV normalised to $10\,\mu$Jy (the corresponding energy flux $f_x \approx 2.4 \times 10^{-11}\,\mathrm{erg/cm^2 s}$), which is a typical level of nonthermal X-ray fluxes reported from shell-type supernova remnants SN 1006 and RX J1713.7-3946.

In Fig. 5.1 the integral fluxes of π^0-decay and inverse Compton γ-rays from a SNR of age 10^3 yr are shown. The π^0-decay γ-ray flux corresponds to the scaling factor $A = 1$. The IC fluxes are calculated by normalising the synchrotron X-ray fluxes to $S_{1\mathrm{keV}} = 10\mu$Jy for ambient magnetic fields of 3μG, 10μG, and 30μG. For both electrons and protons we assume continuous acceleration over 10^3 years with a time-independent injection spectrum with $\Gamma = 2$ and $E_0 = 100$ TeV. Note that for the normalisations used, the results presented in Fig. 5.1 only slightly depend on the source age, unless it is larger than the synchrotron cooling time of multi-TeV electrons,

$$t_{\mathrm{synch}} \approx 1.2 \times 10^3 \, (B/10\,\mu\mathrm{G})^{-2} \, (E_\mathrm{e}/100\,\mathrm{TeV})^{-1} \, \mathrm{yr} \;. \tag{5.8}$$

Examination of the condition given by Eq.(5.7) is of particular interest. If it could be shown that the TeV signals, reported by the CANGAROO collaboration from two shell type SNRs, SN 1006 (Tanimori et al., 1998b) and RX J1713.7-3946 (Muraishi et al., 2000) *cannot be* explained by IC scattering, this would be observational evidence of shock-acceleration of protons in a SNR, because the only alternative for the explanation of TeV emission of these objects are γ-rays of nucleonic origin produced in interactions of accelerated protons with the ambient gas (Aharonian, 1999).

Generally, with the exception of Cas, and perhaps a few other specific objects, the IC radiation of SNRs in the TeV region is dominated by the 2.7 K CMBR seed photons. As long as the gas number density does not significantly exceed 10 cm^{-3}, in this energy regime we can safely ignore the contribution from electron bremsstrahlung. The relative contributions of these radiation components can be estimated from Fig. 5.3 where the electron energy-loss timescales due to different cooling processes are shown. This is true, however, for the stationary (continuous) electron accelerator. After the electron accelerator turns off, the number of electrons with $E \geq 20$ TeV producing ≥ 1 TeV inverse Compton γ-rays diminishes, and bremsstrahlung then may dominate even in the TeV regime. It is interesting to compare also the bremsstrahlung and π^0-decay γ-ray fluxes. The cooling times of both processes slightly (logarithmically) depend on particle energy. The cooling time of electrons due to relativistic bremsstrahlung is comparable with the $p - p$ cooling time of protons, but it is shorter, by a factor of 3 to 5, than the characteristic cooling time of protons through the π^0-decay channel. Therefore the π^0-decay γ-ray flux would dominate over the bremsstrahlung γ-ray flux (at energies above 100 MeV), if the overall energy in accelerated protons exceeds the energy in relativistic electrons by a factor of 10 or more.

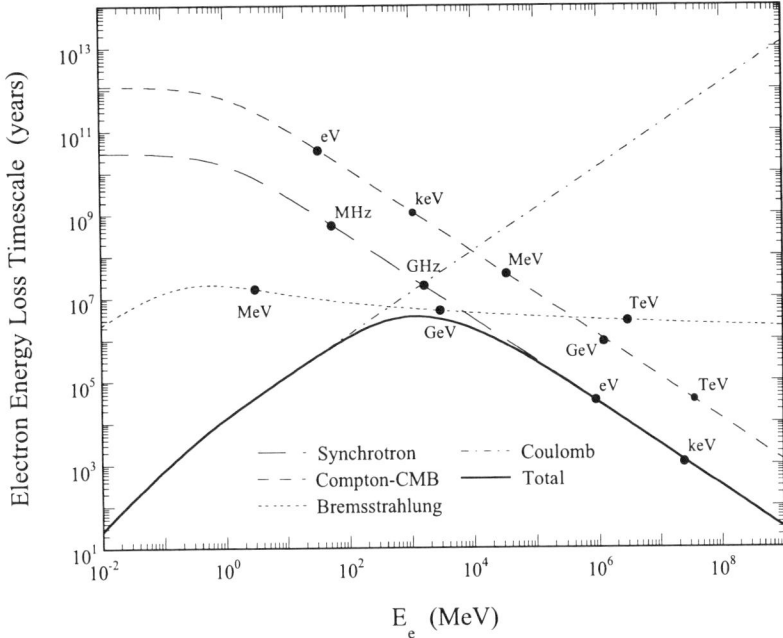

Fig. 5.3 The electron-energy-loss timescales, $t = E/(dE/dt)$, due to synchrotron emission, bremsstrahlung, Compton scattering of the 2.7 CMBR and Coulomb collisions for the case of $n = 4$ cm^{-3} and $B = 20$ μG (it is assumed that the interstellar gas with $n_{\rm ISM} = 1$ cm^{-3} and $B_{\rm ISM} = 5$ μG is compressed by a strong shock by a factor of 4). For these parameters, Compton losses dominate below 100 MeV, bremsstrahlung losses dominate near 1 GeV, and the synchrotron losses dominate above 10 GeV. On each radiative loss curve the typical photon energy emitted by an electron with the given kinetic energy is indicated. This shows that the radio synchrotron and the relativistic bremsstrahlung (as well as π^0-decay) emission components will endure for the longest time after the electron source has turned off, and the X-ray synchrotron, TeV inverse Compton and non-relativistic (sub-MeV) bremsstrahlung emission components will decrease fastest. (From Sturner et al., 1997).

5.3 Synchrotron X-ray Emission of SNRs

The IC origin of TeV emission of SNRs implies the existence of multi-TeV electrons and, therefore, unavoidable synchrotron radiation extending to O/UV and shorter wavelengths. In the galactic plane the magnetic field is estimated to be as large as $B \geq 3$ μG . Since the the energy density of such magnetic fields exceeds the density of the 2.7 K CMBR ($\approx 4 \times 10^{-13}$ erg/cm^3), the energy flux of TeV radiation is expected to be (always) below the energy flux of synchrotron X-rays. Diffusive shock acceleration models predict significant shock-compression of the magnetic field, by a factor of 4 or even more (in the case of the development of nonlinear shocks). Therefore, the energy flux of TeV emission should not exceed 10 percent of the synchrotron X-ray flux, unless the SNR is located well above the galactic plane and we deal with strictly parallel shock acceleration.

In synchrotron models the spectral fit to the X-ray flux is crucial because the fluxes are produced by electrons in the region of the exponential cutoff E_0 between 10 and 100 TeV. For a "power-law with exponential cutoff" type spectra, the power-law index of electrons Γ is derived from the radio data, while the information about the cutoff energy E_0 is contained in the X-ray part of the spectrum. In the particular case of negligible energy losses, the spectrum of electrons $N(E_e)$ repeats the injection spectrum, $N(E_e) \propto Q(E_e)$, therefore in the δ-functional approximation, the differential synchrotron spectrum can be presented in a convenient form

$$J(\varepsilon) \propto \varepsilon^{-(\Gamma+1)/2} \exp[-(\varepsilon/\varsigma\varepsilon_0)^{1/2}], \qquad (5.9)$$

where

$$\varepsilon_0 = h\nu_c \simeq 5.3 \left(\frac{B}{10\mu\text{G}}\right) \left(\frac{E_0}{100\,\text{TeV}}\right)^2 \text{ keV}, \qquad (5.10)$$

is the characteristic synchrotron energy for an electron of energy E_0, and ς is a free parameter introduced to adapt this formula to accurate numerical calculations, which show that the best *broad-band* fit with an accuracy of better than 25% in the cutoff region up to $\varepsilon \sim 20\varepsilon_0$, is provided by $\varsigma = 1$.

In the framework of diffusive shock acceleration model, the synchrotron cutoff energy is determined by the *"acceleration rate=synchrotron loss rate"* condition. The acceleration time (see e.g. Malkov and Drury, 2001) can be written, with accuracy of about 50 per cent, in a simple form $t_{\text{acc}} \approx 10D/v_s^2$, where D is the diffusion coefficient in the upstream region, and v is the upstream velocity into the shock. The diffusion coefficient is generally highly unknown parameter, however if one requires acceleration of electrons to highest energies (thus allowing extension of synchrotron radiation to the X-ray domain), we must assume that the diffusion proceeds in the Bohm regime. Therefore it is convenient to parametrise the diffusion coefficient in terms of the Bohm diffusion coefficient

$$D(E) = \eta \frac{r_g c}{3}, \tag{5.11}$$

where $r_g = 3.3 \cdot 10^{15} (E/10\,\mathrm{TeV})(B/10\,\mu\mathrm{G})^{-1}$ cm is the particle gyroradius, and $\eta \geq 1$ is the gyrofactor; $\eta = 1$ implies the smallest possible diffusion coefficient, and correspondingly the shortest possible acceleration time,

$$t_{\mathrm{acc}} \approx 2.7 \times 10^3 \left(\frac{E_e}{100\,\mathrm{TeV}}\right) \left(\frac{B}{10\,\mu\mathrm{G}}\right)^{-1} \left(\frac{v_s}{2000\,\mathrm{km/s}}\right)^{-2} \eta \; \mathrm{yr}. \tag{5.12}$$

The maximum energy of accelerated electrons in the regime dominated by synchrotron losses, is determined from Eqs.(5.8) and (5.12):

$$E_0 \simeq 67\, \eta^{-1/2} \left(\frac{B}{10\,\mu\mathrm{G}}\right)^{-1/2} \left(\frac{v_s}{2000\,\mathrm{km/s}}\right) \; \mathrm{TeV}. \tag{5.13}$$

From Eqs.(5.10) and (5.13) one can find that the the energy of exponential cutoff in the synchrotron spectrum,

$$\varepsilon_m \simeq 2 \left(\frac{v_s}{2000\,\mathrm{km/s}}\right)^2 \eta^{-1} \; \mathrm{keV}. \tag{5.14}$$

does not depend on either the magnetic field or the age of the source. Since $\eta \geq 1$, and the shock speed in the Sedov phase does not significantly exceed 2000 km/s, the steep featureless X-ray spectrum above 1 keV may serve as a decisive diagnostic tool to identify the synchrotron origin of radiation. On the other hand, the sheer fact of the detection of nonthermal X-radiation from several shell type SNRs (Petre et al., 1999) and its possible synchrotron origin imply that the acceleration of electrons in these SNR proceeds in the regime close to the Bohm diffusion.

5.4 TeV Gamma Radiation of SN 1006 and Similar SNRs

5.4.1 *Inverse Compton models of TeV emission*

The featureless X-ray emission observed by ASCA from the shell of SN 1006 (Koyama et al., 1995) is naturally interpreted as synchrotron emission of electrons accelerated to energies $\sim 100\,\mathrm{TeV}$ at the shock front. Motivated by this fact, several theorists (Mastichiadis, 1996; Mastichiadis and De Jager, 1996; Pohl, 1996; Yoshida and Yanagita, 1997) predicted strong TeV emission produced by the same electrons upscattering off the 2.7 K CMBR. Therefore, the detection of TeV emission from SN 1006, claimed soon after these predictions (Tanimori et al., 1998b) strengthened the belief in the electronic origin of TeV radiation. The π^0-decay contribution to the observed flux is widely considered to be less important. However, it has been argued (Aharonian, 1999) that the simple one-zone synchrotron and IC model applied to SN 1006 might have serious theoretical problems. The interpretation of TeV emission in terms of IC mechanism requires several bold conditions, the validity of which requires thorough theoretical studies and new X-ray and γ-ray observations concerning both the spectral and morphological properties of the source:

Constraint on the magnetic field. Since the 2.7 K CMBR serves as the main target photon field for the IC scattering, the flux of γ-rays is determined, for the given X-ray flux ($S_{1\mathrm{keV}} \sim 20\mu\mathrm{J}$), only by the strength of the magnetic field. Therefore, in the framework of the IC origin of TeV γ-rays the ratio of observed fluxes of X-rays and TeV γ-rays provides an estimate of the the strength of the ambient B-field. In the simplified *one-zone model* which assumes that the synchrotron X-rays and IC γ-rays are produced by same electrons confined in a spatially homogeneous region, Fig 5.1 provides quite accurate estimates for IC fluxes . However, the study of spectral features of the γ-ray emission requires a careful derivation of the spectrum of $\geq 10\,\mathrm{TeV}$ electrons which should be controlled by the spectrum of synchrotron X-rays. In the synchrotron-inverse Compton models the spectral fit to the X-ray flux is crucial because the fluxes are produced by electrons in the region of the exponential cutoff E_0 between 10 and 100 TeV. For the "power-law with exponential-cutoff" type energy distribution of electrons, the resulting synchrotron radiation depends on the spectral index of electrons Γ_e and the parameter $\Pi = E_0 B^{1/2}$ (see Eq.5.10). Strictly speaking, this result is valid for the δ-functional approximation, provided that the deformation of the injection spectrum of electrons due to the energy losses can be ignored (i.e. when $t_{\mathrm{source}} \leq t_{\mathrm{synch}}$). Since the inverse Compton origin of TeV radiation from SN 1006 requires a magnetic field less than 10 μG, and the age of the accelerator is less than 10^3 yr, such an approximation is quite acceptable. Moreover, numerical calculations show that this parameter remains a (quasi) *invariant* even in the general case, although its absolute value could differ somewhat from predictions of the δ-function approximation.

In Fig.5.4 the spectra of synchrotron radiation are shown calculated for the same product

$$\Pi = E\, B^{1/2} = 61.5\,\text{TeV}\,\mu\text{G}^{1/2}\,, \tag{5.15}$$

but for 3 different combinations of E_0 and B: (1) $B = 3\mu\text{G}$, $E_0 = 35.5\,\text{TeV}$; (2) $B = 5\mu\text{G}$, $E_0 = 27.5\,\text{TeV}$; (3) $B = 8\mu\text{G}$, $E_0 = 21.3\,\text{TeV}$

The power-law index Γ of electrons is derived from the radio synchrotron spectrum with spectral index α_r, $\Gamma = 1 + 2\,\alpha_r \simeq 2.15$. In all calculations the same normalisation to the radio fluxes is used. Therefore different magnetic fields require different total energy in relativistic electrons. At the same time, because the target photon field (2.7 K CMBR) for the IC scattering is fixed, there is a strong dependence of the IC γ-ray fluxes on the magnetic field. In particular, in the case of small magnetic fields, $B \leq 10\mu\text{G}$, when the radiative cooling time of 10 to 100 TeV electrons, responsible for the observed X-rays and TeV γ-rays, exceeds the age of the source, we have $F_\gamma \propto B^{-(\alpha_r+1)}$. Note that for $B \leq 3\mu\text{G}$ the radiative losses are dominated by IC on the 2.7 K CMBR.

Fig. 5.4 shows that the interpretation of the observed TeV radiation by the IC mechanism in the framework of a simplified spatially homogeneous (one-zone) model is possible only for an ambient magnetic field in a very narrow range around $5\mu\text{G}$. Magnetic fields $\geq 7\mu\text{G}$ result in a strong reduction of the IC flux, while the assumption of low magnetic fields, $B \leq 4\,\mu\text{G}$, leads to overproduction of γ-rays.

Maximum electron energy. For a magnetic field in the NE rim $B \simeq 5\,\mu\text{G}$ and the limited age of 10^3 yr of SN 1006 a question arises as to which mechanism could be efficient enough to accelerate electrons to energies $E_0 = \Pi/\sqrt{B} \sim 25\,\text{TeV}$ (in fact, to much higher energies, $E \gg E_0$, taking into account that in a magnetic field as low as $5\,\mu\text{G}$ only electrons with energy more than 200 TeV could produce synchrotron X-rays up to 8 keV as detected by ASCA). In the model of diffusive shock acceleration, the absolute maximum energy that an electron can achieve is determined by Eq.(5.12), assuming that the cooling time of electrons given by Eq.(5.8) exceeds the age of the source T_0:

$$E_0 \simeq 35 \left(\frac{B}{10\,\mu\text{G}}\right) \left(\frac{T_0}{10^3\,\text{yr}}\right) \left(\frac{v_s}{2000\,\text{km/s}}\right)^2 \eta^{-1}. \tag{5.16}$$

For $B = 7\,\mu\text{G}$, and assuming that the Sedov phase in SN 1006 started several hundred years after the explosion, from Eqs. (5.16) and (5.15) one finds $v_s \simeq 3500\,\eta^{1/2}\,\text{km/s}$. Thus, even assuming that the acceleration in SN 1006 takes place at maximum possible rate ($\eta = 1$), the lower limit on the shock speed exceeds the estimates of v_s based on the Sedov solution (Willingale et al., 1996; Winkler and Long, 1997).

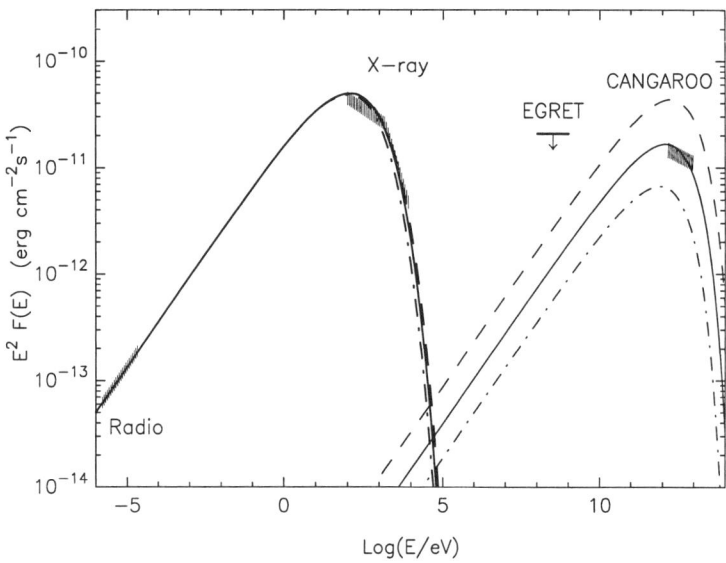

Fig. 5.4 The synchrotron and IC fluxes calculated for the homogeneous source without escape of accelerated particles ("one-zone" model). A power-law injection spectrum of electrons with $\Gamma_e = 2.15$ is assumed in order to fit the radio data. The maximum energy E_0 is determined from the condition given by Eq.(5.15) for 3 different values of the magnetic field: $B = 3\,\mu\text{G}$ (dashed line), $5\,\mu\text{G}$ (solid line), and $8\,\mu\text{G}$ (dot-dashed line).

Diffusion effects. The required small magnetic field, $B \leq 7\,\mu\text{G}$, allows the highest energy electrons produced in the shell to effectively escape from the acceleration region. This will give rise to enhanced IC emission outside of the rim, namely in the interior regions of the remnant where the magnetic field could be as low as the typical interstellar B-field. The escape of electrons from the rim is unavoidable because of the diffusive and convective propagation of particles with a characteristic timescale $\tau_{\text{esc}} = (\tau_{\text{con}}^{-1} + \tau_{\text{dif}}^{-1})^{-1}$. The convective escape time is

$$\tau_{\text{con}}(t) = \Delta r / u_2, \qquad (5.17)$$

where u_2 is the fluid speed downstream of the shock (in its rest frame), thus at present $u_2 = v_s/\rho \sim 500\,\mathrm{km/s}$ for the compression ratio $\rho \sim 4$. The width, Δr, of the NE rim in X-rays as measured by ROSAT does not exceed 20 per cent of the angular radius of the remnant of about 17 arcminutes (Willingale et al., 1996) which corresponds to $r_s = 4.9\,d_{\mathrm{kpc}}\,\mathrm{pc}$. In calculations below a constant $\Delta r/r_s = 0.2$ ratio is assumed throughout the evolution of the remnant. The diffusive escape time is

$$\tau_{\mathrm{diff}}(t) \simeq \frac{(\Delta r)^2}{2\,D(E)}, \tag{5.18}$$

where $D(E)$ is the diffusion coefficient in the shocked region (the rim). In the theory of shock acceleration the diffusion coefficient is generally taken in the form given by Eq.(5.11). The maximum confinement (and therefore the maximum energy) of particles is achieved in the Bohm regime corresponding to $\eta = 1$.

The kinetic equation and the solution to this equation for the two-zone model (Aharonian and Atoyan, 1999), which treats the overall (i.e. integrated over the volumes) fluxes from the rim (zone 1) and outside (zone 2), allows to study the effects associated with the electron escape. In particular, in the case of small magnetic field the electron escape results in comparable contributions of γ-rays from the rim and outside, as demonstrated in Fig. 5.5. The same could be true also for the synchrotron radiation, unless the magnetic field in the rim does not exceed the average field of the nebula. In fact, the ASCA observations show a narrow X-ray rim with sharp edges. Thus, we may conclude that the magnetic field in the rim is significantly enhanced, which can be naturally explained by strong shock compression. Also, from Fig. 5.5 we may draw the conclusion that the magnetic field in the rim should be within the limits $5 \leq B_1 \leq 10\,\mu\mathrm{G}$ (preferably $B_1 \simeq 6 - 8\,\mu\mathrm{G}$), and η close to 1, i.e. the diffusion in the rim should take place essentially in the Bohm limit. In order to provide maximum electron energy, E_0, of order of 30 TeV, the shock speed should exceed 3000 km/s, which implies a large distance to the source, $d \geq 1.5\,\mathrm{kpc}$. For this set of parameters, approximately half of the total TeV emission is contributed from the inner parts of the remnant, $r \leq 0.8r_s$. The narrow range of the required magnetic field in the rim allows a rather accurate estimate of the total energy of relativistic electrons in SN 1006. For example, for the distance $d = 1.8\,\mathrm{kpc}$, the total electron energy in the rim is $W_e \simeq 3 \times 10^{48}\,\mathrm{erg}$, whereas the energy in the electrons which escaped the rim is $W_e \simeq 2 \times 10^{48}\,\mathrm{erg}$.

The total energy of the magnetic field in the remnant with $B_2 = B_1/4 \simeq 2\,\mu\mathrm{G}$ for a distance 1.8 kpc is about 10^{46} erg. A similar amount of magnetic field energy is expected also in the rim since the amplification of the B-field there is compensated by a smaller volume of the rim. Thus, the inverse Compton origin of TeV γ-radiation requires about 1 percent of the total

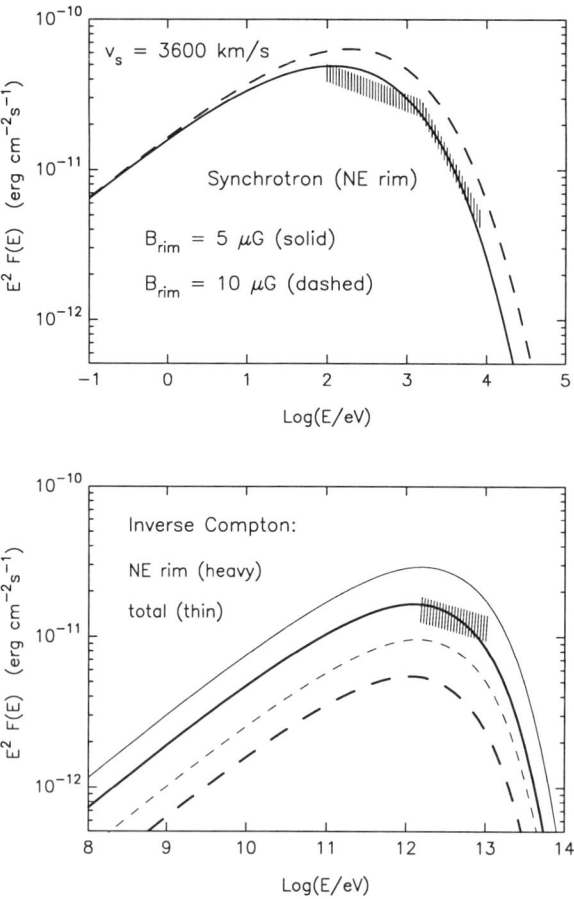

Fig. 5.5 Synchrotron (top panel) and IC γ-ray (bottom panel) fluxes calculated in the framework of the 2-zone model for two values of the magnetic field in the rim: $B_1 = 5\,\mu\mathrm{G}$ (solid lines) and $B_1 = 10\,\mu\mathrm{G}$ (dashed lines). The magnetic field outside of the rim is taken as $B_2 = B_1/4$. The heavy lines show the fluxes from the NE rim, and thin lines correspond to the total fluxes. The distance to the source is assumed to be 1.8 kpc. The maximum electron energy E_0 is calculated from Eq. (5.16) for the gyrofactor $\eta = 1$ for $B_1 = 5\,\mu\mathrm{G}$, and $\eta = 2$ for $B_1 = 10\,\mu\mathrm{G}$. (From Aharonian and Atoyan, 1999).

explosion energy of SN 1006 in relativistic electrons, and implies that the conditions in SN 1006 are far from equipartition between the relativistic electrons and the magnetic field. These two conclusions are intrinsic for the leptonic models of TeV emission of SN 1006. The estimates of energies of relativistic electrons and magnetic fields are very robust and almost independent of the model parameters.

5.4.2 Hadronic origin of TeV emission?

Currently, the nucleonic origin of the observed TeV emission is treated by the community as an inadequate alternative to the IC mechanism, the main argument being the low ambient density of the gas in SN 1006, $n \simeq 0.4\,\mathrm{cm}^{-3}$ (Willingale et al., 1996) as well as the presumed large distance to the source of about 2 kpc (Winkler and Long, 1997). However these arguments are not sufficiently robust to dismiss such an important possibility with far reaching conclusions concerning the origin of the nucleonic component of galactic cosmic rays (Aharonian, 1999).

The very fact of the existence of $\gg 10\,\mathrm{TeV}$ electrons, as follows from the X-ray data, is evidence for a strong shock in SN 1006, and implies a large compression factor, $\rho \simeq 4$ or even more, up to 10, as follows from non-linear studies of shock acceleration in SNRs (e.g. Baring et al., 1999). Therefore it seems not unrealistic to assume a significantly higher density of the gas in the rim region, e.g. $n \simeq 2\,\mathrm{cm}^{-3}$. The current estimates of the distance to the source also contain significant uncertainties, and do not exclude a distance of about 1 kpc or even less. To explain the reported flux of TeV emission, for $\Gamma_\mathrm{p} = 2$ the scaling factor $A \simeq 0.83$ is needed (Fig. 5.6). This requires for the total energy in accelerated protons

$$W_\mathrm{p} \simeq 4 \times 10^{49}\,(n/2\,\mathrm{cm}^{-3})^{-1}\,d_\mathrm{kpc}^2\,\mathrm{erg}\,. \tag{5.19}$$

This is only ~ 10 per cent of the total kinetic energy of explosion estimated for SN 1006 as 5×10^{50} erg. The numerical calculations (Berezhko et al., 2002) within a specific nonlinear kinetic model of particle acceleration in SNRs (Berezhko et al., 2002) confirm that indeed the existing TeV data can be explained by accelerated protons.

The estimate of the scaling factor A derived from a comparison of the calculated and observed TeV fluxes depends significantly on the spectrum of the protons. In Fig. 5.6 the π^0-decay γ-ray fluxes calculated for 3 spectra of protons with spectral indices, $\Gamma_\mathrm{p} =1.8, 2$, and 2.1, are shown. The fluxes are normalised to the reported flux at 3 TeV. The corresponding values of the scaling factor are A=0.44, 0.83, and 1.35, respectively. Even in the case of relatively soft spectrum of protons with $\Gamma_p = 2.1$, the required scaling factor would be still acceptable if more than 20 per cent of the energy of the supernova explosion could be transformed into accelerated protons. Steeper proton spectra with slopes $\Gamma_p \geq 2.1$ do not match the energy budget of the source. Such spectra are excluded also by the EGRET flux upper limit, $F_\gamma \leq 1.7 \cdot 10^{-7}\,\mathrm{ph/cm^2 s}$ at $E \geq 100\,\mathrm{MeV}$.

The γ-ray spectra in Fig. 5.6 are calculated assuming a maximum proton energy of $E_0 = 200\,\mathrm{TeV}$. Although the exact value of E_0 does not significantly change the requirements to the scaling factor A, it has a noticeable impact on the spectral form of γ-rays above 10 TeV. Therefore only future precise spectroscopic measurements in this energy region can provide an important information about E_0. At the same time the existing sub-10 TeV data already tell us that within the framework of nucleonic model of TeV radiation of SN 1006 the cutoff energy E_0 cannot be significantly less than 100 TeV. Such large values of E_0 could be achieved only under the assumption of a large magnetic field in the shock region, $B \geq 100\,\mu\mathrm{G}$ (see Eq. 5.16). This assumption does not leave any chance for the IC mechanism

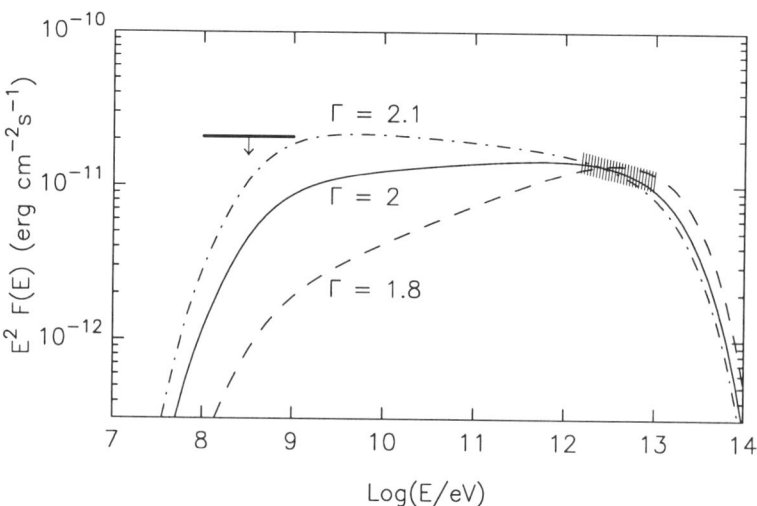

Fig. 5.6 The fluxes of π^0-decay γ-rays calculated for a proton spectrum with $E_0 = 200\,\mathrm{TeV}$, and 3 different power-law indices: $\Gamma_p = 1.8$, 2 and 2.1. The fluxes are normalised to the reported flux at 3 TeV. The corresponding values for the scaling parameter are $A = 0.44$, 0.83, and 1.35, respectively. (From Aharonian and Atoyan, 1999).

to give a noticeable contribution into the observed TeV emission. At the same time, the X-ray emission can be explained by synchrotron radiation for arbitrary magnetic field strengths . In particular, for $B \geq 100~\mu G$, the maximum electron energy is determined by the balance between the acceleration rate and the synchrotron energy loss rate, and therefore the position of the corresponding synchrotron cutoff does not depend on the magnetic field and the age of accelerator (see Eq.5.14). The X-ray spectrum of SN 1006 is satisfactorily fitted by Eq.(5.9) with $\varepsilon_0 \sim 0.1$ keV. This gives an interesting estimate of the acceleration rate. Indeed, for the characteristic shock speed $v_s \approx 2000$ km/s, the parameter $\eta \sim 20$, i.e. the acceleration rate is an order of magnitude slower than the acceleration in the Bohm regime. Then from Eq.(5.16) it follows that in order to accelerate protons to $E_0 \sim 200$ TeV, the magnetic field should be as large as 1 mG. The magnetic field in SNRs can be amplified to such levels through the generation of Alfvén waves by accelerated particles themselves (Lucek and Bell, 2000).

5.4.3 *Distinct features of electronic and hadronic models*

Within the one-zone model for IC γ-radiation from SN 1006, which assumes that all accelerated electrons are confined in the NE rim, the observed X-ray to TeV γ-ray flux ratio requires a very low magnetic field of about 5 μG. A more realistic model of γ-ray production that includes the effect of convective and diffusive escape of electrons allows a somewhat larger magnetic field, up to 10 μG, with preferable values being in a rather narrow range between 6 and 8 μG. Given the short time of particle acceleration available, $t_0 \leq t_{SN1006} \simeq 10^3$ yr, such a small magnetic field would imply a very high shock speed in SN 1006, $v_s \geq 3000$ km/s, in order to provide the acceleration of electrons to energies $E_0 \sim 30$ TeV needed to explain the X-ray and TeV observations.

This model allows definite predictions which could be tested by future observations. A significant, if not dominant, fraction of the IC TeV emission should be produced outside the NE rim. Namely, a γ-ray flux comparable with the flux from NE rim should be expected from the extended inner region of the remnant adjacent to the rim. The size of this region depends on the character of propagation of electrons in the rim. Indeed, even in the case of diffusion of electrons in the Bohm limit in the remnant with a magnetic field of $B_2 \sim 2~\mu G$, the characteristic distance of penetration of electrons towards the central region of the remnant can be estimated as

$$l(E_e) \simeq \sqrt{2\,D(E_e)t_0} \simeq 0.75\,(E_e/10\,\text{TeV})^{1/2}\,\text{pc}. \qquad (5.20)$$

Since the IC production of TeV γ-rays (on the 2.7 K CMBR) with energies less than several TeV takes place in the Thompson regime, and therefore their characteristic energy scales as $E \simeq 1\,(E_e/20\,\text{TeV})^2\,\text{TeV}$, the size of the emission region of TeV gamma rays should be as large as the width of the rim, and increases linearly with the energy of the γ-rays. Moreover, if the propagation of electrons in the remnant is much faster than in the Bohm limit ($\eta \gg 1$), the electrons could fill up practically the entire remnant. In that case the energy dependence of the size of γ-ray emission will be significantly weakened.

Another distinctive feature of the IC origin of TeV radiation is the spectral shape of radiation: (i) very hard, with the photon index $\Gamma \sim 1.5$ below 1 TeV, (ii) flat with $\Gamma \sim 2$ between 1 and 10 TeV, and (iii) very steep above 10 TeV.

The alternative to the IC interpretation is the nucleonic origin of the observed TeV radiation. This interpretation becomes energetically comfortable, which implies no more than 10^{50} erg in accelerated protons and nuclei, if we assume (i) a *small* distance to the source of about 1 kpc, and (ii) a significant enhancement of the gas density in the rim due to strong shock compression. This predicts a compact size of the TeV emission comparable with that of the X-ray rim. In the region below 1 TeV the spectrum of γ-rays of nucleonic origin, with a power-law index of $\Gamma \simeq \alpha_p \sim 2$ is expected to be steeper than the spectrum of IC γ-rays. However a flatter spectrum of π^0-decay γ-rays cannot be excluded if the protons have an acceleration spectrum as hard as $E^{-1.5}$ (see Malkov, 1997).

The shape of the spectrum of π^0-decay γ-rays above 1 TeV depends on the characteristic maximum energy of accelerated protons. Since the flux of TeV γ-rays is no longer connected with the X-ray fluxes, the value of the magnetic field in the acceleration region (NE rim) could be as high as 1 mG. Correspondingly, the energies $E_0 \geq 100\,\text{TeV}$ could be achieved. On the other hand if E_0 is significantly less than 100 TeV, one may expect a turnover in the spectrum of γ-rays above several TeV.

Thus, it is possible that both in the low (sub-TeV) and in the high (multi-TeV) energy regions the IC and π^0-decay γ-rays have similar spectra. Therefore the spatial rather than spectroscopic measurements of γ-radiation above 100 GeV with future IACT arrays will provide decisive information about the origin of very high energy radiation of SN 1006. The IC models predict significant γ-ray fluxes not only from the rim, but also from the inner parts of the remnant. On the other hand, the π^0-decay γ-rays trace the density profile of the gas in the production region, therefore one should expect a rather compact γ-ray source essentially coinciding with the rim.

Because of the small range of magnetic field strength allowed by the IC model of TeV radiation of SN 1006, the energy required for relativistic electrons in this model is predicted with good accuracy: $W_e \simeq 5 \times 10^{48}$ erg. This is at least two orders of magnitude larger than the energy contained in the magnetic field, even if we ignore the energy contained in accelerated protons. Thus, the interpretation TeV radiation in terms of IC scattering would imply that the conditions in SN 1006 are far from the equipartition regime. Also the acceleration of electrons should proceed in the regime close to the Bohm diffusion limit with $\eta = 1$.

The nucleonic model of TeV emission requires a total energy in accelerated protons from 2×10^{50} erg down to 10^{49} erg, depending on the enhancement factor for the gas density in the rim, the spectral index of the accelerated protons, and the distance to the source. In particular, for a moderate assumption for the spectral index $\Gamma_p = 2$, and the gas compression ratio $\rho = 4$ (i.e. $n = 1.6 \, \text{cm}^{-3}$), the energy in protons is estimated to be $\approx 5 \times 10^{49} (d/1 \, \text{kpc})^2$ erg. In order to accelerate protons to energies $E_0 \sim 100$ TeV the ambient magnetic field should exceed 100 μG.

5.4.4 *Concluding remarks*

The synchrotron X-ray emission from SN 1006 is an indication for acceleration of electrons in these objects to energies significantly exceeding 10 TeV. Although the inverse Compton scattering of same electrons unavoidably leads to production of VHE γ-rays, the flux of TeV γ-ray emission from SN 1006 reported by the CANGAROO collaboration exceeds, at least an order of magnitude, the γ-ray flux of inverse Compton origin expected from this supernova remnant, unless one assumes unreasonably small magnetic field in the production region of synchrotron X-rays. The required magnetic of about 5 μG seems to be quite unrealistic not only because we generally expect much stronger field in the compressed shocked regions of the shell, but also because it implies acceleration of electrons in the extreme Bohm diffusion regime by the shock with a speed as large as 3500 km/s. Also, the IC model requires a strong departure from the equipartition condition in the region of particle acceleration, $W_e \geq 100 W_B$.

The γ-ray fluxes predicted by hadronic models are also noticeably below the reported TeV flux, unless one assumes that the acceleration of protons proceeds with efficiency significantly exceeding the nominal value of 10 per cent, and that the gas density in the shell is significantly enhanced due to strong shock compression.

Thus, it is clear that both the leptonic and hadronic models face serious difficulties to explain the TeV fluxes reported by the CANGAROO collaboration from SN 1006. Either the CANGAROO result is wrong (the simplest solution from the point of view of a theorist), or something is conceptually wrong with our current understanding of acceleration, propagation and radiation of very high energy particles in supernova remnants (the favoured outcome from the point of view of an observer). On the other hand, there is little doubt that, if the supernova remnants are indeed effective factories of multi-TeV electrons and protons, TeV γ-ray emission should eventually show up, at one flux level or another.

Although the results of this section were explicitly devoted to SN 1006, they, in fact, can be successfully applied to a *generic* supernova remnant, by considering the acceleration rates of electrons and protons as free parameters. Hopefully, the observations of SNRs with the next generation of detectors with significantly improved sensitivities will comprise a solid observational basis for further phenomenological and theoretical studies. In particular, the extension of measurements both down to GeV energies and beyond 10 TeV will provide the currently missing arguments in favour of (or against) the leptonic or hadronic models. However, key information about the origin of this radiation most probably will be provided by studies of the *spatial distribution* of TeV emission. The IC models require a very small magnetic field, and therefore relatively broad spatial distribution of TeV radiation. The hadronic models assume strong shock compression of the ambient gas, therefore they predict TeV emission produced essentially in the narrow shell. The forthcoming arrays of imaging Cherenkov telescopes have the potential to address these issues in the near future.

5.5 Molecular Clouds Overtaken by SNRs

The reasons which make the detection and identification of π^0-decay γ-rays from SNRs that are located in ordinary (low-density) regions of the ISM rather difficult, are twofold: (i) slow interaction rate ($t_{\rm pp} \simeq 5 \times 10^7 (n/1~{\rm cm}^{-3})^{-1}$ yr), allowing only limited efficiency of conversion of energy of relativistic protons to γ-rays, $L_\gamma \simeq \dot{W}_{\rm p}(3 t_{\rm pp}/t_{\rm SNR})^{-1} \leq 10^{-3} \dot{W}_{\rm p}$;

(ii) significant "contamination" due to the IC component of radiation, especially at TeV energies. Therefore, the best hope to obtain solid evidence of γ-rays of hadronic origin is associated with SNRs in dense environments (Aharonian *et al.*, 1994a). In this regard two scenarios seem to be relevant – "SN explosion inside the GMC" and "GMCs overtaken by the supernova shell". The features of γ-radiation of both cases are discussed in Sec.4.2 (Chapter 4) without specifying the type of particle accelerator. In the particular case of SNR shocks operating as particle accelerators, the scenario of "SN explosion inside the GMC" cannot guarantee effective particle acceleration to very high energies, because of rapid cooling of the shock. Moreover, even if particles are somehow accelerated to very high energies, they may escape the clouds quickly, leading to formation of steep particle spectra inside the cloud. The second scenario gives a more or less passive role to giant molecular clouds (GMCs) which act as targets for protons, significantly enhancing the production rate of π^0-decay γ-rays. This scenario seems the most natural way to amplify the γ-radiation to a level comparable with the sensitivity of EGRET. Therefore, if the the reported associations of several EGRET sources with SNRs (Sturner and Dermer, 1995; Esposito *et al.*, 1996; Torres *et al.*, 2003) are not result of accidental geometrical coincidence, dense GMCs overtaken by SNRs seem the most natural sites of γ-ray production. A SNR in cloudy environments may in fact initiate a cluster consisting of several γ-ray sources associated with it (Aharonian and Atoyan, 1996).

Montmerle (1979) was perhaps the first to suggest that SN explosions occurring in OB associations which are rich in massive molecular clouds may lead to observable high energy γ-ray fluxes. Such evidence has been found by Pollock (1985) who argued that the interaction of the SNR G78.2+2.1 (γ Cygni) with a nearby cloud should be responsible for γ-rays detected by COS B in that direction.

The discovery by EGRET of approximately 80 γ-ray sources at low galactic latitudes, the most of them being still unidentified, initiated new interest in SNRs interacting with nearby clouds. In the last several years the possible correlation of sources from the 3rd EGRET catalog (Hartman *et al.*, 1999) with relatively young galactic SNRs has been extensively explored. These efforts, motivated by a search for counterparts for the unidentified EGRET sources at low galactic latitudes, revealed new evidence of SNRs interacting with molecular clouds. An interesting outcome of these studies was, for example, the discovery of a large ($8° \times 8°$), low brightness shell type SNR in the Capricornus region, the radio structure of which spatially correlates with three unidentified EGRET sources (Combi et al., 2001). Moreover, the 21 cm line observations revealed that two of these sources coincide with HI clouds. Although the possibility of a chance association cannot be ruled out, this cluster of γ-ray sources can be naturally interpreted as radiation of dense clouds overtaken by the remnant. Despite large observational and theoretical uncertainties (e.g. in the ambient gas density and strength of the magnetic field, the age of the remnant, the energy spectra and e/p ratio of accelerated particles, etc.), the interpretation of γ-ray data as the result of interactions of accelerated particles in dense regions ($n \geq 10 \text{ cm}^{-3}$) favours two radiation mechanisms – electron bremsstrahlung and π^0-decay γ-rays. On the other hand it is difficult to estimate the contributions from each of these channels based merely on the low-energy γ-ray observations. Detailed multiwavelength studies of a large sample of SNR-GMC interacting systems are needed to understand the acceleration and radiation processes in these objects.

X-ray observations are of special interest. Hard X-rays of synchrotron origin detected from several shell type SNRs are an unambiguous indicator of multi-TeV electrons accelerated, most likely, by strong SNR shocks. Since the cutoff energy of this component determined by Eq.(5.14) is always expected around or below 0.1 keV, the synchrotron X-radiation of SNRs can be easily recognised by its steep spectrum above 1 keV, as well as by its spatial localisation in the shell, where the acceleration takes place. Besides the X-rays of synchrotron origin, we may expect another component of hard X-radiation due to the nonthermal bremsstrahlung of subrelativistic electrons or protons. Because of the Coulomb-loss-flattened distribution of electrons in high density environments, this mechanism predicts an extremely hard X-ray spectrum with power-law photon index $\Gamma \sim 1$ (Uchiyama et al., 2002a). Therefore, the high energy γ-radiation of either bremsstrahlung or π^0-decay decay origin should be accompanied of subrelativistic X-radiation with a characteristic ε^{-1} type spectrum, detection of which may serve as an indicator of the existence of a large amount of subrelativistic particles in clouds. For a given spectrum of the electron population, the ratio of the X- and γ-ray fluxes depends on the position of the Coulomb break in the electron spectrum, which in its turn depends on the product of the ambient density and the age of the source. It is likely that such radiation has been indeed detected from γ Cygni (Uchiyama et al., 2002a) and RX J1713.7-3946 (Uchiyama et al., 2002b).

5.5.1 Bremsstrahlung X-rays from γ Cygni

The broad-band spectral energy distribution of the so-called Hard X-ray Clump (HXC) detected by ASCA in the supernova remnant γ Cygni is shown in Fig. 5.7. Nonthermal bremsstrahlung from the accelerated electrons is a natural source of the HXC flux, because the shocked dense clouds act as an effective target for energetic electrons. Due to Coulomb interactions the high density gas results in a significant hardening of low-energy electrons below the "Coulomb break", giving rise to the standard ε^{-1} type bremsstrahlung spectrum in the X-ray band, which agrees with the ASCA data. The bremsstrahlung spectrum above the "Coulomb break" essentially repeats the acceleration spectrum of electrons.

Fig. 5.7 presents the results of numerical calculations for 2 sets of parameters which describe the gas density and the acceleration spectrum of electrons, assuming that electron bremsstrahlung is responsible for both the ASCA hard X-ray and the EGRET γ-ray fluxes. For the electron spectrum with the acceleration index $\Gamma_e = 2.1$, the best fit is achieved for a gas density of $n = 34$ cm^{-3}. A steeper acceleration spectrum with $\Gamma_e = 2.3$ requires larger gas density, $n = 130$ cm^{-3}. Note that the adopted acceleration spectra are consistent with the reported radio spectral index $\alpha_r = 0.5 \pm 0.15$. Also, an exponential cutoff in the electron spectrum at 10 TeV is assumed. If the electron distribution with $\Gamma_e = 2.3(2.1)$ extends beyond GeV energies, for the magnetic field 10^{-5} G the calculated radio flux density amounts to about 10%(60%) of the measured radio flux density integrated over the whole remnant. Furthermore, if the electron distribution extends beyond TeV energies, the bremsstrahlung spectrum with $\Gamma_e = 2.1$ exceeds the Whipple upper-limit, whereas the spectral index $\Gamma_e = 2.3$ is still in agreement with the Whipple result. Both combinations of model parameters quite satisfactorily fit the spectral shape and the absolute flux of hard X-rays.

Ignoring the energy losses of electrons would lead to significantly steeper X-ray spectra, and would also result in overproduction of absolute X-ray fluxes (dotted curves in Fig. 5.7). Note that the main contribution to X-rays comes from relatively high energy electrons with energies close to 10 MeV for the acceleration index $\Gamma_e = 2.3$.

Because of its poor angular resolution, EGRET measurements do not provide clear information about the site(s) of production of high energy γ-rays. Nevertheless, it is likely that only a part (perhaps, even only a small part) of the reported high energy γ-ray fluxes originates in the HXC region.

The γ-ray fluxes could be easily suppressed by assuming lower gas densities. Indeed, such an assumption would lead to a shift of the "Coulomb break" energy in the electron spectrum to lower energies, and the predicted high energy γ-ray spectra would appear significantly below the reported EGRET fluxes (the solid curve in Fig. 5.7).

A more likely candidate site for the production of the bulk of high energy γ-rays is the region called DR4 from which most of the radio emission emerges. A massive cloud with a density of ~ 300 cm^{-3} occupying $\sim 5\%$ of the SNR volume has been suggested to exist in the vicinity of DR4 to explain the γ-ray flux. Actually the reported EGRET error circle is somewhat removed from the HXC, but closer to the DR4. A density of ~ 300 cm^{-3} is higher than the upper limit density of the HXC. Such high gas density implies a high (about 50 MeV) "Coulomb break" energy in the electron spectrum, and considerable suppression of the X-ray flux. This

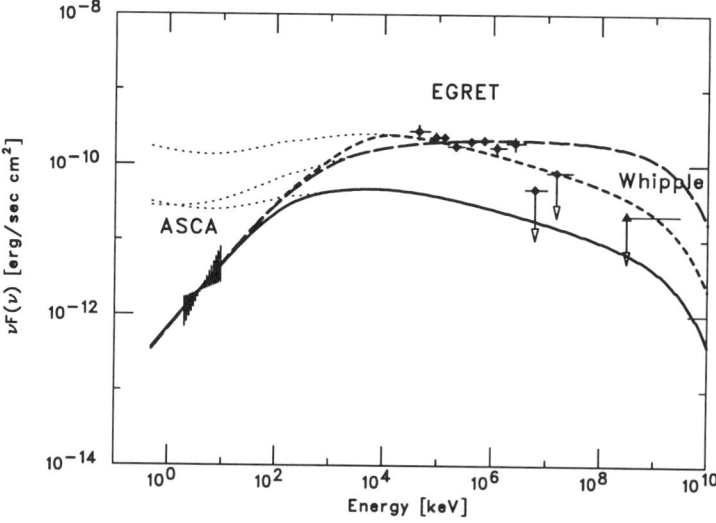

Fig. 5.7 Broadband spectral energy distribution of the HXC region of γ Cygni. The range of the power-law fit of the hard X-ray component is shown together with γ-ray data (≥ 100 MeV) of 2EG J2020+4026 taken from Esposito *et al.* (1996) and the Whipple TeV upper limit from Buckley *et al.* (1998). The bremsstrahlung photon spectra from the loss-flattened electron distribution are calculated for the electron index $\Gamma_e = 2.1$ and the gas density $n = 34$ cm^{-3} (*long-dashed line*), $\Gamma_e = 2.3$ and $n = 130$ cm^{-3} (*dashed line*), and $\Gamma_e = 2.3$ and $n = 10$ cm^{-3} (*solid line*). The *dotted lines* show the bremsstrahlung spectra corresponding to the acceleration spectra of electrons, i.e. ignoring the Coulomb losses of electrons. (From Uchiyama *et al.*, 2002a).

would naturally explain the lack of noticeable hard X-ray flux from the DR4 region, which is bright in radio and perhaps also in γ-rays.

For a gas density of about 100 cm^{-3} in the HXC, the X-ray flux is produced predominantly by electrons with energies of about 10 MeV. The X-ray flux is roughly proportional to the product of the gas density and the number of relativistic electrons, because the relativistic bremsstrahlung cross-section depends only logarithmically on the electron energy. On the other hand, the Coulomb energy loss rate of relativistic electrons, $dE/dt \propto E/\tau_{\text{Coulomb}}$ is proportional to the gas density and almost independent of the electron energy. Therefore the energy loss rate of the bulk of the electrons can be uniquely determined by the hard X-ray luminosity, independent of the density and the shape of the electron energy distribution. The estimated hard X-ray luminosity, $L_X \simeq 1.2 \times 10^{33} D_{1.5}^2$ erg/s, can be converted to an energy loss rate of $L_e \sim 5 \times 10^{37} D_{1.5}^2$ erg/s. The energy released in relativistic electrons then would be estimated as $W_e \sim \tau_{\text{age}} L_e \sim 10^{49} D_{1.5}^2$ ergs. This enormous energy deposition due to Coulomb collisions would heat the emission region of the HXC. Subsequently the heat would be radiated away in the far-infrared band by molecular line emission, if the gas is comprised of molecules. The observed infrared luminosity of γ Cygni is indeed comparable to the above estimate of the Coulomb energy loss rate.

5.5.2 The case of RX J1713.7-3946

This shell type supernova remnant is of great interest, being one of three SNRs so far detected both in synchrotron X-rays (Koyama et al., 1997; Slane et al., 1999, Uchiyama et al., 2003) and TeV γ-rays (Muraishi et al., 2000, Enomoto et al., 2002). Two very massive ($M \sim 10^5 M_\odot$) and dense ($n \sim 500$ cm^{-3}) clouds are found in the vicinity of this SNR, one of which (cloud A) probably borders with the shock-wave region of the remnant (Slane et al., 1999). The results based on observation in ASCA's large FoV revealed an extended X-ray source (Uchiyama et al., 2002b) coincident with cloud A (see Fig.5.8a). The cloud has a positional association also with the unidentified EGRET source 3EG J1714-3857. Butt et al. (2001) claimed that the shock front of RX J1713.7-3946 is interacting with the cloud A, and argued that this would be a natural site of production of ≥ 100 MeV γ-rays, presumably of hadronic origin.

The X-rays from cloud A have, most likely, nonthermal origin. As discussed in the previous section, the unusually flat spectrum (the pho-

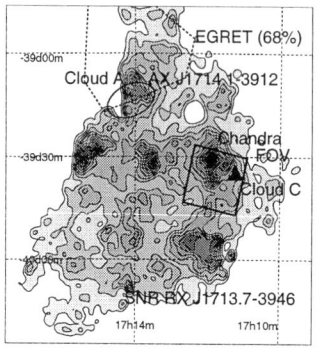 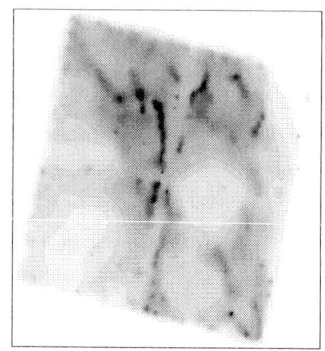

Fig. 5.8 (a) Overall ASCA view of RX J1713.7-3946 in the energy band 5-10 keV (left; from Uchiyama, Takahashi and Aharonian, 2002) and (b) the X-ray image of the northwest rim in the Chandra's $17' \times 17'$ field-of-view in the 3-5 keV band (right; from Uchiyama, Aharonian, and Takahashi, 2003).

ton index $\Gamma = 1 \pm 0.3$) can be best interpreted in terms of characteristic bremsstrahlung emission from the ionization-loss-flattened distribution of either subrelativistic electrons or protons. Although this model explains the observed spectral features of the X-ray emission, it requires huge (marginally acceptable) kinetic energy of about 10^{50} erg in the form of subrelativistic particles, if the estimate for the distance to the source of about 6 kpc (Slane et al., 1999) is correct. The energy requirement is reduced to a quite reasonable level if one assumes that the SNR is located much closer, e.g., at $d \simeq 1$ kpc (Koyama et al., 1997). In any case, the relation of this radiation to RX J1713.7-394 is not quite clear, and in fact is a subject of further studies.

In this regard, the X-ray emission detected from the northwest rim of RX J1713.7-39, as well as the TeV radiation detected from the same direction have more fundamental implications for this supernova remnant. The synchrotron origin of the X-radiation implies the existence of multi-TeV electrons, and contains information about the parameter $B^{1/2}E_0$. In order to disentangle B and E_0, one needs an additional piece of information, and so long the 2.7 K CMBR remains the main reservoir for target photons and the magnetic field does not significantly exceed 10 μG, such information is contained in IC γ-rays. But these observations would have much more fundamental implications, in particular for the origin of galactic CRs, if the detected TeV-radiation was of hadronic origin (e.g. Aharonian, 2002a).

The observations of RX J1713.7-3946 by the CANGAROO collaboration using the new 10 m diameter imaging Cherenkov telescope (Enomoto et al., 2002) not only confirmed the TeV signal reported earlier (Muraishi et al., 2000), but also provided very important information about the shape of the differential spectrum, which between 0.4 TeV and 10 TeV is approximated by a power-law with photon index $\alpha = 2.84 \pm 0.35$. If confirmed, this information would be almost sufficient for a definite conclusion about the origin of TeV γ-rays.

Indeed, the energy flux of TeV radiation of about $5 \cdot 10^{-11}$ erg/cm^{2}s is quite comparable with the flux in synchrotron X-rays (Enomoto et al., 2002), which implies that the IC origin of TeV γ-rays would require a magnetic field as small as $B = (8\pi w_{2.7K})^{1/2} \simeq 3~\mu$G. This requirement is tighter than the case of SN 1006, especially if one takes into account that RX J1713.7-394 is located in the galactic plane. Given the shock compression of the magnetic field in the shell by a factor of 4 or more, this upper limit on the magnetic field seems unrealistic, and consequently the interpretation of TeV data in terms of the IC scattering on the 2.7 K CMBR rather inadequate. Moreover, the observed steep spectrum at relatively low energies does not agree with prediction of the "2.7 CMBR inverse Compton" origin of TeV γ-rays. Indeed, the spectral fit for the synchrotron radio to X-ray data requires $B^{1/2}E_0 \simeq 100~\mu$G$^{1/2}$TeV and the electron spectral index $\alpha_e = 2.08$ (Enomoto et al., 2002). For $B \leq 3~\mu$G, this gives a lower limit on the cutoff energy of electrons, $E_0 \geq 50$ TeV. Since the target photon is fixed, this allows a robust prediction for the spectral shape of the TeV emission. Namely, it can be described by a hard power-law with a photon index of $\Gamma \simeq 1.5$ up to 1 TeV, after which it gradually steepens to $\Gamma \sim 2$ around a few TeV, and $\Gamma \geq 3$ above 10 TeV. Since this is in apparent conflict with the observed single power-law type spectrum with $\Gamma \simeq 2.85$, the IC interpretation can be discarded if the 2.7 K CMBR represents the main target field for electron scattering (see Fig. 2.10).

Formally one may assume that some other seed photon field, different to the 2.7 K CMBR, dominates in the formation of the IC scattering component. In this case the steep γ-ray spectrum can be explained by the Klein-Nishina effect. This implies that the mean energy of these seed photons should be 1 eV or larger. In order to make this IC component dominant, one has to assume an unreasonably high density of optical/UV photons in RX J1713.7-39, one that is not supported by observations. The bremsstrahlung component also can be confidently excluded. Although for

$B \simeq 3$ μG and $n \simeq 300$ cm^{-3} the absolute bremsstrahlung flux around 1 TeV can match the observed γ-ray flux, in order to avoid overproduction of γ-rays at higher energies one must assume a cutoff in the electron spectrum around 1 TeV which, however, contradicts the synchrotron-best-fit condition of $B^{1/2}E_0 \simeq 100$ μG$^{1/2}$TeV.

After excluding all possible γ-ray components of leptonic origin, one may arrive at conclusion that π^0-decay γ-radiation remains the only option which can explain the TeV data without being in conflict with the X-ray data. The X-rays can be still explained by synchrotron radiation of electrons. This of course would require suppression of the IC component of γ-radiation assuming $B \geq 3$ μG. The model calculations presented in Fig. 2.10 demonstrate that the TeV spectral points can be satisfactorily fitted by the π^0-decay mechanism assuming an exponential cutoff in the proton spectrum at $E_0 = 10$ TeV. For the average hydrogen density in the cloud of about $n \simeq 300$ cm^{-3}, the observed TeV fluxes can be explained if the content of relativistic protons in the cloud is about $W_\mathrm{p} = 3 \times 10^{48}(d/1 \mathrm{\ kpc})^2$ erg. The different estimates of the distance to the source vary between 1 kpc (Koyama et al., 1997) and 6 kpc (Slane et al., 1999). The total energy accelerated in protons typically is limited to 10^{50} erg, a significant part of which is may contained in subrelativistic protons or electrons. Also, for any realistic geometry only a fraction of accelerated protons can be captured/confined in the cloud. Therefore, a relatively small distance to the source, close to its lower limit of about 1 kpc, seems preferable on energy grounds.

The hadronic model of the TeV emission of RX J1713.7-39 has been criticised by Reimer and Pohl (2002) and Butt et al. (2002) who argued that this interpretation violates the γ-ray upper limits set by EGRET at GeV energies. This is, however, a rather shaky argument; the "GeV upper limit" problem can be overcome in several natural ways. For example, (i) adopting a slightly harder proton spectrum, it is possible to avoid the conflict with the EGRET data. Even for a proton spectrum steeper than E^{-2}, it is still possible to suppress the GeV γ-ray flux, if one invokes (iii) the effects of energy-dependent propagation of protons while travelling from the accelerator (SNR shock) to the nearby clouds. Moreover, the lack of the GeV γ-rays can be naturally explained by (iii) confinement of low-energy (≤ 10 GeV) protons in the shell, in contrast to the effective escape of high-energy (multi-TeV) protons, assuming that the particle acceleration proceeds in the Bohm diffusion regime. And finally, the GeV γ-ray emission perhaps can be additionally suppressed assuming that (iv) the low energy protons do not freely penetrate the densest regions of the cloud due to quasi-static magnetic mirrors or different type of plasma instabilities caused by the same cosmic rays.

On the other hand, the limited energy budget available for cosmic rays requires TeV γ-ray production in very dense regions, in particular in giant molecular clouds which should significantly override the proton acceleration site(s). Unfortunately, the quality of the CANGAROO data is not sufficient for definite conclusions concerning the exact location of the TeV production region(s). The observations of this source by the H.E.S.S. and CANGAROO-III arrays of Cherenkov telescopes should allow detailed morphology of the TeV production region with a bin size as small as several arcminutes. A similar accuracy for study of spatial distribution of GeV γ-rays can be achieved in future observations of RX J1713.7-39 by AGILE and GLAST. Although hadronic models generally predict similar spatial

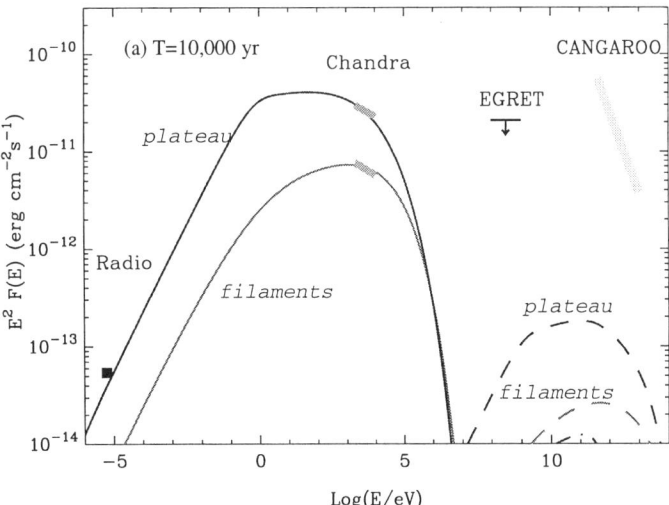

Fig. 5.9 Multiwavelength synchrotron (*solid lines*) and IC (*dashed lines*) radiation spectra from the filaments and the plateau regions of RX J1713.7-39 (from Uchiyama, Aharonian and Takahashi, 2003). It is assumed that the electron acceleration takes place in the filaments. The following parameters sets have been used in calculations: The age and the distance to the source: $T = 10\,000$ yr and $d = 6$ kpc. The acceleration rate, the spectral index and the exponential cutoff of the electron spectrum: $L_e = 2.8 \times 10^{36}$ erg s^{-1}, $\alpha = 1.84$, and $E_0 = 125$ TeV, respectively. The magnetic field in the filaments and in the plateau region : $B_{\rm fil} = B_{\rm pla} = 50\,\mu$G. The convective escape time $\tau_{\rm con} = 1000$ yr, the gyrofactor in the filaments $\eta = 83$.

distributions of GeV and TeV γ-rays, some differences nevertheless may arise due to possible energy dependent effects in the propagation of protons from their acceleration site to the nearby clouds.

The importance of comprehensive studies of both spatial and spectral properties of nonthermal radiation has been demonstrated by the Chandra observations of the northwest rim of RX J1713.7-39, the brightest region of nonthermal X-rays. The Chandra image revealed (Uchiyama et al., 2003) a complex network of X-ray filaments surrounded by a fainter diffuse plateau, and a dark region of circular shape (see Fig. 5.8b). The examination of the individual spectra from different parts of the rim showed that despite significant brightness variations, the X-ray spectra everywhere in this region are similar, being well fitted with a power-law with photon index $\Gamma \simeq 2.3$. Furthermore, the spectra indicate that the synchrotron cutoff is located beyond 10 keV. Both the filamentary structure and the lack of synchrotron cutoff below a few keV may challenge the perceptions of the standard shock acceleration models concerning the production, propagation and radiation of ultrarelativistic electrons.

The X-ray brightness distribution shown in Fig. 5.8b most likely is the result of highly inhomogeneous production and propagation of relativistic

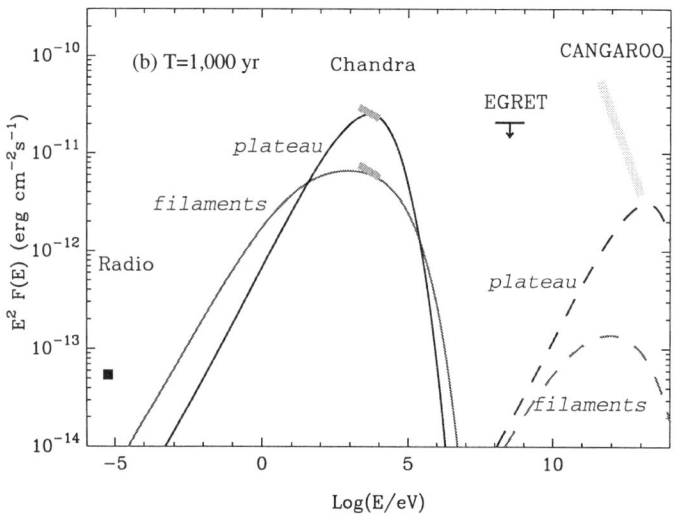

Fig. 5.10 The same as in Figs. 5.9, but for the following model parameters: $T = 1000$ yr, $d = 2$ kpc, $L = 1.6 \times 10^{37}$ erg s^{-1}, $\alpha = 1.95$, $E_0 = 200$ TeV, $B_{\text{fil}} = 20\,\mu\text{G}$, $B_{\text{pla}} = 6\,\mu\text{G}$, $\tau_{\text{con}} = 500$ yr, $\eta = 1$. (From Uchiyama, Aharonian and Takahashi, 2003).

electrons. At first glance, homogeneous distribution of multi-TeV electrons over the rim cannot be ruled out, because in this case the filamentary structure might be referred to a local enhancement of the magnetic field. However, in the regime, when the synchrotron cooling time is shorter than the source age, the synchrotron X-ray luminosity does not depend significantly on the magnetic field. In this regime an equilibrium is quickly established between the electron injection and synchrotron losses. This implies that the synchrotron radiation saturates to the maximum possible rate determined by the electron injection rate.

In this context it seems natural to identify the X-ray filaments as sites of particle acceleration. It is likely that the filaments are bright not only because of the enhanced magnetic field, but also due to high concentration of electrons inside the accelerators. A significant fraction of these electrons may escape from the thin filaments, and fill much larger regions. This will result in synchrotron X-rays from the extended regions (plateau), the brightness of which can be low, but their overall X-ray flux may exceed the total flux from filaments.

The results of a numerical time-dependent treatment, which includes the energy losses and escape of electrons from their accelerators, but do not specify the acceleration mechanism (i.e. assumes an *a priori* acceleration rate and spectrum of electrons extending to $\geq$ 100 TeV), are shown in Figs. 5.9 and 5.10, for two different sets of model parameters (Uchiyama *et al.*, 2003). These results show that the hypothesis of electron acceleration in the filaments and hot-spots does explain the fluxes and spectral shapes of the X-ray emission from both the filaments and the plateau region. On the other hand, the calculated IC fluxes are at least one order of magnitude below the TeV fluxes reported by the CANGAROO collaboration.

5.6 A Special Case: Gamma Rays from Cassiopeia A

The shell type supernova remnant Cassiopeia A is the brightest and one of the best studied radio sources in our Galaxy. The synchrotron radiation of this source continues to submillimeter wavelengths, and perhaps even further to the infrared (Tuffs *et al.*, 1997) and hard X-rays (Allen *et al.*, 1997). The estimates of the mean magnetic field and the energy content in radio emitting electrons based on simple equipartition arguments give 0.16 mG and 5×10^{49} erg, respectively (Anderson *et al.*, 1991). Assuming a larger field, of about 0.3 mG, and taking into account the propagation effects one may reduce the total energy in relativistic electrons to a more comfortable level of about 3×10^{48} erg compared to the estimate of 1.5×10^{50} erg in the kinetic energy of diffuse ejecta of Cas A (Braun, 1987). On the other hand, the TeV γ-ray observations by HEGRA (Aharonian et al., 2001d) constrain the total amount of accelerated protons to 10^{49} erg (see below). Thus, the content of electrons in accelerated particles in this source is significantly enhanced compared to the ratio $e/p \sim 1/100$ observed in local CRs. This implies that Cas A cannot be representative of the source population which provides the bulk of the observed CR fluxes. To a large extent this is not a surprise because Cas A is a unique source in our Galaxy in general, and among SNRs, in particular.

Although most of the radiation of Cas A of both thermal and nonthermal origin comes from a shell enclosed between two spheres with angular radii of 100 arcsec and 150 arcsec, corresponding to spatial radii $R = 1.7$ pc and $R = 2.5$ pc for a distance to the source of $d = 3.4$ km (Reed et al., 1995), the bulk of nonthermal energy may originate not in the shell, through the diffusive shock acceleration of particles (as in most of SNRs), but in the numerous hot spots which appear to be fast moving knots (Scott and Chevalier, 1975) or compact, steep-spectrum radio structures (e.g. Bell, 1977; Cowsik and Sarkar, 1984). The radio structures are of special interest because it is likely that their bright radio emission is not just the result of an enhanced magnetic field, but is (also) caused by the local enhancement of relativistic electrons. In other words, these regions can be the sites of particle acceleration. The calculations of the spectral and temporal evolution of the synchrotron radiation of Cas A within a two-zone model which distinguishes between compact, bright steep-spectrum radio knots and the diffuse "plateau" (Atoyan et al., 2000a), provide further support for this hypothesis. In particular, it has been shown that the energy distributions of electrons in these compact structures becomes significantly steeper than the acceleration spectrum on timescales of the *energy-dependent* escape of electrons into the surrounding diffuse plateau region.

A possible fit to the broad band synchrotron fluxes of Cas A is shown in Fig. 5.11. Magnetic fields $B_1 = 1.2$ mG and $B_2 = 0.3$ mG in the compact bright radio structures and in the diffuse plateau region, respectively, are assumed. The total magnetic field energy in the shell is then $W_{B2} = 3.8 \times 10^{48}$ erg. In the compact zone-1 regions it is significantly less, $W_{B1} = 2 \times 10^{47}$ erg. On the other hand, the contents of relativistic electrons in zones 1 and 2 are comparable – $W_{e1} = 1.6 \times 10^{48}$ erg and $W_{e2} = 1.8 \times 10^{48}$ erg, respectively.

By comparing the bremsstrahlung flux of radio emitting electrons with the flux upper limit $I(> 100\,\mathrm{MeV}) \leq 1.1 \times 10^{-6}\,\mathrm{ph/cm^2 s}$ set by the SAS-2 and COS B detectors, Cowsik and Sarkar (1980) derived a lower limit to the mean magnetic field in the shell of Cas A, $B_0 \geq 8 \times 10^{-5}\,\mathrm{G}$. Such a meaningful constraint on the magnetic field is possible because of the high radio flux of Cas A and effective production of high energy γ-rays via electron bremsstrahlung. This is actually quite an unusual situation compared with other shell type "isolated" SNRs (i.e. remnants in regions free of massive clouds) for which bremsstrahlung is generally a less effective γ-ray production mechanism. The reasons are twofold: a very large amount of electrons with energy extending to at least 10 GeV (the electrons which produce synchrotron radiation at mm wavelengths), and an unusually large density of the gas in the nebula, $n \geq 10\,\mathrm{cm^{-3}}$. In addition, in the heavy, oxygen-rich gas in Cas A, the electron bremsstrahlung is amplified ($\propto Z(Z+1)/A$) by a large factor of about 4 compared to bremsstrahlung in a pure hydrogen gas.

Another interesting feature of the γ-ray production in Cas A is that

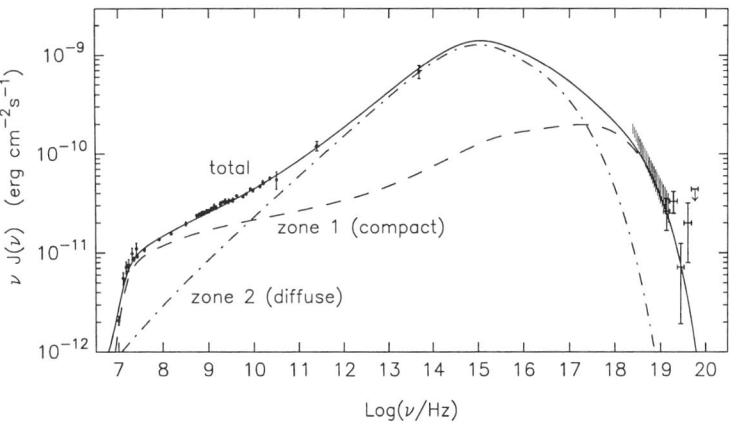

Fig. 5.11 Synchrotron fluxes calculated in the framework of two-zone model for Cas A (from Atoyan et al., 2000a). The acceleration spectrum of the electrons in the zone 1, which is modelled as being composed of 150 compact structures with a mean radius $R = 0.06\,\mathrm{pc}$, is assumed to be a power-law with an index of 2.2 and exponential cutoff at $E_0 = 18\,\mathrm{TeV}$; the escape time of electrons from knots is assumed as $\tau(E) = 28(E/1\,\mathrm{GeV})^{-0.7}\,\mathrm{yr}$. The magnetic fields in the knots (zone-1) and in the plateau region (zone-2) are $B_1 = 1.2\,\mathrm{mG}$ and $B_2 = 0.3\,\mathrm{mG}$, respectively. Dashed and dot-dashed curves show contributions of zone 1 and zone 2 to the overall synchrotron radiation flux (solid curve).

the inverse Compton component is dominated by scattering of electrons on far infrared dust emission with $T \simeq 100$ K (Atoyan et al., 2002). For other SNRs, the 2.7 K CMBR dominates over all other seed photon fields. Even so, because of severe synchrotron energy losses of the electrons in the strong magnetic field, exceeding 100 μG, the IC component of radiation in Cas A is significantly suppressed, and only at TeV energies it may become comparable with the bremsstrahlung flux.

The expected γ-ray fluxes produced by the electrons responsible for the broad-band synchrotron emission presented in Fig. 5.11, are shown in Fig. 5.12. The thin solid line corresponds to the total flux of the bremsstrahlung γ-rays, and the dashed and dot-dashed lines represent the bremsstrahlung fluxes from zone-1 and zone-2, respectively.

Because of the steep decline of the energy distribution of radio electrons in the compact radio structures (zone-1), the intensity of the γ-ray flux at $E \sim 1\,\text{GeV}$ is dominated by the flat-spectrum bremsstrahlung of the plateau region (see Fig. 5.12). It is also important that the contribution of other γ-ray production mechanisms at this energy is not yet significant. Therefore, future measurements of the differential flux of high energy γ-rays from Cas A by GLAST should allow rather accurate determinations of a number of important parameters in Cas A.

If the fluxes of hard X-rays observed at $E \geq 10\,\text{keV}$ have indeed a synchrotron origin (Allen et al., 1997), relativistic electrons in Cas A should be accelerated to energies up to tens of TeV. These electrons should then produce TeV γ-rays. Along with bremsstrahlung, at these energies the principal mechanism for γ-ray production is the inverse Compton (IC) scattering of electrons in the field of thermal dust emission with $T = 97\,\text{K}$ (Mezger et al., 1986).

The fluxes of IC γ-rays expected from Cas A depend very sensitively on the mean magnetic field B_2 in zone-2 (the shell). In particular, in the VHE regime, $F_{\text{IC}} \sim B_2^{-(5+\alpha_2)/2}$, where $\alpha_2 \geq 3$ is the spectral index of TeV electrons in the shell. Thus, the flux of IC γ-rays could be significantly increased assuming smaller values of B_2, and hence also of B_1 fields, since the radio data require the ratio of magnetic fields in two zones to be approximately at the level of $B_1/B_2 \sim 4$. On the other hand, B_2 cannot be significantly less than 0.3 mG, otherwise bremsstrahlung would lead to overproduction of X- and γ-rays, in contradiction with the OSSE/RXTE and EGRET data (see Fig. 5.12).

Formally, the IC γ-ray flux can be increased assuming a more structured model for the magnetic field distribution in the shell. Namely, one may assume that the magnetic field in the shell of Cas A is reduced from the highest value B_1 in the compact zone-1 (the acceleration sites) to a lower value B_2 in the surrounding zone-2, and further on to some $B_3 \leq B_2$ in zone 3. The latter would then represent the regions of the shell with relatively low magnetic field or the regions adjacent to the shell. Relativistic electrons could then escape from zone-2 into zone-3, as they do from zone-1 into zone-2. Because the shell is significantly larger than the compact zone-1 structures, the characteristic timescales for the electron escape from zone-2 into zone-3 should be significantly larger than that the escape time from zone-1 into zone-2.

In Fig. 2.11 the fluxes of IC γ-rays, calculated in the framework of the 3-zone model with $B_3 = 0.1\,\mathrm{mG}$, are shown. Two different values for the magnetic field in zone-1, $B_1 = 1\,\mathrm{mG}$ (solid line) and $B_1 = 1.6\,\mathrm{mG}$ (dashed line) are assumed. The magnetic field in the zone-2 for both cases is fixed

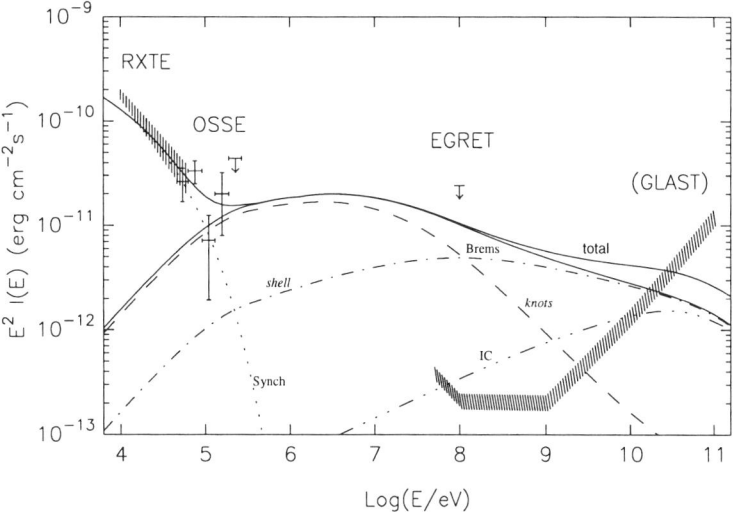

Fig. 5.12 The fluxes of synchrotron (dotted line), inverse Compton (3-dot–dashed line), and bremsstrahlung (solid line) radiations calculated in the framework of the two-zone model of Cas A for the same model parameters as in Fig. 5.11. The broad band overall spectral energy distribution consisting of these three components of radiation is shown by the curve marked as "total". The bremsstrahlung fluxes produced in zone-1 and zone-2 are shown separately by the dashed and dot-dashed curves, respectively. The hatched region shows the expected flux sensitivity of GLAST. The X-ray/soft γ-ray fluxes measured by RXTE and OSSE detectors, as well as the flux upper limit from EGRET, are also shown.

to $B_2 = B_1/4$. Although B_1 (and B_2) in these 2 cases change only by a factor of 1.6, the fluxes of TeV γ-rays drop significantly, by a factor of 6.

For the given magnetic fields B_1 and B_2 in the zones 1 and 2, the 3-zone model allows an increase (compared to the 2-zone model) of the γ-ray fluxes at energies above 1 TeV by a factor of 3. The assumption of the magnetic field B_3 smaller than 0.1 mG does not result in a further increase of TeV γ-ray fluxes, because for such low magnetic fields all electrons up to several TeV are not in the "saturation" regime, therefore variations of B_3 do not affect the electron energy distribution in zone 3.

Thus, the IC fluxes, shown in Fig. 2.11, should be treated as *strict upper limits*. This is an important conclusion which should be taken into account in interpretations of TeV radiation. In this regard, the original interpretation of hard X-rays as synchrotron radiation from multi-TeV electrons (Allen et al., 1997) needs more solid observational support, and deserves further theoretical studies. In particular, this radiation could be explained also in terms of the nonthermal bremsstrahlung of electrons accelerated by plasma waves to subrelativistic energies (Laming, 2001). If so, the need of TeV electrons would disappear, and correspondingly the IC interpretation of the TeV signal from Cas A would become rather shaky and, in fact, somewhat redundant, especially if one takes into account the fact that the TeV flux can be quite comfortably explained by interactions of accelerated protons with the ambient gas. As it can seen from Fig. 2.11, the interpretation of the TeV flux in terms of π^0-decay γ-rays requires only 10^{49} erg in accelerated protons which is only factor of 10 more than the energy in relativistic electrons, and, at the same time, constitutes only $\leq 7\%$ of the total kinetic energy available in Cas A. Note that the π^0-decay γ-ray flux shown in Fig. 2.11 corresponds to a rather modest, from the point of view of the total energy budget, assumption concerning the proton energy distribution with a spectral index $\alpha_p = 2.15$ and $E_0 \gg 1$ TeV. But "economically" less attractive proton spectra, e.g. with $\alpha_p = 2.4$, which would require a factor of 4 larger total energy in protons, are still acceptable.

Actually, the nonlinear diffusive shock-acceleration model applied to Cas A faces just opposite problem. This model, based on parameters adopted to fit the X-ray data in terms of synchrotron radiation, predicts a flat γ-ray flux significantly exceeding the observed TeV flux (Berezhko et al., 2001). Such a discrepancy can be removed assuming that the X-rays are not of synchrotron origin. Then, an alternative (to the diffusive shock-acceleration in the shell) could be proton acceleration in the bright radio structures, e.g. by the second-order Fermi mechanism as discussed by Chevalier et al.

(1978) and Cowsik and Sarkar (1984).

The predictions of the 3-zone model, which was introduced above to estimate the maximum (theoretically possible) flux contributed through the IC channel, just marginally match the TeV flux (see Fig. 2.11). Therefore, at this stage the leptonic origin of the TeV radiation cannot be ruled out. The crucial test for the origin of TeV radiation of Cas A can be provided by future detailed spectroscopic γ-ray measurements in the energy interval between 100 GeV and several TeV. The spectrum predicted by the leptonic model at energies above 1 TeV is very steep, with a photon index $\Gamma \geq 3$. Therefore the detection of a differential γ-ray spectrum harder than E^{-3} at TeV energies, would be a strong argument in favour of the hadronic origin of radiation. At the same time, the detection of a steep TeV spectrum cannot be uniquely interpreted, because introducing an "early" energy cutoff E_0 in the proton spectrum around several TeV one may easily reproduce a flat high energy γ-ray spectrum with strong steepening above 1 TeV. An additional piece of information is contained at low energies. The bremsstrahlung γ-ray fluxes expected above 100 MeV are almost two orders of magnitude above the GLAST sensitivity, unless the magnetic field in the shell does not significantly exceed 1 mG. This, however, seems unlikely, because otherwise it would require an unrealistically large energy, $W_{B2} \geq 3 \times 10^{50}$ erg, to be deposited in the magnetic field of the shell. Thus, we may conclude that GLAST as well as sub-100 GeV threshold IACT arrays should detect γ-ray signals from Cas A. The spectrum of γ-radiation of leptonic origin, which at these energies is dominated by electron bremsstrahlung, can be predicted (from radio data) very robustly. It should be close to a flat power-law, $\propto E^{-2}$. This makes the identification of radiation mechanisms quite difficult, because the proton spectrum (and, consequently, the π^0-decay γ-ray spectrum), cannot be significantly steeper than $E^{-2.2}$, otherwise the hadronic mechanism would require an unacceptably large energy in accelerated particles. In this regard, more promising is the energy region below 100 MeV, where the π^0-decay as well as the IC components are strongly suppressed, and thus the flux measured at low energies would provide an important "calibration point" for extrapolation of the bremsstrahlung contribution to higher energies.

In summary, the γ-ray spectrometry of Cas A with GLAST and future IACT arrays will give definite answers about the energy distribution between the magnetic field and accelerated electrons and protons, and thus will provide a crucial insight into the nature of acceleration processes in one of the most prominent nonthermal objects in our Galaxy.

5.7 "PeV SNRs"

The featureless energy spectrum of cosmic rays, extending as a single power-law up to the *knee* around 10^{15} eV, indicates that the flux of galactic CRs is dominated by a *single* source population. If so, the representatives of this source population should be sources of nonthermal synchrotron X-rays and very high energy γ-rays. Since both radiation components have been reported from 3 prominent SNRs – SN 1006, RX J1713.7-39, and Cas A – one may interpret this as strong evidence of close association of this hypothetical source population with SNRs. On the other hand, the requirement of a *single* source population to dominate in production of the bulk of galactic CRs would then imply that the SNRs should be able to accelerate particles to at least 10^{15} eV ($\equiv$ 1 PeV). Although the "standard" diffusive shock acceleration theory applied to SNRs does not guarantee effective acceleration of protons beyond 10^{14} eV, particle accelerations to $\geq 10^{15}$ eV is possible, at least in principle, in those SNRs (hereafter "PeV SNRs") where the magnetic field exceeds 0.1 mG. Remarkably, the CRs themselves may provide the necessary magnetic field through amplification of the ambient interstellar field by large-amplitude plasma waves (Lucek and Bell, 2000). The increased magnetic field reduces the diffusion coefficient, and correspondingly increases, by an order of magnitude or more, the maximum achievable energy (Bell and Lucek, 2001).

The most straightforward search for sites of such extreme "PeV SNRs" can be performed by ground-based detectors operating effectively in the energy regime $\geq$ 100 TeV. In Fig. 5.13 are shown the fluxes of X- and γ-rays of hadronic origin expected from a "PeV SNR". Both radiation components are initiated by interactions of accelerated protons with the ambient medium with a gas number density $n = 1$ cm^{-3} and magnetic field $B = 0.2$ G. While the γ-rays arise directly from the decay of π^0-mesons, the X-rays are result of synchrotron radiation of secondary electrons, the products of $\pi^{\pm}$-decays. The lifetime of electrons producing X-rays, $t_{\text{synch}} \simeq 1.5 B_{\text{mG}}^{-3/2}(E_X/1 \text{ keV})^{-1/2}$ yr, in a magnetic field exceeding 0.1 mG is very short (≤ 50 yr) compared to the age of the source. Therefore the synchrotron X-radiation actually could be considered as an unavoidable "prompt" radiation component of hadronic interactions, emitted simultaneously with π^0-decay γ-rays. Consequently, both the X- and γ-ray fluxes depend on the total amount of high energy protons currently accumulated in the source, and the rate of p-p interactions, i.e. on the density of the ambient matter. Although approximately the same fraction of energy of the parent protons is transferred to secondary electrons and γ-rays, because the energy of relatively low energy (sub-TeV) electrons is not radiated away effectively, the π^0-decay γ-ray luminosity exceeds the synchrotron luminosity. The L_X/L_γ ratio depends on the proton spectrum as well as on the history of particle injection, but for any reasonable proton spectrum extending to $\gg 100$ TeV, the energy release through the X-ray channel exceeds 20-30 percent of the energy released through the π^0-decay channel. The broadband spectra of hadronic radiation of a "PeV" SNR shown in Fig. 5.13 are calculated for a proton accelerator at a distance $d = 1$ kpc operating during 10^3 yr with a constant rate $L_p = 3 \times 10^{38}$ erg/s; the total energy accelerated in protons is about $W_p = L_p \cdot T \simeq 10^{49}$ erg. The power-law spectrum of protons is assumed to be $\alpha_p = 2.1$. The solid, dashed, and dot-dashed curves correspond to 3 different cutoff energies in the proton spectrum - $E_0 = 10^{15}$, 3×10^{15}, and $E_0 = 10^{16}$ eV, respectively. Remarkably, the spectrum of π^0-decay γ-rays in the corresponding cutoff region at $\sim (1/10) E_0$ almost repeats the shape of the proton spectrum around E_0. Thus the search for ultra-high energy γ-rays from SNR in the energy domain ≥ 100 TeV would lead to the discovery and identification of galactic sources responsible for the CR spectrum up to the knee. Moreover, accurate γ-ray spectroscopic measurements at highest energies would provide extremely important information about the shape of the source (acceleration) spectra of protons. Needless to say, this information could be crucial for identification of the acceleration mechanisms in SNRs, as well as for understanding of the role of different processes (e.g. acceleration *versus* propagation) which are responsible for the formation of the knee in the CR spectrum.

Unfortunately, the prospects of realizing this exciting method in practical terms is quite limited due to the lack of projects of γ-ray detectors with an adequate sensitivity in the energy interval $10^{14} - 10^{16}$ eV. The deep survey by the large PeV air-shower array CASA-MIA, with sensitivity $\sim 10^{-14}$ ph/cm^{2}s above 100 TeV did not reveal any source across a large fraction of the northern sky (McKay *et al.*, 1993). The fluxes shown in Fig. 5.13 require an order of magnitude better detector sensitivity, unless the total energy in accelerated protons exceeds 10^{50} erg and/or the sources are located in dense environments. Unfortunately, an improvement of the CASA-MIA sensitivity by an order of magnitude seems quite unrealistic, given the limited experimental interest in this energy interval, at least at present.

In such circumstances, the search for "PeV SNRs" via hard synchrotron radiation of secondary ($\pi^{\pm}$-decay) electrons is an alternative, and perhaps an even more powerful tool, given the superior potential of X-ray detectors. In particular, Chandra and XMM-Newton have sufficient sensitivity to perform such studies. Remarkably, the spectrum of this radiation in the cutoff region is much smoother that the corresponding cutoff in the γ-ray spectrum. For example, while the energy distribution in the form of an exponential cutoff in the proton spectrum is directly transferred to electrons (but shifted by a factor of 10 to lower energies), the synchrotron radiation in the corresponding cutoff region behaves as $\exp{(-\varepsilon/\varepsilon_0)^{1/2}}$. This important feature, which is seen in Fig. 5.13, should allow a more comprehensive study of the highest energy part of the spectrum of accelerated protons via X-rays - the third generation products of p-p interactions.

An apparent difficulty in this method is connected with the separation and identification of the "hadronic" synchrotron X-rays from other X-ray radiation components, in particular from the "nominal" synchrotron radiation of directly accelerated electrons. These two components can be distinguished if the magnetic field in the X-ray production region exceeds 0.1 mG and the proton spectrum extends to $E \geq 10^{15}$ eV. In the case of a high magnetic field the position of the cutoff of the synchrotron radiation

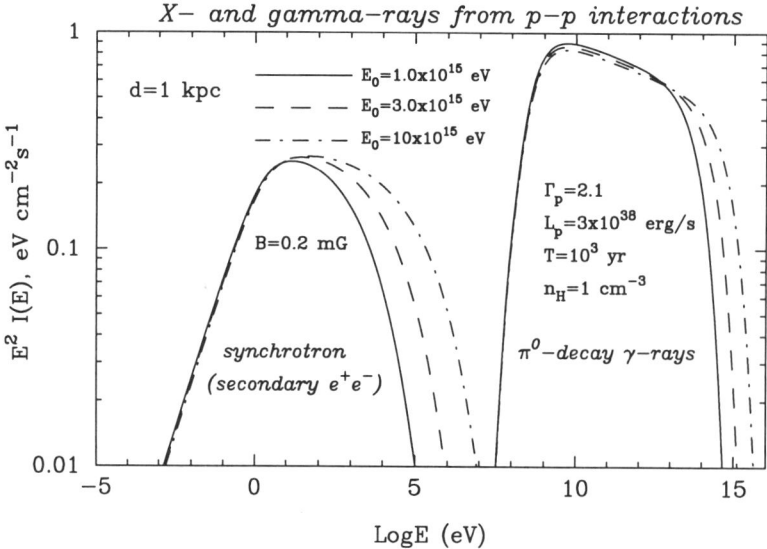

Fig. 5.13 The broad-band spectra of radiation from a "PeV SNR" initiated by interactions of accelerated protons with the ambient gas.

of directly accelerated electrons does not depend on the magnetic field and typically appears in the soft X-ray domain, $h\nu_{m,1} \leq 1$ keV (see Eq.(5.14)). On the other hand, the position of the spectral cutoff of the synchrotron radiation produced by secondary electrons, $h\nu_{m,2}$, is very sensitive to the magnetic field. For example, in the Bohm diffusion regime, the proton cutoff energy is proportional to the strength of the magnetic field, thus $h\nu_{m,2} \propto B \cdot E_0^2 \propto B^3$. Thus, for extreme SNRs, with very large magnetic fields and proton cutoff energies, the synchrotron radiation of secondary electrons may have a relatively hard X-ray-spectrum compared with the synchrotron spectrum of directly accelerated electrons. The search for such a component in the nonthermal X-ray spectra of young SNRs is of great interest. If we are lucky, this may result in the discovery of "PeV SNRs" !

Chapter 6

Pulsars, Pulsar Winds, Plerions

Pulsars – rapidly rotating, magnetized neutron stars – were the first astrophysical sources discovered in high energy γ-rays (for review see e.g. Fichtel and Trombka, 1997) In fact, these objects, with the Vela pulsar as the brightest persistent GeV source in the γ-ray sky, remain the only firmly identified γ-ray source population in our Galaxy.

The existence of neutrons stars had been predicted long before the discovery of pulsars. In their seminal papers published in 1934, Baade and Zwicky argued that the enormous energy release observed in supernovae might be result of transition of an ordinary star into a dense neutron star, consisting mainly of neutrons. The feasibility of formation of such an unusual stellar configuration had been theoretically demonstrated by Landau (1938).

The close association of pulsars with neutron stars was realized by Gold (1968,1969), soon after their discovery in 1967 by Hewish *et al.* (1968). Remarkably, before the discovery of pulsars, Pacini (1968) argued that the rapidly rotating neutron stars with extremely strong magnetic fields should be able to power supernova remnants like the Crab Nebula, and predicted that the rotating, magnetized neutron stars might be observable at radio frequencies.

There is little doubt that the neutron stars are produced in supernova explosions. It should be noted in this regard that the ages of the young radiopulsars in the Crab and Vela supernova remnants, derived from their periods P and period derivatives $\dot{P}$, are in very good agreement with the ages of these remnants. Moreover, it is established that these synchrotron nebulae are powered by pulsars, most likely through relativistic winds terminated at standing reverse shock waves (Rees and Gunn, 1974). It is believed that the pulsar winds basically consist of electrons and positrons which originate in the pulsar magnetospheres. The rotationally induced electric field in the magnetosphere is sufficient to drive pair cascades (Sturrock, 1971) producing these electrons, which then may escape the magnetosphere from the polar cap regions. However, the wind-to-magnetosphere connection remains an unsolved theoretical challenge (see Michel, 1998).

Gamma-radiation from rotation powered pulsars can be produced through several radiation mechanisms in three physically distinct regions: (1) *magnetosphere*, (2) *relativistic wind*, and (3) *synchrotron nebula*.

These sites are schematically shown in Fig. 1.8. Pulsed γ-rays are generated in the pulsar magnetosphere limited by the light cylinder. The relativistic electron-positron *wind* ejected by the pulsar can itself produce very high energy γ-rays. The wind, that carries almost the whole rotational energy of the pulsar, eventually terminates in the surrounding interstellar medium. This results in formation of a nonthermal (synchrotron and inverse Compton) nebula the spectrum of which extends from radio wavelengths to ultra-high energy γ-rays.

6.1 Magnetospheric Gamma Rays

Presently, the number of catalogued radio pulsars – single neutron stars powered by fast rotation – exceeds 1000. Only six of them have been reported by the EGRET team as high energy γ-ray sources (Thompson *et al.*, 1999), with up to five possible associations with unidentified EGRET sources (Torres *et al.*, 2003). Actually, this is not a big surprise, and can be explained, by the limited sensitivities of γ-ray instruments. For example, the minimum detectable γ-ray flux for EGRET of about 10^{-11} erg/cm^{2}s is almost 6 orders of magnitude higher than the detection threshold of pulsars by current radio telescopes at GHz frequencies. This striking difference in sensitivities is compensated, to a certain extent, by much larger energy fluxes of γ-rays compared with the radio fluxes. While the pulsars emit in the radio band a very small fraction ($\leq 10^{-6}$) of their rotational energy (Manchester and Taylor 1977), the γ-ray luminosities of some of the EGRET pulsars above 100 MeV exceeds one per cent of their spin-down luminosities. It is expected that the future major high energy γ-ray detectors like GLAST will increase the number of *radio* pulsars seen also in γ-rays by an order of magnitude (Thompson 2001). Moreover, the radio pulsars, which are only observable if the radio beam crosses the Earth orbit, are believed to be only a fraction of a larger source population called *rotation powered pulsars* (RPP). Remarkably, the γ-ray beams are expected to be significantly wider than the radio beams, and these beams do not fully overlap. If so, the RPPs have more chances of being detected in high energy γ-rays rather than at radio wavelengths. This could be the case of some of the unidentified EGRET sources, a large fraction of which is likely to be radio-quiet pulsars like Geminga. Thus, the searches for pulsed radiation components in the spectra of a large fraction of unidentified EGRET sources (suspected to be pulsars) without invoking information from lower (radio, optical, X-ray) frequency domains, seems to be another high priority objective, because in many cases the periodic signals at lower frequencies could be suppressed. Such studies could be effectively performed by GLAST (McLaughlin and Cordes, 2000), and by future sub-10 GeV imaging Cherenkov arrays like 5@5 (Aharonian and Bogovalov, 2003).

The brightness temperatures of the radio emission of pulsars vary within $10^{23} - 10^{30}$ K (Manchester and Taylor, 1977). Therefore it is likely that the pulsed radio emission is generated coherently by a beam of relativistic electrons. In the IR/optical and X-ray bands the emission is not coherent; it may have a thermal and/or nonthermal origin. The specific mechanism(s) responsible for the radio (e.g. Lyubarsky, 1995) as well as IR/optical and X-ray (Pacini, 1971; Becker and Trümper 1997) remain highly uncertain even for the most prominent and well studied objects like the Crab and Vela pulsars.

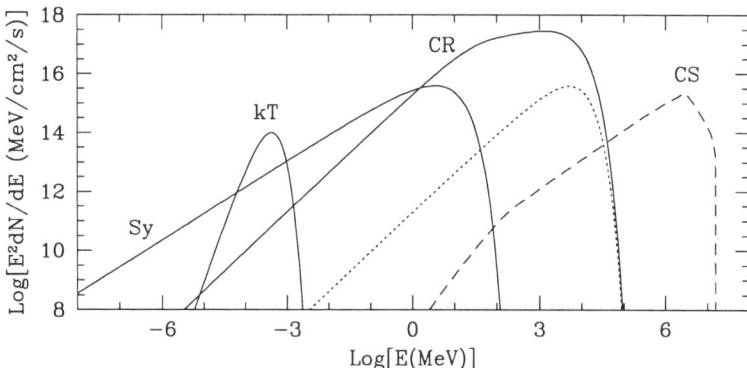

Fig. 6.1 Characteristic phase-averaged spectrum for a young γ-ray pulsar. Solid lines show the curvature (CR), synchrotron (Sy), and the thermal surface flux (kT). The dashed curve (CS) represents the TeV pulsed spectrum from Compton upscattering of the synchrotron spectrum on the primary $e^{\pm}$. (From Romani, 1996).

The situation is somewhat better in the high energy γ-ray domain. Modulation of γ-ray light curves at the rotation periods of these objects indicates that γ-rays are produced near the surface of neutron stars by electrons accelerated either at the polar cap or in the vacuum gaps at the outer magnetosphere. Three γ-ray production mechanisms – the curvature radiation, synchrotron radiation, and inverse Compton scattering – can effectively contribute to the γ-radiation from MeV to TeV energies. An example of relative contributions of these process in the formation of high energy spectra of pulsars expected in the so-called outer gap model is shown in Fig. 6.1. It is also recognised that cascade processes in the magnetosphere, initiated by pair production in the magnetic field, play an important role in the formation of broad-band γ-ray spectra, especially in the *polar cap* models (Daugherty and Harding, 1982).

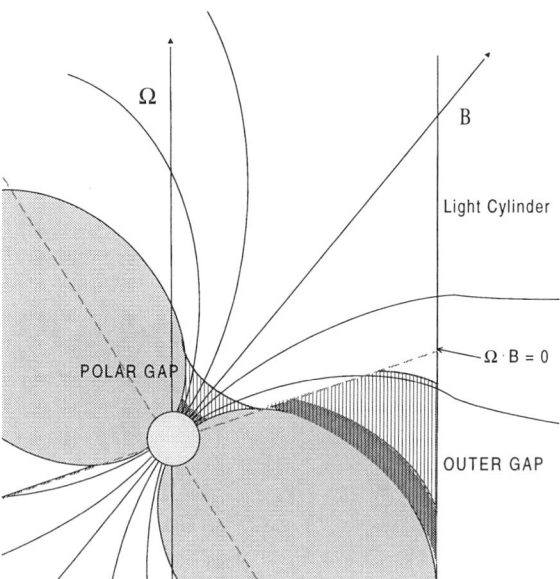

Fig. 6.2 Geometry of polar and outer gaps. Dark regions are thin gaps of younger pulsars. Hatched regions are thick gaps of older pulsars. (From Harding, 2001).

At the same time, one of the key issues of the physics of γ-ray pulsars – the location of γ-ray production region(s) – remains an open question. Presently two basic classes of models are discussed in this regard - the *polar cap* (Daugherty and Harding, 1982, 1996; Usov and Melrose, 1995) and *outer gap* (Cheng et al., 1986; Romani, 1996; Hirotani and Shibata, 2002) models. The geometry of polar caps and outer gaps is schematically presented in Fig.6.2.

In both models primary γ-rays are generated by ultrarelativistic electrons accelerated in the quasi-static electric fields. The charge density in the pulsar magnetosphere is high enough to screen the electric field parallel to the B-field, thus the co-rotation condition $\mathbf{E} \cdot \mathbf{B} = 0$ is maintained everywhere except at a few locations. These regions, where $E_{\|} \neq 0$, can exist close to the surface of the neutron star in the polar caps (polar cap/inner gaps model), or at distances comparable with the light cylinder of the pulsar along the null charge surface defined by the condition $\mathbf{\Omega} \cdot \mathbf{B} = 0$, where the corotation charge density changes sign (outer gap model).

6.1.1 *Polar cap versus outer gap models*

The ideas which constitute the basis for the polar cap models were developed by Sturrock (1971) and Ruderman and Sutherland (1975) who proposed that the particle acceleration and their subsequent radiation occur near the star surface, namely at the magnetic poleş. Pair cascade development is an important component for γ-ray production in all versions of the polar cap model, but the details depend on which radiation process (inverse Compton or curvature radiation) controls the production of pairs responsible for the screening of the accelerating field. In scenarios where the inverse Compton scattering is both the dominant primary γ-ray radiation mechanism and the initiator of the pair cascades (Sturner et al., 1995), the particle energies are limited to Lorentz factors $\gamma_e \sim 10^6$ (Harding and Muslimov, 1998). If the inverse Compton controlled zones are not stable, the particle acceleration continues up to $\gamma_e \sim 10^7$. This model predicts quite sharp (super-exponential) cutoffs at energies typically below 10 GeV. This agrees with the upper limits on the pulsed TeV signals from 6 EGRET pulsars (see Fig.2.1). Fig. 6.3 shows the high energy cutoffs predicted by the polar cap model as a function of the surface field strength calculated for different radii of photon emission (Harding, 2001). The observed cutoff energies of 8 (identified and suspected) γ-ray pulsars versus their fields are also shown (the B-fields are derived from the measured parameters P and $\dot{P}$, assuming $R = 10^6$ cm). It is seen that the low B-fields ($\equiv$ older pulsars) allow higher cutoff energies.

In the outer gap models the γ-ray production occurs at large distances from the star surface, where the local field is reduced by several orders of magnitude compared with the field at the star surface, and so the absorption of γ-rays in the the magnetic field is not sufficient for development of pair cascades and does not significantly distort the γ-ray spectrum. These mod-

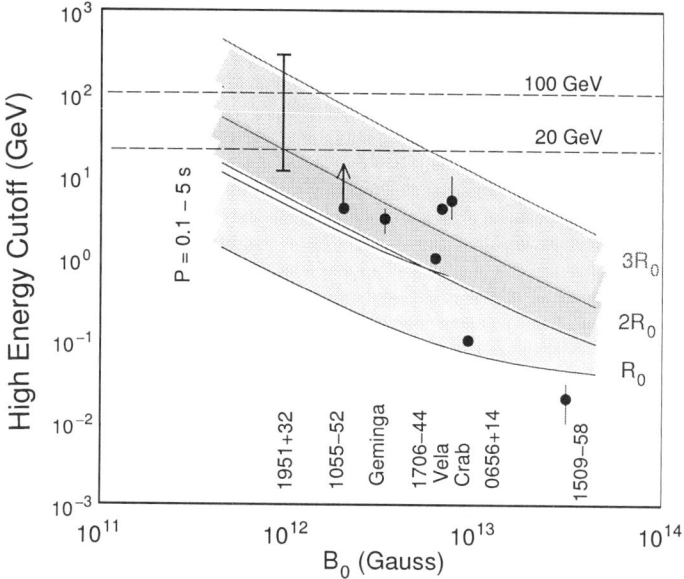

Fig. 6.3 The expected spectral cutoff energies due to magnetic pair attenuation versus the surface field strength for a range of periods at different radii (in units of the neutron star radius) of the emission region. The estimated turnover energies of known γ-ray pulsars are also shown. (From Harding, 2001).

els can reproduce, more or less successfully, both the spectral shape and the light curves of γ-ray pulsars, if it is postulated that the observed γ-rays are dominated by the emission from *one* pole (Romani and Yadigaroglu, 1995). In this model the cutoff energy in the γ-ray spectrum is caused by the maximum energy of accelerated electrons, therefore it is not severely constrained. Moreover, while the polar cap model predicts very sharp, superexponential cutoffs (because the attenuation coefficient itself increases exponentially with energy), in the outer gap model the γ-ray spectrum drops more slowly. Thus detailed spectrometric γ-ray measurements should help to distinguish between these two models (or rule out both). The Vela pulsar seems to be the best target for such a study. Although the EGRET data contain important information about the approximate location of the cutoff energy, they do not allow, however, because of poor photon statistics above 1 GeV, accurate determination of the spectral shape in the cutoff region. In this regard GLAST is much better suited for such a task, as demonstrated in Fig. 2.5.

The need for accurate measurements of pulsars in the multi-GeV region is not limited by detailed spectrometric studies in the cutoff region. The EGRET data show that the light curves in the 100 MeV range and above 5 GeV are significantly different. Namely, for all pulsars except PSR B1706-44, the multi-GeV light curves are dominated by one of the two pulses seen at lower energies (Thompson, 2001). Thus, the detailed study of the temporal characteristics of EGRET pulsars in high energy bands seems to be an equally important objective of future observations. While at lower energies such studies can be effectively performed by GLAST, for the multi-GeV band large photon statistics is a critical requirement which only marginally can be satisfied by GLAST. Ground-based sub-10 GeV imaging Cherenkov telescope arrays like 5@5 would be ideal instruments for this purpose. 5@5 with its enormous detection area should be able to detect a positive γ-ray signal above several GeV from the Vela pulsar with a count rate between 20 and 50 photons per second, depending on the shape of the spectrum above 5 GeV (Aharonian and Bogovalov, 2003).

The high detection rate capability of 5@5 should allow the detection of multi-GeV signals from many other pulsars. For example, unpulsed signals from sources with spectra similar to Vela, but with absolute fluxes as faint as 3 milli Vela ($\simeq 10^{-11}$ erg/cm^{2}s), can be detected by 5@5 in only 3 h observation time. Note that the one-year survey by GLAST will provide a similar sensitivity at energies above 100 MeV for pulsars in the galactic plane, and approximately 10 times better sensitivity for high galactic attitude pulsars (Thompson, 2001). The sensitivities of 5@5 and GLAST are shown in Fig. 6.4 together with the spin-down "fluxes" of radio pulsars – the spin-down luminosities of individual objects (derived from P and $\dot{P}$) divided by $4\pi d^2$. The spin-down flux is a meaningful parameter which, despite uncertainties in both the γ-ray production efficiency and the γ-ray beaming factor, generally may be taken as an objective criterion for the "pulsar observability".

Although Fig. 6.4 indicates the impressive potential of future gamma-ray detectors for discovery of new γ-ray pulsars, the number of detectable pulsars significantly depends on the radio and γ-ray emission models. In accordance with some (perhaps, too optimistic) estimates the number of pulsars detectable by GLAST could be as large as 750 (McLaughlin and

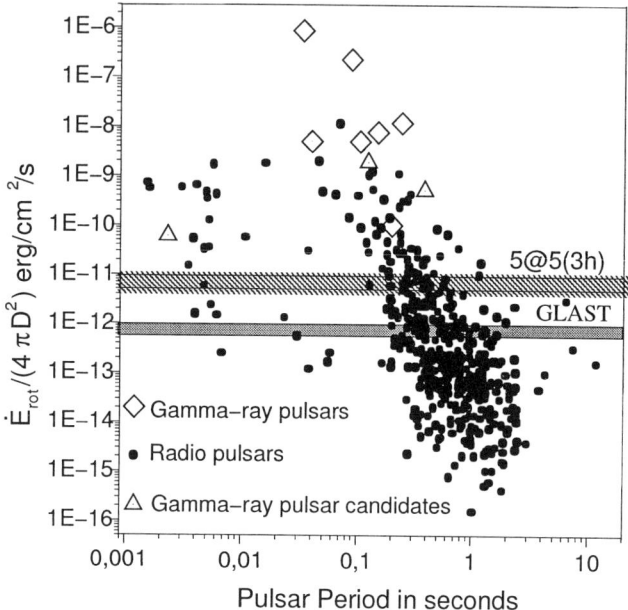

Fig. 6.4 Gamma-ray pulsar observability in terms of the spin-down "flux". The DC (unpulsed) flux sensitivities of GLAST (above 100 MeV) for high galactic latitude pulsars after a one-year sky survey, and of 5@5 (above 5 GeV) for 3 h observation time are also shown.

Cordes, 2000). The simple extrapolation of the "logN-logS" curve constructed on the basis of the EGRET pulsars gives more conservative number, something between 30 and 100 γ-ray pulsars (Thompson, 2001).

In any case, the dramatic increase in the number of γ-ray pulsars should allow an adequate statistical study of the γ-ray luminosities L_γ of pulsars versus their magnetic field B and the period P. Because different models predict significantly different functions $L_\gamma(B_0, P)$, these studies will help to distinguish not only between the polar cap and outer gap models, but also between different scenarios within one of these general models, for example between the "curvature-radiation initiated cascade" (Harding and Zhang, 2001) and "inverse-Compton initiated cascade" (Sturner et al., 1995) polar cap models. The analysis of known γ-ray pulsars shows an approximate dependence $L_\gamma \propto B_0 P^{-2}$ which is better explained by polar cap models (bad news for TeV γ-ray astronomy). But a decisive test can be provided only by future observations. For example, while the polar cap models predict γ-ray emission at some level from *all pulsars*, the outer gap models predict a "dead line". The latter divides the $B_0 - P$ space between young pulsars which are capable of supporting effective pair production in the outer gaps from the older pulsars which cannot (see Harding (2001) and references therein). Thus detection of γ-rays from a pulsar below this line (i.e. a very old pulsar) would give a preference to the polar cap type models.

A significant improvement in the sensitivity of γ-ray instruments is critical also for searching for γ-ray pulsars with unknown pulse periods. The finding of periodic signals in the γ-ray data is not a simple problem (see e.g. Chandler et al., 2001). Note that the pulsed components of all EGRET pulsars have been discovered only after application of the information about the period and phase obtained at other energy bands. However, in many cases the time structure could be essentially different at different wavelengths. Therefore, it is crucial to search for periodic γ-ray signals without relying on observations at other energy bands. In its turn, this requires adequate γ-ray photon statistics.

The statistics of photons increases with an increases in the observation time, and in principle could be very large, in particular for large field of view instruments like GLAST. However, the duration of valuable observation time can not be increased infinitely unless the periodic source is not a perfect clock. Pulsars *are not perfect clocks*. The periods of radio pulsars do *increase* with time due to rotational losses. The phase of the signal varies with time as $\varphi = 2\pi(t\nu + t^2\dot{\nu})$, where $\nu = 1/P$. The variation of the phase due to the change of the period of rotation can be neglected while $t^2\dot{\nu} < 1/2$. Thus, the duration of the time for the search of a periodical signal in absence of an *a priori* information about the period P and period derivative $\dot{P}$ should satisfy the condition $T_{\rm obs} < (1/2\dot{\nu})^{0.5}$. During this time the number of detected γ-ray photons should be large enough to allow us to reveal the period of rotation. Generally, the observation time should not exceed 10 h (Aharonian and Bogovalov, 2003). This perfectly matches the performance of the sub-10 GeV threshold IACT arrays which should be able to detect positive γ-ray signals from pulsars at the level of ≈3 milli-Vela with a photon statistics of about 10^3, just after several hours of continuous observations.

6.1.2 *Magnetospheric TeV gamma rays?*

In the polar cap model we do not expect TeV γ-ray emission from pulsar magnetospheres, because the high energy γ-rays are heavily absorbed before they escape from the magnetosphere. The upper limits on the flux of pulsed TeV γ-ray emission from the EGRET pulsars shown in Fig. 2.1 are in good agreement with predictions of this model. Millisecond pulsars with modest magnetic fields in the range of 10^8 to 10^9 G, could, however, be an exception. In such small magnetic fields the effect of γ-ray absorption is significantly reduced, and therefore high energy radiation can escape from the production region. In these pulsars, the γ-ray emission between 1 MeV and a few hundred GeV is dominated by curvature radiation; at higher energies an additional inverse Compton component is formed with a distinct maximum around 1 TeV (Bulik et al., 2000). Although the predicted absolute γ-ray fluxes are well below the sensitivities of both EGRET and current ground-based instruments, several such objects could be promising candidates for future observations with GLAST as well as with low-energy threshold Cherenkov telescope arrays.

A TeV component of magnetospheric γ-radiation caused by inverse Compton scattering is expected also in the outer magnetosphere gap model of "ordinary" pulsars with strong magnetic fields (Cheng et al., 1986). An interesting feature of this radiation was claimed to be a very hard spectrum below 1 TeV with a sharp cutoff above several TeV (Romani, 1996). If so, the most promising energy region for detection of this component would be a rather narrow interval around 1-3 TeV. In particular, the calculations of pulsed γ-radiation from Vela shown in Fig. 6.1 predict that the energy flux in the pulsed TeV γ-rays could be as large as 1% of the pulsed GeV flux. The upper limit on the TeV flux from the Vela pulsar reported by the CANGAROO collaboration (see Fig. 2.1) is below the flux predicted by Romani (1996), but is consistent with the most recent calculations of pulsed TeV γ-rays within the outer gap model (Hirotani, 2001). Note that the magnetospheric TeV flux predictions strongly depend on the density and the spectrum of infrared pulsed photons within the light cylinder. Although recent observations of optical radiation with the VLT show that the optical emission smoothly extends to 1 eV, no data are available at the far infrared wavelengths which provide the bulk of the seed photons for the production of TeV γ-rays. As long as such information is lacking, the upper limits on the pulsed TeV emission from Vela unfortunately cannot be used for robust conclusions concerning other model parameters.

6.2 Gamma Rays from Unshocked Pulsar Winds

Rotation powered pulsars eject plasma in the form of relativistic winds that carry off most of the rotational energy of the pulsars. The best observed example is the pulsar in the Crab Nebula. This pulsar ejects a relativistic wind which is terminated at a distance of about 0.1 pc by a standing reverse shock. The shock accelerates electrons up to energies of 10^{15} eV, and randomises their pitch angles (Rees and Gunn, 1974; Kennel and Coroniti, 1984; Arons, 1996). This results in the formation of an extended synchrotron source in the region downstream of the shock.

The pulsar winds are characterised by the so-called *magnetization* or σ-parameter – the ratio of the electromagnetic energy flux to the kinetic energy flux of particles in the wind. At $\sigma \geq 1$ the wind is Poynting flux dominated. At $\sigma \leq 1$ the wind is kinetic energy dominated (hereafter, KED wind). The analysis of observations of the Crab Nebula prefers $\sigma \sim 10^{-3}$ (Kennel and Coroniti, 1984; De Jager and Harding, 1992; Atoyan and Aharonian, 1996). The recent observations of the compact nebula surrounding the Vela pulsar, in particular its small size and low luminosity, favour a somewhat larger σ parameter for the wind of the Vela pulsar, but still it does not exceed 1.

On the other hand, all models of electron-positron pair production in the pulsar magnetospheres predict $\sigma \sim 10^3 - 10^4$, i.e. at the base of the wind the plasma is likely to be Poynting flux dominated. Therefore, it is difficult to avoid the conclusion that a phase transition of the wind takes place somewhere between the light cylinder and the wind termination shock. Despite certain theoretical efforts in the past to understand the reasons (and mechanisms) of pulsar wind acceleration, the formation of KED winds remains an unsolved problem in pulsar physics. It is generally believed that the activity in this direction should have pure theoretical character, because the region where almost the whole rotational energy of the pulser is transferred into the kinetic energy of the wind cannot be directly observed. This has a simple explanation. Although the wind electrons may have an energy as large as 10^{13} eV, they move together with the magnetic field and thus do not emit synchrotron radiation. However, this statement is valid only for the synchrotron radiation of the wind. In fact, the unshocked KED wind can be *directly observed* through its inverse Compton radiation (Bogovalov and Aharonian, 2000). The IC γ-radiation of the wind electrons is unavoidable due to the illumination of the wind by external low-energy photons of different origin. These are nonthermal (synchrotron) and ther-

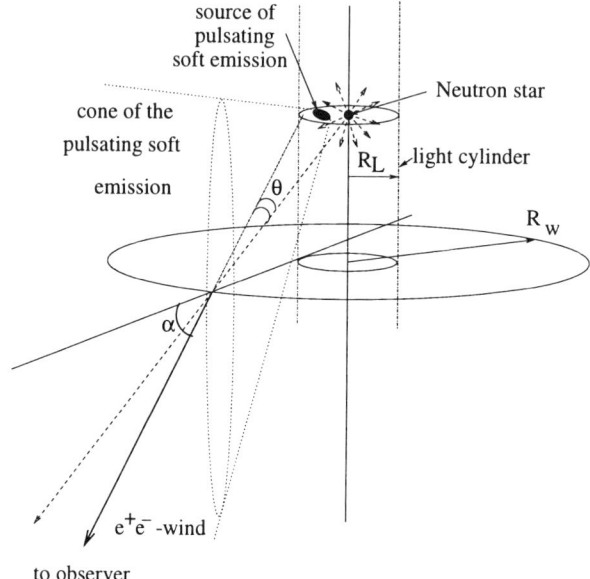

Fig. 6.5 Sketch illustrating the trajectories of the electron-positron plasma after acceleration, and positions of soft photon sources. (From Bogovalov and Aharonian, 2000).

mal (dust) radiation of the nebula surrounding the pulsar, 2.7 K CMBR, optical emission of the companion star (in the case of the binary systems), and finally the soft, thermal and nonthermal emission of the pulsar itself.

6.2.1 Characteristics of the KED wind

The interaction of the KED wind with the soft emission of the pulsar is unavoidable because, even if the soft radiation is emitted radially, the KED wind does not move strictly along radial directions. This feature of the KED winds is a direct consequence of energy and angular momentum conservation laws and does not depend on specific mechanisms of wind acceleration.

The energy and the angular momentum fluxes of the wind consist of two parts: one corresponds to matter and the other corresponds to the electromagnetic field. The total fluxes of the kinetic energy $\dot{E}_{\rm kin}$ and the angular momentum $\dot{L}_{\rm kin}$ can be presented as $\dot{E}_{\rm kin} = \dot{N}mc^2 <\gamma_{\rm w}>$, $\dot{L}_{\rm kin} = \dot{N}m <r\gamma_{\rm w}v_\varphi>$, where $\dot{N}$ is the total injection rate of particles, $<\gamma_{\rm w}>$ is the average Lorentz factor of the wind, r is the distance to the axis of rotation, and v_φ is the azimuthal component of the plasma velocity. Below it is assumed that the wind consists only of electrons and positrons.

Classical mechanics provides a simple relationship between the rates of the pulsar's rotational energy losses, $\dot{E}_{\rm rot} = I\Omega\dot{\Omega}$, and the angular momentum losses, $\dot{L} = I\dot{\Omega}$: $\dot{E}_{\rm rot} = \dot{L}\Omega$, where I is the moment of inertia of the pulsar. In the KED wind the electromagnetic field carries off negligible energy and angular momentum, therefore

$$\frac{v_\varphi}{c} = \frac{R_{\rm L}}{r}. \tag{6.1}$$

The geometry of the flow and of the IC process is shown in Fig. 6.5. The pulsar is located on the axis of rotation and ejects the wind radially. The dash-dotted vertical lines show the light cylinder. The wind is accelerated at $R_{\rm w}$. Particles in the wind move along straight lines without further acceleration beyond $R_{\rm w}$. The equatorial plane of the pulsar is inclined to the observer at the angle α. IC photons move along the direction of motion of the electrons. Therefore, only the particles of the wind directed towards to us can produce observable emission.

According to Eq.(6.1) the projected vector of the plasma velocity after acceleration lies on the line tangential to the light cylinder. The angle θ between the direction of the motion of relativistic particles in the wind and the soft thermal photons emitted from the pulsar depends on the distance r to the axis of rotation, $\sin\theta = (R_{\rm L}/r)\cos\alpha$.

6.2.2 The ejection rate and the particle spectrum

The total rate of the particle ejection can be estimated within the inner electrostatic gap models of $e^\pm$ pair production in the pulsar magnetosphere (Ruderman and Sutherland, 1975), $\dot{N}_{\rm gap} = n_\pm c S_{\rm cap} \lambda$, where $S_{\rm cap} = 2\pi R_*^3 \Omega/c$ is the total area of the polar caps of the neutron star where the roots of the open magnetic field lines are placed, $n_\pm$ is the density of the particles in the primary beam, Ω is the angular velocity of rotation of the pulsar, R_* is the radius of the pulsar and the factor λ takes into account the electron multiplication due to the development of electromagnetic cascades in the magnetosphere. Electromagnetic cascades are initiated by the beam of electrons accelerated up to the Lorentz-factor $\gamma_{\rm gap} \sim 2 \times 10^7$ (Ruderman and Sutherland, 1975). The density of the particles in the beam is of the order of magnitude of the Goldreich-Julian density $n_\pm = n_{GJ}$ determined as $n_{GJ} = (\Omega {\bf B})/2\pi ec$ (Goldreich and Julian, 1969). For the Crab pulsar the electromagnetic cascade in the pulsar magnetosphere increases the number of particles by a factor of $\lambda \sim 10^4$ (Daugherty and Harding, 1982; Gurevich and Istomin, 1985). Correspondingly, the Lorentz-factor of particles is decreased by the same factor. The uncertainties in the model parameters and assumptions do not allow an accurate theoretical estimate of λ, but give a broad range of possible values of λ between 10^3 and 10^5.

The cascade multiplication of the primary electrons accelerated in the pulsar magnetosphere results in the formation of an $e^{\pm}$ plasma with

$$\dot{N} = \frac{\lambda B_0 \Omega^2 R_*^3}{ec^2}, \quad \text{and} \quad \gamma_{w0} = \frac{\gamma_{\text{gap}}}{\lambda}. \tag{6.2}$$

The initial Lorentz-factor of particles is close to $\gamma_{w0} \sim 10^3$. Then, for the Crab pulsar, the kinetic energy flux of the initial wind is $\dot{E}_0 \approx \gamma_{w0} mc^2 \dot{N}$, and $\sigma_0 = \dot{E}_{\text{rot}}/\dot{E}_0 \sim 2.5 \times 10^4$.

We assume that the KED wind with $\sigma \ll 1$ is formed in some "acceleration region" at a distance R_w from the axis of rotation. The average Lorentz-factor of the KED wind is determined as $<\gamma_w> = \dot{E}_{\text{rot}}/\dot{N}$, with a typical value $\sim 10^6 - 10^7$. The KED wind is believed to be cold since the region of the flow of this wind is observed as an "under-luminous" region (e.g. Kennel and Coroniti, 1984). A hot wind would produce noticeable synchrotron emission which would contradict the existence of the under-luminous region.

In calculations presented below we assume that the particle flux in the wind is isotropic. However to be consistent with the Chandra observations (Weisskopf et al., 2000), we assume that the Lorentz-factor of the wind depends on latitude. In the split-monopole model (Bogovalov, 1999), the Lorentz-factor

$$\gamma_w = \gamma_{w0} + \gamma_{\max} \cos^2 \alpha, \tag{6.3}$$

where $\gamma_{\max} = \sigma_0 \gamma_0$ is the maximum Lorentz-factor of the plasma on the equator ($\alpha = 0$); $\gamma_0 \sim 10^2$ and $\sigma_0 \sim 10^4$ are the Lorentz-factor and the magnetization parameter of the wind near the light cylinder. The wind is assumed to be monoenergetic at a given latitude. It follows from Eq.(6.3) that $<\gamma_w> = 2/3 \gamma_{\max}$ with $\gamma_{\max} = \frac{eH_0 R_*}{2mc^2\lambda}(R_*\Omega/c)^2$.

6.2.3 IC Radiation of the pulsar wind in Crab

The fluxes of the IC emission are calculated assuming that the observer detects photons from a monoenergetic beam of electrons of the wind moving towards the observer. Since the plane of the Crab pulsar equator is inclined to the observer at 33^0 (Hester et al., 1995), it is reasonable to take $\gamma_{max} = 3/2\dot{E}_{rot}/\dot{N}mc^2$, with $\gamma_w(\alpha) = \gamma_{max}\cos^2(33^0)$.

The low frequency emission from the Crab pulsar consists of pulsed and unpulsed components of radiation, as shown in Fig. 6.6. The pulsed component detected at IR/optical and X-rays is believed to originate in the magnetosphere. The magnetospheric (nonthermal) emission is likely to contain also an unpulsed power-law optical component. Although we may also expect thermal radiation from the surface of the neutron star, there are only upper limits on the flux of this radiation component. For a radius of the neutron star of 10 km, this gives us an upper limit on the temperature of the surface of 1.9×10^6 K.

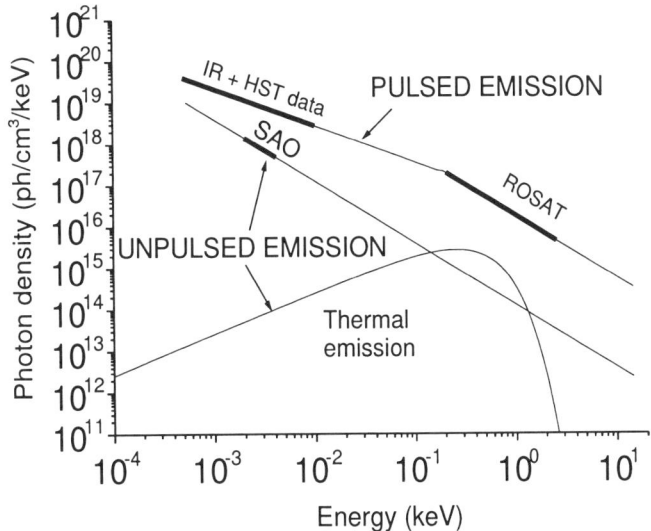

Fig. 6.6 Average radiation densities near the light cylinder of the Crab pulsar. The heavy solid lines show the measured fluxes. Thin solid lines show the interpolations and extrapolations of the observational data. The spectrum of the expected blackbody emission component corresponds to a temperature of 1.9×10^6 K.

(a) target photons of nonthermal origin

The IC γ radiation caused by illumination of the wind by the pulsed soft photon emission should be modulated at the same period of the pulsar. This is true even for the isotropic wind. To calculate the IC fluxes it is assumed that the nonthermal soft emission is generated by relativistic electrons which have v_φ close to zero, as happens in the inner and outer gap models. Therefore, the soft photons move along the radial direction.

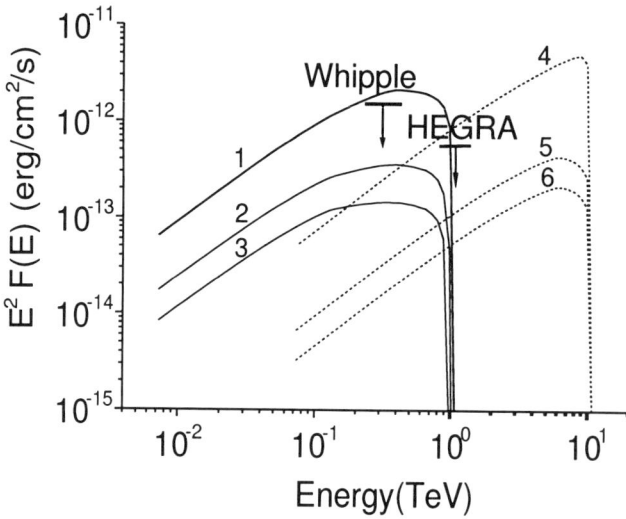

Fig. 6.7 The spectra of *pulsed* γ-ray emission produced by a pulsar wind with $\gamma_{max} = 3 \times 10^6$ (solid lines) and $\gamma_{max} = 3 \times 10^7$ (dashed lines). The curves 1, 2, and 3 correspond to R_w/R_L =30, 50, and 70, and the curves 4, 5 and 6 to R_w/R_L = 50, 70, 100. The upper limits on the pulsed radiation from Crab Nebula, reported by the Whipple group at 300 GeV and by the HEGRA group at 1 TeV, are also shown. The beam of the soft photons is assumed to be radial (from Bogovalov and Aharonian, 2000).

The comparison of the calculated fluxes with the reported TeV upper limits on the pulsed emission from the Crab (Fig. 6.7) set a robust lower limit on the distance at which the wind acceleration takes place. For standard assumptions about the location of sources of soft nonthermal radiation, it should exceed $30R_L$, otherwise it would result in a significant overproduction of VHE γ-rays (compared with the reported upper limits). This conclusion agrees with the wind acceleration model of reconnection of the striped magnetic field in the winds of radio pulsars (Michel, 1982; Coroniti, 1990). However in Crab-like pulsars this mechanism cannot provide an effective transformation of the Poynting flux to the kinetic energy of particles (Lyubarsky and Kirk, 2001). The above constraint on the site of wind acceleration, $\geq 30R_L$, can be avoided if one assumes that the low-frequency pulsed emission is associated with the site of acceleration of the wind. If so, the soft photons move almost parallel to the relativistic electrons, and therefore the effect of Compton scattering becomes negligible. This is, however, a rather drastic (to some extent, provocative) assumption, the implications of which need to be thoroughly studied. In the case of the Crab pulsar, the broad-band nonthermal emission from radio to γ-rays are in phase with each other, therefore are likely to be produced in the same region of the magnetosphere. Now, assuming that the KED wind also originates from the same region, we actually propose that the pulsed radio, optical, X-rays and γ-rays below 10 GeV are produced at the light cylinder, but not in the inner magnetosphere, contrary to the standard predictions of the inner and outer electrostatic gaps models.

Independent of the location of the source(s) of nonthermal soft emission, we still should expect another Compton radiation component of the unshocked wind associated with the *thermal emission* of pulsar.

(b) target photons from thermal isotropic radiation

The γ-ray fluxes expected from the bulk motion Comptonization of the thermal isotropic radiation of the pulsar by the KED wind produced near the light cylinder . are shown in Fig. 6.8.

Solid lines correspond to the fluxes produced by the KED wind produced at $R_\mathrm{w} = R_\mathrm{L}$ and $2R_\mathrm{L}$ and temperature $T_\mathrm{r} = 1.9 \times 10^6$ K. The dashed lines show the spectra of the IC emission for $R_\mathrm{w} = R_\mathrm{L}$ and $T_\mathrm{r} = 0.9 \times 10^6$ K. Note that the differential spectra of IC γ-rays have a specific line-type feature which is a result of the IC scattering in deep Klein-Nishina regime. The comparison of the calculated spectra with the observed TeV γ-ray fluxes of the Crab Nebula apparently rule out the acceleration of the KED wind at the light cylinder with Lorentz factors within the limits $3 \times 10^5 < \gamma_\mathrm{max} < 3 \times 10^7$. This contradiction could be resolved assuming that (1) the thermal radiation from the neutron star is significantly less than the flux for the assumed black-body radiation with $T_r \sim 10^6$ K or (2) the Lorentz factor of the KED wind is larger than 3×10^7 or smaller 3×10^5. Since the assumed surface temperature is close to the theoretical expectations, the second option seems more attractive.

Although there are certain theoretical arguments which give preference to Lorentz-factors of the KED wind larger than 10^6 (Kennel and Coroniti, 1984), we cannot rule out the formation of the KED wind with a larger pair multiplication factor λ and, correspondingly, smaller γ_w. Therefore significantly lower (by a factor up to 10-100) values cannot be excluded. Moreover, such small values of the wind Lorenz factor may have an interesting implication regarding the possible interpretation of the unpulsed GeV γ-ray emission reported by EGRET, which so far does not have an adequate explanation, either by the electrostatic gap models of the Crab pulsar or by the inverse Compton models of the Crab Nebula.

Fig. 6.9 demonstrates that a KED wind with $\gamma_{\max} \sim 4 \times 10^4$ formed at $R_{\rm w} = 3R_{\rm L}$ may produce IC radiation comparable with the EGRET unpulsed γ-ray emission below 10 GeV. Of course the exact values for $\gamma_{\max}$ and $R_{\rm w}$ depend on the temperature of the blackbody radiation emitted from the surface of the neutron star as well as on the geometry of the wind. Nevertheless, the implications of this interpretation are independent of the details of model assumed. If confirmed by future observations with GLAST, this hypothesis would allow the first direct measurement of the Lorentz factor of a pulsar wind. It would also lead to two important conclusions: (i)

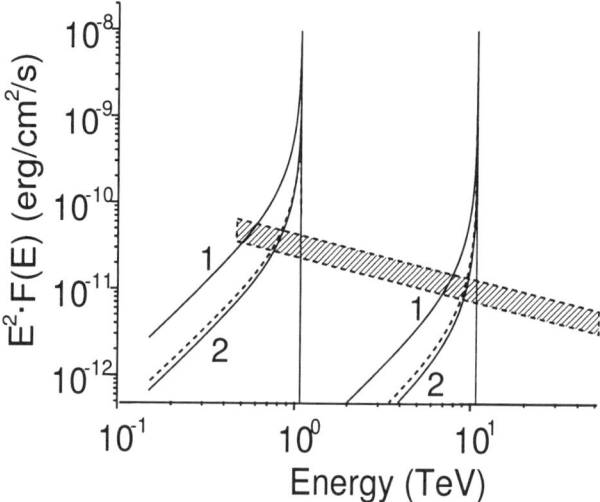

Fig. 6.8 The spectra of *unpulsed* TeV radiation from unshocked wind of the Crab pulsar. It is assumed that the KED wind with $\gamma_{\max} = 3 \times 10^6$ and $\gamma_{\max} = 3 \times 10^7$ is illuminated by the thermal (blackbody) emission from the surface of the neutron star with radius $R = 10$ km. The solid curves correspond to the surface temperature $T = 1.9 \times 10^6$, assuming that the wind is accelerated at $R_{\rm w} = R_{\rm L}$ (1) and $R_{\rm w} = 2R_{\rm L}$ (2). The dashed curves correspond to the emission with a lower temperature, $T = 0.9 \times 10^6$ K, and $R_W = R_L$. The range of the observed unpulsed fluxes of TeV γ-rays from the Crab is shown by the dashed zone. (From Bogovalov and Aharonian, 2000).

unusually small Lorentz-factor of the KED wind, and (ii) production of the pulsed low-frequency radiation of the pulsar in the same region (outside of the magnetosphere) where the KED wind is formed. The second conclusion is required in order to suppress the Comptonization of the wind by the pulsed low-frequency radiation, and thus to avoid the conflict with observations around 10 GeV.

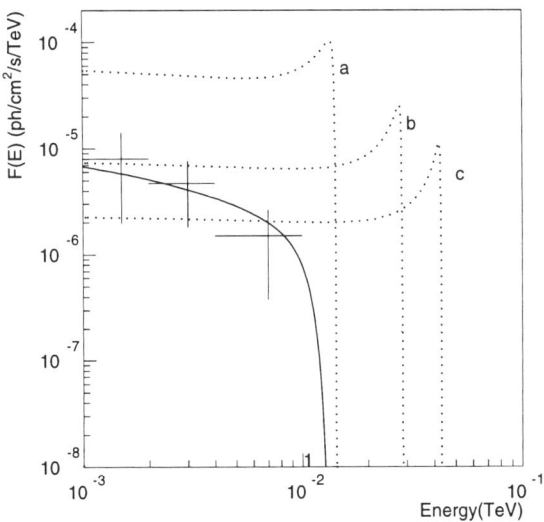

Fig. 6.9 The spectra of unpulsed IC gamma-radiation produced by the wind of the Crab pulsar. The dotted curves correspond to the KED wind formed at the light cylinder with Lorentz-factors 4×10^4 (a), 8×10^4 (b), and 1.2×10^5 (c). The solid curve is calculated for the wind with a Lorentz-factor 4×10^4 formed at $R_\mathrm{w} = 3R_\mathrm{L}$. The error bars present the EGRET fluxes (Nolan et al., 1993) of the unpulsed γ-ray component of radiation from the Crab. (From Aharonian and Bogovalov, 1999).

6.2.4 Gamma rays from winds of PSR B1706-44 and Vela?

The radio pulsar PSR B1706-44, which has a rotation period $P = 102$ ms, characteristic age $T = 1.7 \times 10^4$ yr, and spin-down luminosity $\dot{E} = 3.4 \times 10^{36}$ erg/s, is one of the six high energy γ-ray pulsars detected by EGRET (Thompson, 1999). The pulsed X-ray emission from this young, energetic radio and γ-ray pulsar was resolved recently by the Chandra X-ray Observatory (Gotthelf et al., 2002). This result is consistent with a ROSAT upper limit of less than 18 per cent for the pulsed component

(Becker and Trümper, 1997). Only an upper limit on the flux of optical emission (pulsed or unpulsed) from the pulsar has been reported at the level of 5×10^{29} erg/s (Mignani et al., 1999).

The pulsar PSR B1706-44 is surrounded by a compact X-ray nebula (Becker and Trümper, 1997; Finley et al., 1998) with a size of ≈ 5 arcsec (Dodson and Golap, 2002). However the X-ray source is faint, therefore one cannot that a noticeable fraction of the X-ray flux is due to the unpulsed component of X-radiation that originates from the pulsar. The spectrum of X-ray emission in the 0.1-2.4 keV interval, after correction for absorption, can be approximated by power-law with photon index 2.4 ± 0.6.

The CANGAROO and Durham collaborations have claimed detection of unpulsed TeV γ-ray emission from this source (see Sec. 2.3.2.2). The "standard" interpretation of this radiation – the inverse Compton scattering of relativistic electrons responsible also for the synchrotron X-ray nebula – may have a serious problem related to the TeV luminosity of the source which significantly exceeds the X-ray luminosity. This implies that one has to assume unrealistically small magnetic field in the nebula or invoke strong energy-dependent propagation effects of electrons to overcome the problem (see Sec.6.4.4). Below we discuss a different possible solution to the problem of low L_x/L_γ ratio, assuming that the TeV γ-ray emission is produced by unshocked ultralelativistic wind. Because the wind is supposed to be cold, in this scenario the TeV radiation are not accompanied by synchrotron X-rays.

The reported TeV fluxes are shown in Fig.6.10, together with the spectra of hypothetical IC γ-rays from the unshocked wind, calculated for the wind Lorentz-factor $\gamma_{max} = 2.2 \times 10^6$. We assume that half of the X-ray flux observed from the direction of PSR B1706-44 originates from the pulsar. For the fixed seed photon X-ray luminosity at 1 keV, the resulting IC γ-ray flux strongly depends on the photon index of X-radiation and the site of formation of the KED wind. The calculations are performed under the assumption that we observe the pulsar in the equatorial plane, which increases the flux of γ-rays to the maximum possible value. This is an inescapable assumption because of the low level of IR/optical and X-ray photons from the pulsar/neutron star, which constitute the main target fields for production of inverse Compton γ-rays. The calculated γ-ray flux of the wind can only marginally explain the observed TeV flux, even if the KED-wind is formed very close to the light cylinder. The low-frequency radiation from PSR B1706-44 is strongly dominated by the unpulsed component, therefore the TeV γ-ray emission is expected to be essentially unpulsed as well.

It is seen that the calculated spectra are in a reasonable agreement with the reported fluxes "C" and "D97" for $\alpha_x = 2.4$. If we adopt the higher (revised later by the Durham group) flux of $\geq 300\,\text{GeV}$ γ-rays, the calculated IC flux could match the reported flux "D98" only if we adopt extreme values for both site of formation of the KED wind ($R_w = R_L$) and the spectral index of X-rays ($\alpha_x = 3$). Thus, despite large uncertainties in the estimates of the X-ray spectrum at low energies (due to the absorption of soft X-rays in the interstellar medium), as well as large systematic uncertainties in the reported TeV γ-ray fluxes, the unshocked wind origin of TeV γ-rays requires rather tough assumptions concerning the Lorentz-factor and the geometry of the wind. Namely, the TeV γ-ray emission observed from PSR B1706-44 can be explained within the framework of this hypothesis provided that (i) the wind is produced within the $\leq 2\,R_w$ proximity of the light cylinder, . (ii) it has a Lorentz-factor close to 2×10^6, and (iii) we see the pulsar in the equatorial plane.

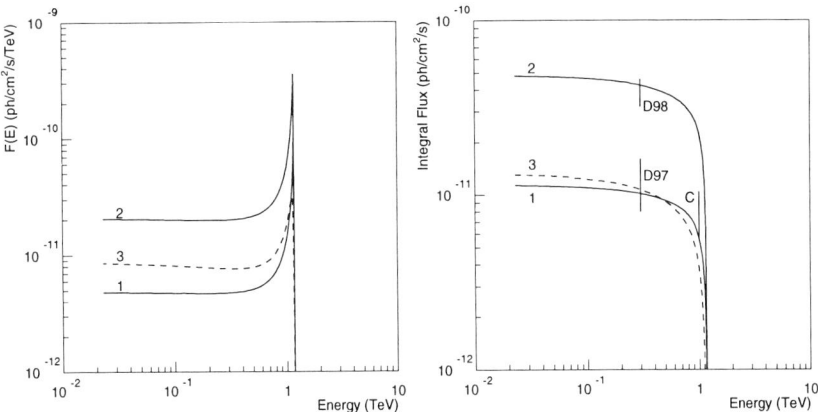

Fig. 6.10 Differential (left) and integral (right) fluxes of IC γ-ray emission of the wind of PSR B1706-44 calculated for $\alpha_x = 2.4$ (curves "1") and $\alpha_x = 3$ (curves "2") and assuming that $R_W = R_L$ (solid curves). The dashed curves correspond to $\alpha_x = 3$ and $R_w = 1.6\,R_L$. The symbol "C" indicates the flux level reported by the CANGAROO collaboration, and the "D97" and "D98" correspond to the initial and revised fluxes reported by the Durham group.

A decisive test for this hypothesis will be possible by study of spectral and angular features of radiation at 1 TeV and below. The unshocked wind model predicts a specific energy spectrum (see Fig.6.10) which can be checked by the 100 GeV threshold IACT arrays. Also, this model predicts a strictly point source structure for the TeV emission, therefore any evidence of extended character of the source would discard the model.

An important part of the search strategy for γ-ray signals from "unshocked pulsar winds" would be observations of several good candidates, in particular the Vela pulsar. The large spin-down luminosity, $L_0 = 7 \times 10^{36}$ erg/s, and relatively well known fluxes of pulsed and unpulsed components of low-energy (optical and X-ray) emission, make this nearby ($d \simeq 250$ pc) pulsar a good target for future ground-based γ-ray observations in general, and for the search for unshocked pulsar winds, in particular. The IC γ-ray fluxes expected for different assumptions concerning the Lorentz factor of the wind, γ_{max}, are shown in Fig.6.11. The spectra are calculated for two sites of formation of the KED wind, $R = 2R_{\mathrm{L}}$ and $R_{\mathrm{W}} = R_{\mathrm{L}}/\cos\alpha$, where $\alpha \simeq 34.6°$ is derived from the toroidal structure of the synchrotron X-ray nebula as measured by Chandra (Helfand et al., 2001). Note that $R_{\mathrm{w}} = R_{\mathrm{L}}/\cos\alpha$ corresponds to the minimum possible value of R_{w}. In Fig.6.11 the flux sensitivities of GLAST and the proposed

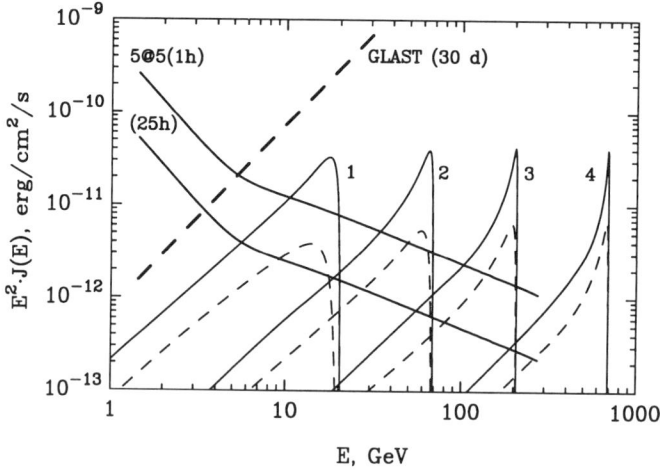

Fig. 6.11 Spectral energy distribution of the IC gamma-radiation from the wind of the Vela pulsar calculated for $\gamma_{\mathrm{max}} = 6 \times 10^4$ (1), 2×10^5 (2), 6×10^5 (3), and 2×10^6 (4). Solid curves correspond to $R = R_{\mathrm{L}}/\cos\alpha$, and dashed curves correspond to $R = 2R_{\mathrm{L}}$. The "5-sigma" sensitivities of GLAST and the IACT array 5@5 are also shown.

IACT array 5@5 (see Chapter 2) are also shown. 5@5 should be able to detect the predicted fluxes for all possible values of γ_{max} in a broad interval from 10^4 to 10^6.

6.2.5 IC γ-rays from the binary pulsar PSR B1259-63

IC γ-ray fluxes from pulsar winds depend significantly on the density of target photon fields. It is clear that the 2.7 K CMBR and diffuse galactic IR/optical photons cannot play significant role in γ-ray production. In the case of isolated pulsars, the thermal and nonthermal radiation components associated with the surface of the neutron star and the pulsar magnetosphere appear sufficient for the production of detectable IC γ-ray fluxes from the winds of pulsars with spin-down fluxes $L_0/4\pi d^2 \geq 10^{-8}$ erg/cm^2, provided that the KED ultrarelativistic wind is formed close to the light cylinder. If, however, the KED winds are formed well beyond the light cylinder, only the winds of strong pulsars like the Crab and Vela can produce (marginally) detectable IC γ-ray emission. In this regard, the winds of binary pulsars are expected to be more prolific IC γ-ray emitters, because these winds encounter an enormous density of photons from companion stars. A perfect example of such a source is the binary pulsar system PSR B1259-63/SS2883 (Ball and Kirk, 2000). Despite the relatively modest spin-down luminosity of the pulsar PSR B1259-63, $L_0 = 8.3 \times 10^{35}$ erg/s, the luminous and hot ($L \simeq 9 \times 10^3 L_\odot$, $T \simeq 2.2 \times 10^4$ K) companion star SS2883 provides copious seed photons for Comptonization of the wind electrons. Indeed, at periastron the companion is only $23R_* \sim 10^{13}$ cm from the pulsar, thus the density around periastron is about 1 erg/cm^3, i.e. 12 orders of magnitude greater than the density of the diffuse galactic background radiation (Ball and Dodd, 2001). The existence of such a dense photon field should result in conversion of a significant fraction of the electron-positron wind power into high energy γ-rays. Assuming that the wind momentum is entirely carried by electron-positron pairs, the IC γ-ray flux should be detectable by future powerful IACT arrays, even if only 10^{-4} part of the energy of the wind of PSR B1259-63 is released in γ-rays ($f_\gamma \simeq 3 \times 10^{-13}$ erg/cm^{2}s for a distance to the source $d = 1.5$ kpc).

Actually, the IC radiation of this source consists of two components associated with the shocked and unshocked wind. Kirk *et al.* (1999) considered IC scattering in the shocked region of the wind, downstream of the termination shock, where the pressure of the pulsar wind is balanced with the companion Be-star wind. Later, Ball and Kirk (2000) realized that the target photons from the companion star should pervade the wind upstream of the termination shock. The effect is so strong that the wind can be decelerated by 'Compton drag' as the electrons lose their energy and momentum while interacting with background photon fields (Ball and Kirk, 2000). The resulting flux of IC γ-rays is expected to be variable because of the pulsar's eccentric orbit. Both the absolute flux and the spectrum of radiation depend on the geometrical parameters shown in Fig. 6.12. When $\theta_p = 0°$, the electrons are travelling straight towards the companion star, and inverse Compton drag has the greatest effect because of the large density of target photons. For large values of θ_p, the drag is less effective because of the small electron–photon collision angle and the drop of the target photon density as the wind propagates away from the companion star.

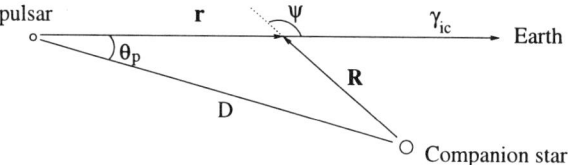

Fig. 6.12 Sketch of the binary system defining angles and distances. (From Ball and Kirk, 2000).

Figure 6.13 shows the spectrum of scattered radiation for three different initial wind Lorentz factors and three values of θ_p. The calculations are performed assuming that the wind momentum is entirely carried by electron-positron pairs The spectrum is broadest, and the integrated flux largest, when θ_p is small and $\gamma_w(0)\epsilon_0 \sim 1$ (where $\epsilon_0 \simeq 3kT \simeq 5$ eV is the mean energy of photons from the companion star). As long as $\gamma_w(0)$ exceeds 10^5, Compton scattering takes place in the Klein-Nishina regime, and thus the IC radiation has a line-type spectrum.

The increase from the minimum to the maximum occurs over 65 days beginning on day -6, just before periastron. The subsequent decrease occurs far more slowly over the remaining 1170 days of the orbital period. This dependence, together with the changes in the photon density due to the changing binary separation, results in a regular variation of γ-radiation.

All the spectra shown in 6.13 are calculated assuming that the wind is not terminated by the pressure balance. The impact of the termination of the wind on the scattered emission from the unshocked wind has been studied by Ball and Dodd (2001) who showed that this effect can decrease the overall level of IC radiation of the unshocked wind and change the spectrum. Also, the termination shock almost eliminates the asymmetry of the light curve before and after the periastron. This implies that unfortunately it is not easy to distinguish between the contributions from the unshocked and shocked regions of the pulsar wind. Nevertheless, even if the wind of the companion optical star dominates that of the pulsar (terminating it in a shock that wraps around the pulsar), the IC γ-ray emission of the unshocked wind should be still detectable for a wide range of wind parameters. The detection of a positive γ-ray signal and its identification with the IC radiation of the unshocked pulsar wind would provide unique information about the structure of the pulsar wind. Even an upper limit on the γ-ray flux, would lead to important conclusions, in particular to revision of our current belief that the pulsar winds are particle dominated cold ultrarelativistic outflows consisting mainly of electrons and positrons.

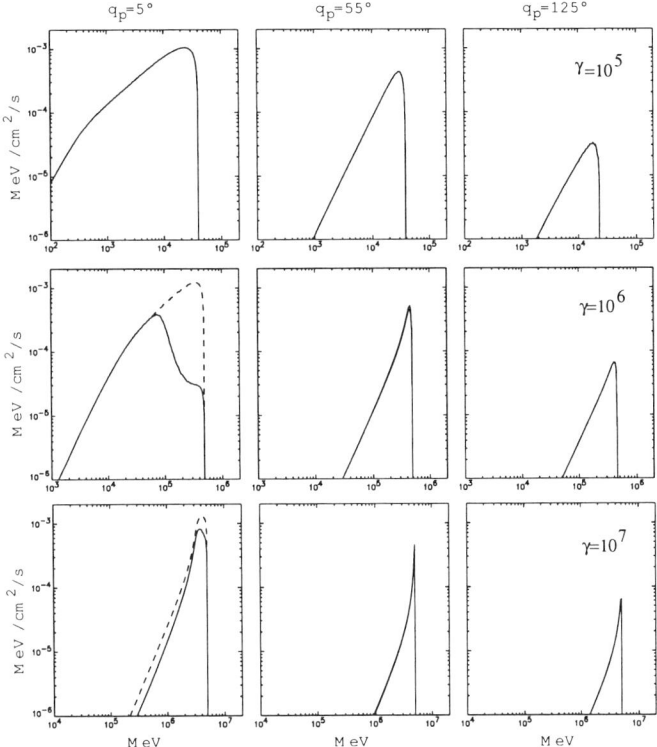

Fig. 6.13 The spectral energy distribution of inverse Compton scattered photons from the unshocked wind of PSR B1259-63 at periastron as a function of photon energy – at angles $\theta_p = 5°$ (left), 55° (middle), and 125° (right). The top row of spectra are calculated for a pulsar wind with an initial Lorentz factor $\gamma_w(0) = 10^5$, the middle for $\gamma_w(0) = 10^6$, and the bottom row for $\gamma_w(0) = 10^7$. The solid curves include the effects of the pair production optical depth. (From Ball and Kirk, 2000).

6.3 Gamma Rays from Pulsar Driven Nebulae

Pulsars lose their rotational energy by driving ultrarelativistic winds of electrons, positrons and, possibly, ions. If a pulsar is surrounded by a supernova remnant, its winds are thought to terminate at a collisionless shock front, the location of which is determined by the balance between the wind ram pressure and the total pressure in the nebula caused by accumulation of the energy injected in the nebula over the time of the pulsar (Rees and Gunn, 1974). This condition, as well as an assumption that the energy density of magnetic field is approximately half of the overall pressure (so-called "equipartition condition"), gives quite accurate estimates of the position of the shock and the average nebular magnetic field, in particular for the most prominent pulsar-driven nebula or *plerion* (filled supernova remnant), the Crab Nebula.

The MHD wind model developed by Kennel and Coroniti (1984) satisfactorily explains many basic features of the Crab Nebula. In this model the pulsar wind is terminated by a standing reverse shock which accelerates electrons up to 10^{16} eV and randomises their pitch angles. This leads to formation of a bright synchrotron source in the region downstream of the shock. A synchrotron nebulae is formed when the ultrarelativistic, kinetic energy dominated wind is confined in a slowly (nonrelativistically) expanding shell of the supernova remnant. Thus, the major fraction of the rotational energy of the pulsar is eventually released in nonthermal synchrotron radiation which extends to the hard X-ray or even γ-ray energy domain. Inverse Compton scattering of the 2.7 K CMBR and (in the Crab Nebula) the synchrotron radiation of the plerion by the same ultrarelativistic electrons lead to the formation of more energetic γ-radiation extending to multi-TeV energies. Although in many cases the energy release in IC γ-rays may constitute only a small fraction of the synchrotron luminosity of the plerion, the high energy γ-radiation provides a unique channel of information about the basic parameters of the nebula.

6.3.1 *Broad-band nonthermal radiation of the Crab Nebula*

The Crab Nebula is a unique cosmic laboratory with an unprecedentedly broad spectrum of the observed nonthermal radiation which extends over 21 (!) decades of frequencies – from radio wavelengths to very high energy γ-rays (see Fig. 2.7). This emission is dominated by two major mechanisms connected with interactions of relativistic electrons with the magnetic and photon fields of the nebula. While the synchrotron component is responsible for the radiation from radio to relatively low energy γ-rays ($E < 1\,\text{GeV}$), the inverse Compton (IC) scattering of electrons is thought to be the most probable mechanism for TeV γ-rays.

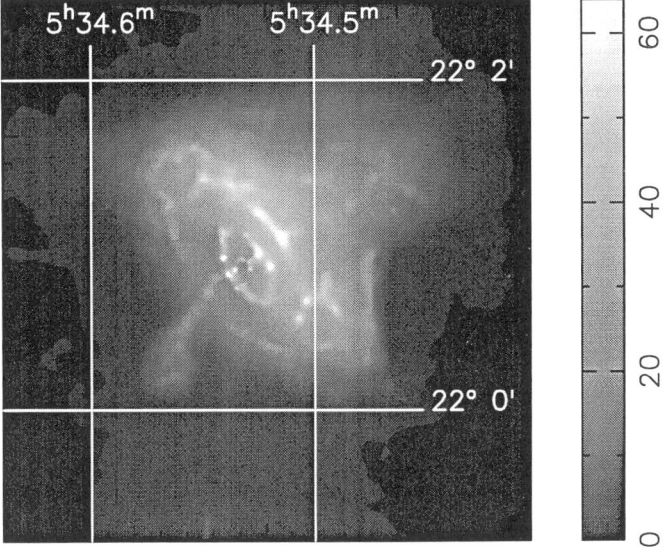

Fig. 6.14 Adaptively smoothed Chandra HETG-ACIS-S image of the central $200'' \times 200''$ region of the Crab Nebula (From Weisskopf et al., 2000).

The typical energies of electrons responsible for production of synchrotron photons in different energy bands of the Crab spectrum are indicated in Fig. 2.7. Note that although the conclusion about highest energy electrons is still based on a model (although well justified) assumption about the synchrotron origin of the hard X-rays/low-energy γ-rays, the detection of γ-rays well above $10^{13}\,\text{eV}$ is the first *unambiguous* evidence of effective acceleration of particles beyond $10^{14}\,\text{eV}$.

It is commonly accepted that the synchrotron nebula is powered by the relativistic wind of electrons generated at the pulsar and terminated by a standing reverse shock wave at a distance $r_s \sim 0.1\,\text{pc}$ (Rees and Gunn, 1974). Relativistic MHD models, even in their simplified form (e.g. ignoring the axisymmetric structure of the wind and its interaction with the optical filaments), successfully describe the general characteristics of the synchrotron nebula, and predict realistic distributions of relativistic electrons and the magnetic field in the downstream region behind the shock (Kennel and Coroniti, 1984).

While the synchrotron and IC mechanisms seem to provide a reasonable explanation of the overall nonthermal radiation of the Crab Nebula (see Fig. 2.7), one cannot exclude possible deviations in different frequency domains from the simplified picture of the outer nebula described by the spherically symmetric MHD models. Indeed, the imaging of the Crab Nebula by the Hubble Space Telescope (Hester et al. 1995) revealed a very rich and complex structure of the inner part of the nebula, on scales down to 0.2″. It consists of features like wisps, jets, knots, *etc.*, and exhibits cylindrical symmetry. The recent imaging of the Crab Nebula in X-rays by Chandra (Weisskopf *et al.*, 2000) with subarcsecond resolution confirms the striking richness of the inner nebula (Fig. 6.14). In particular, these observations revealed, for the first time, an X-ray inner ring within the X-ray torus (Aschenbach and Brinkmann, 1975), as well X-ray hot spots along the inner ring.

The distinct axisymmetrical structure of the inner nebula is strong evidence that the most of the rotational energy of the pulsar is released in the form of a wind which flows along the pulsar equator. But, after 30 years of extensive theoretical efforts, astrophysicists are still far from a full understanding of the important details of the physics of interaction of the pulsar wind with the synchrotron nebula. This prominent source remains a great challenge for future theoretical work.

6.3.1.1 *Synchrotron and IC radiation*

To calculate the nonthermal radiation of the Crab Nebula, one has to specify the spatial distribution of the magnetic field, the acceleration site(s) and the injection spectrum of the relativistic electrons, as well as the character of their propagation in the nebula. The calculations presented below are based on the MHD model of Kennel and Coroniti (1984). Although this model assumes a spherical geometry, which significantly deviates from the picture shown in Fig. 6.14, it provides a quite accurate estimates for IC γ-ray fluxes integrated over the size of the nebula.

The MHD model of Kennel and Coroniti (1984) provides all necessary parameters, in particular it allows self-consistent calculations of the spatial and spectral distribution of high energy electrons, freshly accelerated at the wind termination shock and injected into the nebula (*wind* electrons).

Meanwhile, the radio emission of the nebula requires an additional low energy ($E \leq 100$ GeV) component of electrons (*radio* electrons) accumulated, most probably, during the whole history of the Crab Nebula. Although the origin and site(s) of this "relic" component are not yet established, it can be easily incorporated into the calculations of the broad-band synchrotron spectrum with a minimum number of assumptions based on radio observations (Atoyan and Aharonian 1996).

Fig. 2.7 demonstrates the good agreement of calculations with the observed spectrum of the Crab Nebula up to hard X-rays, and a reasonable explanation of the γ-ray fluxes up to 1 GeV by the synchrotron radiation. The best fit is reached by the following combination of the spectra of the *radio* and *wind* electrons: (1) $n_{\rm re}(E) \propto E^{-\alpha_{\rm re}} \exp(-E/E_{\rm c})$ with $\alpha_{\rm re} = 1.52$ and $E_{\rm c} = 150$ GeV, and (2) $n_{\rm we}(E) \propto (E_* + E)^{-\alpha_{\rm we}} \exp(-E/E_1)$ with $\alpha_{\rm we} = 2.4$, $E_* = 200$ GeV, and $E_1 = 2.5 \times 10^{15}$ eV. The presentation of the spectrum of the *wind* electrons $n_{\rm we}(E)$ in such a form provides a the necessary (and natural in the framework of MHD model) flattening of the spectrum below $E_* \sim 200$ GeV. The transition from the hard to steep power-laws in the total (*radio* + *wind*) electron spectrum at energies around 100 GeV accounts for the sharp steepening of the spectrum at IR/optical wavelengths. Therefore detailed spectroscopic measurements in this energy region would allow us to specify more precisely the values of $E_{\rm c}$ and E_* which define the degree of "smoothness" of transition from *radio* to *wind* electrons. Independent information about the electrons in this transition region is contained in the 10 to 100 GeV γ-rays produced by the same electrons upscattering the ambient low-frequency radiation. Meanwhile, determination of the high energy cutoff E_1 in $n_{\rm we}(E)$ is contingent on measurements of γ-rays in the 1 MeV to 1 GeV energy band.

The existence of ultra-relativistic electrons in the synchrotron nebula provides production of detectable TeV γ-ray fluxes through IC scattering (Gould, 1965; De Jager and Harding, 1992; Aharonian and Atoyan, 1995; De Jager *et al.*, 1996; Atoyan and Aharonian, 1996; Hillas *et al.*, 1998). Since the target radiation fields which play a major role in the production of IC γ-rays in the Crab Nebula (synchrotron, thermal infrared, and 2.7 K CMBR) are well known, the flux of IC γ-rays can be calculated with good accuracy. For the given flux of synchrotron X-rays, the number of TeV electrons strongly depends on the nebular magnetic field. Therefore the IC γ-ray fluxes are very sensitive to the average magnetic field: $I(\geq 1\,{\rm TeV}) \simeq 8 \times 10^{-12}(\bar{\rm B}/0.3\,{\rm mG})^{-2.1}\,{\rm ph/cm^2 s}$. Thus the comparison of the predicted and observed TeV γ-ray fluxes allows determination of the magnetic field in the central $r \sim 0.5\,\mathrm{pc}$ region (where the bulk of X-rays and TeV γ-rays are produced) with accuracy $\Delta B/\bar{B} \simeq 0.5(\Delta F/F)$. Although the statistical significance of γ-ray observations of the Crab Nebula is very high, the systematic uncertainties in the flux estimates remain rather large, $\Delta F/F \sim 1$. Even with the present uncertainties, the TeV observations favour the magnetic field in the X-ray production region to be $\bar{B} \sim 0.2 - 0.3\,\mathrm{mG}$, which is in agreement with the estimated equipartition field (Marsden et al., 1984).

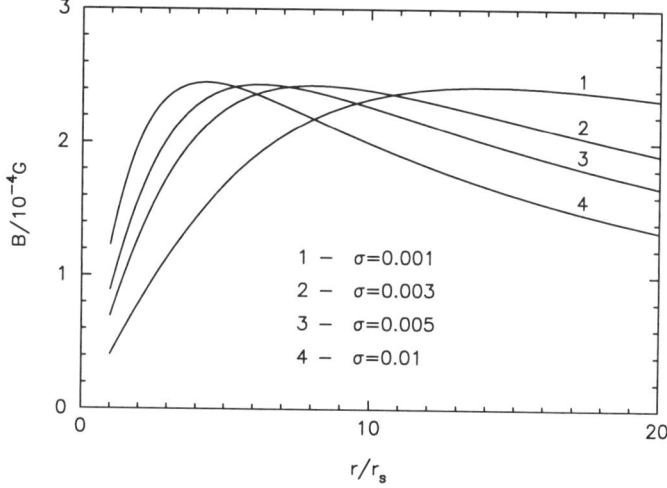

Fig. 6.15 The radial distribution of the magnetic field in the Crab Nebula expected within the framework of the MHD model of Kennel and Coroniti (1984) for different values of σ-parameter. $r = r_\mathrm{s}$ corresponds to the distance of the wind termination shock from the pulsar.

The shape of the spectrum of IC γ-rays does not depend much on the basic parameters of the nebula, in particular, it is almost independent of the parameter σ (the ratio of the electromagnetic energy flux to the particle energy flux at the shock) which in the framework of MHD model defines the spatial distribution of the magnetic field $B(r)$ at $r \geq r_\mathrm{s}$. Although the calculations in Fig. 2.7 correspond to $\sigma = 0.005$, the measured synchrotron fluxes can be equally well fitted with σ anywhere between 0.001 and 0.01. Indeed, since at distances $\geq 3r_\mathrm{s}$, where the bulk of synchrotron and IC photons are produced, the magnetic field $B(r)$ calculated with the MHD model of Kennel and Coroniti (1984) depends rather weakly on σ (Fig. 6.15), only minor changes in the assumed injection spectrum of electrons are required to provide the same synchrotron spectrum for different σ within 0.001-0.01. Consequently, as far as the target photon fields for IC γ-rays are fixed, there is no room for strong dependence of the calculated IC fluxes on σ.

In Fig. 2.7 are shown the synchrotron and IC components of radiation, calculated in the framework of spherically symmetric MHD model, which describes the flux of the Crab Nebula over the whole range of observed frequencies fairly well. At the same time, in some specific energy intervals, in particular in the 1-10 MeV, 1-10 GeV, and perhaps also at ≥ 10 TeV bands the observed energy spectra cannot be easily explained within the simplified synchrotron-Compton model. Below we discuss possible ways to account for these features.

6.3.1.2 *Second High Energy Synchrotron Component*

The spectral measurements of the unpulsed radiation of the Crab by COMPTEL revealed an unexpected flattening of the spectrum at energies 1-10 MeV (van der Meulen *et al.*, 1998) which follows the well established steepening of the spectrum above 100 keV. Since such sharp feature could hardly be attributed to peculiarities in the injection spectrum of shock accelerated electrons, a more natural interpretation of this spectral feature can be given assuming the existence of an additional radiation component. Explanation of this radiation excess in terms of nuclear γ-ray line emission is not supported by observations (van der Meulen *et al.*, 1998), and more importantly, it contradicts the total luminosity of the Crab Nebula since only $\xi \leq 10^{-5}$ of the energy losses of nonrelativistic protons and nuclei is released in prompt γ-ray lines, while the main part goes to heating of the ambient gas, and thus should show up in the form of thermal radiation with an unacceptably high luminosity $L_\mathrm{thermal} \sim \xi^{-1} L_\mathrm{obs}(1-10\,\mathrm{MeV}) \geq 10^{41}$ erg/s.

While remaining in the framework of the hypothesis of the synchrotron origin for the radiation up to 1 GeV, the steepening above 100 keV implies an exponential cutoff in the injection spectrum of the *wind* electrons at energies smaller than $E_1 = 2.5 \times 10^{15}\,\mathrm{eV}$ used in Fig. 2.7 . If so, the flat spectrum observed by COMPTEL requires a second population of high energy electrons. In Fig. 6.16 a possible fit of the observed fluxes up to 1 GeV by two-component synchrotron emission is presented. The first component is attributed to the same *wind* electrons as in Fig. 2.7 but with $E_1 = 5 \times 10^{14}\,\mathrm{eV}$. For the second component a very hard acceleration spectrum, for example of Maxwellian type, $n_2(E) \propto E^2 \exp(-E/E_2)$ with

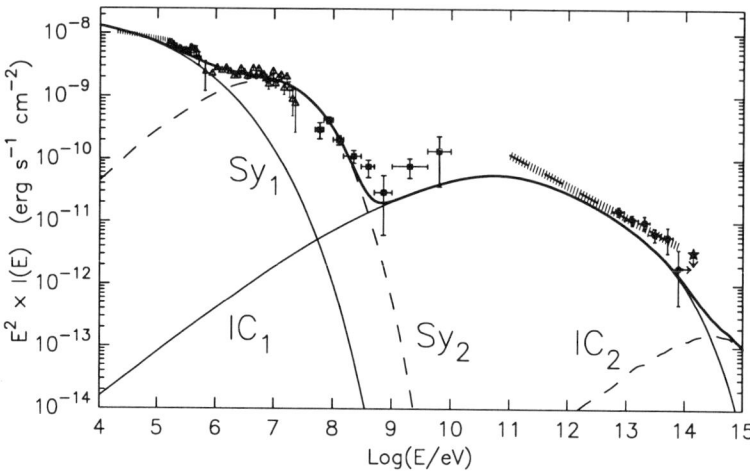

Fig. 6.16 Synchrotron and IC radiation components produced by the first (solid) and second (dashed) populations of electrons (see text). The heavy solid line shows the total flux. The hatched region corresponds to $I(E) = (2.5 \pm 0.4)(E/1\,\mathrm{TeV})^{-2.5}\,\mathrm{cm}^{-2}\,\mathrm{s}^{-1}\,\mathrm{TeV}^{-1}$ which generally describes the reported fluxes from 300 GeV to 70 TeV. (From Aharonian and Atoyan, 1998a).

$E_2 = 5.5 \times 10^{14}$ eV, has been assumed. Although here $E_1 \approx E_2$, actually the mean energy of the second population $\bar{E} = 3E_2 \simeq 1.6 \times 10^{15}$ eV is larger that the highest energy particles $E \sim E_1$ effectively present in the first population. Note that the required acceleration power in the second electron population is only $P_2 \sim L_{\mathrm{obs}}(1 - 10\,\mathrm{MeV}) \sim 10^{36}$ erg/s, i.e. less than 1% of the first (main) population, $P_1 \sim 3 \times 10^{38}$ erg/s. This explains the small contribution of the second electron population in the total IC γ-ray fluxes up to energies 10^{14} eV (see Fig. 6.16).

The possible sites of acceleration of the second electron population could be the peculiar compact regions such as wisps, knots, etc. Since the equipartition magnetic field in these regions is estimated to be as high as few mG (in that case E_2 is reduced to $\simeq 10^{14}$ eV) the highest energy electrons could not escape the acceleration sites due to severe synchrotron losses. Although these variable structures, with typical size 0.2″ (Hester, 1995), are not resolvable by low-energy γ-ray instruments, the detection of variability of the 1-100 MeV emission would be direct proof for the synchrotron origin of the "MeV bump".

6.3.1.3 Bremsstrahlung and π^0-decay gamma rays?

The second feature seen in Fig. 6.16 is the deficit in the predicted IC fluxes compared to the reported fluxes at GeV energies. The expected IC γ-ray flux in this energy region is estimated to be, within 20 per cent accuracy, $\approx 10^{-8}(\bar{B}/0.3\,\mathrm{mG})^{-1.3}\,\mathrm{ph/cm^2 s}$. For the equipartition magnetic field of $B \simeq 0.3\,\mathrm{mG}$ this flux is a factor of 5 below that measured by EGRET (Nolan et al., 1993, Fiero et al., 1998). In order to explain the measured fluxes in this energy region, an additional component of GeV radiation was suggested in De Jager et al. (1996), which the authors in their best-fit "synchrotron+IC" model call a *second IC power-law component*. To increase the IC flux one has to suppose that the magnetic field in the radio nebula (where the low energy γ-rays are produced) is $B \sim 10^{-4}\,\mathrm{G}$. However this value does not agree with the well defined flux of TeV radiation. Also, such a low magnetic field would imply that the energy in radio electrons is $W_\mathrm{e} \simeq 7 \times 10^{48}\,\mathrm{erg}$, while the energy in the magnetic field is a factor of 30 smaller. This would make the confinement of electrons in the nebula rather problematic. Thus, if the high γ-ray flux above 1 GeV reported by EGRET originates outside the pulsar, i.e. in the nebula, one may need to invoke additional radiation mechanism.

Bremsstrahlung gamma-rays. For a mean gas density in the nebula of $\bar{n} \approx 5\,\mathrm{cm}^{-3}$, the flux of the bremsstrahlung γ-rays cannot exceed 15 % of the flux of IC γ-rays. In fact, in the Crab Nebula the gas is concentrated mainly in dense filaments where $n \sim 10^3\,\mathrm{cm}^{-3}$ (Davidson and Fesen, 1985). In the case of a uniform distribution of relativistic electrons throughout the nebula the *effective* gas density is defined by the mean density of the nebula, $n_\mathrm{eff} \approx \bar{n}$. However, if electrons are trapped, at least partially, in the regions of high density, i.e. if they propagate slower inside the filaments than outside, then $n_\mathrm{eff} \gg \bar{n}$. The fluxes of bremsstrahlung γ-rays calculated for $n_\mathrm{eff} = 50\,\mathrm{cm}^{-3}$ are shown in Fig. 6.17. The contribution of the "amplified" bremsstrahlung flux not only could explain the measured GeV γ-ray fluxes, but also would significantly modify the spectrum at very high energies (Atoyan and Aharonian, 1996). Indeed, in the energy range between 100 GeV and 10 TeV the superposition of the IC and the "amplified" bremsstrahlung components results in an almost power-law spectrum with an index $\alpha_\gamma \simeq (2.5-2.7)$ in contrast to the curved IC γ-ray spectrum *alone*, which is hard at $E \simeq 100\,\mathrm{GeV}$ ($\alpha_\gamma \simeq 2.0$), but becomes significantly steeper at higher energies ($\alpha_\gamma \simeq 2.7$ at $E \simeq 10\,\mathrm{TeV}$).

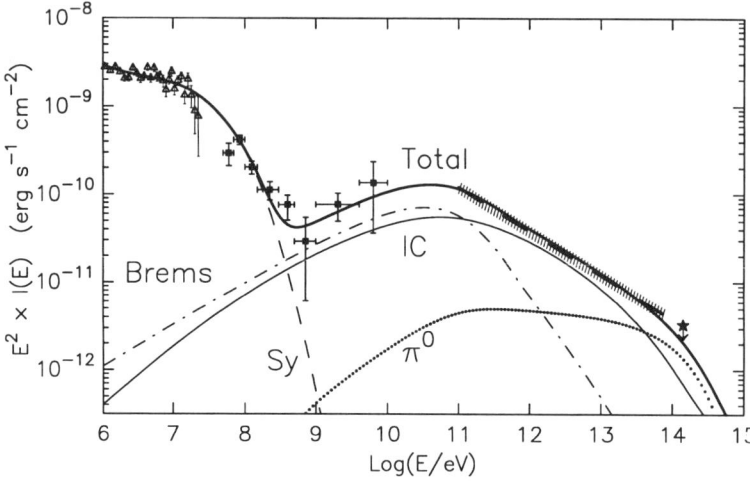

Fig. 6.17 The contributions of different γ-ray production mechanisms to the total nonthermal radiation of the Crab Nebula. The Synchrotron and IC components are the same as in Fig. 6.16. The bremsstrahlung and π^0-decay γ-ray fluxes are calculated for $n_{\rm eff} = 50\,{\rm cm}^{-3}$. (From Aharonian and Atoyan, 1998b).

π^0-decay gamma-rays. Interactions of the nucleonic component of accelerated particles (which may in principle acquire a significant part of the power of relativistic wind at the reverse shock; see Arons, 1996) with the ambient gas lead to the production of γ-rays through secondary π^0-decays. However, since the average gas density in the nebula is low, the contribution of this mechanism to the γ-radiation could be detectable only in the case of partial confinement of relativistic particles in the filaments, so that $n_{\rm eff} \gg \bar{n}$. If so, the π^0-decay γ rays may show up (on top of the steep IC spectrum) at TeV energies and beyond (Atoyan and Aharonian, 1996; Bednarek and Protheroe, 1997). In this regard, the detection of up to 50 TeV γ-rays from the Crab as reported by the CANGAROO (Tanimori et al., 1998a) and HEGRA groups (Horns et al., 2003) may have significant implications concerning the content of the wind and propagation/interaction of accelerated particles in the filaments. Although the reported fluxes do not contradict, within the uncertainties in the nebular magnetic field, an IC origin of the radiation, the estimated differential power-law spectra from 1 TeV to 50 TeV with power law index close to 2.5-2.7 seems to be significantly harder than the predicted IC spectrum ($\alpha_\gamma \simeq 3$ at 30 TeV).

In Fig. 6.17 the spectrum of π^0-decay γ-rays is shown, calculated for a power-law differential spectrum of accelerated protons with $\alpha_p = 2.1$, exponential cutoff at $E = 10^{15}\,\text{eV}$ and significant flattening below $E \sim 1\,\text{TeV}$, as is expected from wind acceleration models. For $n_{\text{eff}} = 50\,\text{cm}^{-3}$ used in Fig. 6.17, the π^0-decay fluxes shown correspond to a total energy in accelerated protons of $W_p = 1.5 \times 10^{48}\,\text{erg}$, a quite acceptable amount from the point of view of the energy budget of the Crab. It is interesting to note that for the chosen parameters the superposition of 3 components of radiation – "IC+bremsstrahlung+π^0" – results in the power-law spectrum with $\alpha_\gamma \simeq 2.5$ over the entire energy range from 100 GeV to 100 TeV. This spectrum significantly differs from the pure IC spectrum, and provides a better fit to the reported data, the compilation of which is presented in Fig. 6.16 and 6.17 by the hatched zone. However, given the large uncertainties in the reported γ-ray fluxes, one cannot at present make a strong statement about the role of the bremsstrahlung and π^0 signatures in the Crab spectrum.

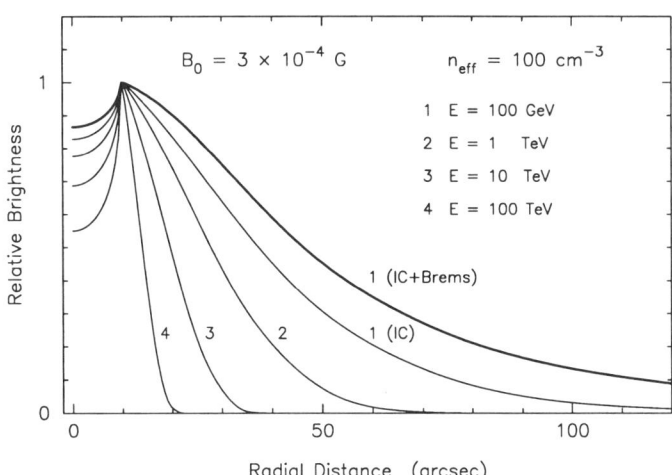

Fig. 6.18 Predicted brightness distributions of high energy γ-rays in the Crab Nebula at different energies. The heavy solid curve corresponds to the brightness distribution at $E = 100\,\text{GeV}$ expected in the case of enhanced bremsstrahlung contribution. All other curves are for the pure IC γ-rays. (From Atoyam and Aharonian, 1996).

6.3.1.4 *The objectives of future gamma ray studies*

The multiwavelength observations of the Crab Nebula have already provided important information about the nonthermal energy in the form of magnetic fields and relativistic electrons. However, many details remain still unresolved which need to be addressed by future observations. Importantly, the fluxes of the source exceed, by at least one order of magnitude, the sensitivities of the current or planned instruments at practically all frequencies of the observed spectrum. This ensures further significant progress in understanding the complex processes in the interaction of the relativistic pulsar wind with the nebula. Below we outline some issued to be addressed by new observations in different energy bands.

Probing electrons and B-fields in the synchrotron nebula. The most informative frequency band to probe the acceleration site(s) and the character of propagation of the ultrarelativistic electrons in the nebula is the X-ray domain. The recent studies of of the spatial and spectral structure of the X-ray emission of the Crab Nebula by Chandra (Weisskopf *et al.*, 2000) provide essential material for comprehensive modelling of synchrotron radiation in the inner part of the nebula. Since the cooling time of electrons responsible for the X-radiation, $t_{\rm sy} \approx 50(B/10^{-4}\ {\rm G})^{-3/2}(\epsilon/1\ {\rm keV})^{-1/2}$ yr, is very short compared to the age of the source, the total flux of synchrotron X-rays is a perfect *calorimetric* measure of the (quasi)-continuous acceleration rate of ultrarelativistic electrons, $\dot{W}_{\rm e} \approx L_{\rm x}$, and thus also the power of the kinetic energy dominated wind. At the same time, since the luminosity of synchrotron X-rays emitted in this regime is almost independent of the magnetic field, the X-ray measurements do not tell us much about the strength and spatial distribution of the magnetic field. Such information is contained in TeV γ-rays produced by the same electrons, because the fraction of energy of electrons released in IC γ-rays is determined by the ratio of energy density of magnetic field $(\propto B^2)$ to the energy density of target photon fields.

Although the limited angular resolution of γ-ray detectors does not allow, at least presently, mapping of the source on subarcmin scales, the measurements of integral fluxes of IC γ-rays at different energies, being coupled with synchrotron radiation in the relevant energy bands, could compensate for this disadvantage. Such an analysis is possible due to the relationship between the energies of synchrotron and IC photons produced by the same electrons. Thus, the spatial distribution of synchrotron radiation in different bands provides important (although not completely model-independent) information about the regions of production of γ-rays at different energies. The brightness distributions of γ-rays shown in Fig. 6.18 are calculated within the framework of the spherically symmetric MHD model of Kennel and Coroniti (1984). Although more realistic spatial distributions of the TeV electrons seen in Chandra images would modify these symmetric distributions, the results presented in Fig.6.18 quite correctly describe the average extensions of γ-ray production regions in different energy bands. For example, since the TeV γ-rays are produced by IC scattering of the electrons responsible for the observed keV X-rays, the estimate of the magnetic field based on keV/TeV data relates to the central $r \sim 0.5$ pc region of the Crab Nebula. A model-independent estimate of the magnetic field in the outer parts of the (optical) nebula can be provided by measurements of γ-ray fluxes at $E \sim 100$ GeV. Similarly, the fluxes of $E \geq 10$ TeV γ-rays, combined with hard X-ray data ($E \geq 100$ keV), could allow determination of the magnetic field in the vicinity of the wind shock front at $r \sim 0.1$ pc. Remarkably, as the target photon fields for IC γ-rays are well known, the accuracy of the estimate of the magnetic field could be better than 25%, provided that the γ-ray fluxes are measured with an accuracy of better than 50%.

While at energies between 1 and 10 TeV IC scattering dominates over all other possible radiation mechanisms, at energies below 1 TeV and above 10 TeV other processes connected with interaction of the relativistic particles with the nebular gas might contribute to the production of γ-rays as much as the IC does. Therefore determination of the magnetic field based on γ-ray fluxes in these energy regions requires separation of the IC contribution from the possible contamination due to γ-rays of other origin. The shape of the spectrum of IC γ-rays does not vary much with the basic parameters of the nebula. While at GeV energies the IC spectrum is very hard, with a power-law index of $\Gamma_\gamma \approx 1.5$, in the VHE region the spectrum gradually steepens from $\Gamma_\gamma \approx 2$ at $E \sim 100$ GeV to $\Gamma_\gamma \approx 2.5$ at $E \sim 1$ TeV, and $\Gamma_\gamma \approx 2.7$ at $E \sim 10$ TeV. Confirmation of this spectral shape is one of the important issues to be addressed by new observations, in particular by the stereoscopic systems of imaging atmospheric Cherenkov telescopes which provide good spectrometry with energy resolution ≤ 20 per cent and high localisation precision for point VHE sources with subarcmin accuracy. Although this still is not sufficient for adequate study of the angular structure of the VHE γ-ray production region in the Crab, in combination with expected large photon statistics it could provide an answer as to whether the observed γ-rays are produced in nebula, or should they be attributed to the pulsar or unshocked wind.

Interaction of accelerated particles with filaments. Although the uncertainties in the reported fluxes at energies 1-10 GeV do not allow us to draw definite conclusions about the conflict between the observations and the predicted IC γ-ray fluxes, the confirmation of high EGRET fluxes by future observations, e.g. by GLAST, would require more effective mechanism(s) responsible for γ-ray production in this energy region. The bremsstrahlung of radio electrons seems to be an intriguing possibility. In particular, it implies an effective confinement of the relativistic particles in dense filaments to provide sufficiently high effective gas density ("seen" by relativistic particles) $n_{\rm eff} \geq 50\,{\rm cm}^{-3}$. Another possible explanation of the EGRET excess flux is IC radiation of the unshocked pulsar wind, provided that the Lorentz factor of the wind is smaller by an order of magnitude than the "nominal" value, $\Gamma \sim 10^6$, required by the model of Kennel and Coroniti (1984).

The enhanced rate of interactions of relativistic particles with the gas due to their confinement in high density regions, opens the possibility of probing the content of the pulsar wind by searching for π^0 γ-rays in the spectrum of the Crab at very high energies. Indeed, for the effective gas density $n_{\rm eff} \geq 50\,{\rm cm}^{-3}$, the total energy of accelerated protons is estimated as $W_{\rm p} \leq 2 \times 10^{48}$ erg. This is quite a reasonable amount for the energy budget of the Crab, which however would be sufficient for significant modification (hardening) of the resulting "IC+π^0" spectrum. If future accurate spectrometric observations of the Crab show that the power-law γ-ray spectrum measured at TeV energies extends, without noticeable steepening, well beyond 10 TeV, it would be strong evidence of acceleration of a nucleonic component of cosmic rays in the Crab up to energies of 10^{15} eV, as well as a high concentration of relativistic particles in dense gas regions. Interestingly, since this will result in enhanced bremsstrahlung γ-rays as well, the total "IC+bremsstrahlung+π^0" radiation spectrum is expected to be almost a single power-law for over three decades in energy, from 100 GeV to 100 TeV.

Origin of hard (MeV) synchrotron radiation. The steepening in the synchrotron spectrum above 100 keV followed by hard MeV radiation ("MeV bump") can be naturally interpreted as the superposition of two different radiation components, both, most probably, of the synchrotron origin. If the first component is attributed to the diffuse emission of the synchrotron nebula, the second component could originate in one or a few compact structures, e.g. in knots or in the jet, where the strong magnetic field ($B \geq 10^{-3}$ G) could create favourable conditions for both effective acceleration of highest energy electrons and production of synchrotron radiation up to ~ 100 MeV. The limited angular resolution of hard X-ray/soft γ-ray detectors does not allow direct identification of these compact structures. On the other hand, since the cooling time of electrons producing MeV synchrotron radiation does not exceed the light crossing time of these compact structures, the detailed study of the spectrum and flux variability of radiation in this energy domain by INTEGRAL and GLAST missions, could compensate, to some extent, for the lack of spatial information.

Indirect information about the site(s) of the synchrotron "MeV bump" is also contained in low energy X-rays. Indeed, even in the case of an extremely hard injection spectrum (e.g. monoenergetic or Maxwellian type) of electrons responsible for the "MeV bump", the energy flux of this component at 1-10 keV range cannot be less than 10^{-10} erg/cm²s, so it perhaps could be resolved by Chandra even if this flux is due to the superposition of a large number of subarcsecond structures. Note, however, that the detection of these structures in X-rays will not be sufficient to identify them as the sites of "MeV bump". The important criteria for such identification would be the energy spectrum of this component which is expected to be essentially harder than the X-ray spectrum of the surrounding diffuse synchrotron nebula.

6.4 High Energy Gamma Rays from Other Plerions

The existence of the bright synchrotron nebula around the Crab pulsar very often is interpreted as a crucial condition for effective production of IC γ-rays. In fact, the *strong* magnetic field in the Crab Nebula produced by the *strong* wind only reduces the γ-ray production efficiency. Indeed, the energy density of the B-field exceeds by almost three orders of magnitude the density of the photon fields, thus only ~ 0.1 per cent of the energy of accelerated electrons is converted to the IC γ-rays, the rest being emitted away by synchrotron photons. The low efficiency of γ-ray production is compensated by the powerful injection of relativistic electrons, thus the Crab Nebula is a copious γ-ray emitter not because of *the existence of the surrounding nebula* but as a result of *very high injection rate of relativistic electrons* by the pulsar into the nebula, $\dot{L}_e \simeq 4 \times 10^{38}$ erg s^{-1}. In fact, the efficiency of IC γ-ray production would be expected significantly higher if the magnetic field were weaker. In other plerions with significantly lower pulsar spin-down luminosities, and consequently weaker pulsar winds, the magnetic fields are expected to be more than one order of magnitude smaller. This makes these objects more effective γ-ray emitters in which the radiative energy loss of electrons is shared between synchrotron and IC channels more evenly, $L_\gamma/L_X = w_{\rm MBR}/w_{\rm B} \simeq 1\,(B/3\mu{\rm G})^{-2}$.

Low ambient magnetic fields can be realized, in particular, if the energetic electrons are injected into the ISM by a pulsar which has already left its supernova remnant. Another interesting possibility for the effective formation of an intense IC γ-ray nebulae around pulsars can be attributed to old pulsars which continue to inject relativistic electrons at late stages as well, when the remnant has almost disappeared.

The energy of relativistic electrons, if injected by an isolated pulsar into the conventional regions of the ISM with a magnetic field $B_{\rm ISM} \sim 3-5\,\mu{\rm G}$, is released mainly in the form of extended synchrotron (optical to X-ray) and IC (γ-ray) nebulae. The X-ray luminosity of the nebula around an isolated pulsar is expected to be much less than the X-ray luminosity of the Crab Nebula due to both lower spin-down luminosities and low ambient magnetic fields in the vicinity of these pulsars. Nevertheless, modern X-ray detectors like Chandra an XMM-*Neuton* should be able to see faint X-ray nebulae around many pulsars.

The energy spectrum and the angular distribution of both the X-ray and γ-ray components depend essentially on the character of propagation of electrons. The parameter $L_0/4\pi d^2$, where L_0 is the pulsar spin down luminosity and d is the distance to the source, might be an indicative of the level of the overall (integrated over the entire extended emission region) X-ray and γ-ray fluxes, provided that the significant fraction of the pulsar's spin down luminosity (e.g. $\eta \geq 50\%$ like in the Crab pulsar) is transferred to relativistic electrons, and the age of the pulsar exceeds the cooling time of $\geq 10\,{\rm TeV}$ electrons. Then, fluxes of TeV γ-rays at the level of 10 mCrab can be predicted from the direction of pulsars having a parameter $S_0 = L_{37}/d_{\rm kpc}^2 \geq 10^{-3}$ (where $L_{37} \equiv L_0/10^{37}\,{\rm erg/s}$ and $d_{\rm kpc} \equiv d/1\,{\rm kpc}$), provided that approximately 1 per cent of the wind power ($\approx L_0$) goes into TeV γ-rays. This corresponds to the maximum allowed magnetic field strength $B \approx (8\pi \times 100 w_{2.7{\rm K}})^{1/2} \approx 30\,\mu{\rm G}$. Tens of pulsars with $S_0 = L_{37}/d_{\rm kpc}^2 \geq 10^{-3}$ are found in our Galaxy. This permits optimism that many IC γ-ray nebulae surrounding these pulsars could be detected by forthcoming IACT arrays with predicted flux sensitivities at TeV energies better than 10 mCrab.

There are only a few parameters which define the fluxes of the synchrotron and IC radiation observable over more than 20 decades of frequencies from 10^7 Hz to $\nu \geq 10^{27}$ Hz, the principal ones being the magnetic and ambient photon fields, as well as the "prompt" (current) spectrum of relativistic electrons. The latter in its turn depends on the electron injection spectrum and the history of evolution of the source.

6.4.1 Time-evolution of electrons

The spectrum $N \equiv N(E_e, t)$ of relativistic electrons at an instant t is defined by the equation (e.g. see Ginzburg and Syrovatskii, 1964)

$$\frac{\partial N}{\partial t} = \frac{\partial}{\partial E_e}[PN] - \frac{N}{\tau} + Q, \tag{6.4}$$

where $Q \equiv Q(E_e, t)$ is the electron injection rate, $P \equiv P(E_e)$ is the energy loss rate, and $\tau \equiv \tau(E_e)$ is the escape time of the electrons from the nebula. Assuming a δ-function injection at instant t_1, i.e. $Q(E_e, t) = Q_1(E_e)\delta(t-t_1)$, one finds the Green-function solution $G(E_e, t, t_1)$ for an arbitrary injection spectrum $Q_1(E_e) \equiv Q(E_e, t_1)$, if considering Eq.(6.4) at $t > t_1$ with initial condition $G(E_e, t = t_1, t_1) = Q_1(E_e)$. The complete Green-function solution in the case of energy-independent escape, $\tau(E_e) = const$, is given in Ginzburg and Syrovatskii (1964). For an arbitrary $\tau(E_e)$ and arbitrary $Q_1(E_e)$ the Green-function solution to Eq.(6.4) at $t \geq t_1$ is

$$G(E_e, t, t_1) = \frac{P(\zeta_t) Q_1(\zeta_t)}{P(E_e)} \exp\left(-\int_{t_1}^{t} \frac{dx}{\tau(\zeta_x)}\right), \tag{6.5}$$

where ζ_t depends on $t - t_1$ and E_e, and is found from the equation

$$t - t_1 = \int_{E_e}^{\zeta_t} \frac{dE}{P(E)}. \tag{6.6}$$

The variable ζ_t corresponds to the initial energy of an electron at instant t_1 which is cooled down to given energy E_e by the instant t. Note that $\zeta_{t_1} = E_e$, and $d\zeta_t/dt = P(\zeta_t)$. Integration of $G(E_e, t, t_1)$ over dt_1 gives the general solution to Eq.(6.4):

$$N(E_e, t) = \frac{1}{P(E_e)} \int_{-\infty}^{t} P(\zeta_t) Q(\zeta_t, t_1) \exp\left(-\int_{t_1}^{t} \frac{dx}{\tau(\zeta_x)}\right) dt_1. \tag{6.7}$$

In the numerical calculations below continuous injection of electrons with a "power-law with exponential cutoff" spectrum, $Q(E) = Q_0 E^{-\Gamma_e} \exp(-E/E_0)$, is assumed. An example of electron spectra formed in a nebula after 10^4 yr stationary injection of relativistic electrons by a pulsar into the surrounding medium is shown in Fig. 6.19.

6.4.2 Target photon fields

In order to demonstrate the impact of uncertainties of the background photon fields on the IC γ-ray fluxes, in Fig. 6.20 the contributions of 4 different target photon fields into the total IC flux are shown assuming that a pulsar of age t_0 powers, through termination of the relativistic wind, the surrounding medium with $B = 3 \times 10^{-5}$ G by injecting relativistic electrons with a constant rate $L_e = 10^{37}$ erg/s.

Formally, the target photon fields can be subdivided into 2 groups, namely, (i) the *external* galactic and extragalactic photon fields, and (ii) the *internal* radiation produced by the source itself. With the exception of the Crab Nebula, the internal radiation fields do not play significant role in production of IC γ-rays. The most important extragalactic field for production of IC γ-rays is the 2.7 K CMBR. The contribution of the galactic

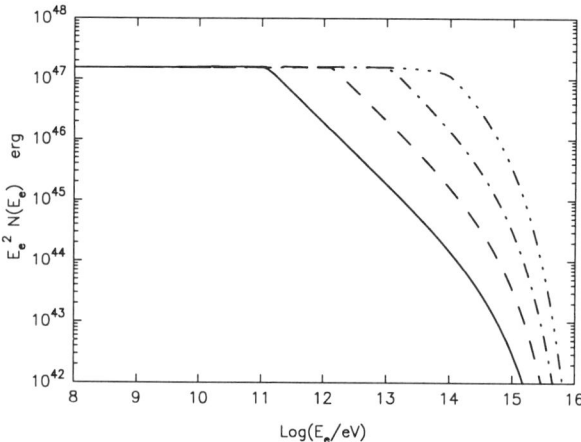

Fig. 6.19 The electron spectra in a nebula at the present epoch, assuming that a 10^4 yr old pulsar continuously injects relativistic electrons into the surrounding medium with a spectrum $Q(E) \propto E^{-2} \exp(-E/10^3 \text{ TeV})$ and power $L_e = 10^{37}$ erg/s. The curves shown are calculated for the following magnetic fields: $B = 10^{-4}$ G (solid line), $B = 3 \times 10^{-5}$ G (dashed line), $B = 10^{-5}$ G (dot-dashed line), and $B = 3 \times 10^{-6}$ G (fancy line).

background radiation is essentially due to the dust and starlight photons with peak intensities at $\lambda \sim 100$ μm and $\lambda \sim 1$ μm, respectively. While the density of the 2.7 CMBR is universal with $w_{\mathrm{MBR}} = 0.25\,\mathrm{eV/cm^3}$, the density of the both galactic background fields varies from site to site, with average values being $w_{\mathrm{dust}} \sim 0.05\,\mathrm{eV/cm^3}$ and $w_{\mathrm{sl}} \sim 0.5\,\mathrm{eV/cm^3}$ (Mathis et al., 1983). Since plerions are located near the galactic plane, this may lead, in principle, to uncertainties of the calculated IC γ-ray fluxes. Fortunately these uncertainties turn out not to be crucial. Indeed, as it follows from Fig. 6.20, even for densities of the dust FIR and starlight radiations $w_{\mathrm{FIR}} = 0.5\,\mathrm{eV/cm^{-3}}$ and $w_{\mathrm{sl}} = 1\,\mathrm{eV/cm^{-3}}$, which are respectively 10 and 2 times

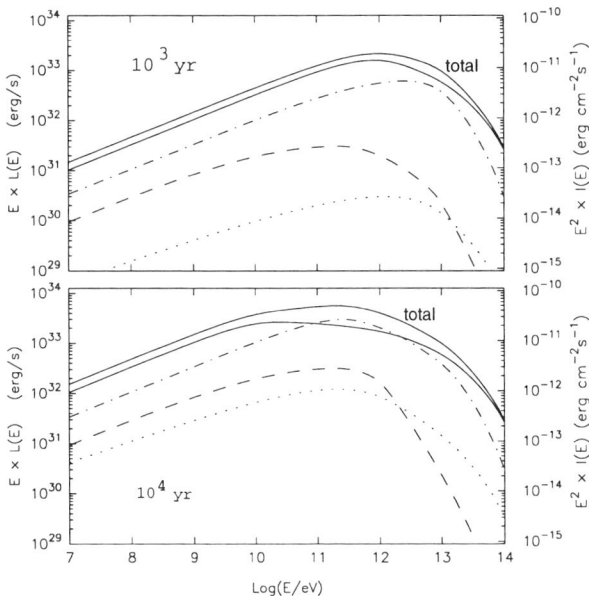

Fig. 6.20 The contributions of different target photon fields to the production of IC γ-rays : 2.7 K CMBR (solid line); FIR with $w_{\mathrm{FIR}} = 0.5\mathrm{eV/cm^3}$ and $T = 100$ K (dot-dashed line); NIR/optical with $w_{\mathrm{sl}} = 1\mathrm{eV/cm^3}$ and $T = 5000$ K (dashed line); synchrotron-self-Compton radiation produced by relativistic electrons for the source size $r_0 = 2\,\mathrm{pc}$ and magnetic field $B = 3 \times 10^{-5}$ G (dotted line). The overall spectrum of IC γ-rays is also shown. The injection spectrum of relativistic electrons with $\Gamma_e = 2$, $E_0 = 100\,\mathrm{TeV}$, and total injection rate $L_e = 10^{37}\,\mathrm{erg/s}$ are supposed. Ages for the pulsar of $t_0 = 10^3$ yr (upper panel) and $t_0 = 10^4$ yr (lower panel) are considered. The left-hand-side axis indicates the γ-ray flux luminosities, the right-hand side axis is scaled to the energy fluxes expected for a distance to the pulsar of $d = 1\,\mathrm{kpc}$. (From Aharonian et al., 1997b).

higher than the average galactic disk values, the IC scattering of electrons on the 2.7 K CMBR dominates at all γ-ray energies. Since it is rather difficult to speculate with such dramatic gradients of the diffuse galactic background even in very bright regions of the galactic disk, one may neglect with rather high confidence the contribution of these two target photon fields. In Fig. 6.20 the synchrotron-self-Compton component of radiation calculated for a source size 2 pc is also shown. Its contribution does not exceed a few per cent of the total γ-ray flux. Thus, in the production of IC γ-rays the 2.7 K CMBR dominates over all other seed photon fields. This fact significantly reduces the uncertainties of predicted γ-ray fluxes.

6.4.3 Effects of B-field, electron energy, and pulsar age

The fluxes and the spectral shape of both synchrotron and IC radiation at a given time t_0 depend on the prompt spectrum of electrons $N(E_e, t_0)$ given by Eq.(6.7).

For magnetic fields comparable with the strength of ISM magnetic field, $B_{\rm ISM} \sim (3-5) \times 10^{-6}$ G, the synchrotron and IC processes contribute equally to the total energy losses. Correspondingly, in this regime the energies released through the synchrotron and IC radiation channels are comparable. But when the magnetic field in the source noticeably exceeds $B_{\rm ISM}$, the energy dissipation of electrons is determined by synchrotron losses. This leads to a strong imbalance between the energies released in the synchrotron and IC channels. This is seen in Fig. 6.21a, where the calculated luminosities of a nebula with different magnetic fields are shown. We assume that a 10^4 yr old pulsar continuously injects into the nebula relativistic electrons with luminosity $L_e = 10^{37}$ erg/s. In order to show clearly the effect of steepening of the electron spectrum due to the radiative energy losses, in this figure a very large value of $E_0 = 10^3$ TeV is assumed. The relevant electron spectrum $N(E_e, t = 10^4 \, {\rm yr})$ is shown in Fig 6.19.

Because of cooling, the prompt spectrum of electrons above energies $E \geq E_* \simeq 12(B/10^{-5} \, {\rm G})^{-2}(t_0/10^4 \, {\rm yr})^{-1}$ TeV (defined from the condition $t_{\rm r} = t_0$) becomes steeper than the injection spectrum, $\Gamma_{\rm e} + 1$. An increase of the magnetic field leads to a decrease of the high energy part of the electron spectrum, and correspondingly to suppression of the IC γ-ray luminosity above $E \sim h\nu(E_*/m_e c^2)^2$ as $L_\gamma \propto B^{-2}$. At the same time, since the synchrotron emissivity is proportional to B^2, the suppression of the number of high energy electrons does not result in a reduction of the synchrotron luminosity.

Moreover, since at $B \geq 10^{-5}$ G the Compton losses can be neglected, the synchrotron luminosity saturates at its maximum value (a factor of 2 higher than at $B = 3 \times 10^{-6}$ G). A higher magnetic field has a strong impact on the low frequency spectrum of the synchrotron radiation. Indeed,

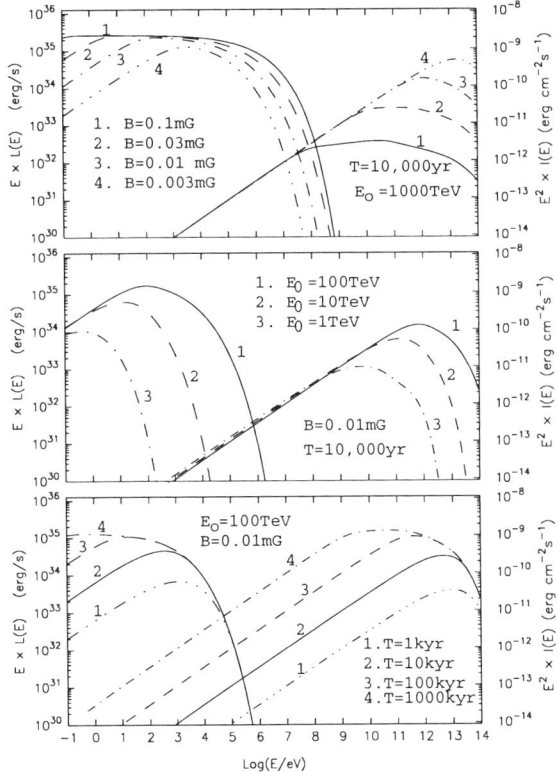

Fig. 6.21 Synchrotron and inverse Compton radiation components from a nebula surrounding a pulsar injecting electrons with a power-law spectrum, $\alpha_e = 2$, at a constant rate $L_e = 10^{37}$ erg/s. **a** (top panel): The age of the pulsar $T = 10^4$ yr and the cutoff energy of electrons $E_0 = 1000$ TeV are fixed. The curves correspond to $B = 10^{-4}$ G (solid lines), $B = 3 \times 10^{-5}$ G (dashed lines), $B = 10^{-5}$ G (dot-dashed lines), and $B = 3 \times 10^{-6}$ G (3dot-dashed lines). **b** (middle panel): $T = 10^4$ yr and $B = 10^{-5}$ G are fixed. The curves correspond to $E_0 = 100$ TeV (solid line), $E_0 = 10$ TeV (dashed line) $E_0 = 1$ TeV (dot-dashed line). **c** (bottom panel): $B = 10^{-5}$ G and $E_0 = 100$ TeV are fixed. The curves correspond to $T = 10^3$ yr (fancy line), $T = 10^4$ yr (solid line), $T = 10^5$ yr (dashed line), $T = 10^6$ yr (dot-dashed line). The left-hand-side axes indicate the γ-ray flux luminosities, the right-hand side axes are scaled to the fluxes expected for $d = 1$ kpc. (From Aharonian et al., 1997b)

with the increase of the magnetic field, the modified ($\propto E_e^{-\Gamma_e-1}$) part of the electron spectrum extends to lower energies, which eventually results in the transition of the radiation spectrum from $E^{-1.5}$ to E^{-2}.

Larger magnetic fields lead to a shift of the spectrum of the synchrotron radiation to higher energies ($\epsilon \propto BE_e^2$). However this cannot prevent the cutoff of radiation caused by the exponential cutoff in the injection spectrum of electrons. The corresponding cutoff is seen also in the spectrum of IC γ-rays. The impact of E_0, for the fixed value of the magnetic field $B = 10^{-5}$ G, is demonstrated in Fig. 6.21b. From this figure we may conclude that if the power-law spectrum of electrons continues beyond 10 TeV, the IC γ-ray luminosity peaks at TeV energies and synchrotron luminosity peaks at keV energies, with $L_\gamma/L_x = w_{\rm ph}/w_{\rm B} \simeq 0.1(B/10^{-5}\,{\rm G})^{-2}$. The X-ray fluxes would be dramatically suppressed if E_0 is less than 10 TeV. In particular, at $E_0 \leq 1$ TeV the X-rays disappear from the synchrotron spectrum completely. This leads also to strong suppression of TeV γ-ray fluxes. However, the γ-ray fluxes below 100 GeV still remain at a rather high level, $f(\geq 100\,{\rm GeV}) \sim 3 \times 10^{-12}\,{\rm erg\,cm^{-2}s^{-1}}$, detectable by future atmospheric Cherenkov telescopes with 100 GeV threshold energies. Thus, in such cases we should expect VHE γ-ray *nebulae surrounding pulsars without accompanying X-ray nebulae*.

Since in the case of continuous injection of electrons, the energy $E_* \propto (B^2 t_0)^{-1}$ is the only parameter which determines the spectral shape of the electrons, the spectrum of the IC radiation is an invariant to the product $B^2 t_0$. Obviously, the spectrum of the synchrotron radiation has a more complicated dependence on B and t_0. In particular, since the characteristic energy of synchrotron photons is proportional to the the BE_e^2, the transition region from $E^{-1.5}$ to E^{-2} occurs at photon energies defined by the combination $B^{-3}t^{-2}$.

The extended X-ray nebula around pulsars with $t_0 \geq 10^3$ yr can be seen at the level of $f_x \geq 10^{-12}\,{\rm erg/cm^2 s}$ if the above introduced "spin-down" energy flux $S_0 = L_{37}/d_{\rm kpc}^2 \geq 0.01$, provided that a significant part of the spin-down luminosity is transformed into the wind of relativistic electrons with spectra extending beyond 10 TeV. This conclusion is almost independent of the ambient magnetic field, therefore one may foresee the detection of many faint synchrotron nebulae surrounding pulsars by Chandra and XMM-*Neuton*. The crucial test for this hypothesis would be the detection of accompanying TeV IC γ-radiation from these objects.

6.4.4 Synchrotron and IC nebulae around PSR B1706-44

The PSR B1706-44 is one of the γ-ray pulsars detected by EGRET above 100 MeV (Thompson et al. 1999) and a possible TeV source (Kifune et al., 1995; Chadwick et al., 1998a). The possible association of the TeV emission with the radiation of the unshocked pulsar wind was proposed in Sec.6.2.4. Below a more conventional scenario for TeV γ-ray production is discussed assuming that the reported TeV emission is due to inverse Compton scattering of electrons which are (partly) responsible for the compact X-ray nebula around the pulsar.

The compact X-ray nebula surrounding PSR B1706-44 was discovered by ROSAT (Becker and Trümper, 1997) and confirmed recently by Chandra (Dodson and Golap, 2002). The X-ray morphology supports the association of PSR B1706-44 with the supernova remnant G343.1-2.3 (Dodson and Golap, 2002). The high resolution VLA images at GHz frequencies revealed also a compact (≈ 3 arcmin) spherical nebula of synchrotron radio emission (Giacani et al., 2001). The observations with the ATCA telescope show extended synchrotron radio emission on much larger scales (Dodson and Golap, 2002).

The X-ray nebula is very compact with a size of 5 arcsec, but since the source is faint one cannot exclude that non-negligible X-ray emission is produced also at larger distances from the pulsar. In accordance with the ROSAT data, the soft X-ray emission in the range 0.1-2.4 keV, corrected for the absorption, can be fitted by power-law spectrum with differential photon index $\alpha_x = 2.4 \pm 0.6$ (Becker et al., 1995) and luminosity $L_{0.1-2.4\text{keV}} \approx 1.2 \times 10^{33}$ erg/s. The ASCA results in the 2-10 keV band give a similar result, $\alpha_x = 2.3 \pm 0.3$ (Finley et al., 1998) and $L_{2-10\text{keV}} \approx 3 \times 10^{33}$ erg/s.

PSR B1706-44 has been reported to be a VHE γ-ray source (Kifune et al., 1995; Chadwick et al., 1998a). The inverse Compton origin of this radiation by relativistic electrons accelerated at the wind termination shock seems the most natural interpretation of the TeV radiation (Harding, 1996; Aharonian et al., 1997b; De Jager and Harding, 1998). However, a simple one-zone approach to the interpretation of X-ray and TeV γ-ray data faces a serious difficulty. The problem here is related to the unexpectedly small L_x/L_{TeV} ratio. The reported γ-ray flux (Kifune et al., 1995) $J(\geq 1\,\text{TeV}) \simeq 8 \times 10^{-12}\,\text{cm}^{-2}\text{s}^{-1}$ implies a VHE γ-ray luminosity exceeding 10^{34} erg/s, i.e. an order of magnitude larger than the X-ray luminosity. Since the luminosities of X- and γ-rays are proportional to the energy densities of the magnetic field and the 2.7 K CMBR, we immediately arrive at the conclusion that the magnetic field should be uncomfortably (for a pulsar driven nebula) small, $B \leq 3$ μG. Note that this is smaller than the field in conventional parts of the interstellar medium. Also, the estimate of the equipartition B-field in the radio nebula around PSR B1706-44 gives an order of magnitude larger value, $B \sim 20$ μG (Giacani et al., 2001). Since there is a little gap between the typical energies of electrons responsible for both X-rays and TeV γ-rays, we cannot exploit the dependence of the $L_\mathrm{x}/L_\mathrm{TeV}$ ratio on the shape of the electron spectrum. The assumption about higher seed photon fields around PSR B1706-44 does not help much either (Aharonian et al., 1997b). A more effective way to overcome this problem could be to assume that the bulk of electrons have streamed out to regions outside the wind nebula. This implies that the X-rays are mainly produced in the nebula where the magnetic field is high, while TeV γ-rays are produced everywhere (because of the universal distribution of the main seed photon field, 2.7 K CMBR). In case of fast escape of electrons, the bulk of IC γ-rays is formed outside the nebula. The fact that the X-ray luminosity of the nebula constitutes only a small fraction of the wind power ($\approx$ spin-down luminosity of the pulsar), $L_\mathrm{x}/L_0 \sim 10^{-3}$, is a good argument in favour of this assumption. Note that in the Crab Nebula, where the relativistic wind is likely to be confined by a slowly expanding shell, the spin-down power of the pulsar is dissipated in the nebula, and finally is radiated away in the form of synchrotron radiation with very high efficiency, $L_\mathrm{x} \sim L_0$.

In Fig. 6.22 broad band-spectra of PSR B1706-44 are shown, calculated for a magnetic field profile approximated as $B_\mathrm{x} = 2 \times 10^{-5}$ G inside and $B_\gamma = 3 \times 10^{-6}$ G outside the compact X-ray nebula. It is assumed that the escape of electrons from the compact X-ray nebula is described by the time $\tau(E_\mathrm{e}) = 10\,(E/20\,\mathrm{TeV})^{-\delta}$ yr. If $\delta \neq 0$, the electron spectrum in the X-ray nebula is modified due to the energy-dependent escape. This effect is important at low energies, below 1 TeV, where the radiative losses are negligible. For the energy-independent propagation of electrons ($\delta = 0$), the differential spectra of synchrotron radiation below 0.1 keV and IC γ-rays below 100 GeV have similar shapes with photon index $\alpha_\mathrm{r} = (\Gamma_\mathrm{e}+1)/2 = 1.5$ (solid lines). For demonstration of the effect of energy-dependent escape or steep injection spectrum of electrons, in Fig. 6.22 the fluxes calculated for $\delta = 0.5$ and $\Gamma_\mathrm{e} = 2.5$ are shown. The radiation fluxes calculated for single power-law injection spectra of electrons require an electron injection rate $L_\mathrm{e} \simeq 5 \times 10^{36}$ erg/s which slightly exceeds the spin-down luminosity of PSR B1706-44. The energy requirement could be significantly softened if one assumes low-energy cutoff in the electron injection spectrum, which would suppress the low-frequency synchrotron radiation to a non-observable level, without a noticeable impact on the synchrotron X-ray and TeV IC γ-ray fluxes (Fig. 6.22). Note, in this regard, that the low-energy cutoff at $E \sim 1$ TeV is not an unrealistic assumption. In fact, the hypothesis of the relativistic magnetised wind from a pulsar implies a monoenergetic flux of electrons. The redistribution of these electrons to a power-law spectrum could occur later on the wind termination shock, as in the Crab Nebula (Kennel and Coroniti, 1984). However, for less powerful pulsars the formation of the wind termination shock, may not be as effective as in the Crab. As a consequence, the spectrum of electrons injected into the nebula may not significantly broadened.

Fig. 6.22 demonstrates that the simplified two-zone model with high and low B-fields inside and outside of the compact X-ray nebula, gives quite

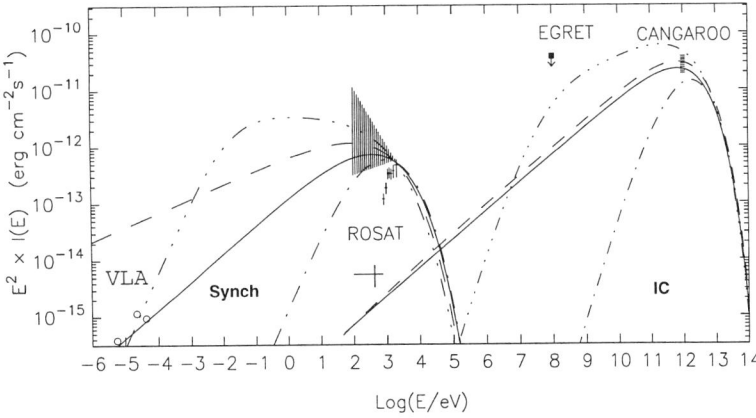

Fig. 6.22 The expected fluxes of the broad-band synchrotron radiation produced in the region of the compact X-ray nebula around PSR 1706-44 with angular size less than 1 arcmin and the IC gamma-radiation produced in more extended region, $\gg 1$ arcmin. The magnetic field profile is approximated by $B_x = 2 \times 10^{-5}$ G and $B_\gamma = 3 \times 10^{-6}$ G inside and outside of the compact X-ray nebula. The escape time of electrons is given by $\tau(E_e) = 10\,(E/20\,\mathrm{TeV})^{-\delta}$ yr. The injection spectrum of electrons is assumed to be a power-law with index $\Gamma_e = 2$ and exponential cutoff at $E_0 = 20$ TeV. The solid curves are for $\delta = 0$, and the dashed curves for $\delta = 0.5$. Two other curves are calculated for $\delta = 0.5$, assuming for the electrons injection spectrum $\Gamma_e = 2.5$ and a low energy cutoff at $E^* = 10$ TeV (dot-dashed) and 100 GeV (3dot-dashed). The fluxes at radio (VLA), X-ray (ROSAT) and TeV γ-rays (CANGAROO) are also shown. (From Aharonian et al., 1997b).

reasonable fits to the X-ray and γ-ray data, provided that the multi-TeV electrons escape the compact nebula on timescales ~ 10 yr. If the propagation of electrons proceeds in the diffusion regime, the diffusion coefficient should be close to the Bohm limit. It is clear that the two-zone scenario discussed above requires comprehensive tests based on detailed morphological and spectrometric studies at both X-ray and TeV energies. While the spectra of X-rays and γ-rays above 1 keV and 1 TeV weakly depend on details concerning the injection spectrum and the character of propagation of multi-TeV electrons, the fluxes of synchrotron radiation at radio, IR and optical frequencies and the fluxes of IC γ-rays below 100 GeV strongly depend on both the electron injection spectrum and the propagation effects.

Broad-band studies of different radiation components of the nebula surrounding PSR B1706-44 should provide important insight into the physics of relativistic electron-positron winds in pulsars. While Chandra and XMM-*Neuton* are well suited for studies of this source in the X-ray band, the observations at optical and IR frequencies are very difficult, given the large (~ 1 arcmin) size of the synchrotron nebula. In this regard, observations in the γ-ray domain seem to be more promising. The fluxes of IC γ-rays at GeV and especially at TeV energies are above the sensitivities of GLAST and IACT arrays. The expected angular extensions and fluxes of this and (hopefully, many) other similar objects match the performance of the stereoscopic Cherenkov telescope arrays like CANGAROO-III, H.E.S.S. and VERITAS to map the moderately extended ($\leq 0.3°$) γ-ray production regions on angular scales ~ 1 arcmin, and provide detailed spectral studies at energies between 100 GeV and 10 TeV.

Chapter 7

Gamma Rays Expected from Microquasars

7.1 Do We Expect Gamma Rays from X-Ray Binaries?

In one of the first attempts to classify the potential VHE γ-ray emitters, X-ray binaries were attributed to the category of serendipitous sources, i.e. objects which "...do not have any firm *a priori* basis for their selection as candidate VHE γ-ray sources" (Weekes, 1992). Nevertheless, this very class of objects played an important role in γ-ray astronomy, in particular initiating in the 1980s a renewed interest in ground based γ-ray observations. But ironically, the same sources ultimately raised questions about the credibility of the results of ground-based γ-ray astronomy from this period (see Chapter 2). Consequently, now most experts treat these data with a (healthy) scepticism, especially after the failure of new generation of ground-based instruments to confirm the early claims of detection of TeV signals from X-ray binaries and cataclysmic variables. As a result, since the early 1990s X-ray binaries have not been considered as primary targets for VHE γ-ray observations. This pessimistic view was largely supported also by the belief that X-ray binaries should be treated, first of all, as *thermal* sources effectively transforming the gravitational energy of the compact object (a neutron star or a black hole) into thermal X-ray emission radiated away by the hot accretion plasma.

However, since the discovery of galactic sources with relativistic jets, dubbed *Microquasars* (Mirabel and Rodriguez, 1994), the general view on the role of nonthermal processes in X-ray binaries has significantly changed. It is now established that non-thermal processes do play a non-negligible role in these accretion-driven objects. Approximately 20 per cent of the ~ 250 known X-ray binaries show synchrotron radio emission, and observations in recent years have revealed the presence of radio jets in several classes of X-ray binary sources (see e.g. Spencer, 1998; Fender, 2001). The high brightness temperature and the polarization observed in the radio emission from X-ray binaries are indicators of the synchrotron origin of radiation. It is now established that the non-thermal power of synchrotron jets (in the form of accelerated electrons and kinetic energy of the relativistic outflow) during strong radio flares could be comparable with, or even exceed, the thermal X-ray luminosity of the central compact object, most likely a stellar mass black hole (for a review see Mirabel and Rodriguez, 1999). If the acceleration of electrons proceeds at a very high rate, the spectrum of synchrotron radiation of the jet can extend to hard X-rays/soft γ-rays (Atoyan and Aharonian, 1999; Markoff et al., 2001). Moreover, the high density photon fields provided by the accretion disk around the compact object and by the companion star, as well as produced by the jet itself, create favourable conditions for effective production of inverse Compton γ-rays inside the jet (Levinson and Blandford, 1996; Atoyan and Aharonian, 1999; Georganopoulos et al., 2002). Generally, this radiation is expected to have an episodic character associated with strong radio flares in objects like GRS 1915+105. Besides, the electrons accelerated by a shock created by the jet propagating through the supersonic wind driven by the companion star (Atoyan et al., 2002) would result in a (quasi) stationary high energy IC γ-rays, the optical target photons being copiously supplied by the companion star (Paredes et al., 2000). The shocks generally should accelerate protons as well. But the γ-ray production through $p-p$ interactions may be effective only in high gas density regions. In particular, the old *"atmospheric target"* (bombardment of the normal star's atmosphere by a relativistic beam of particles accelerated at the compact object – Berezinsky, 1976; Eichler and Vestrand, 1984; Hillas, 1984b) or *"target crosses beam"* (interaction of a moving gas target with a beam of relativistic particles – Aharonian and Atoyan, 1991a; 1996b) scenarios can provide copious production of high energy γ-rays of hadronic origin[1]. And finally, besides the γ-rays produced in small-scale (sub-pc) jets of microquasars, one may expect persistent high energy γ-radiation from extended synchrotron lobes formed by electrons accelerated at the interface between the relativistic jet and the interstellar medium (Aharonian and Atoyan, 1998b), or by protons interacting with dense molecular clouds (Heinz and Sunyaev, 2002).

---

[1] Although these models had been inspired by reports of the detection of VHE signals from Cyg X-3, Her X-1 and some other X-ray binaries, which were later discredited, the models themselves still present a certain interest as viable scenarios of efficient γ-ray production.

7.2 Nonthermal Phenomena in Microquasars

The discovery of microquasars opened new possibilities for exploration of the phenomenon of relativistic jets common on a larger scale in AGN (see Chapter 9). Both type of objects are believed to be powered by accretion of matter by a collapsed object, by a stellar mass (typically 10 $M_\odot$) black hole in a microquasar and by a super-massive (up to 10^9 $M_\odot$) black hole in an AGN. Due to their proximity and many orders of magnitude smaller masses of the central black holes, the microquasars offer an opportunity for monitoring the jets on much shorter spatial and temporal scales than it is possible for AGN. Moreover, the "Eddington flux", $L_{\rm Edd}/d^2$ ($L_{\rm Edd} \simeq 1.3 \times 10^{38} (M/M_\odot)$ erg/s is the so-called Eddington luminosity – an indicator of the *potential* power of the source, and d is the distance to the source), is typically larger by 2 to 4 orders of magnitude for microquasars than for AGN. This enables detection of microquasar jets with Doppler factors even as small as $\delta \leq 1$, in contrast to the case of the *inner* jets in AGN, the detection of which requires strong Doppler boosting. It is important that for microquasar jets with aspect angles of $\sim 90°$, as in the superluminal jet sources GRS 1915+105 and GRO J1655-40, the Doppler factors of the approaching and receding components are comparable, therefore *both* are detectable, providing valuable information obtained by radio observations.

A large fraction of microquasars is associated with the so-called galactic black-hole candidates (BHC). The spectra of these highly variable objects extend to the domain of very hard X-rays. OSSE observations of a number of transient BHCs (see Grove *et al.*, 1998) indicate that there are at least two states which characterise the energy spectra of these objects - (i) the *hard/low* state has a typical power-law photon index ~ 1.5 with a cutoff around 100 keV; in some sources like Cyg X-1 a new component of radiation extending to MeV energies is possible, (ii) the *soft/high* state with a single, relatively steep power-law spectrum with photon index 2.5-3, but without indication of a spectral break up to 1 MeV or so. These two spectral states can be seen in Fig. 7.1 where the energy spectra of seven transient BHCs are shown.

Observations of the microquasars GRS 1915+105 and GRO J1655-40 and the prominent black-hole candidate Cyg X-1 (likely also a microquasar; see Stirling *et al.*, 2001) by the COMPTEL instrument on the Compton GRO show that the spectra of these sources may extend up to 10 MeV (Iyudin, 2000; McConnell *et al.*, 2002). For any reasonable temperature of the accretion plasma, models of thermal Comptonization cannot explain the MeV radiation, even when one invokes the so-called bulk-motion Comptonization (e.g. Laurent and Titarchuk, 1999). The MeV radiation requires a new, most likely nonthermal component of radiation. It is interesting to note that a significant excess above 300 keV in the spectrum of Cyg X-1 was found by HEAO-1 long ago (Nolan *et al.*, 1981). For explanation of this excess Aharonian and Vardanian (1985) have proposed a model which the X/γ spectrum of Cyg X-1 is formed as superposition of two components – (i) the thermal Comptonization component with a conventional temperature of

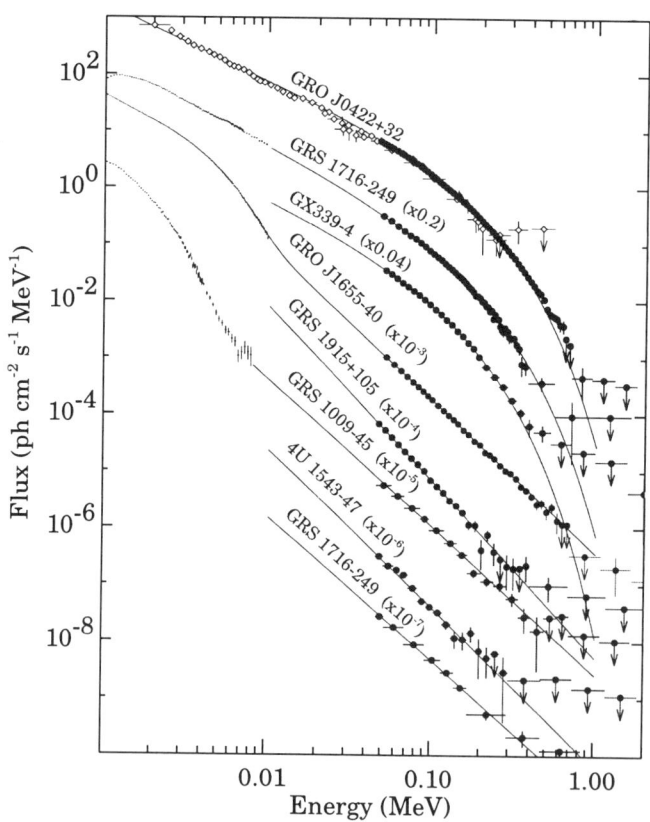

Fig. 7.1 Energy spectra of seven transient BHCs (from Grove *et al.*, 1998). For clarity of the figure, the fluxes are scaled by arbitrary factors as indicated at the curves. The transient GRS 1716-249 is shown twice to demonstrate the difference between the spectra detected at different states.

the accretion plasma $kT_e \sim 20-30$ keV (Sunyaev and Titarchuk, 1980) and (ii) a nonthermal high energy component produced during the development of a linear pair cascade initiated by relativistic particles in the accretion plasma surrounding the black hole. A modern version of that model, called the "hybrid thermal/non-thermal Comptonization" model (for a review see Coppi, 1998), provides reasonable spectral fits in the energy interval from 10 keV to 10 MeV for both the hard/low and the soft/high states observed by the Compton GRO detectors (McConnell et al., 2000). This type of model contains an *ad hoc* assumption about existence of a relativistic electron population in the accretion plasma, either as a result of direct electron acceleration or through pion-production processes in the two-temperature accretion disk with $T_i \sim 10^{12}$ K (e.g. Eardley et al., 1978, Eilek, 1980; Mahadevan et al., 1997). However, despite good agreement with the observed spectra, these models do not explain the tight temporal correlation observed between the low/hard X-ray state and the radio emission (Corbel et al., 2000).

An alternative site for production of γ-rays could be the synchrotron jets. In particular, it has been proposed that the synchrotron emission of microquasars might extend to X-ray energies, either in the extended jet structure (Atoyan and Aharonian, 1999) or close to the base of the jet (Markoff et al., 2001). The hard X-ray/low energy γ-ray emission of microquasars can be explained also in terms of Compton scattering of external photons (from the accretion disk and the companion star) by the relativistic electrons in the persistent jets of objects like Cyg X-1 and XTE J1118+480 (Georganopoulos et al., 2002). The interpretation of γ-rays in terms of nonthermal radiation either via 'synchrotron-self-Compton' or 'external Compton' scenarios is similar to the models proposed for TeV blazars (see Chapter 10).

The possible sites for production of different components of broad-band radiation in a microquasar are schematically shown in Fig. 7.2.

Generally, the microquasars provide more information, compared to the AGN jets, for quantitative study of nonthermal radiation processes in the jets. In particular, the two sided moving jets observed in galactic sources remove some ambiguities typical for extragalactic jets (e.g. the jet Doppler factors). The time scales of energy generation and dissipation processes in the accretion disks which launch (in one or another way) the jets, are proportional to the mass of the central black hole. Therefore, microquasars give us a unique chance to study many dynamical processes in relevant time-scales (from seconds to days), while the corresponding time scales of analogous processes in AGN exceed the human life time-scales (Sam et al., 1996). For example, the simultaneous observations of GRS 1915+105 led to the exciting discovery that the bulk of the X-ray emitting accretion plasma around the black hole disappears in less than several seconds of formation of the (nonthermal) radio jets (see Mirabel and Rodriguez (1999) and references therein). This is a clear indication of the close relation between the accretion and ejection phenomena. At the same time, many questions concerning the origin, content, energy budget and the structure of jets in microquasars remain open. Are they dominated by kinetic energy of particles or by Poynting flux? Is the kinetic power carried by e-p plasma or by electron-positron pairs? Moreover, we do not know whether the jets consists of discrete plasmons or whether they are sub-relativistic continuous flows with internal shocks. Although in the radio images of GRS 1915+105 one clearly sees discrete structures, it does not yet exclude models of quasi-

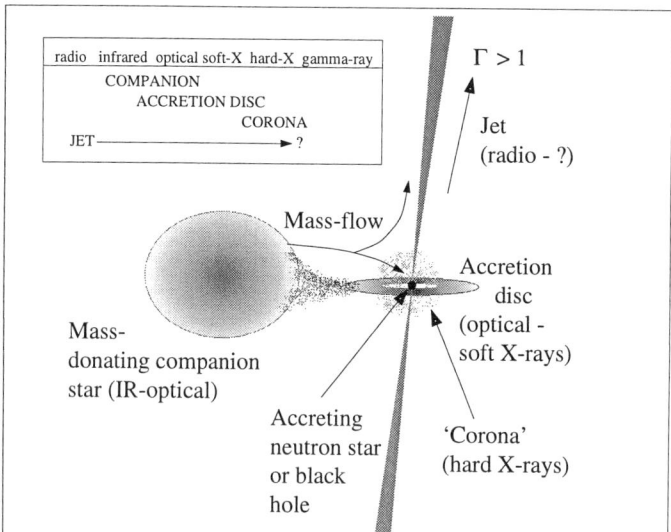

Fig. 7.2 The sites of formation of broad band radiation of a typical microquasar (from Fender, 2002). The normal companion star is a source of optical photons. The soft and hard X-rays are produced in the accretion disk and in the corona. All these radiation components have thermal origin. Low energy γ-rays can be produced due to inverse Compton scattering of a hypothetical nonthermal electron population in the corona. The synchrotron radiation from the jets is observed from radio to mm and infrared wavelengths. It is possible that the radiation of the jet extends to X-rays (due to the synchrotron and/or IC components), and even high energy γ-rays (through inverse Compton scattering).

continuous flow with internal shocks. It is quite possible, in fact, that we are dealing with more than one population of jets. For example, the existence of emission lines in the optical spectrum of the jet in SS 433 is an indicator of the nucleonic content of the outflow in this object. At the same time, we cannot exclude that in other sources the jets are dominated by electron-positron pairs. The jets can two- or one-sided, quasi-continuous or consisting of discrete plasmons, *etc.* The differences of jet structures in three famous galactic objects - GRS 1915+105, SS 433 and Cyg X-3 are demonstrated in Fig. 7.3.

For conclusive predictions about the γ-ray fluxes expected from microquasars, detailed modelling of the radiation processes in relativistic outflows

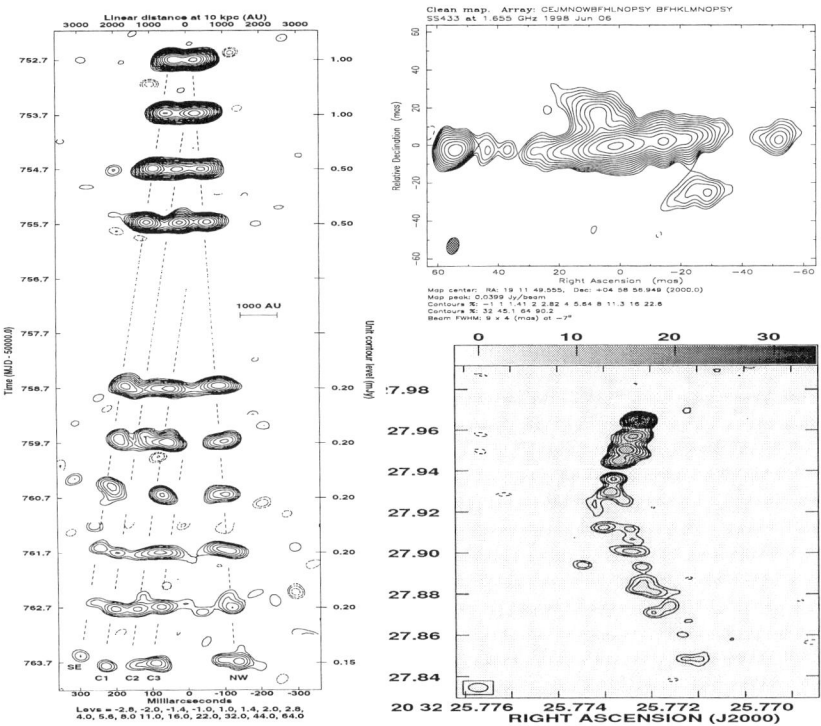

Fig. 7.3 Radio images of three famous galactic sources with relativistic jets (from Fender, 2002). Left: ejections from GRS 1915+105 observed with MERLIN (Fender *et al.*, 1999); Top right: a continuous two-sided image of the jet of SS 433 observed with the VLBA (Paragi *et al.*, 2001); Lower right: a VLBA image of a one-sided curved jet from Cyg X-3 (Mioduszewski *et al.*, 2001).

is obviously needed. The γ-ray fluxes strongly depend on several model parameters characterising physical conditions in the jet. While some of these parameters can be fixed quite confidently by modelling the radio flares of microquasars, for others, like the maximum energy of accelerated particles, we have to introduce certain *ad hoc* assumptions. It is remarkable that radio emission from X-ray binaries like Cyg X-1 and Cyg X-3 has been treated within the model of radiation from a conical jet long before the of discovery of microquasars (Hjellming and Johnston, 1988). Presently a large amount of observational data has been accumulated which allows comprehensive study of synchrotron radio flares of microquasars based on *quantitative* time-dependent treatment of injection, propagation and radiation of electrons in two conceptually different models. The first class of models assumes that the synchrotron radiation is produced by *discrete expanding plasmons* (clouds containing relativistic electrons and magnetic fields) in an optically thin regime. Different modifications of the basic approach of expanding clouds for interpretation of variable radio sources have been suggested in a number of earlier works (Kardashev, 1962; van der Laan, 1966; Band and Grindlay, 1986). A formalism for the quantitative study of evolution of the energy distribution of relativistic electrons in an expanding magnetised medium for an arbitrary time-dependent injection rate, taking into account both adiabatic and radiative energy losses, as well as losses caused by the escape of electrons in *energy-dependent* and *time-dependent* form, has been developed by Atoyan and Aharonian (1999). This formalism is described in Appendix B. It has been shown that the well studied radio flare of GRS 1915+105 during the March 1994 is in good agreement with the synchrotron emission of relativistically expanding clouds.

Kaiser *et al.* (2000) have developed an alternative model for radio flares of microquasars based on the assumption of quasi-continuous jet ejection during outbursts. The shock fronts travelling along the jet accelerate relativistic particles which emit the observed synchrotron radiation. The underlying physical processes of this model are similar to the internal shock model proposed for description of radiation features of gamma-ray bursts (Rees and Meszaros, 1994) as well as for interpretation of X- and γ-ray emission of inner jets in blazars (Ghisellini, 1999).

Despite certain conceptual differences between the "discrete plasmons" and "internal shocks" approaches, the differences in fact are not too large for calculations of the radio emission. The main difference between these approaches actually is that the "discrete plasmon" model does *not* specify the energy reservoir for accelerated electrons, and implies a quasi-isotropic electron distribution in the source rest frame (this corresponds to an assumption that the *bulk* speed of relativistic plasma in the source frame is small). Because in the rest frame of a strong relativistic shock the downstream speed of the plasma is sub-relativistic at best, the difference between these two approaches cannot be dramatic. In fact, this is confirmed by the modelling by Kaiser *et al.* (2000) of the evolution of radio fluxes of the 19 March 1994 outburst of GRS 1915+105 in terms of relativistic internal shocks propagating in the continuous jets, with the parameters for the radio source rather similar to those derived in the 'discrete plasmon' approach (Atoyan and Aharonian, 1999). Note, however, that the model of Kaiser *et al.* (2000) does not allow an energy-dependent escape of radio electrons from the jet, therefore in order to explain the spectral evolution (*steepening*) of the radio fluxes, they have to assume a sharp cutoff in the injection spectrum of electrons at low (GeV) energies. Obviously, in that case one could neither account for the synchrotron flares in the IR (except for very early stages, during the first ≤ 15 min, of the outbursts; see discussion in Kaiser *et al.*, 2000), nor expect high energy γ-ray fluxes.

7.3 Modelling of Radio Flares of GRS 1915+105

Radio flares in microquasars are associated with ejection of *pairs* of relativistic outflows which contain relativistic electrons and magnetic fields. Study of the 19 March 1994 outburst of GRS 1915+105, when such a pair of relativistic ($v \approx 0.92c$) radio jets were first discovered in a galactic source (Mirabel and Rodriguez, 1994), is still of special interest, because this particular flare remains one of the most powerful and long-lived events detected from this source, with rather detailed information available on the evolution of the spectra and fluxes of *both* ejecta resolved and monitored by VLA during > 40 days after the outburst.

An important and general feature of the evolution of radio flares of microquasars is the rapid (≤ 1 d) rise time of the fluxes $S_\nu \propto \nu^{-\alpha_\mathrm{r}}$, with a transition from an optically thick ($\alpha_\mathrm{r} \leq 0$) to optically thin ($\alpha_\mathrm{r} > 0$) synchrotron spectra, which is a characteristic signature of an expanding radio source. Fast expansions of synchrotron sources are observed also in smaller scales flares where during ≤ 1 hr the synchrotron spectrum gradually becomes transparent at progressively longer wavelengths from IR to radio domains (Mirabel *et al.*, 1998).

Another important feature of the evolution of radio flares is that the initially hard spectral index of $\alpha_r \sim 0.5$ around flare maximum later steepens, typically during less than a few days, to $\alpha_r \geq 1$. Because of rather short time scales involved, such fast spectral modifications at photon frequencies $\nu \leq 10\,\mathrm{GHz}$ cannot be attributed to the "synchrotron aging" of radio emitting electrons, which would be possible only on time scales $t_{\mathrm{syn}} \sim 150\,(\nu/10\,\mathrm{GHz})^{-1}(B/1\,\mathrm{G})^{-3/2}$ days. In fact, these fast modifications in the radio spectra can be explained as a result of an *energy-dependent* escape of radio electrons from the emission region.

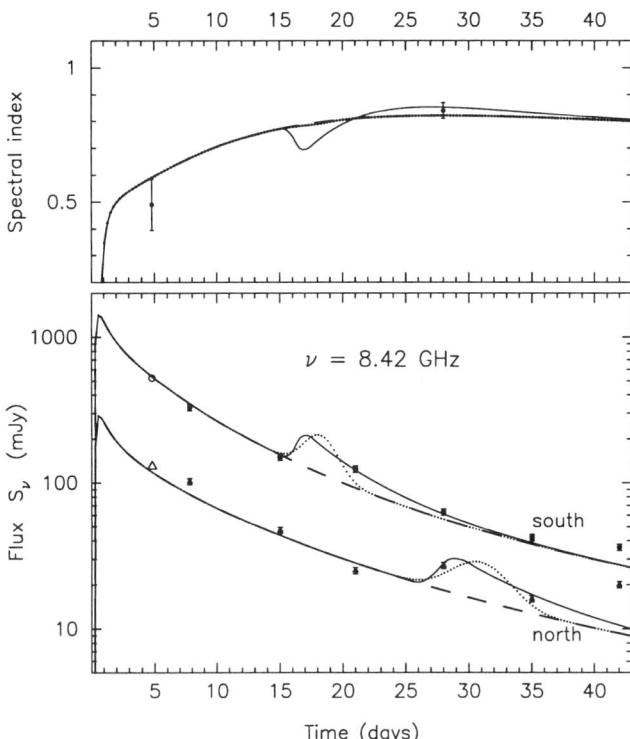

Fig. 7.4 Model fits to the time evolution of the radio fluxes (bottom panel) and spectral indices (top panel) at 8.42 GHz of the southern and northern jets of the 19 March 1994 outburst of GRS 1915+105. The fluxes expected in the case of a smooth time profile of an injection rate of the electrons into the ejecta are shown by dashed curves, and the solid curves show the fluxes in the case of a short-term ($\leq 1\,\mathrm{d}$) increase of the electron injection rate into both clouds at the same intrinsic times $t'_a = t'_r \simeq 9\,\mathrm{d}$. The dotted line corresponds to the case of a similar peculiarity in the behaviour of the magnetic field. For details see Atoyan and Aharonian (1999).

These are two principal features of the model in which the pair of jets of the 19 March flare are approximated as relativistically moving and spherically expanding plasmons. Although the model contains a number of free parameters, the large amount of the observational information available for this particular flare places rather tight limits on the possible range of variations of the principal parameters, such as the expansion speed, the evolution of the magnetic field, the power-law index of the injection spectrum of electrons, the energy dependence of the electron escape from the jet, the characteristic time profile of the *continuous* injection rate of electrons (e.g. through their shocked front surfaces) into the clouds, *etc.* (Atoyan and Aharonian, 1999).

The expansion of the cloud. In order to explain the clouds' opacity ($\tau_\nu \leq 1$) at $t_0 = 4.8$ days after the ejection, the size of the cloud should exceed $R_0 \sim (0.5-1) \times 10^{14}$ cm, (depending on the magnetic field B_0), and consequently an initial speed of expansion v_0 between $0.1\,c$ and $0.2\,c$ is needed. This agrees with the expansion speed deduced directly from observations. Also, to allow energy-dependent escape of electrons from the cloud (e.g. as $\tau_{\rm esc}(E) \propto E^{-1}$), which is assumed for explanation of the observed steepening of the radio spectrum at later stages of the flare, we should require a deceleration of the cloud's expansion, e.g. as $v_{\rm exp} = v_0(1 + t/t_*)^{-k}$ with $t_* \sim 1-3$ days (in the frame of the jet), and $k \sim 0.7-1$.

The relativistic electrons and magnetic fields. The power of injection/acceleration of electrons and the magnetic field in the cloud are strongly coupled. At $t_0 = 4.8$ days after formation of the jet (Mirabel and Rodriguez, 1994), the estimates of the equipartition magnetic field and the injection rate of relativistic electrons give $B_{\rm eq} \simeq 0.2$ G, and $L_e \sim 10^{38}$ erg/s, respectively (note, however, that in the expanding cloud the magnetic field could be lower than its equipartition value). To explain the observed rate of decline of the flare one has to assume: (i) reduction of the magnetic field with the radius of the cloud as $B \sim R^{-m}$ with $m \sim 1$, and (ii) decrease of the injection rate of the relativistic electrons at late stages, e.g. as $q_e(t) \sim q_0(1 + t/t^*)^{-p}$ with $p \sim 2k$. Under these assumptions, the shape of the injection spectrum of electrons is approximated as $Q_e(E,t) = q_e(t) E^{-2} \exp(-E/E_c)$. Since the synchrotron radio emission of the jet requires electrons with energy $\leq$ several GeV, the radio observations do not tell us much about the cutoff energy in the electron spectrum. Meanwhile, the possible synchrotron origin of the IR emission observed from GRS 1915+105, especially at late stages of evolution, when the magnetic field does not exceed 0.1 G, implies that electrons with energy up to at least several tens of GeV should be present in the jet.

The resolution of both approaching and receding jet components of the 19 March flare gives the flux ratio of the ejecta $S_a/S_r = 8 \pm 1$, whereas $S_a/S_r \simeq 12$ is expected for a pair of radio clouds moving in opposite directions with speeds $v_a = v_r = 0.92c$ (Mirabel and Rodriguez, 1994). This feature may be interpreted as an indication of shocks propagating with speeds $v_{sh} \simeq 0.92c$ in the continuous jets of relativistic plasma flowing with a bulk velocity $v_j \simeq 0.7c$ (Bodo and Ghisellini, 1995). However, the observed flux ratio can be readily explained also in terms of a pair of discrete plasmons, if one allows only a small difference (i.e. an asymmetry) between the intrinsic parameters of the counter ejecta, such as a few per cent difference in the speeds of their propagation v_a and v_r (Atoyan and Aharonian, 1997).

The ability of the "discrete plasmon" model to provide good fits to the time evolution of radio emission of both ejecta of March 19 1994 event is demonstrated in Fig. 7.4. It should be noted that it is possible to explain the observations with an accuracy comparable with the high ($\sim 5\,\%$) accuracy of the measurements reported by Mirabel and Rodriguez (1994), if one assumes a short-term impulsive increase ('after-impulse') of the injection rate of relativistic electrons into both clouds at the same intrinsic times $t'_a = t'_r \simeq 9\,\mathrm{day}$ after ejection. If true, such a synchronous 'after-impulses' would prove that the ejecta, even far from their ejection site, still continue to be fed with energy from the central engine through continuous flow of relativistic winds/jets, which remain otherwise invisible before colliding with radio ejecta and terminating in shocks.

7.4 Expected Gamma Ray Fluxes

The observations of synchrotron emission in the near IR band (e.g. Mirabel *et al.*, 1998), and especially the detection of an IR synchrotron jet in GRS 1915+105 a few weeks after ejection (Sams *et al.*, 1996; Eikenberry and Fazio, 1997), when the magnetic field in the expanding cloud hardly exceed 0.1 G, tell us that the spectrum of accelerated electrons in the jets should extend at least up to several tens of GeV. If the electrons in the jets are accelerated further beyond TeV energies, we could expect also X-rays and even MeV γ-rays of a synchrotron origin from microquasars.

The bulk of X-rays observed from GRS 1915+105, with the peak luminosity during strong flares exceeding 10^{39} erg/s, is believed to be produced in the thermal accretion plasma around the black hole. However, since the observed X-ray spectra are rather steep, with a typical power-law photon index $\alpha_x \sim 3$ in the region of tens of keV (see Fig. 7.1), the synchrotron component may become visible above 100 keV where the thermal emission is suppressed.

Direct evidence for acceleration of relativistic electrons beyond TeV energies in the jets of microquasars can be provided by a detection of IC γ-rays at energies $E_\gamma \geq 100\,\mathrm{GeV}$. The calculations in the framework of the synchrotron-self-Compton model show that during the strong flares one may indeed expect significant fluxes of VHE γ-rays. The detectability of these γ-rays essentially depends on physical conditions in the expanding ejecta, especially on the maximum energy of accelerated electrons and on the level of the magnetic field.

Fig.7.5 shows the fluxes of synchrotron and IC radiations of GRS 1915+105 that could be expected from a flare similar to the 19 March 1994 outburst at an early stage, $t = 2.4$ h after the ejection. The calculations are done for different combinations of the two principal (for production of high energy γ-radiation) parameters, the maximum electron energy, E_c, and the magnetic field, B_0, in the clouds at an instant $t_0 \simeq 4.8$ d when the clouds reach the size $R_0 \simeq 10^{15}$ cm. All other model parameters, including the power-law index $m = 1$ that describes the evolution of the magnetic field with time, are the same as the ones used in Fig. 7.4 for modelling of the 19 March 1994 radio flare of GRS 1915+105.

The calculations shown in Fig.7.5a correspond to the magnetic field $B_0 = 0.1$ G (at $t = 4.8$ d), and 3 different cutoff energies in the electron injection spectrum: $E_c = 50$ GeV, 1 TeV, and 20 TeV. It is seen that while the synchrotron spectrum up to IR frequencies (~ 1 eV) and IC spectrum up to 1 MeV are insensitive to E_c (provided that it exceeds 50 GeV), detectable synchrotron X-ray and very high energy IC γ-ray fluxes could be produced (e.g. at the level of 10^{-11} erg/cm^{2}s) only when $E_c \gg 1$ TeV.

The acceleration of electrons beyond 1 TeV is a necessary, but not sufficient, condition for visibility of the jet in VHE γ-rays. The second condition is related to the magnetic field which, as the calculations show, should not exceed its equipartition value, close to $B_0 \sim 0.2$ G, otherwise only a small part of the energy of electrons is released in γ-rays through the IC channel. Indeed, as is seen in Fig. 7.5b, for $B_0 = 0.2$ G the fluxes of both synchrotron X-rays and IC γ-rays are rather low, whereas the assumption of $B_0 \leq 0.1\,\mathrm{G}$ results in a significant increase of both of these fluxes. Such a profound variation of high energy radiation with magnetic field is a result of the strong dependence of the synchrotron radiation on the magnetic field, $S_\nu \propto B^{1+\alpha_\mathrm{r}}$. In order to provide the same flux at radio frequencies, a decrease of the magnetic field, say, by a factor of $a = 2$ should be compensated by an increase of the injection rate of electrons by a factor of $a^{1+\alpha_\mathrm{r}} \sim 3$.

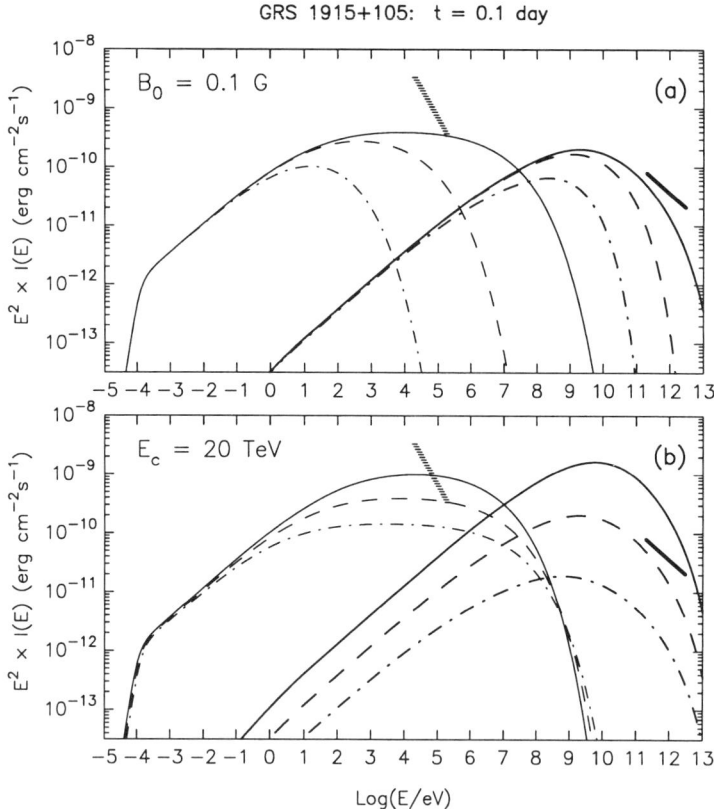

Fig. 7.5 The synchrotron (thin curves), and IC (heavy curves) energy fluxes which could be expected from a flare like the 19 March 1994 event of GRS 1915+105 at 0.1 day after the outburst (from Atoyan and Aharonian, 1999). Top panel: $B_0 = 0.1\,\mathrm{G}$ and $E_\mathrm{c} = 20$ TeV (solid), 1 TeV (dashed), and 50 GeV (dot-dashed). Bottom panel: $E_\mathrm{c} = 20$ TeV and $B_0 = 0.05$ G (solid), 0.1 G (dashed), and 0.2 G (dot-dashed). All other model parameters are the same as those used in Fig. 7.4 to explain the 19 March 1994 radio flare. The hatched region shows the level of hard X-ray flux typical for the strong flares, and the heavy solid line indicates the level of the VHE γ-ray flux detected from the Crab Nebula.

Different magnetic fields B_0 actually require different injection rates L_e for relativistic electrons. Thus, for calculations in Fig. 7.5b the electron injection rates are $L_e \simeq 4 \times 10^{38}$ erg/s, $1.2 \times \times 10^{39}$ erg/s and 3.6×10^{39} erg/s for the fields $B_0 = 0.2$ G, 0.1 G and 0.05 G, respectively. This difference in the injection rates has the strongest impact on the fluxes of synchrotron radiation at high energies which are produced, *unlike* the radio fluxes, in the so called saturation regime (when the electrons radiate away all their initial energy during the time period which is significantly shorter than the typical dynamical timescales characterising the source).

It is seen from Fig. 7.5b that during the first several hours after ejection, when the cloud is still opaque at the radio frequencies, the synchrotron flux at energies ≥ 100 keV may dominate over the extrapolation of the thermal component, and may result in a noticeable hardening of the overall spectrum even below 100 keV. The hard X-ray spectra of microquasars are usually explained by the Comptonization in the high temperature accretion plasma around the black hole. However in those cases when the hard power-law tails extend beyond several 100 keV, and in some sources even beyond 1 MeV, a simple thermal Comptonization approach seems insufficient, and synchrotron MeV γ-rays could then be a possible alternative.

The magnetic field strength in the jets has a stronger impact on the intensity of IC γ-rays than on the synchrotron flux (see Fig.7.5b). For a given field of soft photons, the increase of the magnetic field by factor of a results in the decrease of the ratio of the photon to the magnetic field energy densities, $w_{\rm rad}/w_{\rm B}$, by a factor of a^2. Since the IC and the synchrotron emissivities are proportional to $w_{\rm rad}$ and $w_{\rm B}$, respectively, with an increase of the magnetic field in the ejecta not only the electron injection power becomes smaller, but additionally the fraction of that power which is channelled (in the "saturated emission" regime) into the IC γ-rays is considerably reduced.

Fig.7.6 demonstrates the evolution of integral fluxes of synchrotron and IC γ-rays expected during a flare like the outburst of 19 March 1994, assuming a low B-field ($B_0 = 0.05$ G) and different exponential cutoff energies E_c in the spectra of accelerated electrons. It is seen that in the case of acceleration of electrons in the ejecta beyond 10 TeV the synchrotron radiation may dominate over the flux of IC γ-rays at least up to 10 MeV

The γ-ray fluxes above several 100 GeV during the first few hours of the flare could be as strong as the persistent flux of the Crab Nebula (see Fig. 7.6), provided that the electrons are accelerated to 20 TeV and beyond. If so, observations during the early stage of the flare (≤ 1 h) could result in detection of a statistically significant high energy γ-ray signal from the jet. During next 1-3 days the IC γ-ray fluxes drop significantly, but they remain still at the detectable '0.1 Crab' level. Afterwards the signal drops beyond the reach of contemporary detectors.

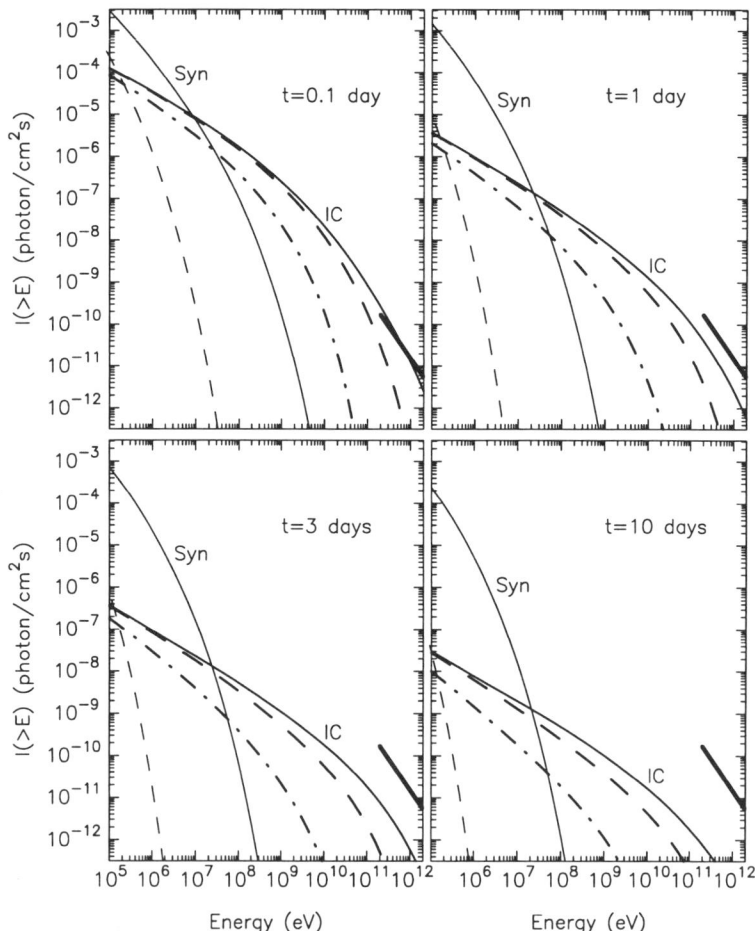

Fig. 7.6 Gamma-ray fluxes expected at different times t during a powerful flare like the 19 March 1994 outburst of GRS 1915+105, calculated for $B_0 = 0.05$ G assuming 3 different exponential cutoff energies: $E_c = 20$ TeV (solid), $E_c = 1$ TeV (dashed), and $E_c = 30$ GeV (dot-dashed). The heavy solid line corresponds to the level of the VHE γ-ray flux of the Crab Nebula. (From Aharonian and Atoyan, 1998b).

7.5 Searching for Gamma Ray Signals from Microquasars

The results of the previous section show that for flares as strong as the March 1994 outburst of GRS 1915+105, there are reasonably high expectations for detection of γ-ray signals from microquasars. The predicted fluxes are rapidly declining with the flare evolution, and depend strongly on the level of magnetic field in the ejecta as compared with the energy density of relativistic electrons w_e. Thus, significant IC γ-ray fluxes could be expected for a magnetic field in the ejecta evolving with power-law index $m = 1$ and $B_0 = 0.05$ G (when the cloud expands to a size $R_0 \simeq 10^{15}$ cm). Formally, a lower magnetic field would result in higher fluxes. However, since already for $B_0 \leq 0.03$ G the required electron injection power, $L_e \geq 10^{40}$ erg/s, is close to the bulk kinetic power of the jet of GRS 1915+105 (Gliozzi et al., 1999), we should not expect γ-ray fluxes significantly exceeding the ones shown in Fig. 7.6.

On the other hand, for a magnetic field close to the equipartition level, $B_0 \simeq 0.2$ G, the IC fluxes are reduced to the level which can only marginally be detected by current instruments. For the given radio intensity at GHz frequencies, the flux of synchrotron X-rays decreases as well. But since the X-rays are produced almost in the *saturation* (synchrotron-loss dominated) regime, the impact of the magnetic field on the flux of synchrotron X-rays is less dramatic than for IC γ-rays. Comparison of the synchrotron X- and IC γ-ray fluxes at different stages of evolution of the flare should allow decoupling of the magnetic field and the energy content in relativistic electrons, and thus would result in robust estimates of the magnitudes and evolution of these important parameters during expansion of the ejecta.

Observations of the synchrotron flares from microquasars in the IR region imply acceleration of electrons at least up to several tens of GeV. Thus we should expect production of at least high energy γ-rays $E_\gamma \geq 100$ MeV in the jets. In this energy domain, during first few hours of a powerful outburst the fluxes of IC γ-rays may exceed the 10^{-6} ph/cm^{2}s level. While such fluxes could be revealed by the EGRET in the best case only marginally, the future GLAST instrument, with a much larger detector area and better angular resolution, should be able to detect GeV γ-ray flares at the flux level 10^{-8} ph/cm^{2}s, i.e. even 1 day after the onset of the flare (see Fig. 7.6).

For VHE γ-rays, the narrow field of view of IACTs and the limited time available for observations (no more than several hours during moonless nights) require a rather thorough strategy for detection of flares with maximum duration $\leq$ few days. In order to be considered as a reasonable target for the ongoing observations, a particular microquasar not only should satisfy relevant zenith angle constraints, but should first of all be in an active state which may typically last from several weeks up to several months. In particular, for GRS 1915+105 the "active state" implies (see e.g. Mirabel and Rodriguez, 1999):

(i) high and strongly variable X-ray fluxes, typically exceeding ≥ 0.5 'Crab', and corresponding to the *high/soft* state;

(ii) increase of the radio fluxes to the level of $S_\nu \sim 50-100$ mJ, so called *plateau* state, on top of which episodes of *radio flares* with durations from several days to a few weeks are superimposed.

Formation of jets is still poorly understood and remains rather unpredictable, therefore a straightforward strategy could consist of daily monitoring of the source during its active state; An alternative, time-consuming approach could consist in observations of the source only after receiving a relevant signal about the ejection of a powerful jet. Since γ-ray fluxes are expected to be sufficiently high only during the first 1-3 days of the jet evolution, a timely trigger from observations of the flare at wavelengths sensitive to the early epochs of the jet evolution would be essential.

Because of the appreciable time (1-2 days in the case of powerful flares) needed for an expanding cloud to become optically thin with respect to synchrotron self absorption, the radio flares could appear too late to serve as a useful trigger for Cherenkov telescopes. In this regard, detections of powerful flares in the X-ray band may be very useful. However, not all of the observed X-ray flares, which are dominated by thermal emission of the accretion plasma, are accompanied by production of powerful non-thermal jets. A much more appropriate trigger, which may be sufficiently fast for early confirmation of a powerful non-thermal flare, can be provided by monitoring an active microquasar in the IR band. At last, with the advent of GLAST with its wide field of view detection ability, the most adequate trigger for IACT observations could be given by this instrument.

Note that the main factor which defines and *limits* the ability of different gamma-ray telescopes to detect the flares is connected with the γ-ray photon statistics that could be collected during a short time of the flare evolution. It is expected that the next generation of IACT arrays, with detection areas as large as 10^5 m^2 and energy thresholds as low as 50 GeV, should be able to conduct an effective search for γ-ray flares at the level of 0.1 Crab in several hours observations, provided, of course, that the electrons in the microquasars are accelerated beyond 1 TeV. In this context, because the acceleration of electrons in microquasar jets to several tens of

GeV is indicated by detections of synchrotron IR flares, detection of γ-ray flares from microquasars can be guaranteed for ground-based detectors with energy thresholds approaching 10 GeV. In this regard, the proposed 5@5 instrument (see Chapter 2) with its potential to detect a sub-10 GeV γ-ray signal at the energy flux level of $\leq 10^{-11}$ erg/cm^{2}s just for a few hour of exposure time, can serve as an ideal tool to study the nonthermal processes in microquasars.

7.6 The Case of Microblazars

A characteristic feature of the two superluminal microquasars, GRS 1915+105 and GRO J1655-40, from which both the approaching and the receding jet components have been detected, is that the jets propagate at a very large angle to the observer, $\theta \geq 70°$ (Mirabel and Rodriguez, 1999). As a result, given that the Lorentz factors Γ_j of both these components should be approximately the same, $\Gamma_a \sim \Gamma_r$, the relativistic Doppler factors $\delta_{r,a} = [\Gamma_j(1 \pm \beta_j \cos\theta)]^{-1}$ of the counter jets in those sources are not very much different, $\delta_a/\delta_r < 2$. The brightness of the receding jet component is then still sufficient for detection.

At the same time, for both GRS 1915+105 and GRO J1655-40, with $\Gamma_j \sim 2.5$ or perhaps even larger (Mirabel and Rodriguez, 1999), the large θ results in the Doppler factors $\delta_j < 1$ even for the approaching jets. Now, taking into account that the observed energy flux of the jet $F_{\rm obs} \propto L_{\rm int}\delta_j^4/d^2$, where $L_{\rm int}$ is the intrinsic luminosity of the source, it is clear that at *all* wavelengths the energy fluxes of these objects are Doppler-*suppressed* by more than one order of magnitude. This is just the opposite case to *blazars* – AGN with jets directed to us typically with $\delta_j \geq 10$, which, despite their large distances $d \geq 100$ Mpc (and therefore much smaller "Eddington flux" $L_{\rm Edd}/d^2$ than for microquasars), are prominent high energy γ-ray sources both at GeV and TeV energies. We see γ-rays from these objects essentially because of the Doppler boosting of nonthermal radiation produced in relativistic jets.

By analogy with AGN we may expect a sub-population of microquasars characterised by jets directed to us (or more specifically, with $\delta_a > 1$) - the *microblazars*. Although, there could be little doubt in existence of microblazars in principle, the probability of finding of a microblazar is proportional to $\sin\theta$, and therefore quite limited (Mirabel and Rodriguez, 1999). A characteristic feature of such objects should be a profound *one-sidedness* of radio jets if these jets would have relativistic speeds $\beta \geq 0.9$ typical for other galactic BH-candidate microquasars. In this regard, the recent report (Mioduszewski et al., 2001) of a one-sided highly relativistic jet from Cyg X-3 is of great interest. The ratio of the flux density in the approaching jet to that in the (undetected) receding jet exceeds 300. If this asymmetry is due to the Doppler boosting, then the implied jet speed is $\geq 0.8c$. The precessing jet model, together with the assumption of an intrinsically symmetric jet, constrains the jet inclination to the line of sight to $\leq 14°$ and the cone opening angle to $\leq 12°$ (Mioduszewski et al., 2001). The rapid and extreme radio variability, as well as the fact that Cyg X-3 is the most luminous X-ray binary at radio wavelengths, provide additional (indirect) support for the hypothesis of the microblazar origin of this source. If the electrons in this object are accelerated to very high energies, then the synchrotron radiation of this source could extend to X-rays or even γ-rays. In particular, in case of the electron acceleration at the maximum (theoretically possible) rate, the synchrotron peak is expected at $h\nu_c \simeq 150\delta_j$ MeV (see Chapter 3), i.e. it appears in the γ-ray domain. In addition, we may

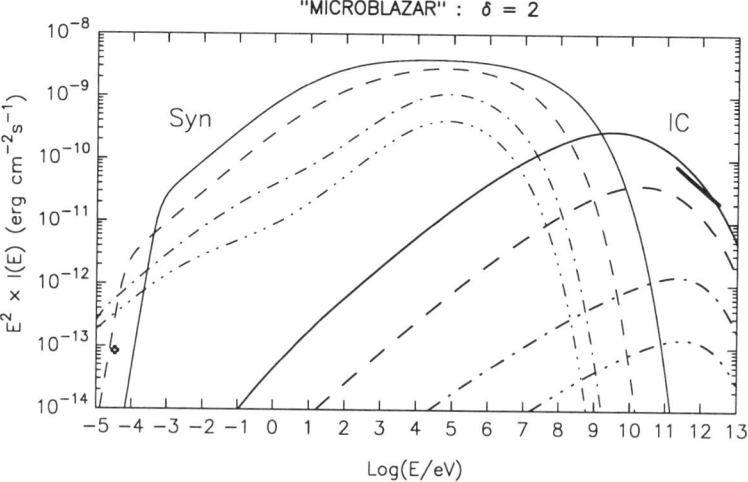

Fig. 7.7 The synchrotron (thin curves), and IC (heavy curves) energy fluxes expected from a hypothetical *microblazar* at the distance $d = 10$ kpc, with the jet position angle $\theta = 30°$, speed $v_0 = 0.9c$ and electron injection rate $L_0 = 10^{38}$ erg/s. The fluxes are shown at different times after the ejection: $t = 20$ min (solid), 0.1 day (dashed), 1 day (dot-dashed), and 3 days (three dot-dashed). The cross corresponds to the flux $S_\nu = 1$ Jy at $\nu = 8.5$ GHz, and the heavy solid line indicates the level of VHE γ-ray flux detected from the Crab Nebula.

expect IC γ-rays with a spectrum extending well beyond GeV energies. Interestingly, high energy, γ-rays have been indeed detected from the direction of Cyg X-3 by EGRET (Mori et al., 1997; Lamb and Macomb, 1997), although the association of these energetic photons with Cyg X-3 remains questionable.

While the detection of the receding jets of microblazars in radio observations is very difficult, the Doppler boosting of the approaching jets with $\delta_a > 1$ would make their detection in γ-rays possible even for objects with total nonthermal power much less than that of the strong flares of GRS 1915+105. Another important consequence of the jets in microblazars would be a significantly shorter timescale for the flare evolution, as well as a shift (by a factor δ_a) of the intrinsic energy spectrum of the jet towards higher energies.

Fig.7.7 shows the synchrotron and IC fluxes of a *hypothetical microblazar* at a distance $d = 10$ kpc with the jet position angle $\theta = 30°$ and propagation speed $\beta_j = 0.9\,c$ (i.e. for a relatively modest Doppler factor $\delta_j \simeq 2$). Calculations are done for the model parameters similar to the ones used for March 1994 flare of GRS 1915+105, except for the electron injection power which is taken to be a factor of 40 smaller, $L_0 = 10^{38}$ erg/s, and assuming for the magnetic field $B = 0.05$ G at the second day after the outburst. The fluxes are shown for 4 different stages (20 min, 0.1 day, 1 day, and 3 days) of evolution of the approaching cloud, which expands with a speed 0.15c. It is seen that during the first day the cloud becomes optically thin at 10 GHz, with the fluxes reaching the 10 Jy level, i.e. an order of magnitude larger than the maximum flux level observed from GRS 1915+105, but in agreement with the radio fluxes observed during the strong flares of Cyg X-3. Meanwhile, the fluxes of VHE γ-rays drop quickly, from 1 Crab to ≤ 0.1 Crab, during the first day of the flare.

7.7 Ultraluminous Sources as Microblazars?

X-ray studies based on the archival data obtained by the *Einstein, ROSAT* and *ASCA* satellites, as well as recent *Chandra* observations have revealed a new class of variable, and therefore compact, X-ray sources in nearby galaxies with luminosities in the range of $10^{39} - 10^{40}$ erg/s (e.g. Colbert and Mushotzky, 1999; Makishima et al., 2000) called *ultraluminous X-ray* (ULX) sources. Some of these objects show variability and spectral transitions which support the idea that they are powered by accretion.

However, given the range of observed luminosities, which is intermediate between the classical X-ray binaries ($10^{35} - 10^{48}$ erg/s) and AGN ($10^{42} - 10^{46}$ erg/s), this implies super-Eddington X-ray sources for conventional stellar-mass black holes with $M \sim 10\, M_\odot$, or *intermediate-mass* black holes with $M \sim 10^2 - 10^4 M_\odot$ (Colbert and Mushotzky, 1999). While the first option requires quite an unusual mode of accretion (e.g. Begelman, 2002), the second alternative simply postulates the existence of black holes in an unusual mass range. While both the *stellar-mass* and super-massive black holes have natural explanations (collapse of massive stars and primordial clouds, respectively), presently there is no viable idea for a process/scenario which could be responsible for the formation of intermediate mass black holes. This has led King *et al.* (2001) to the conclusion that the emission may, in fact, be beamed. If so, because the possibility of the detection of jets directed to the observer is quite small ($\propto \sin\theta$), this should be a rather rare class of objects in the near Universe. Roberts and Warwick (2000) have catalogued 28 sources with $L_x \geq 10^{39}$ erg/s in a sample of 83 galaxies, indicating that indeed only one in five of galaxies surveyed hosts one or more ULX. This view is supported also by the X-ray population synthesis study by Körding *et al.* (2002) who showed that the distributions of detected ULX are consistent with black-hole masses $M \leq 10\, M_\odot$ and bulk Lorentz factors for jets $\Gamma \sim 5$. And finally, the idea that ULXs are related to X-ray binaries is supported by optical observations suggesting that the ultraluminuous X-ray source NGC 502 X-1 has an O star companion (Roberts *et al.*, 2001), and by a possible X-ray periodicity observed in an ULX in the spiral galaxy IC 342 (Sugiho *et al.*, 2001).

A natural interpretation of the X-ray beaming seems to be production of X-rays in the jet. The Synchrotron-self-Compton (SSC) model of a microblazar which assumes acceleration of electrons to TeV energies in a jet oriented at a small angle to the observer (see Chapter 10), may readily provide the observed apparent X-ray luminosities of ULX. If the jet parameters do not deviate much from the ones used in Fig. 7.7, but assuming a larger Doppler factor, $\delta_{\rm jet} \geq 5$, one can achieve the $10^{39} - 10^{40}$ erg/s apparent luminosity of a *typical* ULX. Remarkably, within this model one should also expect high energy γ-rays from ULX, typically at the level of apparent luminosities $L_\gamma \sim (0.01-0.1)L_x \sim 10^{37} - 10^{39}$ erg/s. This implies γ-ray fluxes between 10^{-13} to 10^{-12} erg/cm^{2}s for a source at a distance of 1 Mpc. Thus, ULX sources located in galaxies within 1 Mpc, may present potentially interesting targets for observations with GLAST and the new generation of ground-based γ-ray instruments. The failure to detect such energetic γ-rays would imply either high magnetic fields in the jets which would suppress the Compton channel of radiation, or inability of the jets to accelerate electrons to TeV energies. While the flat X-ray spectra of ULX sources in low/hard state are consistent with the extension of electron acceleration to TeV energies, the steep spectra in high/soft state with photon index ≥ 2.5 favour an "early cutoff" in the electron spectrum below 100 GeV. This would cause the VHE γ-rays to vanish, although GeV γ-rays could be still present in the IC spectrum (see Fig. 7.5a).

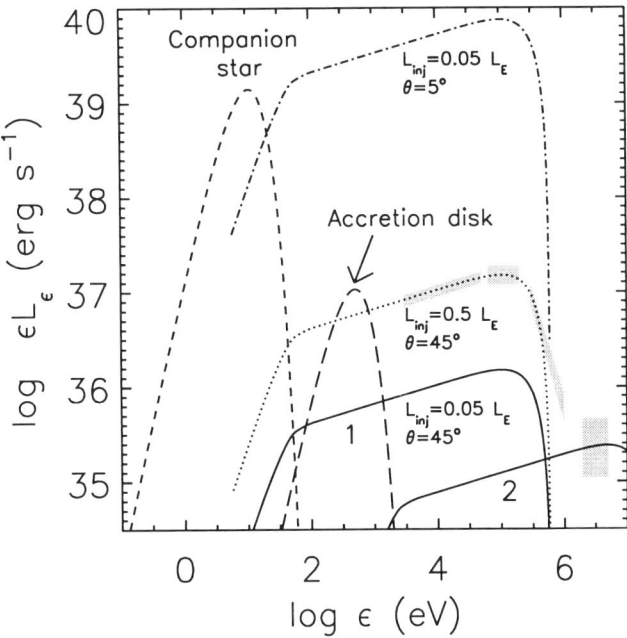

Fig. 7.8 X-ray spectra of a jet in a system similar to Cyg X-1. The IC emission due to the star (1) and the accretion disk (2) seed photons is plotted for an observing angle $\theta = 45°$ and $L_j = 0.05 L_{\rm Ed}$ (solid lines). The IC contribution due to the companion star is also plotted for $\theta = 45°$ and $L_j = 0.5 L_{\rm Ed}$ (dotted line), and $\theta = 5°$ and $L_j = 0.05 L_{\rm Ed}$ (dot-dashed line). The Lorentz factor of the flow is $\Gamma = 3$. The shaded area corresponds to observations (McConnel et al., 2000, Di Salvo et al., 2001). The short and long dashed lines represent the black body radiation components with characteristics similar to the companion star and accretion disk, respectively (From Georganopoulos, Aharonian and Kirk, 2002).

X-rays of ULXs can be produced also by low energy $E_e \sim 100$ MeV electrons through the inverse Compton channel (Georganopoulos et al.,

2002). Such electrons are present in (quasi) persistent radio jets of X-ray binaries. If the acceleration of electrons in the jet takes place within the binary system (i.e. at distances $l \leq R_s$, where R_s is the separation between the companion star and the black hole), and especially close to the accretion disk, the photons from the companion star and/or the accretion disk upscattered by $E_e \sim 100$ MeV electrons, would be Compton boosted to hard X-rays.

The prototype of this class could be Cyg X-1, but with a jet directed to the observer. Cyg X-1 has a super-giant companion star with luminosity $\simeq 10^{39}$ erg/s and mean photon energy ~ 3 eV. The X-ray spectrum of this source in the low/hard state has a photon index $\simeq 1.8$ and peak luminosity at 100 keV, $L_x \simeq 1.5 \times 10^{37}$ erg/s. The power spectrum of Cyg X-1 peaks at $f \sim 0.1$ s^{-1} (Sunyaev and Revnivtsev, 2000). Therefore the source size should not exceed $R \sim c/f \sim 3 \times 10^{11}$ cm. In Fig. 7.8 the X-ray fluxes of IC radiation of the jet in Cyg X-1 are plotted, calculated under different assumptions concerning the jet power in the form of electrons (and positrons). The bulk Lorenz factor of the jet is set to $\Gamma_j = 3$ and viewing angle $\theta = 45°$, in agreement with observations (Stirling et al., 2001). This corresponds to the jet Doppler factor $\delta_j = 1$. The solid curves are calculated for the electron acceleration power $L_e = 0.05 L_{Ed}$. The power-law index of accelerated electrons is assumed to be 1.6, and the maximum energy at 80 MeV. It is seen that the model X-ray flux is an order of magnitude lower than the observed flux (curve 1). Thus, if we want to attribute the observed X-ray luminosity solely to the IC component of radiation of the jet, then we should require 10 times higher jet power (dotted curve in Fig. 7.8). However, the strong reflection features in the low/hard state spectrum of Cyg X-1 suggest that the X-rays should be dominated by the disk component. This constraints the electron luminosity, $L_j \leq 0.5 L_{Ed}$. At the same time, for a plausible value $L_j \leq 0.05 L_{Ed}$, the MeV "excess" can be naturally explained by IC scattering of the accretion disk photons (curve 2 in Fig. 7.8).

Now assuming that a Cyg X-1 like source has a jet directed toward us, the luminosity of the IC radiation would be amplified due to the Doppler boosting by a factor of δ_j^5 for a continuous jet. Thus, for the relativistic jet parameters similar to Cyg X-1 ($\Gamma_j = 3$, $\delta_{max} \simeq 5.8$), the flux amplification factor could be as large as 7×10^3, thus such a source could reach an apparent X-ray luminosity $\sim 2 \times 10^{40}$ erg/s. This is illustrated in Fig. 7.8 for the jet with $\theta = 5°$ and $L_j = 0.05 L_{Ed}$.

If ULXs are microquasars similar to Cyg X-1, both their soft and hard states cannot be dominated by thermal (accretion plasma) component, since the accretion cannot produce, even at Eddington luminosity, more than 10^{39} erg/s for a $\sim 10 M_\odot$ black hole. This is in agreement with the proposed picture for the beamed hard state, where the thermal emission, including the reflection features are swamped by the jet component of emission. Importantly, the lack of observed reflection features in ULXs is confirmed not only for hard but also for the soft state (Makishima et al., 2000). This is an important fact, because the "microblazar" model for ULX sources requires existence of a relativistic jets both in high and low states of ULX sources. The jet parameters in these states should be different, reflecting the different state of the central engine producing the jet. In particular soft/hard spectra in high/low states can be interpreted in terms of the maximum energy of electrons. For typical energies of seed photons ($\sim 1 - 3$ eV) provided by the companion star, the spectrum of resulting X-rays is very flat if $E_{\rm max} \geq 50$ MeV. For the soft spectra in the high state, with photon indices 2.5-3, the maximum energy of electrons should not exceed $E_{\rm max} \sim 10$ MeV. This implies that, if in the low/hard state we may hope, at least in principle, to see also high energy γ-rays (provided that $E_{\rm max} \geq 100$ GeV), then in the high/soft state such a possibility is excluded.

7.8 Persistent Gamma Ray Emission from Extended Lobes

Besides the γ-rays produced in small-scale (sub-parsec) jets of microquasars within the synchrotron-self-Compton (SSC) scenario, one may expect high energy radiation from the extended synchrotron lobes produced by the relativistic electrons accelerated at the interface between the relativistic jets and the interstellar medium, similar to the scenario suggested for the extragalactic FRII radio sources. This could be the case for SS 433, a prominent galactic jet source associated with the supernova remnant W50. The large-scale jet of SS 433, which travels at a speed $0.26c$, extends to ≥ 50 pc, and terminates at $\simeq 1°$ from the central object.

Assuming that the relativistic jet termination shock accelerates electrons to extremely high energies, we may expect extended emission of synchrotron X-rays and IC γ-rays caused by VHE electrons interacting with the ambient interstellar magnetic field and with the 2.7 K CMBR, respectively. The possible synchrotron origin of the extended X-ray source coincident with the eastern radio "ear" $\sim 1°$ from SS 433 (Safi-Harb and Ogelman,

1997), implies the existence of relativistic electrons of TeV energies in the regions associated with the terminal shock of the eastern jet. The X-ray emission of the inner parts of the jets of SS 433 seems to be of nonthermal (synchrotron or inverse Compton origin) as well (Band and Grindlay, 1986; Yamauchi et al., 1994). However, while the inner X-ray lobes represent emission originating in the jet, the outer eastern X-ray lobe is attributed to the shocked interstellar medium (Safi-Harb and Ogelman, 1997). The very steep X-ray spectrum also indicates the synchrotron origin of this radiation.

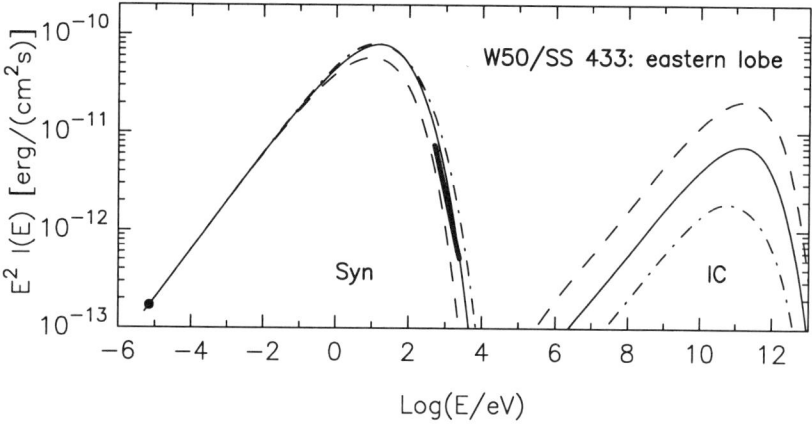

Fig. 7.9 Energy fluxes of the synchrotron and IC radiation expected from the eastern radio "ear" of W50/SS 433 for 3 different ambient magnetic fields: $5\,\mu$G (dashed), $10\,\mu$G (solid), $20\,\mu$G (dot-dashed). The full dot indicates the 10 Jy radio flux measured at $\nu = 1.7$ GHz, and the heavy line indicates the level of the observed X-ray fluxes. (From Aharonian and Atoyan, 1998b).

Fig. 7.9 shows the model synchrotron and IC fluxes assuming that the electrons, accelerated in the hot spot at the end of the jet, give rise, during their propagation in the interstellar medium, to an extended synchrotron and IC nebulae. The radio spectral index of ≈ 0.5, as observed from the eastern "ear" (Dowens et al., 1986), requires an electron acceleration spectrum $\propto E^{-2}$. Also, since the very steep spectrum of X-rays from approximately the same region, with photon index $3.7^{+2.3}_{-0.7}$ (Safi-Harb and Ogelman, 1997) cannot be explained by energy losses of electrons (at least in the framework of continuous electron acceleration), the nonthermal interpretation of this radiation implies that the synchrotron X-rays are produced by electrons from the "exponential tail" of the spectrum. On the other hand, since for any reasonable ambient interstellar magnetic field ($B \leq 20\mu\mathrm{G}$) the X-rays can be produced by electrons with energy exceeding several tens of TeV, we should assume a sharp, e.g. exponential cutoff, in the electron spectrum at energies below 10 TeV. Indeed, it is seen from Fig. 7.9 that the acceleration spectrum $E^{-2}\exp(-E/7\,\mathrm{TeV})$ of electrons continuously injected into the surrounding medium during the age of the source, about $T = 10^4$ yr, fits reasonably both the radio and X-ray fluxes, the assumed magnetic field being between $5\mu\mathrm{G}$ and $20\mu\mathrm{G}$. Since the calculated spectra are normalised to the radio flux emitted by the low energy electrons for which the energy losses are negligible, the acceleration rate strongly depends on the assumed magnetic field: $L_e = 2.8 \times 10^{38}$, 9.9×10^{37}, and 3.5×10^{37} erg/s for $B = 5$, 10, and $20\,\mu\mathrm{G}$, respectively. Note that even for the lowest possible magnetic field, the required L_e does not exceed 10% of the minimum kinetic power of the jet (e.g. Mirabel and Rodriguez, 1999).

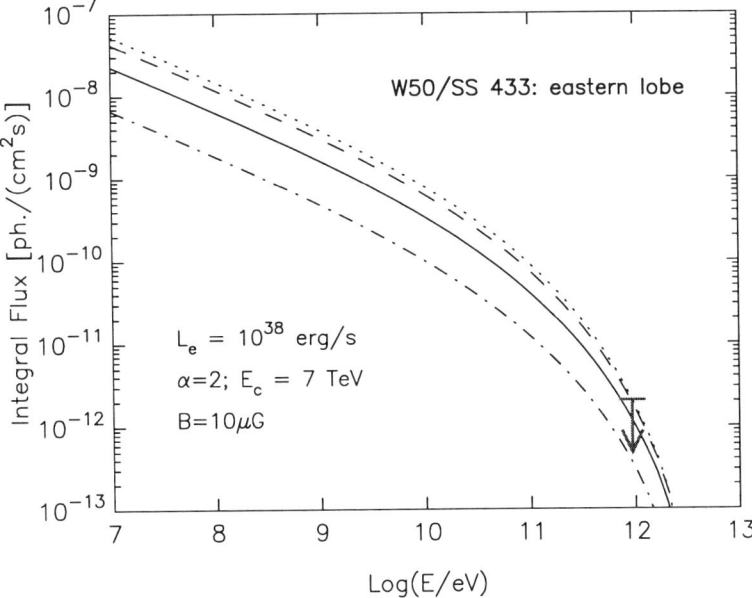

Fig. 7.10 The integral fluxes of γ-rays expected from the direction of the eastern radio "ear" of W50/SS 433 within different opening angles: $0.1°$ (dot-dashed), $0.25°$ (solid; the size of the "ear"), $0.5°$ (dashed), and $2°$ (dots). The upper limit on the TeV flux set by the HEGRA IACT system is also shown. (From Aharonian and Atoyan, 1998b).

The angular distribution of radiation depends on the combination of the age of the source t_0 and the propagation speed (or the diffusion coefficient D_{diff}, if the electron propagation proceeds in the diffusion regime) of electrons. Interestingly, for $T = 10^4$ yr, the angular radius of the eastern "ear" $\simeq 0.25°$ (Safi-Harb and Ogelman, 1997) requires $D_{\text{diff}} \sim 10^{28}$ cm^2/s, a rather typical value for propagation of cosmic rays in the galactic disk.

In Fig. 7.10 the integral fluxes of IC γ-rays are shown, calculated for $B = 10\,\mu\text{G}$ within different angles from the center of the eastern "ear". Note that while the predicted very hard flux of γ-rays above 100 MeV can be detected only by GLAST, the sensitivities of the current ground-based γ-ray instruments are already sufficient for searches for a IC γ-ray nebula surrounding the eastern "ear". Motivated by this prediction, the HEGRA collaboration used their system of Cherenkov telescopes to observe the SS 433/W50 system at very high energies. The analysis of various extended sources of location guided by the *ROSAT* X-ray images revealed no evidence for TeV emission above 1 TeV. The HEGRA flux upper limit (Rowell, 2002) is consistent with the flux predictions shown in In Fig. 7.10 . This implies a lower limit on the magnetic field within the region of the eastern "ear" of $B \geq 10\ \mu\text{G}$. Note that a larger magnetic field by a factor of two would reduce the γ-ray fluxes by a factor of four. With their predicted sensitivities between 100 GeV and 10 TeV, the next generation of ground based detectors should provide a deeper probe of the magnetic field and the energy budget of TeV electrons in the jet termination region. Finally, other microquasars with extended lobes, like the two persistent sources of relativistic outflow, 1E 1740.7-2942 and GRS 1758-58, and perhaps also GRS 1915+105, could be promising targets for observations with current and planned ground-based and satellite-borne γ-ray detectors.

Chapter 8

Large Scale Jets of Radio Galaxies and Quasars

In this and the following Chapters we discuss properties of high energy nonthermal radiation observed or expected from the largest structures in the Universe - the so-called large-scale AGN jets and clusters of galaxies. There is an apparent link between these two source populations. The clusters of galaxies contain many AGN with powerful jets which energise the intracluster medium through the termination shocks accompanied by particle acceleration and magnetic field amplification. Particle acceleration takes place also inside the jets. The most energetic particles eventually escape the jets, and make a non-negligible contribution, among other important channels like the SN explosions, galactic winds, accretion discs, *etc.*, to the energy budget of the host galaxy cluster. Large scale AGN jets and clusters of galaxies are believed to be potential sites of cosmic ray acceleration up to energies $\sim 10^{20}$ eV, therefore both source populations are of special interest for the cosmic ray community (see Fig. 1.6).

While the acceleration of the highest energy cosmic rays in AGN jets and clusters of galaxies is still a theoretical conviction, the presence of nonthermal particles of lower energies in these objects is an established fact. Moreover, it is likely that the formation and radiation of large-scale extragalactic jets is a phenomenon strongly dominated by nonthermal processes. Relativistic electrons with Lorentz factors $\gamma_e \sim 10^3 - 10^5$ in these objects are "seen" through their synchrotron radio emission. The possible synchrotron origin of nonthermal X-rays observed from both source populations would imply that the energy spectra of electrons extend well beyond 1 TeV. It is difficult to imagine a scenario when the acceleration of electrons is not accompanied by acceleration of protons and nuclei. Note that the proton-to-electron ratio in galactic cosmic rays is close to 100. For typical conditions characterising the energy losses and propagation of cosmic rays in the interstellar medium, the large observed p/e ratio reflects the *acceleration rates* of relatively low energy (less than 1 TeV) protons and electrons in our Galaxy. Obviously this ratio cannot be mechanically applied to AGN jets and clusters of galaxies where the sources and acceleration mechanisms could be quite different. On the other hand, the p/e ratio in these objects is expected to be very large even for similar proton and electron acceleration rates. Indeed, the ratio of the *current* densities of protons and electrons is related to the ratio of corresponding acceleration rates as $w_p/w_e = (\dot{w}_p \cdot t_{cool}^{(p)})/(\dot{w}_e \cdot t_{cool}^{(e)})$, where $t_{cool}^{(p)}$ and $t_{cool}^{(e)}$ are the characteristic energy loss-times of protons and electrons. In clusters of galaxies both the inelastic interaction time and the escape time of protons exceed, at least at energies $E \leq 10^3$ TeV, the cluster age, the latter being comparable with the the Hubble time $1/H_0 \simeq 1.5 \times 10^{10}$ yr. Therefore, clusters of galaxies serve as unique "cosmological stockpiles" of hadronic cosmic rays. On the other hand, because of inevitable interactions with the 2.7 K CMBR, the lifetime of radio electrons with Lorentz factors $\geq 10^4$ does not exceed 2×10^8 yr. Thus, assuming that particle acceleration has started at the early stages of cluster formation, and continues until now with a more or less constant and similar rates for electrons and protons, $\dot{w}_p/\dot{w}_e \sim 1$, the p/e ratio would be larger than $t_H/t_{cool}^{(e)} \sim 100$ around several GeV, and increases linearly at higher energies. If the nonthermal activity at the first stages of cluster formation was higher, we may expect even larger proton-to-electron ratio. While protons from these remote epochs survive without significant losses, we can see radio electrons produced only during the last 10^8 yr. A similar conclusion is true also for the kpc-scale structures in AGN jets. While the characteristic lifetime of electrons responsible for the optical and X-ray synchrotron emission cannot significantly exceed 100 yr, the minimum proton escape time from these structures (assuming that the protons propagate rectilinearly) exceeds several thousand years. Thus, the search for signatures of the hadronic component in the observed nonthermal radiation, and the estimates of the accumulated total proton energy is of great interest, and not only for studies of specific particle acceleration mechanisms. The detection and identification of nonthermal radiation associated with the hadronic component of cosmic rays has more general implications concerning the role of cosmic rays in the energy and pressure balance in these powerful objects.

8.1 Synchrotron and IC Models of Large Scale AGN Jets

Chandra observations have revealed various bright X-ray features like knots and hot spots in large-scale extragalactic jets (for a review see e.g. Harris, 2001). The undisputed synchrotron origin of the radio and optical radiation of the resolved jet structures in several powerful AGN and radiogalaxies implies the presence of relativistic electrons with a typical Lorentz factor $\gamma_e \simeq 10^3 \, B_{mG}^{-1/2} \nu_{GHz}^{1/2}$, where $B_{mG} = B/10^{-3}$ G is the magnetic field in units of milli-Gauss and $\nu_{GHz} = \nu/1$ GHz is the frequency of synchrotron radiation in units of GHz. The inverse Compton scattering of the same electrons leads to the second component of nonthermal radiation. The 2.7 K CMBR provides a universal photon target field for inverse Compton emission. The electrons boost the energy of the seed photons to $E \simeq 4kT\gamma_e^2$; this gives a simple relation between the energies of inverse Compton and synchrotron photons produced by the same electron: $E \simeq 1 \, \nu_{GHz} B_{mG}^{-1}$ keV. Thus, for the characteristic magnetic field, which in the knots and hot spots is believed to be between 0.1 mG and 1 mG, the inverse Compton radiation of radioelectrons appears in the Chandra energy domain. Yet, the energy density of 2.7 K CMBR is too low to provide significant X-radiation. Indeed, the inverse Compton luminosity constitutes only a small part of the emission of radioelectrons, $L_X \approx 10^{-5} B_{mG}^{-2} L_R$, unless we assume that the radio emitting regions are relativistically moving condensations with a Lorentz factor of bulk motion $\Gamma \geq 10$, thus the energy density of 2.7 K CMBR "seen" by electrons in the jet's frame is enhanced by a factor of Γ^2 (Tavecchio et al., 2000; Celotti et al., 2001). This hypothesis automatically implies that the jet features are moving towards the observer, otherwise we would face unacceptably large requirements for the energy budget of the central source.

The emissivity of inverse Compton radiation could be significantly enhanced also due to additional target photon populations, in particular due to the jet's own synchrotron radiation. The density of this radiation in a knot at a distance d from the observer is estimated as $w_r = (d/R)^2 c^{-1} f_{R-O} \simeq 0.9 \, f_{-12} \theta^{-2}$ eV/cm^3, where $f_{-12} = f_{R-O}/10^{-12}$ erg/cm^{2}s is the observed radio-to-optical flux normalised to 10^{-12} erg/cm^{2}s, R is the source radius and $\theta = R/d$ is the source angular size in arc-seconds. For example, the fluxes of 1 to 1000 GHz radio synchrotron emission of two distinct hot spots ("A" and "D") of the radiogalaxy Cygnus A, both of $\sim 0.''5$ angular size, are very high, $f \sim 10^{-11}$ erg/cm^{2}s (Wilson et al., 2000), thus the density of synchrotron radiation exceeds the density of 2.7 K CMBR by two orders of magnitude. Therefore, the X-ray emission of these hot spots can be readily explained by the synchrotron-self-Compton (SSC) model, assuming $B \sim 0.2$ mG, and taking into account that the energy flux at radio frequencies exceeds the X-ray flux by 1.5 orders of magnitude. This is not the case, however, for the majority of jet features resolved by Chandra for which we observe, in fact, just the opposite picture. The X-ray fluxes, for example, from the jets in Pictor A (Wilson et al., 2001), 3C 120 (Harris et al., 1999), 3C 273 (Röser et al., 2000; Sambruna et al., 2001; Marshall et al., 2001), and PKS 0637-752 (Schwartz et al., 2000; Chartas et al., 2000) are at least a factor of 10 larger than the radio and optical fluxes.

For X-ray emission of many resolved extragalactic knots and hot spots the electron synchrotron radiation is considered 'the process of choice' (Harris, 2001). Synchrotron radiation is indeed an extremely effective mechanism converting the kinetic energy of relativistic electrons into X-rays with an almost 100 per cent efficiency, provided, of course, that the electrons are accelerated to multi-TeV energies. The single power-law spectrum smoothly connecting the radio, optical and X-ray fluxes with spectral index $\alpha \simeq 0.76 \pm 0.02$ observed from the knot A1 in the jet of 3C 273 (Röser et al., 2000; Marshall et al., 2001) formally could be used as an additional argument for the synchrotron origin of X-rays. However, this often used argument for a single power-law distribution of electrons is a quite naive simplification. Detailed calculations show (see e.g. Brunetti 2002a) that the electron spectra established in jets have quite complex forms, and only in narrow energy intervals they can be approximated by power-law functions. Also, the observed spectral steepening (or cutoff) at optical frequencies (but with yet hard X-ray spectra) in the so-called "problem sources" (Harris, 2001) like the knot D/H3 in the same source, 3C 273 (Röser et al., 2000, Marshall et al., 2001), as well as the knots in 3C 120 (Harris et al., 1999), PKS 0637-752 (Schwartz et al., 2000) and in the hot spot of Pictor A (Wilson et al., 2000), tell us that we deal with a rather complex picture with multiple electron populations or a single source population with a complex spectrum. An interesting feature in the electron spectrum may appear due to the Klein-Nishina effect. Indeed, if the energy of accelerated electrons extends to ≥ 100 TeV, and the energy losses of electrons are strongly dominated by IC scattering, the Klein-Nishina effect would result in spectral hardening of the electron spectrum at highest energies. Consequently, this would lead to the hardening of the spectrum of synchrotron X-rays (Dermer and Atoyan 2002). The effect becomes significant when

$\Gamma_{\rm jet} \gg 10(B/30~\mu{\rm G})(1+z)^{-2}$. Therefore, a bulk motion Lorentz factor of $\Gamma_{\rm jet} \geq 10$ is required to provide an adequate energy density of 2.7 K CMBR in the frame of the jet ($\propto \Gamma_{\rm jet}^2$) and/or significantly lower magnetic fields compared to the typical values assumed in standard models, especially for sources with $z \ll 1$. The injection of extremely high energy electrons with $E \geq 100$ TeV is another key assumption of this model.

Generally, the high radiation efficiency is treated as a crucial component of any successful model/mechanism of luminous nonthermal sources. However, for the large-scale extragalactic jets this highly desired feature ironically leads to certain difficulties for the synchrotron X-ray emission of electrons. The process, in fact, seems to be "over-efficient". This, at first glance paradoxical, statement has a simple explanation. The cooling time of electrons responsible for the radiation of an X-ray photon of energy E, $t_{\rm synch} \simeq 1.5\, B_{\rm mG}^{-3/2}(E/1~{\rm keV})^{-1/2}$ yr is small (for any reasonable magnetic field $B \geq 0.01$ mG), even compared with the minimum available time – the light travel time across the source, $R/c \sim 3 \times 10^3$ yr. This simple estimate has the following interesting implications.

(i) The acceleration of electrons should take place throughout entire volume of the hot spot, otherwise the short propagation lengths of electrons would not allow formation of diffuse X-ray emitting regions on kpc scales as it is observed by Chandra. However, operation of huge quasi homogeneous 3-dimensional accelerators of electrons with linear size of 1 kpc or so, seems to be a serious theoretical challenge. Possible alternatives could be (1) the superposition of many compact accelerators within the observed knots, or (2) a scenario in which the cloud of TeV electrons, accelerated in a relatively compact region, expand at a speed of light, for example in the form of a (quasi) symmetric relativistic "hot" wind (i.e. a relativistically expanding cloud in the frame of which the electrons have a broad relativistic momentum distribution enabling their synchrotron radiation).

(ii) The electron-synchrotron model of X-ray jet structures implies prompt formation of the spectrum of (radiatively) cooled electrons with power-law index $\Gamma_e = \Gamma_0 + 1$. It is important to remember that this relation is valid only for the acceleration spectra with $\Gamma_0 \geq 1$. Otherwise $\Gamma_e = 2$, independent of Γ_0. As long as the synchrotron energy losses (${\rm d}E_e/{\rm d}t \propto E_e^2$) play a dominant role in formation of the electron spectrum, the latter cannot be harder, independent of the initial (acceleration) spectrum $Q(E_e)$, than $N(E_e) \propto E_e^{-2}$. Indeed, since the cooled steady-state spectrum of electrons is determined by Eq.(3.5), even in the extreme case of an abrupt low-energy cutoff in the injection spectrum (i.e. $Q(E_e) = Q_0 E_e^{-\Gamma_0}$ at $E_e \geq E_*$, and $Q(E_e) = 0$ at $E_e \leq E_*$), the spectrum of cooled electrons has a broken power-law shape with $N(E_e) \propto E_e^{-(\Gamma_0+1)}$ at $E_e \geq E_*$, and $N(E_e) \propto E_e^{-2}$ at $E_e \leq E_*$. Correspondingly, the spectral index of the synchrotron radiation ($S_\nu \propto \nu^{-\alpha}$) is $\alpha = \Gamma_0/2$ for $\Gamma_0 \geq 1$, and $\alpha = 0.5$ for *any* electron acceleration spectrum flatter than E^{-1}. Thus, the spectrum of synchrotron emission of radiatively cooled electrons cannot be harder than $S_\nu \propto \nu^{-0.5}$.

Typically, the spectra of the knots of extragalactic jets are very hard with $\alpha \leq 1$. In particular, the power-law index of X-radiation of the knot A1 of 3C 273 was reported to be close to $\alpha_x \approx 0.6$ (Marshall et al., 2001). This requires an electron acceleration spectrum with $\Gamma_0 = 1.2$ which would challenge any (in particular, shock) acceleration scenario. Moreover, it is difficult to reconcile the reported optical-to-X-ray spectral index $\alpha \leq 0.35$ of the so-called 25" knot of 3C 120 (Harris et al., 1999). Actually, spectral indices $\alpha \leq 0.5$ could be typical for the optical-to-X-ray spectra of jet structures of some other AGN as well, for which a systematic spatial displacement between the radio, optical and X-ray regions have been reported.

It is possible to have very hard optical-to-synchrotron X-ray spectra assuming that the energy losses of electrons are dominated by relativistic expansion or by Compton losses in the *Klein-Nishina regime*. The first process keeps the original (acceleration) spectrum of electrons unchanged because the adiabatic cooling time does not depend on energy. The second process makes it even harder compared to the acceleration spectrum because in the Klein-Nishina regime the energy loss rate has a very weak (logarithmic) dependence on energy (see Chapter 3). Since the magnetic field in the jet structures could hardly be less than $10\mu G$, the Compton losses in the Klein-Nishina regime can dominate over the synchrotron losses if the energy density of background radiation significantly exceeds $B^2/8\pi \simeq 2.5$ eV/cm^3. The energy density of the 2.7 K CMBR is smaller by a factor of 10. Moreover, the energy of the 2.7 K CMBR photons is not high enough for the Compton scattering of electrons to proceed in the deep Klein-Nishina regime. In order to get the required density and photon energy (in the jet frame), one may invoke relativistic bulk motion of the jet with Lorentz factor $\Gamma_{\rm jet} \gg 10$. Alternatively, an additional source of target photons at the near infrared or higher frequencies would be required. In this regard, the *non-linear* SSC scenario seems an interesting possibility, and deserves future theoretical investigation.

In conclusion, the synchrotron, inverse Compton and Synchrotron-self-

Compton models of nonthermal X-ray emission in the context of *directly* accelerated/reaccelerated electrons are generally considered by the "AGN community" as standard models capable of describing the main features of nonthermal emission reported from large scale AGN jets. Even so, many observed morphological and spectral features of jet structures cannot be easily accommodated by these models, at least in their simplified (homogeneous, one-zone) versions. Future studies will show whether it will be possible to escape these difficulties by assuming more sophisticated models. At the same time, it is important to explore other, non-conventional scenarios concerning the origin of electrons, as well as to study other radiation mechanisms related to the hadronic component of accelerated particles.

8.2 Ultra High Energy Protons in Jets

The jets of powerful radiogalaxies and AGN are one of a few potential sites in the Universe where protons can be accelerated to the highest observed energies of about 10^{20} eV (see e.g. Hillas, 1984; Cesarsky, 1992; Rachen and Biermann, 1993; Ostrowski, 1998; Henri *et al.*, 1999, Aharonian *et al.*, 2002). Dissipation of bulk kinetic energy and/or the Poynting flux of jets results in, most likely through strong terminal shocks, generation/amplification of magnetic fields, heating of the ambient plasma, and acceleration of particles in the knots and hot spots.

8.2.1 *Secondary electrons*

The undisputed synchrotron origin of the nonthermal radio radiation of many knots and hot spots extending to (at least) optical wavelengths is clear evidence for acceleration of electrons to TeV energies. Besides, relativistic electrons are produced also in interactions of high energy protons with ambient gas and photon fields. Biermann and Strittmatter (1987) have proposed that the secondary component of electrons produced by extremely high energy protons interacting with local photon fields may play an important role in formation of the nonthermal radiation of radio jets. This idea has been further developed within the so-called PIC (Proton Induced Cascade) model by Mannheim *et al.* (1991) for interpretation of radiation of some prominent radio jets' features like the west hot spot in Pictor A and the knot A1 in 3C 273. The intrinsic feature of this model is that it produces a distinct maximum in the spectral energy distribution

(SED) νS_ν at MeV energies, and a standard hard X-ray spectrum with a spectral index $\alpha_x \sim 0.5$ at keV energies. The recent Chandra observations of Pictor A (Wilson et al., 2001) give a flux which at 1 keV appears almost 3 orders of magnitude below the PIC-model predictions. The predictions of this model for the knot A1 of 3C 273 also fall well below the reported Chandra flux at 1 keV. Although it is formally possible to increase the X-ray fluxes, assuming significantly larger power in accelerated protons or speculating with denser target photon fields, this raises serious problems with the required nonthermal energy budget. Moreover, the flat X-ray spectrum of the hot spot of Pictor A, with a spectral index $\alpha = 1.07 \pm 0.11$ excludes the PIC model which predicts a much harder X-ray spectrum. The X-ray spectral index of the knot A1 in 3C 273, $\alpha_x = 0.60 \pm 0.05$, is closer to the PIC model predictions, but in this source the efficiency of the process is too low to explain the detected X-ray flux.

For a broad spectrum of target photons, the photo-meson cooling time of protons can be estimated as $t_{p\gamma} = (c <\sigma f> n(\epsilon^*)\epsilon^*)^{-1}$, where $<\sigma f> \simeq 10^{-28}$ cm^2 is the photo-meson production cross-section weighted by inelasticity at the photon energy ~ 300 MeV in the proton rest frame and $\epsilon^* = 0.03 E_{19}^{-1}$ eV, $E_{19} = E_p/10^{19}$ eV is the energy of the proton in units of 10^{19} eV (see e.g. Mücke et al., 1999). The fluxes of jet features most relevant for photo-meson production, in the frequency band 10^{12}-10^{14} Hz, typically are poorly known. Therefore it is convenient to present the low-frequency photon flux density in the form $S_\nu = S_0 \nu_{\rm GHz}^{-\alpha}$, where S_0 is the flux density at 1 GHz in units of Jy. This gives

$$t_{p\gamma} \simeq 10^9 c_\alpha S_0^{-1} E_{19}^{-\alpha} \theta^2 \text{ yr}, \tag{8.1}$$

where $c_\alpha \approx 0.42$; 3.9 and 36 for $\alpha = 0.5$, 0.75 and 1.0, respectively; $\theta = R/d$ is the angular radius of the source in arcseconds. For example, in the A1 knot of 3C 273 with $\theta \sim 0.5 - 1$ arcsecond, $S_0 \simeq 0.05$Jy and $\alpha \simeq 0.75$, the characteristic photo-meson production time appears more than 2×10^9 yr even for 10^{20} eV protons. This is too large compared with the escape time of highest energy protons from the X-ray production region. Indeed, the particle escape cannot be longer than the time determined by diffusion in the Bohm limit, $t_{\rm esc} \approx R^2/2D$ with the diffusion coefficient $D(E) = \eta r_g c/3$, where $r_g = E_p/eB$ is the gyroradius, and $\eta \geq 1$ is the so-called gyro-factor (in the Bohm limit $\eta = 1$). Thus

$$t_{\rm esc} \simeq 4.2 \times 10^5 \eta^{-1} B_{\rm mG} R_{\rm kpc}^2 E_{19}^{-1} \text{ yr}. \tag{8.2}$$

and, representing the source radius as $R = \theta \cdot d \simeq 4.8\theta(d/1 \text{ Gpc})$ kpc, for 3C 273 with redshift $z = 0.158$ one finds (assuming for the Hubble constant $H_0 = 60$ km/s Mpc):

$$t_{\rm esc}/t_{\rm p\gamma} \simeq 8 \times 10^{-5} \eta B_{\rm mG} E_{19}^{-0.25}. \tag{8.3}$$

This ratio implies a very limited efficiency for transformation of the proton kinetic energy to nonthermal radiation. Note that both characteristic times, $t_{\rm esc}$ and $t_{\rm p\gamma}$ are proportional to R^2, consequently the ratio $t_{\rm esc}/t_{\rm p\gamma}$ does not depend on θ, and thus we cannot increase the photo-meson production efficiency by assuming that the acceleration and radiation take place in smaller (but numerous) regions inside the knot. The ratio also depends slightly on the energy of protons. Although the photo-meson production time decreases with energy, the escape time decreases even faster, thus the efficiency of the process $(t_{\rm esc}/t_{\rm p\gamma})$ becomes, in fact, less at higher energies. The direct (Bethe-Heitler) production of (e^+, e^-) pairs cannot significantly enhance the efficiency of $p\gamma$ interactions either. Although this process requires lower energy protons, which are confined more effectively, for a broad-band target radiation this process is typically slower than the photo-meson production.

The fluxes of mm radiation in some hot spots, e.g. in the radiogalaxies Pictor A and Cygnus A are 2 or 3 orders of magnitude higher than in the knot A1 of 3C 273, therefore the efficiency of $p\gamma$ interactions in these objects could approach a few per cent, provided that particle escape takes place in the Bohm regime. Even so, the PIC remains a rather ineffective scenario for the production of X-rays, because it allows less than 1 per cent of the overall luminosity of nonthermal radiation to be released at keV energies.

Secondary electrons can also be produced also at interactions of relativistic protons with the ambient gas. The characteristic cooling time of this process is given by Eq.(3.13). Because the X-ray synchrotron radiation in the jet could be produced by ≤ 10 TeV electrons, the pp interactions require relatively low energy protons, $E_{\rm p} \leq 100$ TeV, the escape time of which may exceed the jet age. On the other hand the pp cooling time of protons is quite large, therefore this attractive mechanism, which produces relativistic electrons throughout the entire jet structure, may have an impact on the overall nonthermal radiation only when the ambient gas density in the knots and hot spots exceeds $0.1 - 1$ cm^{-3}, or the jet is moving relativistically towards the observer with Doppler factor $\gg 1$. This possibility for 3C 273 in discussed in Sec. 8.2.4.

The low-frequency radiation converts, through photon-photon interactions, a fraction of high energy γ-rays, before they escape from the knots, into (e^+, e^-) pairs. Thus, these interactions could, in principle, initiate electromagnetic cascades in the jet. For a power-law spectrum of low-frequency radiation with the normalisation defined above, the optical depth characterising the efficiency of the cascade development at γ-ray energy E is estimated as

$$\tau_{\gamma\gamma} \approx \tau_\alpha S_0 \theta^{-2} R_{\rm kpc}(E/1 \text{ TeV})^\alpha \,, \qquad (8.4)$$

with $\tau_\alpha \simeq 4 \times 10^{-3}; 1.8 \times 10^{-4}$ and 8.2×10^{-6} for α =0.5, 0.75 and 1.0, respectively. For example, in the knot A of 3C 273 with $\theta \sim 0.5$ arcsec, $\alpha \simeq 0.75$ and $S_0 \simeq 0.05$ Jy, the optical depth exceeds 1 only at $E \geq 10^{18}$ eV. Thus, the cascade in this specific knot could be initiated only in the presence of very high energy γ-rays. At the same time, absorption of γ-rays propagating along the jets of tens or even hundreds of kpc could be quite significant, albeit the mean density of photons in the overall jet is noticeably less than in the knots. Moreover, if the γ-ray energy exceeds 10^{15} eV, the $\gamma\gamma$ pair production in such large scale jets becomes inevitable due to interactions with photons of the 2.7 K CMBR. This would lead to the generation of electron-positron pairs over the length of the jet through effective development of electromagnetic cascades initiated by ultra-high energy γ-rays injected in the jet from the central object. We discuss this interesting effect in Sec. 8.3.

8.2.2 *Synchrotron radiation of protons*

In this section we discuss a new mechanism for production of X-ray emission in large scale jets. Namely, we assume that X-ray emission of jets has a synchrotron origin but is produced by *protons* (Aharonian, 2002b). At first glance this mechanism seems to be quite inefficient for the production of X-rays. However, adopting magnetic fields in jet structures somewhat larger than the field assumed in the standard electron synchrotron or IC models, and speculating that protons are accelerated in the jet structures to energies $E_{\rm p} \sim 10^{18}$ eV or more, it is possible to construct a model which allows effective cooling of protons via synchrotron radiation on very "comfortable" timescales, namely on timescales comparable with the (diffusive) escape time of protons and/or the age of the jet. This provides effective propagation of protons over the entire jet structures on kpc scales, and thus can naturally explain the diffuse character of X-ray emission, as well as the broad range of spectral indices for X-ray emission observed from different objects. Yet, as long as the *proton synchrotron cooling time*, the particle *escape time* and the *age of the jet* are of the same order of magnitude, the proton synchrotron model offers a quite high (from 10 to almost 100 per cent) efficiency for transformation of the kinetic energy of protons to hard X-rays. This makes the proton synchrotron radiation an attractive and viable mechanism for interpretation of nonthermal X-ray emission from large-scale extragalactic jets.

During the jet lifetime most of the energetic particles escape the knots. In particular, in the Bohm regime, when the particles tend to drift away most slowly ($\eta = 1$), the break in the initial proton spectrum takes place at $E_\mathrm{p} \simeq 4.2 \times 10^{17} B_\mathrm{mG} R_\mathrm{kpc}^2 (\Delta t/10^7 \mathrm{\ yr})^{-1}$ eV. Since the observations limit the size of a typical knot or a hot spot to $R \leq$ few kpc, the only possibility to move the break point to higher energies is the increase of the magnetic field, $B \geq 1$ mG. However, at such large magnetic fields the proton synchrotron radiation becomes an additional, or even more important limiting factor with a characteristic cooling time

$$t_\mathrm{synhc} \simeq 1.4 \times 10^7 B_\mathrm{mG}^{-2} E_{19}^{-1} \mathrm{\ yr} \;, \tag{8.5}$$

Remarkably, the radiation of the highest energy protons is released in the X-ray domain:

$$h\nu_\mathrm{m} = 0.29 h\nu_\mathrm{c} \simeq 2.5 \times 10^2 B_\mathrm{mG} E_{19}^2 \mathrm{\ keV}. \tag{8.6}$$

Correspondingly, the characteristic time of radiation of a proton-synchrotron photon of energy E is

$$t(E) \simeq 2.2 \times 10^8 B_\mathrm{mG}^{-3/2} (E/1 \mathrm{\ keV})^{-1/2} \mathrm{\ yr} \;. \tag{8.7}$$

Taking into account that 1 keV photons are produced by protons with energy $E_{19} \simeq 0.063 \, B_\mathrm{mG}^{-1/2}$, the radiation time of synchrotron X-ray photons becomes less than the time of particle escape in the Bohm regime if $B_\mathrm{mG}^3 R_\mathrm{kpc}^2 \geq 30$. For a typical size of jet knots of about 1 kpc, this would require $B \geq 3 \times 10^{-3}$ G and correspondingly the total energy contained in the magnetic field $W_\mathrm{B} = B^2 R^3/6 \geq 4 \times 10^{58}$ erg. Interestingly, in such an extreme regime, when almost the whole kinetic energy of accelerated protons is released in synchrotron X-rays for a typical time of about $\leq 4 \times 10^7$ yr, the luminosity of synchrotron radiation at 1 keV is expected to be at the level $L_\mathrm{X} \simeq 3 \times 10^{43} \kappa$ erg/s, assuming equipartition between the magnetic field and accelerated protons, $W_\mathrm{p} = W_\mathrm{B}$. The parameter κ is the kinetic energy in protons (responsible for synchrotron X-rays, i.e. $E_p \geq 10^{17}$ eV) as a fraction of the total energy. For an E_p^{-2} type proton spectrum extending beyond 10^{18} eV, $\kappa \sim 0.1$. Note that in the regime dominated by proton synchrotron losses, the total luminosity of the proton-synchrotron radiation does not depend on the magnetic field, and is determined simply by the acceleration power of highest energy protons.

Below we present detailed numerical calculations for several prominent extragalactic jet features. We assume that during the lifetime of the jet Δt, protons are injected into a homogeneous and spherically symmetric region, with acceleration spectrum $Q_p(E_p) = Q_0 E_p^{-\Gamma_p} \exp(-E_p/E_0)$ and at a (quasi) continuous constant rate $L_p = \int Q_p(E_p) E_p dE_p$. The kinetic equation for the energy distributions of relativistic particles which takes into account the radiative and escape losses, and the time-dependent solutions to this equation, are described in Appendix B. In this study the interactions of protons with the magnetic field (synchrotron radiation), photon fields (photo-meson- and Bethe-Heitler pair production) and the ambient gas ($pp \to \pi \to e^+e^-\gamma$), as well as the synchrotron radiation and inverse Compton scattering of secondary electrons are included in calculations. The fast synchrotron cooling of secondary electrons in relatively strong irregular magnetic fields allow us to ignore the cascade processes (initiated by γ-γ interactions) without a significant impact on the accuracy of calculations.

8.2.3 Pictor A, PKS 0637-752, and 3C 120

The luminosities of proton-synchrotron radiation calculated for 3 prominent jet features – the western hot spot in Pictor A, the radio knot at 25" distance from the nucleus of the Seyfert galaxy 3C 120, and the bright X-ray region in the inner western jet of PKS 0637-752 (at $\sim 8''$ from the core) - are shown in Fig. 8.1. The shape of the resulting proton spectrum primarily depends on the index of acceleration spectrum Γ_0, the magnetic field B and the size R. The total power in the accelerated protons L_p determines the absolute X-ray flux.

Pictor A. The western X-ray hot spot $4'.2$ from the nucleus of this nearby ($z = 0.035$) powerful radiogalaxy coincides with the radio jet and has a lateral extent of ~ 2 kpc (Wilson et al., 2001). The spectacular Chandra image of the jet and hot spot of Pictor A in X-rays is shown in Fig. 8.2. The X-ray emission from the hot spot is well described by a flat power-law spectrum with a photon index of $\alpha + 1 = 2.07 \pm 0.11$. This implies that within the proton-synchrotron model the spectral index of radiating protons should be close to 3. Thus, for the shock acceleration spectrum with canonical power-law index $\Gamma_0 \sim 2$ we must allow an increase of the spectral slope by 1. Such a steepening can be naturally provided by intense synchrotron cooling and/or by fast energy-dependent escape of protons from the source with $\tau_{\rm esc}(E) \propto 1/E$. In the synchrotron-loss dominated regime the corresponding break in the proton-synchrotron spectrum (the point where $S_\nu \propto \nu^{-0.5}$ is changed to $\propto \nu^{-1}$) takes place at

$$(h\nu)_b \simeq 5\, B_{\rm mG}^{-3} \Delta t_8^{-2} \text{ keV} , \tag{8.8}$$

where $\Delta t_8 = \Delta t/10^8$ yr is the duration of the particle acceleration in units of 10^8 yr. For a typical lifetime of the jet (the period of activity of the central engine) $\Delta t \leq 10^8$ yr, the flat ($\nu S_\nu = const$) spectrum of the hot spot above 1 keV would require magnetic field larger than 1 mG. In the regime dominated by particle escape, the break in the spectrum depends on the propagation character of protons. In particular in the regime of Bohm

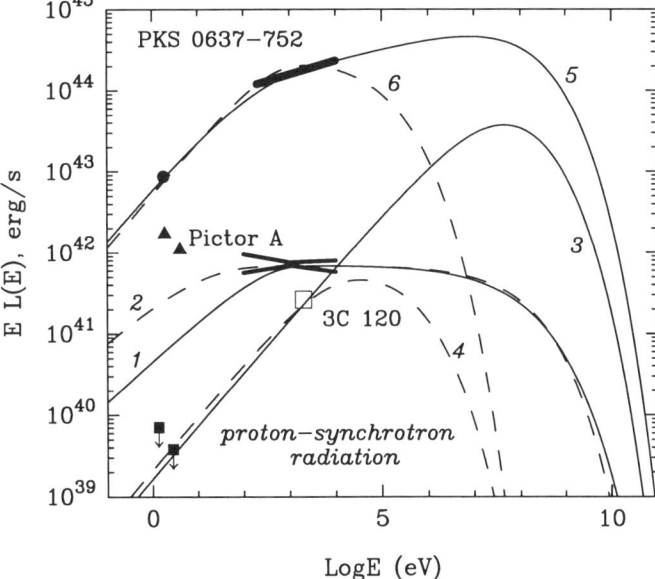

Fig. 8.1 Proton synchrotron luminosities of jets in Pictor A (western radio hot spot), 3C 120 (25" radio knot), and PKS 0637-752 (the outer jet component). The optical and X-ray data are from Wilson *et al.* (2001) for Pictor A; Harris *et al.* (1999) for 3C 120; Schwartz *et al.* (2000) for PKS 0637-752. The model parameters are described in the text. (From Aharonian, 2002b).

diffusion the break in the proton-synchrotron spectrum appears at

$$(h\nu)_b \simeq 0.004 \, B_{mG}^3 \Delta t_8^{-2} R_{kpc}^4 \eta^{-2} \text{ keV} . \qquad (8.9)$$

Note that the characteristic times of synchrotron cooling and the particle escape in the Bohm regime have the same $\propto 1/E$ dependence, therefore the dominance of synchrotron losses does not depend on the specific energy interval, and is determined by the following simple condition:

$$B_{mB}^3 R_{kpc}^2 \eta^{-1} \geq 30 . \qquad (8.10)$$

In Fig. 8.1 we show two "proton-synchrotron" fits to the spectrum of the hot spot of Pictor A calculated (a) for a regime dominated by *synchrotron losses* with two key model parameters $B = 3$ mG and $R = 2$ kpc (curve 1), and (b) for a regime dominated by *particle escape* with $B = 1$ mG and $R = 1$ kpc (curve 2). In both cases the following additional model parameters were assumed: gyro-factor $\eta = 1$ (Bohm diffusion); power-law index $\Gamma_0 = 2$ and maximum energy of accelerated protons $E_0 = 10^{20}$ eV. And finally, to match the absolute X-ray fluxes we assume that the proton acceleration takes place in a (quasi) continuous regime during last $\Delta t = 10^8$ yrs with rates $L_p = 4.5 \times 10^{43}$ and 1.35×10^{45} erg/s for the cases (a) and (b), respectively.

Although both spectra satisfactorily fit the reported X-ray data above 0.5 keV, case (a) is more attractive on energetic grounds – it requires ~ 30 times less power in accelerated protons. Note that in the regime dominated

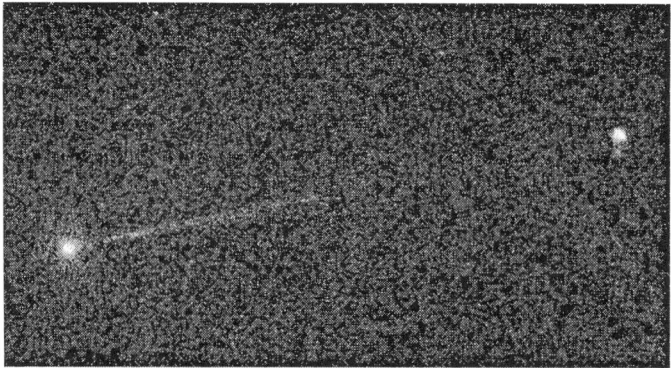

Fig. 8.2 The Chandra X-ray image of Pictor. The perfectly rectilinear jet emanates from the radiogalaxy and extends over 100 kpc towards the bright hot spot at a projected distance of 250 kpc. (From Wilson *et al.*, 2001).

by synchrotron losses, the efficiency of X-ray production is close to 100 per cent, the luminosity being almost independent of the magnetic field, $L_{x-\gamma} \sim L_p$. In the particle-escape dominated regime, $L_{x-\gamma} \propto B^2$.

The spectra of relativistic protons trapped in the hot spot, $W_p(E_p) = E_p^2 N(E_p)$, are shown in Fig. 8.3. Like Fig. 8.1, the curves 1 and 2 correspond to cases (a) and (b), respectively. The spectra are similar with an original (acceleration) shape $\propto E_p^{-2}$ below the break energy at which the escape and synchrotron cooling times exceed the age of the source, and $\propto E_p^{-3}$ at higher energies. At the same time, while in case (a) the total energy in protons is rather modest, $W_p^{(tot)} = \int W(E_p)dE_p \simeq 1.1 \times 10^{59}$ erg, in case (b) it is significantly higher, $W_p^{(tot)} \simeq 2.9 \times 10^{60}$ erg. Thus, while in case (a) the conditions are close to the equipartition between the protons and magnetic filed ($w_p \simeq 1.3 \times 10^{-7}$ erg/cm^3, $w_B \simeq 3.6 \times 10^{-7}$ erg/cm^{3}s), in case (b) the pressure of protons exceeds the B-field pressure by 3 orders of magnitude.

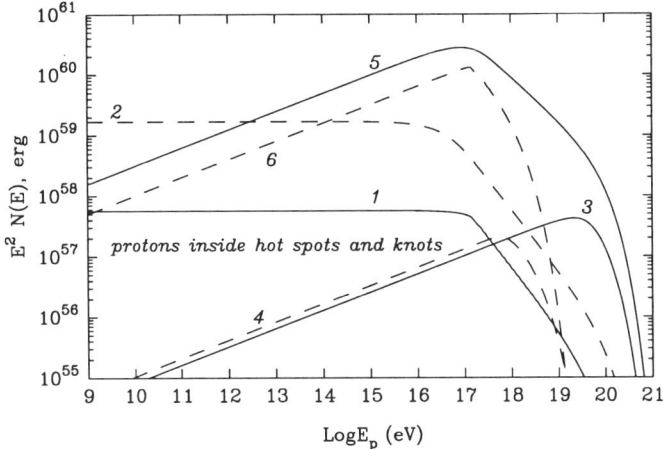

Fig. 8.3 Total energy of relativistic protons confined in X-ray emitting regions of radio jets: the western hot spot in Pictor A (curves 1 and 2), 25″ knot of 3C 120 (curves 3 and 4), and the outer jet component in PKS 0637-752 (curves 5 and 6).

Finally we note that the proton synchrotron radiation cannot be responsible for the observed radio and optical fluxes. The two optical/UV points (triangles) in Fig. 8.1 confirm the break in the spectrum at $\nu \sim 10^{14}$ Hz. Both the slope of the spectrum connecting the radio and optical fluxes, as well as the spectral break above 10^{14} Hz can be naturally explained by synchrotron radiation of electrons. We note, however, that this radiation component is produced, most probably, in region(s) separated from the X-ray production region, both in *space* and *time*.

The calculations presented in Fig. 8.1 include interactions of protons with 2.7 K CMBR. The synchrotron radiation of the secondary electrons peaks at GeV energies, with luminosities at 1 keV of about 3.5×10^{36} erg/s for case (a) and by a factor of 20 larger for case (b) (not seen in Fig. 8.1). In the hot spot of Pictor A, the interactions of protons with the local (synchrotron) radiation at $\nu \sim 0.1 - 100$ GHz are more (by two orders of magnitude) frequent, especially if we assume a small, sub-kpc size of the hot spot. Even so, this process is not sufficiently effective to reproduce the observed X-ray fluxes. Moreover, the synchrotron radiation of secondary (from $p\gamma$ interactions) electrons at 1 keV has, independent of the proton spectrum, a standard spectrum with spectral index $\simeq 0.5$, i.e. significantly harder than the spectrum detected by Chandra.

3C 120. The X-radiation from a radio knot at a distance of 25″ from the nucleus of the Seyfert galaxy 3C 120 found in the ROSAT HRI data (Harris *et al.*, 1999) deserves special interest. The flux predicted by the SSC models appears several orders of magnitude below the observed X-ray flux. The two-component synchrotron model which postulates two different populations of relativistic electrons responsible for the radio-to-optical and X-ray emissions, at first glance seems a natural interpretation. However, the reported flux limit at optical frequencies (Harris *et al.*, 1999) challenges this model as well. Indeed, the upper limit at $\nu = 6.8 \times 10^{14}$ Hz ($S_\nu \leq 0.18\mu$Jy) combined with the X-ray flux at 2 keV (0.018μJy) (see Fig. 8.1) requires an unusually hard spectrum between the optical and X-ray band, $S_\nu \propto \nu^{-\alpha}$ with $\alpha \leq 0.35$. This implies that the spectrum of radiating electrons should be flatter than $N(E_\mathrm{e}) \propto E_\mathrm{e}^{-1.7}$. Although such a spectrum deviates from predictions of the canonical shock acceleration theory, formally we cannot exclude other, more effective particle acceleration scenarios. However, as discussed above, as long as the synchrotron energy losses ($\mathrm{d}E_\mathrm{e}/\mathrm{d}t \propto E_\mathrm{e}^2$) play a dominant role in formation of the electron spectrum, the latter cannot be harder, independent of the initial (acceleration) spectrum $Q(E_\mathrm{e})$, than $N(E_\mathrm{e}) \propto E_\mathrm{e}^{-2}$. The acceleration spectrum $Q(E_\mathrm{e})$ can be sustained unchanged, assuming fast, at the speed of light, escape or adiabatic losses of electrons. However, even with such a dramatic assumption, the electron escape time, $t_\mathrm{esc} = R/c \simeq 3 \times 10^3 R_\mathrm{kpc}$ yr, in the knot with a radius 2-3 kpc appears longer than the typical lifetime of electrons responsible for synchrotron radiation of optical and X-ray photons, $t_{\text{synch}} \simeq 1.5 B_{\text{mG}}^{-3/2}(E/1 \text{ keV})^{-1/2}$ yr, unless the magnetic field is less than $\sim 10^{-5}$ G. Alternatively, we should assume that the X-ray flux is a result of contributions from short-lived (transient) compact regions where the energy-independent escape losses could dominate over the synchrotron losses.

We do not face such a problem with the synchrotron radiation of protons. For a reasonable set of model parameters, the proton escape could dominate over the synchrotron cooling, but yet the luminosity of the proton-synchrotron radiation could be maintained at a rather high level. Obviously, quite similar to the electron synchrotron model, the comparison of the X-ray and optical fluxes tells us that the spectrum of the radiating protons should be harder than $E^{-1.7}$. This implies that for even hardest possible acceleration spectra, $\propto E^{-1.5}$ (Malkov, 1999), the escape of particles should be essentially energy-independent ($t_{\text{esc}} \propto E^{-\beta}$ with $\beta \leq 0.2$). In Fig. 8.1 we show a possible version of the proton-synchrotron radiation calculated for the following model parameters: the magnetic field $B = 1.5$ mG, radius of the X-ray production region $R = 3$ kpc, acceleration spectrum of protons with $\Gamma_0 = 1.7$ and $E_0 = 10^{20}$ eV, energy-independent escape of protons with $t_{\text{esc}} = 5 \times 10^5$ yr. For the diffusive propagation of particles, the latter corresponds to the diffusion coefficient which at proton energies $\geq 10^{17}$ eV (responsible for production of 0.1-10 keV synchrotron photons) is much larger than the Bohm diffusion coefficient. This provides the dominant escape loss, and thus does not allow deformation of the primary (acceleration) spectrum of protons up to the highest energies (see curve 3 in Fig. 8.3). Then, the efficiency of the proton-synchrotron radiation, which is characterised by the ratio of the escape time to the synchrotron cooling time, at 1 keV is about ≈ 0.4 per cent. The efficiency is much higher (close to 100 per cent) for 100 MeV photons (see curve 3 in Fig. 8.1) produced by the highest energy ($\sim 10^{20}$ eV) protons. The corresponding flux of 100 MeV γ-rays $J_\gamma \sim 5 \times 10^{-8}$ ph/cm^{2}s is sufficiently large to be detected by GLAST.

To reproduce the observed X-ray flux at 1 keV we must assume a rather high injection rate of accelerated protons during 10^8 yr operation of the jet, $L_p = 1.65 \times 10^{45}$ erg/s. Note, however, that $\geq 10^{18}$ eV protons do not contribute, for the given magnetic field of about 1.5 mG, to the production of ≤ 1 keV photons. Therefore we may somewhat (by a factor of 4) soften this requirement by reducing the high energy cut-off in the proton spectrum down to $E_0 = 10^{18}$ eV. Further significant reduction of the required acceleration power could be achieved assuming more effective particle confinement. In Fig. 8.1 we show the luminosity of the proton-synchrotron radiation (curve 4) calculated for $t_{\rm esc} = 5 \times 10^7$ yr and $E_0 = 5 \times 10^{18}$ eV, but keeping all other model parameters unchanged. In spite of the increase of the escape time by two orders of magnitude, which becomes comparable with the age of the source $\Delta t = 10^8$ yr, the synchrotron cooling remains a slower process, compared with the particle escape, until $E_{\rm p} = 1.2 \times 10^{18}$ eV. Therefore the break in the synchrotron spectrum occurs only at ~ 5 keV. At the same time, the required proton acceleration power becomes rather modest; now only $L_{\rm p} \simeq 1.0 \times 10^{43}$ erg/s is needed to explain the observed X-ray flux.

PKS 0637-752. This quasar at a redshift $z = 0.651$ has the largest and most powerful X-ray jet detected so far (Chartas *et al.*, 2000, Schwartz *et al.*, 2000). The X-ray luminosity of the ≥ 100 kpc jet of about $L_{\rm X} \sim 4 \times 10^{44}$ erg/s is contributed mostly by bright condensations between $7''.5$ and $10''$. This region contains three 3 knots resolved in radio and optical wavelengths. The enhanced X-ray emission is associated, most probably, with these knots, although its profile does not exactly repeat the radio and optical profiles of the jet. Therefore the combined flux of these three knots, $S_\nu = 0.57$ μJy at an effective frequency 4.3×10^{14} Hz (Schwartz *et al.*, 2000), should be considered as an upper limit for the optical flux from the X-ray emitting regions.

Because the individual X-ray knots are not clearly resolved, here we consider a simplified picture, namely treating the overall emission from the X-ray enhanced region as radiation from a single source. The X-ray luminosity based on the measured flux density at 1 keV, 5.9×10^{-14} erg/cm^{2}s keV and spectral index $\alpha = 0.85$ (Schwartz *et al.*, 2000) is shown, together with the model calculations, in Fig.8.1. Curve 5 is obtained for the following parameters: $B = 1.5$ mG, $\Gamma_0 = 1.75$, $E_0 = 10^{20}$ eV, $R = 5$ kpc. Also it was assumed that particles propagate in a "relaxed" Bohm diffusion regime with gyro-factor $\eta = 10$. Under these conditions the escape losses dominate over the synchrotron losses, and become important above 10^{17} eV. This results in the steepening of the proton spectrum, from the initial $E_{\rm p}^{-1.75}$ to $E_{\rm p}^{-2.75}$. The latter continues up to the exponential cutoff in the acceleration spectrum at $E_0 = 10^{20}$ eV (curve 5 in Fig.8.3). The corresponding proton-synchrotron spectrum nicely fits the observed X-ray spectrum and, at the same time, agrees with the optical flux reported from that part of the jet. Note, however, that the optical flux shown in Fig.8.3 perhaps should be considered as an upper limit. Correspondingly, if the intrinsic optical flux of the X-ray emitting region is significantly less than this upper limit, we must assume a flatter proton acceleration spectrum. It should be noticed also that the extrapolation of the proton synchrotron spectrum down to radio wavelengths appears significantly below the reported radio fluxes at 4.8 GHz and 8.6 GHz (Chartas et al., 2000), which most probably are due to the synchrotron radiation of directly accelerated electrons.

For the chosen set of model parameters, the absolute X-ray flux requires a huge acceleration power $L_\mathrm{p} = 3 \times 10^{46}$ erg/s, which however agrees with the estimates of the jet power (in the form of kinetic energy of particles and of Poynting flux) which in most luminous extragalactic objects could be as large as $10^{47} - 10^{48}$ erg/s (see e.g. Sikora, 2001, Ghisellini and Celotti, 2001). We may nevertheless reduce the energy requirements assuming somewhat different model parameters which would minimise the escape losses. An example of the X-ray luminosity calculated for a favourable set of model parameters is presented by curve 6 in Fig.8.1. In particular, compared with curve 5 the magnetic field is increased to 3 mG, the size is reduced to 3 kpc, and it is assumed that the particle diffusion proceeds in the Bohm regime ($\eta = 1$). This implies that $B_\mathrm{mB}^3 R_\mathrm{kpc}^2 \eta^{-1} \approx 250$, i.e. the proton losses are strongly dominated by synchrotron cooling. In order to reduce the required proton acceleration power by another factor of ~ 3, an "early" exponential cutoff in the acceleration spectrum is assumed at $E_0 = 2 \times 10^{18}$ eV. The resulting proton spectrum shown in Fig.8.3 (curve 6) could be treated as an optimum spectrum designed to maximise the production of X-ray at 1 keV, as is seen in Fig.8.1. The required proton power now is 10 times less than in the previous case, $L_\mathrm{p} = 2.9 \times 10^{45}$ erg/s. Further significant reduction of L_p is almost impossible given the fact that the X-ray luminosity in the interval 0.2-10 keV already is huge, $L_\mathrm{x} \simeq 5 \times 10^{44}$ erg/s, unless we assume that the jet is moving with a bulk Lorentz factor $\Gamma_\mathrm{j} \gg 1$ towards the observer.

8.2.4 The case of 3C 273

3C 273 is the brightest quasar on the sky with broad band distribution of the central source extending from radio to high energy γ-rays (see e.g. Courvoisier 1998) and with a nonthermal large scale jet extending over ≈ 80 kpc. The highly variable emission of the central source and the extension of the spectrum to ≥ 1 GeV γ-rays (see Fig.2.3) implies that this radiation is beamed, and most likely originates in a sub-pc relativistic jet (otherwise severe absorption of the γ-rays by X-rays would be inevitable, see Lichti et al., 1995). The relatively small distance to the source ($z = 0.158$) allows the detection of γ-rays up to energies of about 100 GeV without significant intergalactic absorption. Thus 3C 273 may serve as an ideal laboratory to study the link between multi-kpc and sub-pc jets of powerful radiogalaxies and quasars.

The Chandra image of the radio jet of 3C 273 is shown in Fig.8.4. Approximately 40% of the total X-ray luminosity of the jet is contributed by the brightest knot, called A1 (Marshall et al., 2001). This knot is unique also in the sense that its radio, optical and X-ray fluxes are consistent with a single power-law spectrum (Röser et al., 2000, Marshall et al., 2001). The spectral energy distribution (SED) of knot A1 is shown in Fig.8.5. The observed fluxes at radio (MERLIN array), optical (*HST*), and X-ray (Chandra) bands are from Marshall et al. (2001). The overall slope, with spectral index $\alpha_{r-x} \simeq 0.75$, agrees well with the local slope at optical wavelengths, and matches the X-ray flux at 1 keV, although the local slope at X-rays is slightly different, $\alpha_x = 0.60 \pm 0.05$ (Marshall et al., 2001).

Fig. 8.4 The Chandra X-ray image of 3C 273 (from Marshall et al., 2001). The spectacular jet that emanates from the quasar and extends over 80 kpc, contains several bright condensations. The brightest of them is the so-called knot A1.

The similar polarisation properties observed at radio and optical frequencies, as well as the power-law behaviour of the spectrum imply that a single electron population can be responsible for radiation between these two bands. Formally, the X-rays also can be attributed to the same electron population (e.g. Röser et al., 2000). However, as discussed above, the power-law extrapolation of the hard synchrotron spectrum to X-ray energies is quite problematic given the severe energy losses of electrons at high energies. The recent report of the detection of an excess of near-ultraviolet emission from this region also suggests that we are dealing with a two-component origin of the overall nonthermal emission of the jet (Jester et al., 2002). Also, because of severe synchrotron losses, the X-ray emitting electrons could not propagate far from their birthplace/accelerator, thus we should expect a point-source type morphology rather than diffuse emission. Below two alternative possibilities are discussed, both connected with the accelerated protons by *direct* synchrotron radiation or through synchrotron radiation of secondary electrons. For a knot size of about 1 kpc and for an ambient magnetic field $B \geq 1$ mG, the synchrotron-cooling and the escape times of protons are comparable with the age of the jet. Thus, with certain assumptions about the acceleration spectrum and the propagation of protons, we can satisfactorily fit the fluxes as well as to explain the kpc size of the observed diffuse nonthermal emission.

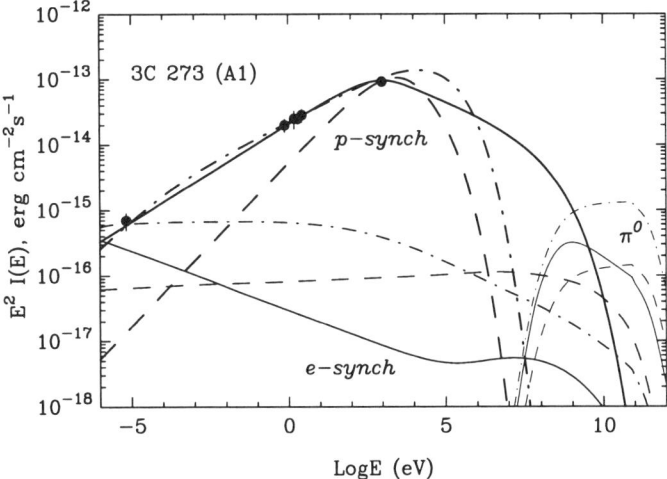

Fig. 8.5 The spectral energy distribution of knot A1 in 3C 273. The radio, optical and X-ray fluxes are from Marshall et al. (2001). The solid, dashed and dot-dashed curves correspond to 3 different sets of model parameters discussed in the text. The *heavy, standard and thin* lines represent the fluxes of (1) synchrotron radiation of protons, (2) synchrotron radiation of secondary electrons produced in $p\gamma$ and pp interactions, and (3) π^0-decay γ-rays produced at pp interactions, respectively.

Proton synchrotron radiation. In Fig.8.5 three spectra of proton-synchrotron radiation modelled for knot A1 are shown. For all 3 curves it is assumed that continuous injection of relativistic protons proceeds at a constant rate over the last 3×10^7 yr. The solid curve is calculated for $B = 5$ mG and $R = 1$ kpc, and assuming that the protons propagate in the Bohm diffusion regime with $\eta = 1$ (model I). This implies that synchrotron losses dominate over the particle escape, but yet, for the assumed source age and the magnetic field, the synchrotron losses become important only for the protons responsible for the X-ray flux above 1 keV ($E_p \geq 3 \times 10^{17}$ eV). Since the escape losses are also small and do not change the acceleration spectrum of protons, in order to explain the entire range of nonthermal emission from radio to X-rays with spectral index $\alpha_{r-x} \simeq 0.7$, we must assume a steep proton acceleration spectrum with $\Gamma_0 = 2\alpha_{r-x} + 1 = 2.4$. The solid curve in Fig.8.5 corresponds to such a power-law index of accelerated protons. The disadvantage of this fit is that it requires an uncomfortably large acceleration rate $L_p = 1.2 \times 10^{47}$ erg/s, with the total energy of protons deposited in the knot during 3×10^7 yrs, $W_p^{(tot)} \simeq 10^{62}$ erg (see Fig.8.6).

Consequently, the energy density of cosmic rays appears 3 orders of magnitude larger than the energy density of the B-field. Note, however, that this huge total energy is contributed by low energy protons which, from the point of view of production of radiation above ≥ 1 GHz, are in fact "vain" particles. Therefore assuming a low-energy cutoff in the acceleration spectrum, e.g. at 10^{13} eV, we can significantly reduce the energy requirements. Another, perhaps a more natural way to avoid the low-energy protons can be achieved assuming a hard acceleration spectrum coupled with an energy-dependent escape of particles. The dot-dashed curve in Fig.8.6 corresponds to the case of the canonical shock acceleration spectrum with $\Gamma_0 = 2$, and the escape time taken in the form $t_{esc}(E) = 1.4 \times 10^7 (E/10^{14} \text{ eV})^{-1/2}$ yr (model II). The resulting proton spectrum inside the knot in the most relevant energy region above 10^{13} eV becomes steeper, $E^{-2.5}$. The assumed large magnetic field $B = 10$ mG, as well as the upper energy cutoff at $E = 10^{18}$ eV in the acceleration spectrum, allow an additional reduction of the acceleration power to 1.05×10^{46} erg/s. The corresponding energy of protons confined in the knot now is 4.9×10^{60} erg. The spectrum of proton synchrotron radiation calculated for this set of parameters is shown in Fig. 8.5 by the dot-dashed curve.

Further significant reduction of the required energy budget is still possible, provided that the proton-synchrotron radiation is responsible only for the X-ray emission, with the radio and optical fluxes being related to other emission components, e.g. to the synchrotron radiation of electrons. The dashed curve in Fig.8.5 corresponds to such a scenario (model III). It is obtained assuming $B = 3$ mG, $R = 2$ kpc, $E_0 = 10^{18}$ eV, $\Gamma_0 = 2$. For these parameters the proton acceleration power is reduced to the level of $L_p = 10^{45}$ erg/s.

Synchrotron radiation of secondary electrons. In Fig.8.5 are shown the spectra of secondary electrons produced at interactions of protons with the ambient plasma and radiation fields. The corresponding synchrotron radiation of these electrons is shown in Fig.8.5. The energy density of mm and sub-mm radiation, which is the most relevant band of electromagnetic radiation for interactions of highest energy protons, is estimated from Fig.8.5, $w_r = \nu S_\nu (d/R)^2/c \approx 10^{-13}$ erg/cm^3, almost an order magnitude less than the energy density of the 2.7 K CMBR at the epoch $z = 0.158$.

The interactions of protons with the 2.7 K CMBR do not contribute significantly to the production of electrons with energy less than 100 TeV, i.e. to the production of electrons responsible for the synchrotron radiation at radio to X-ray energies. The first generation electrons appear with

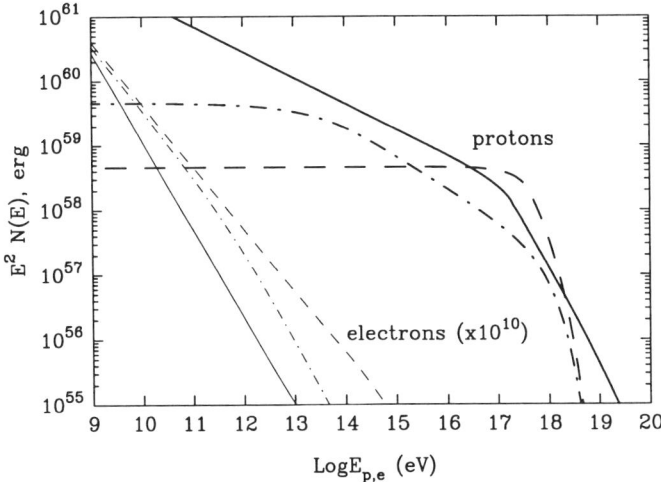

Fig. 8.6 Total energy of relativistic protons trapped in the knot A1 of the quasar 3C 273. The heavy solid, dashed, and dot-dashed curves correspond to 3 different combinations of the model parameters as in Fig.8.5. The fluxes of secondary $p\gamma$ and pp electrons multiplied by 10^{10} are also shown (thin lines).

energies exceeding $E_e \geq 10^{15}$ eV. On the other hand, the large magnetic field and small γ-γ optical depth prevent effective electromagnetic cascade development, which would allow production of lower energy electrons. As a result, the only signature of $p\gamma$ interactions seen Fig.8.5 is the flattening of the spectrum of the electron synchrotron radiation at MeV/GeV energies, caused by electrons produced through the Bethe-Heitler pair production.

For a relatively high gas density in the knot, the interactions of accelerated protons with the thermal plasma contribute more effectively, through decays of $\pi^\pm$ mesons, in the production of relativistic electrons and positrons. Because of the poorly known gas density in the jets, the calculations of secondary "pp" electrons contain large uncertainties. On the other hand, important information about the gas density could be provided by synchrotron radiation of these electrons; the latter obviously should not exceed the observed fluxes at radio, optical and X-ray bands. For the knots in 3C 273, the most informative upper limits are contained in the very low radio fluxes at GHz frequencies. The curves in Fig.8.5 correspond to the gas density $n = 10^{-5}$ cm^{-3} for the model I, and 10^{-3} cm^{-3} for the models II and III. It is seen that for the steep acceleration spectrum of protons, $\propto E^{-2.4}$, the radio data only marginally agree with the calculated synchrotron radio flux by secondary electrons (solid line in Fig.8.5), and therefore $n \leq 10^{-5}$ cm^{-3}. Such a strong constraint is a result of a very large amount of low energy, ≤ 10 GeV, protons (solid curve in Fig.8.6) - the parents of the secondary electrons producing the synchrotron radio emission. Since the gas density in knots is likely to be significantly larger, this constraint could be considered as an independent argument against the steep proton acceleration spectrum assumed in the model I.

The models II and III with flat E^{-2} type acceleration spectrum of protons allow much higher gas densities in the knot, namely, 10^{-3}, and 10^{-2} cm^{-3}, respectively. Because these numbers are quite close to the densities expected in the environments of large scale extragalactic jets, it seems an attractive idea of referring a fraction of, or even the entire, synchrotron spectrum of the knot to the synchrotron radiation of secondary "pp" electrons.

The fluxes of synchrotron radiation of secondary "pp" electrons calculated in the continuous injection regime over the last 10^7 yr are shown in Fig.8.7 for two combinations of model parameters: (a) $R = 1$ kpc, $B = 0.1$ mG, $\eta = 1$; and (b) $R = 0.25$ kpc, $B = 0.03$ mG, $\eta = 10$. In both cases the same "power-law with exponential cutoff" acceleration spectrum for protons is assumed with $\Gamma_0 = 1.5$ and $E_0 = 10^{18}$ eV. These parameters imply essentially different escape times of protons with corresponding breaks at $E_b \simeq 3.5 \times 10^{16}$ eV and 6.5×10^{13} eV for the cases (a) and (b), respectively. The resulting breaks in the synchrotron spectra of secondary electrons appear at 1 MeV and 1 eV, respectively. It is seen that case (a) explains the observed nonthermal spectrum from radio to X-rays quite well, while case (b) can explain only the radio and optical fluxes. Note that for chosen parameters the highest energy protons quickly escape the knot. This effect, combined with the low magnetic field, results in a dramatic drop of emissivity of the proton synchrotron radiation.

The synchrotron radiation by secondary "pp" electrons in radiogalaxies and AGN is not a new idea. The basic problem of this hypothesis - the deficit of an adequate target material - was clearly recognised in the early 1960s (see e.g. Burbidge et al., 1963). The above calculations face the same problem. In order to match the absolute fluxes of the observed nonthermal radiation from knot A1, we must assume a very large product of the gas density and the proton injection power, $nL_p = 8.5 \times 10^{46}$ erg/s cm^3. Thus

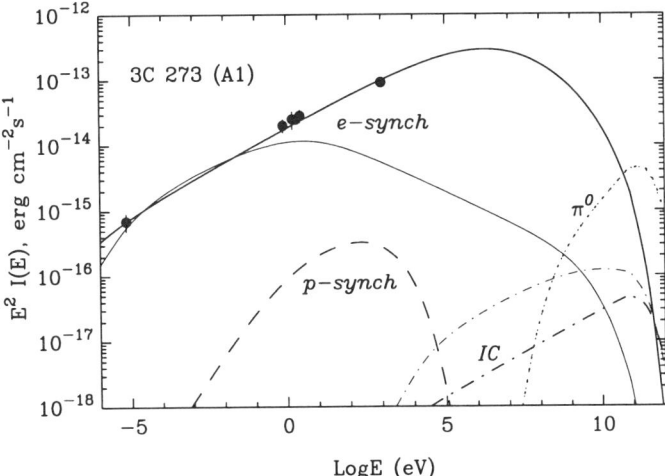

Fig. 8.7 Broad-band ninthermal emission of the knot A1 of 3C 273 produced directly by accelerated protons via the synchrotron radiation (dashed lines) and through the $pp \to \pi^0 \to \gamma$ channel (dotted lines), as well as by secondary electrons through the synchrotron (solid lines) and inverse Compton (dot-dashed lines) radiation. The heavy lines correspond to the case of slow escape (model I) and thin lines correspond to fast escape (model II). The model parameters are discussed in the text. The synchrotron radiation of protons for model II is too weak and is beyond the figure frames. The contributions of π^0-decay γ-rays for the models I and II are almost identical, and therefore cannot be distinguished. (From Aharonian, 2002b).

for any reasonable acceleration rate of protons $L_\mathrm{p} \leq 10^{47}$ erg/s, the plasma density in the knot should be close to 1 cm^{-3}. Such high densities are not supported, however, by depolarisation studies of the radio emission. In principle, this problem could be essentially softened if the knot is moving towards the observer with a relativistic Doppler factor $\gg 1$.

Obviously, the hypothesis of relativistic beaming has a universally "positive" effect for all radiation models. In particular it helps to reduce significantly the energy requirements for both the *proton synchrotron* and the *secondary-electrons synchrotron* models discussed above. Indeed, if the jet maintains the relativistic speed up to kpc scales, for a Doppler factor of about $\delta_\mathrm{j} \sim 5$, these requirements become so relaxed that we can successfully explain the entire radiation of the knot A1 by proton synchrotron radiation alone. At the same time, the relativistic beaming makes the interpretation of the nonthermal emission of the knot by secondary "pp" electrons an attractive alternative with a quite reasonable product of the gas density, the jet lifetime, and the proton acceleration rate $(n/0.1 \text{ cm}^{-3}) \, (\Delta T/10^7 \text{ yr}) \, (L_\mathrm{p}/10^{45} \text{ erg/s}) \sim 1$.

Gamma-ray emissivity of the knots. 3C 273 is a powerful γ-ray emitter with a peak in the spectral energy distribution at 1-10 MeV at the level of $\simeq 4 \times 10^{-10}$ erg/cm^{2}s; the energy flux at 1 GeV is about $\simeq 2 \times 10^{-11}$ erg/cm^{2}s. These fluxes exceed by several orders of magnitudes the flux of the proton synchrotron radiation (see Fig.8.5). Below we briefly discuss two other potential mechanisms for γ-radiation related to interactions of relativistic protons.

Besides the cooling through synchrotron radiation, the secondary electrons release their energy also through inverse Compton radiation which may extend to very high energies. However, for the parameters favourable for the proton synchrotron radiation, in particular for the magnetic field $B \geq 1$ mG, the contribution of the IC component is negligible, and therefore it does not appear in Fig.8.5. For a smaller magnetic field, $B \leq 0.1$ mG, the IC fluxes are higher (Fig.8.7), but still well below the sensitivity of γ-ray detectors. This is also true for "direct" π^0-decay γ-rays produced in pp interactions. The fluxes of this component are shown in Fig.8.5 and Fig.8.7. Note that the sharp drops of γ-ray fluxes above 100 GeV are caused by intergalactic photon-photon absorption. In calculations presented in Figs. 8.7 and 8.7, we adopted one of the recent models of the diffuse extragalactic background of Primack *et al.* (2001) which satisfactorily describes the observational data at near infrared and optical wavelengths - the most important band from the point of view of intergalactic absorption of γ-rays above 100 GeV (see Chapter 10)..

Although the flux of π^0-decay γ-rays is proportional to the ambient gas density n and the total amount of accumulated relativistic protons, $W_{\rm p} \simeq \Delta t L_{\rm p}$, it cannot be arbitrarily increased assuming larger values for the product $n \Delta t L_{\rm p}$. The π^0-decay γ-rays are tightly coupled with the synchrotron radiation of secondary ($\pi^{\pm}$-decay) electrons. Therefore the fluxes of π^0-decay γ-rays are (unavoidably) limited by the energy fluxes of synchrotron radiation of secondary electrons at low frequencies. In particular, since both GeV γ-rays and GHz radio photons are initiated by the same (10 to 100 GeV) primary protons, the flux of γ-rays around 1 GeV cannot exceed 10^{-15} erg/cm^{2}s, as is clearly seen in Fig.8.5 and Fig.8.7. Obviously this upper limit is insensitive to the relativistic beaming effects. This excludes any chance to explain the observed high energy γ-radiation from 3C 273 by pp interactions in the the large-scale jet. The observed MeV/GeV γ-radiation originates, most probably, in the inner jet.

8.3 Large Scale Jets Powered by Gamma Rays

In this section we discuss an interesting scenario when relativistic electrons are introduced throughout the length of the jet through the development of electromagnetic cascades in the 2.7 K CMBR and (perhaps) other local photon fields, initiated by extremely high energy γ-rays injected into the jet from the central object (Neronov et al., 2002). This scenario provides a natural and economic way to power the jets up to distances of 100 kpc and beyond.

The mean free path of γ-rays in the field of 2.7 K CMBR at $z \ll 1$ has a minimum $\Lambda_\gamma \approx 8$ kpc at $E_\gamma \simeq 10^{15}$ eV. At both lower and higher energies Λ_γ increases – sharply (exponentially) at $E_\gamma \ll 10^{15}$ eV, and slowly (almost linearly with energy) at $E_\gamma \gg 10^{15}$ eV, $\Lambda_\gamma \approx 14.6\, E_{\gamma,16} T_{2.7}^{-2}[1 + 0.7 \ln (E_{\gamma,16} T_{2.7})]^{-1}$ kpc, due to the decrease of the cross-section with the parameter $E_\gamma kT \gg m_{\rm e}^2 c^4$. Hereafter, $E_{\gamma,16} = E_\gamma/10^{16}$ eV, $T_{2.7} = T/2.7$ K. Thus, a γ-ray beam with a broad spectrum extending to 10^{18} eV can supply the jet with "desirable" VHE electrons along the jet, and thus power the jet up to distances ~ 100 kpc or even more. At cosmological epochs with $z \geq 1$, the mean free path is reduced by a factor of $(1+z)^2$.

Below we assume that VHE photons are injected into a cylindrical jet. Interacting with the 2.7 K CMBR as well as with the low-frequency radiation of the jet itself, these γ-rays initiate electromagnetic cascade supported by γ − γ pair-production and inverse Compton scattering. The synchrotron radiation causes degradation of energy of the electrons, and thus reduces the efficiency of the cascade development.

High resolution radio observations show that the magnetic field in large scale jets may be dominated by a regular component parallel to the jet axis, although there could be anomalous regions near the knots with oblique field components. Here we adopt a simplified picture assuming that the field consists of two - random and regular (aligned with the jet axis) - components, B_0 and B_r, respectively. Generally, if the first generation electrons appear at small angles to the regular magnetic field (i.e. the γ-ray beam from the central engine is directed along the jet), they will spiral with a small pitch angle θ in the magnetic field B_0 and at the same time get gradually deflected by the random magnetic field B_r. As long as the deflection of electrons does not exceed a few degrees, the synchrotron losses of electrons are dominated by the random magnetic field. At larger angles the synchrotron radiation in the regular field becomes more important. Moreover, for $\theta \geq 30°$ the latter also dominates over Compton losses. At the same time, because of the Klein-Nishina effect, the electrons of extremely high energies, $E \geq 10^{17}$ eV are basically cooled through the synchrotron radiation, even when they move at small pitch angles.

If the deflection of electrons in the jet is described as diffusion in pitch angle, the diffusion length Λ_{diff} can be estimated as $\Lambda_{\text{diff}} \approx 5\ (B_r/0.1\mu\text{G})^{-2}(E_e/100\text{TeV})^2 l_{\text{pc}}^{-1}(\theta_0/3°)^{-2}$ kpc, where l_{pc} is the correlation length of the random B-field in parsecs and θ_0 is the initial opening angle of the primary photon beam. When electrons cool down to ∼ 100 TeV, their propagation length is determined by Compton losses, $\Lambda_e = 0.3(T_{2.7})^{-4}(E_e/100\text{TeV})^{-1}$ kpc. On the other hand, when Λ_{diff} becomes comparable to Λ_e, the trajectories of the cascade electrons are effectively randomised by the irregular B-field at energies below $E_{\text{crit}} = 40(B_r/0.1\mu\text{G})^{2/3} l_{\text{pc}}^{1/3} T_{2.7}^{-4/3}(\theta_0/3°)^{2/3}$ TeV. These electrons are cooled through synchrotron and Compton losses, and form a bright "shell" around the "main stream" of the cascade.

Because the spectra of Chandra jets extend beyond 1 keV, for a reasonable value of $E_{\text{crit}} \sim 100$ TeV, which should be treated as a free parameter, the regular B-field must exceed 10 μG. One needs a sufficiently strong magnetic field in order to explain the X-ray data. The random field cannot play this role because it would destroy the cascade from the very beginning. On the other hand, effective cascade development in many cases is desirable in order to have a more or less homogeneous distribution of electrons, both in density and energy spectrum, throughout the jet. Assuming that the maximum energy of synchrotron radiation from the jet lies in the X-ray energy domain, we find the following relation between the random and regular magnetic fields: $(B_r/0.1\mu G)^2(B_0/10\mu G)^{3/2}l_{pc} \approx 7T_{2.7}^4(\theta_0/3°)^{-2}(h\nu_{max}/1keV)^{3/2}$. An example of evolution of photon and electron spectra in the "main stream" and in the "shell" are shown in Fig.8.8.

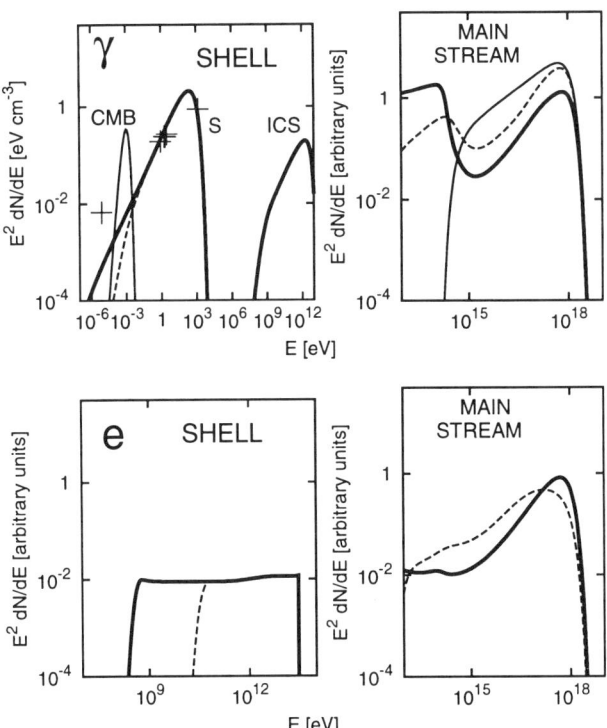

Fig. 8.8 Evolution of energy spectra of γ-rays (upper panel) and electrons (lower panel) in the "main stream" (right) and in the "shell" (left) of the jet. Upper panel (right): thin line – initial photons (at the base of the jet), dashed lines – after 50 kpc, thick lines – 500 kpc. Upper panel (left): the synchrotron/IC spectra of the "shell" (dashed line – after 10^6 yr, thick solid line – after 10^8 yr). For comparison the Chandra points of the knot A of the quasar 3C 273 are also shown, assuming 1 kpc for the knot size. The 2.7 K CMBR flux is shown by the thin solid line. Lower panel: evolution of electron spectra in the "main stream" and in the "shell". (From Neronov et al., 2002).

Generally, in most parts of the jet, the internal radiation fields are negligible compared to the external 2.7 K CMBR background. However, in the knots and hot spots the density of internal (synchrotron) radiation can be comparable or even exceed the density of the 2.7 K CMBR. This is the case, for example, in knot A1 of 3C 273. Therefore, interactions of the highest energy γ-rays with the radio synchrotron radiation may well dominate, at least in the bright knots. Besides, the random B-field in the knots can be stronger than in the rest of the jet, which would result in an increase of the energy $E_{\rm crit}$ at which the electron trajectories are randomised. To demonstrate this effect, in Fig.8.9 the evolution of the electron production rate in the knot A1 is shown. For comparison, the detected profile of X-rays is also shown. In order to avoid large energy release at small distances, a very hard, $E^{-1.5}$ type, γ-ray spectrum with exponential cutoff at 10^{19} eV is assumed.

It is seen that in the initial part of the jet, where synchrotron background is quite low, the injection rate of photo-produced electrons into the shell is small. But in knot A1 the rate increases rapidly. An observer who detects the jet from the side and does not see the "main stream" of the cas-

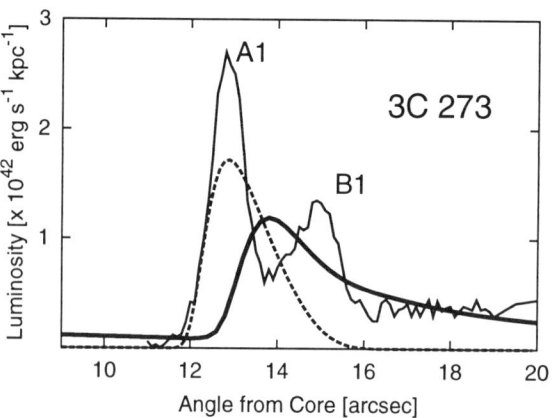

Fig. 8.9 Evolution of the jet's luminosity. The thin solid line represents X-ray data for 3C 273, the dashed line is intensity profile of the synchrotron background used in numerical calculation. The thick solid line is the injection rate of electrons with energy $E \leq E{\rm crit} = 10^{14}$ eV from the "main stream" of the electromagnetic cascade, i.e. the rate of "visible" electrons from the "shell". For the primary spectrum of γ-rays we assumed $E^{-1.5}$ with an exponential cut off at 10^{19} eV. The luminosity of the primary photon beam is 10^{44} erg/s. (From Neronov et al., 2002).

cade, may conclude that electrons are effectively accelerated in the knot. Thus, this model can give a reasonable explanation of the often observed effect of disappearance of the jet and its appearance again at larger distances in the form of knots and hot spots. Such a picture is clearly seen in the one sided jet of the radiogalaxy Pictor A (see Fig.8.2). The perfectly rectilinear jet stemming from the central sources disappears approximately half way to the hot spot located at a projected distanced of 250 kpc. Within this model the luminosity of the main body of the jet is explained by the interaction of $10^{15} - 10^{17}$ eV γ-rays with the 2.7 K CMBR, with the follow-up cascading and synchrotron radiation of electrons deflected from the "main stream" by the random magnetic field. Beyond 100 kpc the "fuel" in the form of $10^{15} - 10^{17}$ eV γ-rays is significantly reduced, but the photons of higher energy survive and continue to interact with the 2.7 K CMBR up to distances of about 1 Mpc. But since the secondary electrons have very large energies, they cannot be effectively deflected by the random magnetic field component, which in addition at such large distances could be significantly reduced. Therefore this part of the jet remains invisible for us. This smooth picture is dramatically changed when the $\geq 10^{17}$ eV γ-rays meet the hot spot. Due to the large density of low-frequency synchrotron radiation, these γ-rays are completely absorbed in the hot spot. The secondary electron-positron pairs immediately cool down in the strong magnetic field of the hot spot of about 10^{-4} G. The resulting synchrotron radiation continues up to 10^{15} eV, but if the jet's aspect angle is large, this radiation cannot be observed because the highest energy electrons do not change their direction until they are cooled down to energies ≤ 100 TeV. The quasi-isotropic (observable) synchrotron radiation of these electrons appears at X-ray and optical frequencies.

The formation of knots in the suggested model is a nonlinear process. The increase of synchrotron luminosity at some point of the jet would lead to an increase of the rate of ejection of electrons from the "main stream". This, in turn leads to a further increase of synchrotron luminosity and the formation of bright knots. The energy losses of shell electrons are dominated by the regular magnetic field which establish a standard $dn_e/dE_e \propto E_e^{-2}$ type spectrum of electrons. Correspondingly, the spectrum of synchrotron radiation has the form $\nu F_\nu \propto \nu^{0.5}$ with a maximum at $h\nu_{\max} = 1.6(B_0/10\mu G)(E_{\text{crit}}/100\text{TeV})^2$ keV. Note, however, that although the radio luminosity of the jet is directly related to the injection rate of electrons in the "shell", and therefore to the VHE γ-ray luminosity of the central source, it depends on some other factors as well, like the time of "operation" of the central source, the strength of the magnetic field in the jet, particle escape, *etc.* Also, the effect of energy-dependent diffusion makes the radio spectrum somewhat steeper. While the X-ray data tell us about the VHE γ-ray luminosity of the central source at the *present epoch*, the radio data rather reflect the history of its evolution.

One of the key questions to be addressed in this model is whether photons with energies larger than 10^{16} eV can be *effectively produced* and *freely escape* the dense photon environments in AGN. The central engines of AGN are believed to be powerful particle accelerators up to 10^{20} eV (see e.g. Kafatos *et al.*, 1981). The accelerated protons can produce VHE γ-rays interacting with the ambient photon fields (supplied, for example by the accretion disk around the massive black hole) through the photo-meson process. Since we need a *beamed* γ-radiation emitted at a small angle to the MHD jet, the protons should cross the photon field almost rectilinearly. Therefore, the condition of a high *proton-to-gamma* conversion efficiency in the production region of a linear size R implies $\tau_{p\gamma} = \sigma_{p\gamma} R\, n_{ph} \geq 1$. On the other hand, the produced γ-rays can effectively escape the production region if $\tau_{\gamma\gamma} = \sigma_{\gamma\gamma} R\, n_{ph} < 1$. Thus, the production region is transparent only for those γ-rays for which $\sigma_{\gamma\gamma} < \sigma_{p\gamma}$. This is possible

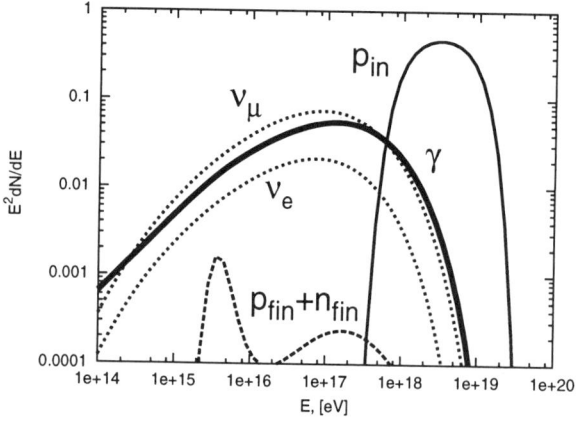

Fig. 8.10 The spectra of secondary photons (thick solid line), nucleons (dashed line) and neutrinos (dotted lines) after they escape the region of proton accelerator close to the base of the jet. The initial proton spectrum is shown by thin solid line. It is assumed that a beam of protons is injected into the region of size 10^{15} cm filled with blackbody radiation with temperature 10^4 K. Total energy emitted in the VHE γ-ray beam is 27 % of the energy contained in the primary photon beam normalised to $\int (E dn/dE) dE = 1$.

for very energetic γ-rays in a "hot" ambient photon gas. As an example, Fig.8.10 shows the spectra of protons and γ-rays emerging from a source filled with thermal radiation with $T = 10^4$ K which serves as a target for protons within a narrow energy interval between $\sim 10^{18}$ and 10^{19} eV. The photon-photon pair-production cross-section is characterised by the product of energies of interacting photons, $EkT/m_e^2 c^4 \sim 10^3 (E/100\text{TeV})$. When this parameter exceed 10^6 (at $E \geq 10^{17}$ eV), the pair-production cross-section is suppressed by several orders of magnitude (see Chapter 3), and becomes less than the (almost energy-independent) photo-meson production cross-section, $\sigma_{p\gamma} \simeq 10^{-28}$ cm^2. Thus, the $E \geq 10^{17}$ eV γ-rays are not only effectively produced at photo-meson interactions, but also are able to escape the source without catastrophic losses. The numerical calculations shown in Fig.8.10 demonstrate this possibility. A broader, e.g. power-law type, proton spectrum would result in effective production of less energetic, $E \leq 10^{16}$ eV, γ-rays as well. However, due to the increase of the photon-photon pair production cross-section, only a small fraction of these photons can escape the source.

The cascade can develop effectively in the jet if the strength of the random B-field does not exceed 1 μG, otherwise, the synchrotron radiation of electrons would dominate over the inverse Compton losses. On the other hand this component of the field should not be significantly below 1 μG in order to force the electrons to escape the "main stream" (otherwise the synchrotron radiation cannot be observed). At very large distances from the central source, the ambient (random) magnetic field can be reduced to a very low level, $B \sim 10^{-9}$ eV or less, therefore the cascade continues almost rectilinearly until the 10-100 TeV γ-rays start to interact effectively with the diffuse infrared background photons. These interactions would result in the formation of giant electron-positron Pair Halos with specific angular and energy distributions (see Chapter 11). If the AGN is located inside a rich cluster of galaxies, where the magnetic field could be as large as 1 μG, then the photo-produced electrons in the intracluster medium would result in nonthermal synchrotron radiation at EUV and X-ray bands (see Chapter 9). And finally, if the central source is a blazar, i.e. the jet is pointed to the observer, we may expect beamed γ-ray emission with the characteristic $E^{-1.5}$ type spectrum extending to 100 TeV. However, due to significant intergalactic absorption, the γ-rays will arrive with significantly distorted spectra. A possible implication of this mechanism for TeV blazars is discussed in Chapter 10.

8.4 Concluding Remarks

Although there is no alternative to the nonthernal origin of radiation of large scale AGN jet, it remains a theoretical challenge to explain the variety of morphological and spectroscopic peculiarities observed from these objects. The recent exciting discoveries by Chandra have added much to our knowledge of structures of these fascinating components of powerful radiogalaxies and quasars. Nevertheless they have not solved the old problems, and, in fact, have brought new puzzles.

The standard models that relate the nonthermal X-ray emission of distinct jet features to the synchrotron or inverse Compton radiation of directly accelerated electrons, are not free of problems. The inverse Compton or synchrotron-self-Compton models typically fail on energetic grounds, unless one assumes that the X-ray emitting regions are jet structures moving relativistically towards the observer. This interesting approach can be applicable for the jets in some objects, but needs further study based on larger source statistics. In any case, it seems unlikely that this mechanism can be applied to all Chandra jets.

While the limited efficiency of the inverse Compton models originates from the lack of sufficiently dense photon target fields, the electron-synchrotron model has just the opposite problem. It is "over-efficient" in the sense that the TeV electrons due to severe radiative losses have very short propagation lengths, and thus can hardly form diffuse X-ray structures on kpc scales. A possible solution could be that the electron acceleration takes place throughout entire volume of a knot or a hot spot. The operation of huge, kpc size accelerators in the jets seems, however, a non-trivial theoretical challenge.

The secondary origin of TeV electrons, produced more or less homogeneously in the knots by relativistic protons interacting with the ambient gas, is an interesting possibility. But this hypothesis requires an unacceptably large density of the ambient thermal plasma and very high proton acceleration rate, unless we assume that the X-ray emitting regions are relativistically moving jet structures with Doppler factor $\delta_j \gg 1$.

Alternatively, the secondary electrons can be produced in photo-meson interactions of protons with ambient photon fields. This process requires acceleration of protons to 10^{20} eV, and therefore makes the idea rather attractive in the context of the origin and sources of the highest energy particles observed in cosmic rays. However, this mechanism does not appear sufficiently effective to achieve the observed X-ray luminosities, as well as to match the spectra of most of the Chandra jets.

The proton-synchrotron radiation mechanism is a more effective channel for X-ray production associated with the hadronic component of accelerated particles. For a certain combination of parameters characterising the acceleration, propagation and radiation of very high energy protons, this model can provide effective cooling of protons via synchrotron radiation on quite comfortable timescales of about $10^7 - 10^8$ yr. This allows effective propagation of protons in the jet over kpc scales, and thus production of extended X-ray structures. Yet, the model allows high radiation efficiencies, and demands quite reasonable proton acceleration rates. Although these rates are comparable with the electron acceleration rates required in the electron-synchrotron models, the proton-synchrotron model implies much higher energy densities in the form of nonthermal particles and magnetic fields. For relativistic jets aligned with the line of sight the energy requirements can be reduced by two or three orders of magnitude. The success of the proton-synchrotron model largely relies on 3 principal assumptions: (*i*) acceleration of protons to energies at least $E_\mathrm{p} = 10^{18}$ eV; (*ii*) a strong ambient magnetic field, $B \geq 1$ mG; (*iii*) slow propagation of protons in the knots in the regime close to the Bohm diffusion with $\eta \leq 10$. A failure of any of these conditions would shift the characteristic energy range of synchrotron radiation towards lower frequencies, and, more importantly, would reduce dramatically the radiation efficiency to an unacceptably low level. And *vice versa*, any observational evidence in favour of the proton-synchrotron origin of the large-scale structures of the Chandra jets would imply a large magnetic field and acceleration of protons to extremely high energies.

Finally, an attractive alternative to all above models is the scenario when the jet is powered by external γ-rays through their interactions with the 2.7 K CMBR, and subsequent development of electromagnetic cascades penetrating through the regular and random fields in the the jet. Generally, nonthermal phenomena are associated with acceleration of relativistic particles. In this model, we have exactly the opposite picture when the external γ-radiation serves as "primary substance" for production of ultrarelativistic electrons. Consequently, there is no need for particle acceleration immediately in the jet. At the same time, this unconventional hypothesis contains certain components of standard scenarios. Indeed, because of its prime motivation to explain X-rays by the electron synchrotron radiation, this model can be considered as a "leptonic" model. On the other hand, this scenario is related to "hadronic" models, because the "primary" γ-rays of energy $10^{15} - 10^{19}$ eV can be produced only in hadronic interactions, most likely in the vicinity of massive black hole. The latter provides a direct link between the large scale jet and the central engine. The beam of ultra high energy γ-rays can be considered as an alternative to the Poynting flux assumed in the standard AGN models for extraction of energy from the rotating black holes. While the transformation of the Poynting flux to the kinetic energy of the outflow, and eventually (through termination shocks) to relativistic particles, remains an unsolved problem, the transformation of γ-rays to relativistic electrons can be realised effectively through the well known mechanism of γ-γ pair production.

Like many other nonthermal phenomena, the γ-rays from large-scale jets may provide straightforward answers to many questions concerning the origin and nature of particles responsible for the observed X-ray emission of the Chandra jets. Unfortunately, the energy fluxes of X-rays in these structures are quite small, below 10^{-12} erg/cm^2, therefore detection of the accompanying γ-radiation is not a simple task. For production of X-rays the inverse Compton models require only low energy electrons. Therefore, generally the γ-ray production is not an important issue in these models.

The model of synchrotron X-radiation by multi-TeV electrons automatically implies high energy γ-ray production due to the inverse Compton scattering on the 2.7 K CMBR. The magnetic field in these models significantly exceeds 1 μG, therefore the energy flux of γ-rays cannot exceed the flux of synchrotron X-rays. Because of the enhanced target photon density, the γ/X flux ratio could be significantly larger in the jets with large bulk motion Lorentz factors. For some combinations of model parameters, the inverse Compton losses can dominate over the synchrotron losses of electrons not only in the Thompson, but also in the Klein-Nishina regime. This would lead to the hardening of the synchrotron spectrum at X-rays, and production of IC γ-ray fluxes which could be marginally detected by GLAST and arrays of atmospheric Cherenkov telescopes. Since this model is proposed for interpretation of X-ray data from the Chandra jets, it automatically implies small viewing angles of jets, otherwise the Doppler factor would be unacceptably small.

Within the framework of the model of jets powered by very high energy γ-rays, we may expect larger γ fluxes from the Chandra jets directed close to the line of sight. Indeed, although in this model comparable fractions of the energy of primary γ-rays are transformed to the cascade γ-rays and synchrotron X-rays, the beamed γ-ray fluxes would be significantly larger compared to the synchrotron radiation of isotropised electrons. Remarkably, while for the jets with large viewing angles the synchrotron X-ray fluxes are not changed, the γ-ray fluxes would be dramatically reduced.

Finally, γ-rays are predicted by the proton-synchrotron model as a continuation of the proton synchrotron X-radiation to higher energies. The energy ranges and fluxes of γ-rays are determined by the strength of the magnetic field, the maximum energy of accelerated protons, and the level of turbulence in the plasma providing slow particle diffusion and escape. The fluxes of γ-rays can reach marginally detectable levels if the magnetic field is sufficient to provide the dominance of synchrotron losses over the particle escape. But in any case the γ-ray spectrum cannot exceed 1 GeV. At the same time, if a significant fraction of the nonthermal energy released in knots and hotspots eventually escapes the jet in the form of relativistic protons, one may expect interesting γ-ray phenomena in the cluster environments harbouring the runway protons. The relevant observable effects are discussed in the next chapter.

Chapter 9

Nonthermal Phenomena in Clusters of Galaxies

9.1 Nonthermal Particles and Magnetic Fields

Galaxy clusters, the largest gravitationally bound structures in the Universe, contain hundreds and thousands of member galaxies enveloped by massive diffuse hot gas with a temperature close to 10^8 K. These are huge reservoirs of thermal energy with X-ray luminosities of optically thin plasma ranging between 10^{43} and 10^{45} erg/s. Our information about the energy content in other forms – in the magnetic field and relativistic particles – is less certain. The recent Faraday Rotation Measure (RM) of a large sample of low-redshifted ($z \leq 0.1$) galaxy clusters, in combination with X-ray bremsstrahlung data, revealed unexpectedly strong magnetic fields with an average value in the inner 0.5 Mpc spheres of about 5 μG (Clarke *et al.*, 2001). The corresponding total magnetic energy $E_{\rm B} = B^2 R^3/6 \simeq 10^{61}$ erg is by two orders of magnitude less than the total thermal energy in the same cluster volume, $E_{\rm th} \simeq 10^{63} n_{-3} T_8$ erg, where $n_{-3} = n_{\rm e}/10^{-3} {\rm cm}^{-3}$ and $T_8 = T/10^8 K$ are the number density and temperature of the intracluster medium (ICM). In the densest known ICM environments within 150 kpc of rich cluster cores (also known as "cooling flow" regions) the magnetic fields could be as strong as 40 μG (Kronberg *et al.*, 2001).

The RM probes claiming large intracluster magnetic fields (ICMFs) have been intensively debated over the last several years, in particular in the context of the origin of nonthermal extreme-ultraviolet (EUV) and hard X-ray (HXR) "excess" emission claimed to detected from some galaxy clusters (Lieu *et al.*, 1996; Bowyer *et al.*, 1996; Boweyer and Berghöfer, 1998; Fusco-Femiano *et al.*, 1999, Bonamente *et al.*, 2001, Rephaeli and Gruber 2002, *etc.*). Magnetic fields larger than 1 μG would imply relatively small energy content in the form of relativistic electrons responsible for the synchrotron radio emission observed from the so-called radio halos. These diffuse structures that permeate the cluster centers with Mpc sizes, are characterised by a steep radio spectrum and low surface brightness with luminosities typically between $10^{40} - 10^{42}$ erg/s. The total energy in radio electrons is estimated as $W_e \simeq L_R t_{synch}$, where the characteristic time of synchrotron emission depends on the frequency and the magnetic field, $t_{synch} \simeq 7.3 \times 10^8 (\nu/1 \text{ GHz})^{-1/2} (B/1 \text{ }\mu\text{G})^{-3/2}$yr. For example, in a cluster with $L_R \simeq 10^{41}$ erg/s, the total energy of radio electrons $W_e \simeq 10^{57}$ erg. Although this is 4 orders of magnitude less than the magnetic energy, it does not yet imply strong departure from the equipartition condition. The deficit can be readily compensated by the hadronic component of cosmic rays. If the ratio of the proton and electron acceleration rates in galaxy clusters is close to that observed in the galactic disk (assuming, for example, that in both cases shock acceleration is responsible for production of cosmic rays), i.e. approximately 100:1, the proton-to-electron flux ratio at the present epoch in the ICM could be as large as 10^4, given the fact that during the source age, $t \sim t_H \simeq 1.5 \times 10^{10}$ yr, the protons are accumulated without significant collisional or escape losses, while the lifetime of radio electrons suffering synchrotron and Compton losses is limited to $10^7 - 10^8$ yr.

The energy content in relativistic electrons could be much higher if one interprets the EUV and HXR "excess" emission observed from galaxy clusters as inverse Compton scattering of electrons on the 2.7 K CMBR. For example, in the Coma cluster the total energy of electrons needed to explain the EUV "excess" should exceed 10^{61} erg. The IC interpretation of the EUV and HXR "excess" emission puts a robust upper limit on ICMF at the level of $0.1 - 0.2$ μG, i.e. an order of magnitude lower than inferred from RM probes. If so, the corresponding magnetic energy cannot significantly exceed 10^{59} ergs. This implies a significant deviation from equipartition, even without including in the balance the pressure contributed by the hadronic component of cosmic rays. Thus, the question of the nature of the EUV/HXR "excess" cannot be reduced to the problem of identification of radiation mechanism(s). It has more fundamental implications concerning the origin of intracluster magnetic fields, the balance between the thermal and nonthermal pressures in the ICM, the structure and strength of intracluster shocks, *etc.* If the ICMF does indeed exceed several μG, the *thermonuclear energy* of stellar sources would be not sufficient to maintain it on ~ 1 Mpc scales. Then, *gravity* seems to remain the only option for the primary source of intracluster fields (Clarke *et al.*, 2002), with the powerful

AGN/accretion disks as the most likely candidates to inject the magnetic field into the ICM (Kronberg et al., 2001). The fact that individual galaxies could inject up to 10^{61} erg of magnetic energy into the IGM was realized by G. Burbidge as early as the mid 50s. Moreover, soon it became clear that gravitation was the only viable source of this energy (Hoyle et al., 1964). Large-scale jets of radiogalaxies can deposit huge amounts of thermal and nonthermal energy through shock-heating and particle acceleration in the ICM (Inoue and Sasaki 2001). For ICMFs of ~ 0.1 μG, other sources, for example supersonic galactic winds (e.g. Völk and Atoyan 2000), can also play non-negligible role in generation of intracluster magnetic fields.

The discrepancy between the estimates of the magnetic field derived from Faraday rotation measures and required by the inverse Compton model of EUV and X-radiation could be to some extent tolerated by more sophisticated IC models and by a thorough treatment of observational selection effects. Apparently, independent channels of information about the ICMF are needed to resolve this important question. Such key information can be provided by IC γ-rays produced in the ICM.

The interest in γ-rays from clusters of galaxies is not limited to inverse Compton models and related topics. An equally important objective is connected with "hadronic" γ-rays expected from interactions of cosmic ray protons and nuclei with the ambient gas and the 2.7 K CMBR. There is little doubt that the relativistic electrons seen through their synchrotron emission should be accompanied by protons. As noted above, since protons do not suffer significant collisional or escape losses, the proton-to-electron ratio could be as large as 10^4, i.e. much larger than in our Galaxy. Actually, such a large energy component in relativistic protons in the ICM is highly desirable. Indeed, if the ICMF exceeds a few μG, as follows from RM probes, one perhaps should postulate an adequate energy density in cosmic ray protons, $w_{\rm CR} \sim 1$ eV/cm^3, in order to avoid uncomfortable departure from equipartition (see e.g. Brecher and Burbidge, 1972). This implies that the total energy of protons trapped in the ICM can be as large as 10^{62} erg. Assuming that this energy is accumulated during the age of the cluster, one finds the required average rate of proton acceleration and their injection into the ICM, $L_{\rm p} = W_{\rm p} H_0 \simeq 2 \times 10^{44}$ erg/s. This is a rather modest rate that can be readily provided in different particle acceleration scenarios. Moreover, as discussed in Chapter 8, the jets of individual powerful AGN and radiogalaxies can accelerate and inject relativistic protons into the ICM at a rate an order of magnitude higher. If such objects appear in a cluster more frequently than once per 10^7 yr (the typical lifetime of a strong AGN), the total energy in relativistic protons in the cluster could approach 10^{63} erg – significantly larger than the magnetic energy, but still comparable with the thermal energy of a standard galaxy cluster. The above estimate is valid if both the energy loss and particle confinement times exceed the source age. This is indeed the case for galaxy clusters.

The characteristic cooling time of relativistic protons due to inelastic nuclear interactions with the ambient gas is given by Eq. (3.13). The average number density of the intracluster gas in the inner ($\leq$ Mpc) parts of rich clusters, like Coma or Perseus, is estimated between 10^{-4} to 10^{-3} cm^{-3}, thus the cooling time of protons in ICM significantly exceeds the source age ($\approx$ Hubble time). On the other hand, the strong magnetic fields in these clusters provide effective confinement of protons up to 10^{15} eV or even higher energies (see below) during the source age. This allows us to estimate the efficiency of γ-ray production in the ICM,

$$\kappa = \frac{L_\gamma}{\dot{W}_p} = \frac{1}{3}(t_{pp}/t_H)^{-1} \simeq 0.03\ H_{60}\ n_{-3}. \qquad (9.1)$$

(the factor 1/3 takes into account that the energy lost by protons per interaction is equally distributed between π^0, π^+ and π^- mesons). Correspondingly, the integrated γ-ray energy flux from a cluster at a distance $d_{100} = d/100$ Mpc is expected to be at the level

$$f_\gamma = \frac{L_\gamma}{4\pi d^2} \approx 2.6 \times 10^{-11}\ n_{-3}\dot{W}_{45}H_{60}d_{100}^{-2}\ \text{erg/cm}^2\text{s}, \qquad (9.2)$$

where $\dot{W}_{45} = \dot{W}_p/10^{45}$ erg/s is the injection rate of protons in the ICM. An important consequence for the effective particle confinement in ICM is that the spectrum of protons accumulated over the Hubble time does not deviate from the acceleration spectrum. Therefore, the π^0-decay spectrum of secondary γ-rays simply repeats the proton acceleration spectrum. For a "standard" E^{-2} type particle acceleration spectrum, we should expect similar energy fluxes of γ-rays released per energy decade, i.e. by a factor of $\log(E_{\text{cut}}/m_p c^2) \geq 5$ less than the total flux given by Eq.(9.2), assuming that the acceleration extends to $E_p \geq 100$ TeV. Thus, for an average (over the Hubble time) injection rate of $\sim 10^{45}$ erg/s of protons into the ICM with density $\sim 10^{-3}$ cm^{-3}, the γ-ray flux per decade could be as large as several times $10^{-12}d_{100}^{-2}$ erg/cm^{2}s. Such fluxes can be best detected at the GeV and TeV energy bands from clusters located between 100 and $\simeq 300$ Mpc. For clusters beyond 300 Mpc, the absolute γ-ray fluxes become less that the detection threshold by modern γ-ray instruments. On the other hand, the detection of π^0-decay γ-rays from nearby clusters is hard because of the large angular extent $\simeq 0.5 d_{100}^{-1}$ degree of γ-ray emission. This concerns, however, gamma-rays produced in the central ≤ 1 Mpc region of the ICM.

It is equally interesting to search for hadronic γ-rays from the so-called "cooling flow" regions within 150 kpc of cluster cores. These regions are not only dense, $n \geq 10^{-2}$ cm^{-3}, but also are characterised by very large magnetic fields, $B \geq 40\mu G$ (Kronberg et al., 2001). The first factor increases the γ-ray production rate, while the strong magnetic field allows effective confinement of protons, albeit with a smaller linear size of the region. These two factors together provide good conditions for copious γ-ray production. On the other hand, the angular extensions of these regions remain quite small even for nearby clusters. This increases dramatically the chances for the detection of high energy γ-rays from the cores of nearby clusters. The core of the Virgo cluster at a distance of about 15 Mpc with the famous radiogalaxy M87 at its center is a perfect candidate for such studies.

Because of limited flux sensitivities, the detection of γ-rays would be very difficult at other, in particular at MeV, energies - both from the core and from the region ≤ 1 Mpc. At the same time, one may expect significant synchrotron fluxes from the secondary $\pi^{\pm}$-decay electrons and positrons (e.g. Blasi and Colafrancesco 1999). Since during the Hubble time these electrons are effectively cooled down to ≤ 1 GeV, the synchrotron radio fluxes of clusters of galaxies allow independent constraints on the total energy budget of hadronic cosmic rays.

The spectra of cosmic rays accelerated in individual galaxies, e.g. in AGN jets or by large-scale intracluster shocks, can extend to 10^{18} eV and beyond. At these energies the interactions of protons with the 2.7 K CMBR become more important than with the intracluster gas. The cooling time of $\geq 10^{18}$ eV protons caused by Bethe-Heitler pair-production is comparable with the source age. Moreover, at highest energies approaching to 10^{20} eV, the protons start to interact with the 2.7 K CMBR through photo-meson processes on timescales $\leq 10^8$ yr, thus the energy of these protons is effectively released in the form of electromagnetic radiation, albeit the increasingly large non-radiative losses caused by the particle escape. The interference of these processes results in formation of several nonthermal radiation components with quite complicated spatial and spectral distributions which carry important information about the highest energy particles, as well as about the intracluster and intergalactic magnetic fields on scales from several 100 kpc to ≥ 10 Mpc.

9.2 Inverse Compton and Bremsstrahlung Models

9.2.1 *Inverse Compton models*

High energy γ-ray production in the ICM through ultrarelativistic electrons upscattering the 2.7 K CMBR is a very effective radiation channel. At GeV and higher energies the Compton cooling of electrons dominates over other radiative and non-radiative energy losses, thus this process proceeds with almost 100 per cent efficiency. This can be seen in Fig. 9.1 where the characteristic energy loss timescales of electrons in a "typical" galaxy cluster are shown. Above 1 GeV, the IC cooling time becomes less than both the age of the source (supposed to be close to the Hubble time) and the diffusive escape time of electrons from the cluster. Thus, even for a strong ICMF of about 3μG, half of the electron energy is released through the IC channel, the other half being radiated away through synchrotron radiation.

The currently most popular interpretation of EUV and hard X-rays by IC scattering on the 2.7 K CMBR requires relativistic electrons with Lorentz factors between 10^2 to 10^4. For an ICMF of $0.1 - 1$ μG, this energy interval partly overlaps with the lower edge of energy of electrons responsible for radio emission observed at frequencies between 10 MHz and 1.4 GHz. This allows us to compare the expected synchrotron and IC fluxes of these electrons with the observed nonthermal emission at radio and X-ray bands, and thus derive a *model-independent* estimate of the ICMF. For example, the radio and X-ray fluxes observed from the Coma cluster, $f_{\rm rad} \simeq 10^{-14}$ and $f_{\rm X} \simeq 10^{-11}$ erg/cm^{2}s, give $B \simeq \sqrt{8\pi w_{2.7\rm K}(f_{\rm rad}/f_{\rm X})} \simeq 0.1$ μG.

Detailed numerical calculations within the one-zone (single-electron population) model performed for the Coma cluster by Atoyan and Völk (2000) are shown in Fig. 9.2. It is assumed that the electrons are continuously injected into the ICM during the last 3×10^9 years. The calculations are done for a fixed value of the magnetic field of $B = 0.12$ μG and for electron injection spectra with 3 different power-low indices - $\alpha_{\rm e} = 2.2, 2.3, 2.6$ and an exponential cutoff at $E_0 \simeq 1$ TeV. The total energy in electrons is defined from the condition required to match the radio flux: $W_{\rm e} = 6.3 \times 10^{61}$ ergs and 1.8×10^{62} ergs for $\alpha_{\rm e} = 2.3$ and 2.6, respectively.

For explanation of the radio fluxes up to ~ 1 GHz the power-law electron spectrum should continue to 100 GeV, but not much beyond. The steepening (or cutoff) of the synchrotron spectrum above 1 GHz seen in Fig. 9.2 requires a cutoff in the electron spectrum below 1 TeV. While the cutoff energy in the electron spectrum does not have an impact on the IC spectrum at EUV and X-ray energies (the highest energy point at 80 keV requires electrons with energy ≤ 10 GeV), its location is crucial for IC γ-ray fluxes above 1 GeV. In particular, the upper limit on the cutoff in the electron spectrum at 1 TeV implies that the γ-ray flux above $\sim 4kT(E_0/m_ec^2)^2 \simeq 5$ GeV is suppressed. This is seen in Fig. 9.3 where the IC fluxes are calculated for two different combinations of magnetic field and the power-law index of the electron injection spectrum (Atoyan and Völk, 2000). Although both combinations match the EUV and X-ray fluxes well, they predict essentially different γ-ray fluxes. In the same figure the contributions from the electron bremsstrahlung calculated for the average intracluster gas density $n = 10^{-3}$ cm^{-3} are also shown. Note that the

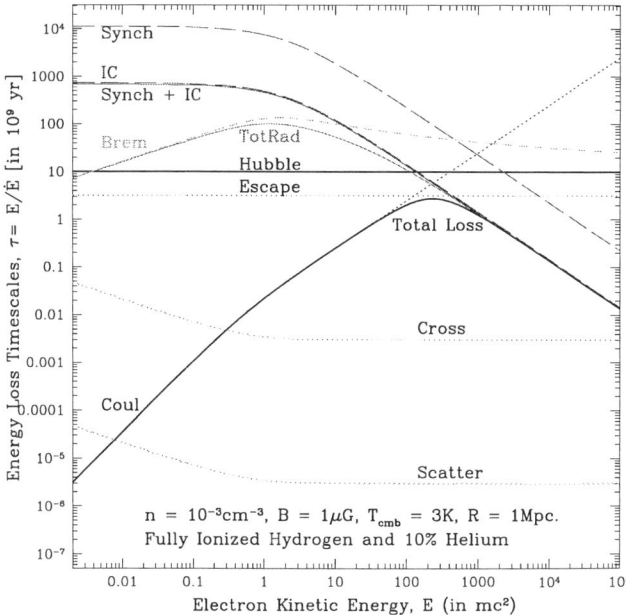

Fig. 9.1 The electron energy loss timescales in a typical galaxy cluster calculated for an ICM with magnetic field 1μG and a plasma average number density $n = 10^{-3}$ cm^{-3} (from Petrosian 2001). The interaction times are shown by solid and dashed curves. The dotted curves represent the electron crossing time $T_{\rm cross} \sim R/(c\beta)$ across a region of size $R \sim 1$Mpc, the scattering time $\tau_{\rm scat} \sim \lambda_{\rm scat}/(c\beta)$ for a constant scattering mean free path $\lambda_{\rm scat}$, and the energy-independent escape time $T_{\rm esc} \sim T_{\rm cross}^2/\tau_{\rm scat}$.

energy of bremsstrahlung photons is comparable to the energy of parent electrons ($E_\gamma \sim 1/2 E_e$), therefore the energy region of electrons responsible for the EUV and X-ray fluxes perfectly overlaps with the electrons that produce bremsstrahlung MeV and GeV γ-rays. Therefore the predictions of bremsstrahlung fluxes at γ-ray energies based on the EUV and X-ray data are quite robust, the only free parameter being the number density of the intracluster gas. The results presented in Fig. 9.3 show that the bremsstrahlung γ-rays should be detected by GLAST for any reasonable density of the intracluster gas exceeding $n \geq 10^{-4}$ cm^{-3}. Thus, GLAST should be able to provide a crucial test for the hypothesis of IC origin of the EUV and X-ray "excess" emission from rich galaxy clusters. And *vice versa*, any detection of ≥ 100 MeV γ-rays by GLAST from the Coma cluster at the flux level $\simeq 10^{-12}$ erg/cm^2 would prove the nonthermal origin of the "excess" EUV emission.

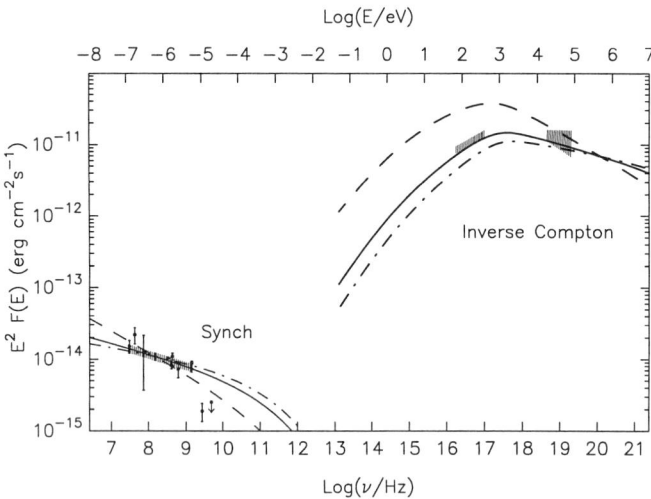

Fig. 9.2 Synchrotron and IC fluxes from the Coma cluster, assuming that the relativistic electrons are continuously injected into ICM over the last 3×10^9 yr (from Atoyan and Völk, 2000). Calculations are performed for 3 different power-law indices of the injection spectrum: $\alpha_{\rm inj} = 2.2$ (dot-dashed curves), $\alpha_{\rm inj} = 2.3$ (solid), and $\alpha_{\rm inj} = 2.6$ (dashed). For all three cases an exponential cutoff is assumed at $E = 1$ TeV. The assumed intracluster magnetic field is $B = 0.12\,\mu{\rm G}$. The absolute injection rates of electrons are determined from the normalisation to the observed radio flux at 400 MHz. The compilation of radio data is from Deiss *et al.* (1997). The hatched regions in the EUV and X-ray domains are the fluxes reported by Lieu *et al.* (1999) and Fusco-Femiano *et al.* (1999).

As discussed above, the radio emission from the central parts of clusters sets an upper limit on the expected flux of high energy IC γ-ray above a few GeV. At larger scales, significantly exceeding 1 Mpc, where the magnetic field could be below 0.1 μG, the γ-ray luminosity is no longer limited by radio observations. Therefore IC γ-ray fluxes from extended regions of galaxy clusters becomes an interesting issue linked to fundamental questions regarding the formation of large scale cosmological structures and gravitational shocks in clusters and superclusters. In structure formation models, the ICM is heated by shocks produced when the baryonic gas falls into the gravitational potential wells of hierarchically merging dark matter. Some workers (e.g. Loeb and Waxman 2000, Miniati, 2003) have recently argued that the same shocks can accelerate electrons up to TeV energies and that the IC radiation of these electrons, upscattering photons of the 2.7 K CMBR, can be best observed at γ-ray energies from regions beyond 3 Mpc, where the accretion shocks occur (see, however, Gabici and Blasi, 2003a). The superposition of contributions of individual galaxy clusters may constitute a major fraction of the diffuse extragalactic γ-ray background (Loeb and Waxman 2000; see however Miniati, 2002; Gabici and Blasi, 2003b). The most attractive candidates to be detected seem to

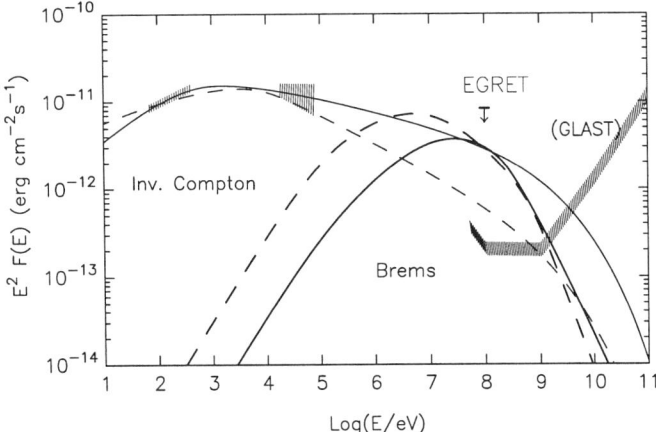

Fig. 9.3 Bremsstrahlung and IC radiation fluxes calculated for the injection of electrons with $\alpha_{\rm inj} = 2.3$ over the last $\Delta t_{\rm inj} = 3\,{\rm Gyr}$ assuming $B = 0.1\,\mu{\rm G}$ (solid curves), and $\alpha_{\rm inj} = 2.6$, $\Delta t_{\rm inj} = 1\,{\rm Gyr}$ assuming $B = 0.15\,\mu{\rm G}$ (dashed curves). A mean gas density of $n_{\rm gas} = 10^{-3}\,{\rm cm}^{-3}$ in the ICM is assumed. The flux upper limit of EGRET (Sreekumar et al., 1996), as well as the flux sensitivity of GLAST are also shown. (From Atoyan and Völk, 2000).

be nearby rich galaxy clusters like the Coma cluster. Given the large angular size of the γ-ray production region, the minimum detectable flux by GLAST and by Cherenkov telescopes exceeds 10^{-11} erg/cm²s. This implies that the γ-ray luminosity, and therefore the acceleration rate of electrons by accretion shocks in these extended structures can be probed on levels down to 10^{43} erg/s. The quantitative estimates of the maximum energy of accelerated electrons, $E_{\max} \simeq 20(kT/1\text{ keV})^{1/2}(\text{B}/0.1\mu\text{G})$ TeV (Loeb and Waxman 2000), as well as more detailed calculations of accretion shocks (Miniati, 2003) show that this radiation cannot extend beyond 1 TeV.

From the discussion of the previous section it follows that there is no room to accommodate strong ($\geq 1\ \mu$G) intracluster magnetic fields in models interpreting the hard X-ray "excess" as emission by IC scattering of moderately energetic (GeV) electrons on the 2.7 K CMBR. Although, this discrepancy could be somewhat reduced within more sophisticated IC models and by adequate treatment of observational selection effects (Petrosian, 2001; Rephaeli and Gruber, 2002), it is important to explore alternative approaches and mechanisms for explanation of the EUV/HXR excess. Fortunately, there are not many options in this regard assuming that the thermal origin of the X-ray "excess" can be firmly excluded. These are X-ray production mechanisms associated with either subrelativistic (keV) or ultrarelativistic (TeV) electrons.

9.2.2 *Nonrelativistic bremsstrahlung*

Motivated by the discrepancy between magnetic field estimates derived from RM observations and those required for the IC model of X-rays, the nonthermal bremsstrahlung (NTB) of subrelativistic electrons has been suggested as a possible radiation mechanism (Enßlin *et al.*, 1999, Sarazin and Kepner 2000, Blasi 2000). However, this explanation suffers a major flaw because it requires an unacceptably large energy input (Petrosian 2001). As discussed in Chapter 3, in the nonrelativistic regime NTB is quite an inefficient radiation mechanism and can hardly be applied to objects/phenomena in which the radiation luminosity constitutes an appreciable fraction of the maximum available power, determined by $W_{\text{total}}/t_{\text{dyn}}$, where W_{total} is the total energy budget and t_{dyn} is the characteristic dynamical time - typically $\leq 10^{64}$ erg and $\sim 10^{10}$ yr in the case of galaxy clusters. At first glance the characteristic time of NTB at subrelativistic energies is comparable to the Compton cooling time of GeV electrons producing X-rays in the same energy interval. However, there is an obstacle to this model caused by a dissipative process acting in parallel - Coulomb interactions which proceed at a much higher rate. The yield of the nonrelativistic NTB is a well-defined quantity independent the details of model parameters. Namely, only a fraction $\leq 10^{-5}$ of the kinetic energy of electrons is radiated via bremsstrahlung, while the rest goes to the Coulomb heating of the gas (see Fig. 9.1). Given the luminosity of nonthermal X-rays from Coma, $\sim 10^{43}$ erg/s, this model seems to be in trouble. One may argue that the injection of nonrelativistic electrons into ICM is associated with a relatively recent event which occurred $t \leq t_{\rm Coul}(\leq 100 \text{ keV}) \simeq 10^6$ yr ago (see Fig. 9.1), so that the time was too short for heating of the ICM. Even so, it is not easy to explain the injection of suprathermal electrons with total energy of $L_{\rm X} t_{\rm NTB} \simeq 10^{61}$ erg over just the last 10^6 yrs.

Nevertheless, one cannot firmly exclude nonrelativistic bremsstrahlung as a source of the hard X-ray "excess" emission. The above conclusion about very low radiative efficiency is true if the energy of suprathermal electrons is well above the temperature of ambient plasma. This is not the case of clusters of galaxies. The energy in suprathermal electrons for production of hard X-rays is less than 100 keV, i.e. not far from the average energy of thermal particles with temperatures of about 10 keV. This reduces the Coulomb interaction rate, and correspondingly increases the efficiency of bremsstrahlung. Thus, even slight deviations of electron distribution from the Maxwellian tail would result in a "quasi-thermal" bremsstrahlung excess. Recently, Liang *et al.*, (2002) have discussed the formation of such a tail of quasi-thermal particles due to *in situ* acceleration of particles through turbulence. If this model is correct, the energy output in quasi-thermal electrons needed to explain the hard X-ray "excess" from Coma, could be pushed down to a quite reasonable level of the order of 10^{45} erg/s (Liang *et al.*, 2002). Note, however, that the "excess" EUV radiation cannot be explained, by definition, by any modification of the bremsstrahlung model of suprathermal particles.

9.3 Synchrotron X- and γ-rays of "Photonic" Origin?

Synchrotron radiation of multi-TeV electrons is a prolific X-ray production mechanism. It works perfectly in various astrophysical environments, from pulsar-driven nebulae and shell-type supernova remnants to small (sub-pc) and large (multi-kpc) jets of AGN. Therefore, it is natural to invoke this mechanism also for interpretation of the reported nonthermal EUV and X-ray "excess" emission of galaxy clusters, especially because of above discussed difficulties associated with the IC and NTB models. At first glance this seems to be a reasonable assumption, given that the characteristic time of X-ray production, $t_{\rm sy} \approx 5 \times 10^4 (E_{\rm x}/1~{\rm keV})^{-1/2} (B/1~\mu{\rm G})^{-3/2}$ yr, is shorter than any radiative or non-radiative timescale characterising energy losses of electrons in these objects. On the other hand, application of this radiation mechanism to galaxy clusters in a standard manner encounters several problems . The energy of electrons needed to produce a synchrotron photon of energy E, is $E_{\rm e} \simeq 250 (E_{\rm x}/1~{\rm keV})^{1/2} (B/1~\mu{\rm G})^{-1/2}$ TeV. Thus, in order to explain the reported nonthermal X-ray spectrum of the Coma cluster extending to 80 keV, one must require extremely large values of the maximum acceleration energy, $E_{\rm max} \sim 10^3$ TeV. Although very high energy electrons can be effectively accelerated by strong accretion or merger shocks (Miniati 2003), even for the most favourable conditions the maximum electron energy does not reach 10^3 TeV; for any reasonable shock-acceleration rate, the lifetime of electrons is too short to boost them to such high energies. Also, the short lifetime does not allow electrons to propagate away from the acceleration sites by more than 1 kpc. Thus, the diffuse character of the observed UV and hard X-rays cannot be explained, unless we assume continuous (in space and time) acceleration of electrons throughout the cluster. Finally, since the intracluster shock acceleration generally implies production of shocks starting at the very beginning of cluster formation, the electrons will be cooled down to 100 MeV (this energy corresponds to the intersection point of the curves relevant to the Hubble time and the radiative cooling times shown in Fig. 9.1). Consequently, we cannot avoid overproduction of radio emission, given the fact that the energy flux of radio emission, for example from the Coma cluster, is three orders of magnitude below the EUV and hard X-ray fluxes.

To overcome the first and second difficulties, we should assume *in situ* particle acceleration with a rate significantly higher than the standard diffusive shock acceleration can provide. In order to suppress the radio emission, one must assume a low-energy cutoff in the electron injection spectrum E_*. However, this assumption does not help much, because the synchrotron and IC losses result in formation of a standard E^{-2} type differential spectrum down to $E \ll E_*$. Thus, even for the cutoff energy E_* as large as 10 TeV, we can avoid the overproduction of radio emission only by assuming that the electron injection was a relatively recent ($\leq 10^7$ yr) event.

It is clear that these are not "innocent" assumptions. They can hardly be accommodated in the current models, therefore any attempt to explain the EUV and X-ray "excess" emission by electron synchrotron radiation requires a non-standard approach to the problem.

These problems can be loosened to a large extent, if we assume that the ultrarelativistic electrons have a secondary origin, namely being products of interactions of hadronic cosmic rays with the ambient matter or photon fields. Protons do not suffer severe radiative losses, therefore the spectra of these particles, accelerated by large-scale shocks or injected into the ICM by individual radiogalaxies, can be extended to 10^{17} eV, or perhaps even higher energies. Both the attractive features and the shortcomings of hadronic models are discussed in the next section. Here we discuss a quite unusual, but perhaps more effective scenario, assuming that the secondary electrons have "photonic" origin (Timokhin et $al.$, 2003). This, at first glance "exotic" hypothesis, can nevertheless be realized, if the individual member-galaxies produce and inject into the ICM γ-rays with spectra extending to 100 TeV and beyond. It is somewhat similar to the scenario discussed in Sec.8.3 in the context of possible energisation of large-scale AGN jets by very high energy γ-rays. Production of such energetic photons is not a trivial processes, and requires a specific combination of very special conditions. At the same time, we know that the γ-ray spectra of TeV blazars *do extend* to energies of 20 TeV (see Chapter 2). Although detection of higher energy photons from these objects is prevented by intergalactic photon-photon absorption, the γ-ray spectra corrected for intergalactic absorption, e.g. from Mkn 501, do not show any tendency for a cutoff up to 20 TeV. In addition, γ-rays of higher energies can be produced in other scenarios that might take place in the central engines and large-scale jets of AGN, as discussed in Sec.8.3. Another (possible) channel for production of multi-TeV γ-rays could be synchrotron radiation of extremely high energy, $E \geq 10^{20}$ eV, electrons and positrons associated with the decay-products of hypothetical super-heavy particles from the Dark Matter halos that provide the major contribution to the gravitational potential of galaxy clusters. Below we will not discuss the origin of the very high energy γ-rays – parents of electrons and positrons in ICM – but simply hypothesise their existence. These γ-rays can be only formally called *primary*, because they themselves are products of other processes. Actually, these are intermediate particles through which the primary energy released in the cores of AGN, at interactions of highest energy protons with 2.7 K CMBR (within a galaxy cluster), or at decays of heavy Dark Matter particles, can be effectively converted to ultrarelativistic electrons *everywhere* in the ICM, and eventually be radiated away in the form of EUV and X-rays.

The energy of the secondary, pair-produced electrons, and therefore the typical energy of their synchrotron radiation strongly ($\propto E^2$) depends on the high energy end of the spectrum of primary gamma-radiation. Below we discuss 2 different cases when spectra of primary γ-rays extend to (i) relatively modest energies of about 10^{15} eV or less, and (ii) to ultra-high energies exceeding 10^{16} eV. Although in both cases the proposed mechanism works with almost 100 per cent efficiency transforming the energy of primary γ-rays to synchrotron radiation of pair-produced electrons, the resulting synchrotron radiation appears in essentially different energy bands. We assume that a source or an ensemble of sources in the central part of the cluster radiate γ-rays with a constant rate and with an energy spectrum given by a "power-law with quasi-exponential cutoff": $\dot{N}_\gamma(E) \propto E^{-\Gamma} \exp[-(E/E_0)^{-\beta}]$. In calculations the flux of the diffuse extragalactic background close to the model 1 in Fig. 10.1 (Chapter 10) is assumed. Then distributions of electrons and positrons (injection spectra) produced by γ-rays from the central source(s) are calculated in each point of the cluster. The time-dependent spectra of pair-produced electrons are obtained taking into account their synchrotron and IC energy losses. Finally, the spectra of synchrotron and IC radiation of these electrons are calculated. It is assumed that the secondary, pair-produced electrons are immediately isotropized. The propagation effects of these electrons are ignored, i.e it is assumed that the electrons "die", due to severe synchrotron losses, not far from their birthplace. For any reasonable intracluster magnetic fields this is a quite acceptable approximation. The propagation effect may become important for low-energy electrons responsible for radio emission. Within the assumed model we do not attempt to explain the radio emission, but rather assume that radio emission is due to relatively low energy electrons associated with nonthermal phenomena, that took place long time ago (Enßlin and Sunyaev, 2002). At the same time, radio emission produced by the cooled low energy (MeV/GeV) electrons should not exceed the observed fluxes. This is an important condition which sets robust upper limit on the active time of operation of γ-ray sources and on their energy spectra.

In Figs. 9.4 and 9.5 the synchrotron and IC spectra of pair produced electrons in the Coma cluster are shown, calculated for two values of the average field: (a) $B = 6\ \mu\text{G}$ and (b) $B = 0.3\ \mu\text{G}$. For both cases the following parameters of the initial γ-ray spectrum are assumed: $\Gamma = 2, \beta = 1/2, E_0 = 700$ TeV. It is assumed that injection of γ-rays with a quasi-constant rate into ICM have been started 10^7 yr ago. The fluxes shown are corrected for the intergalactic absorption due to interactions with the diffuse extragalactic background, assuming that the source is located at a distance of 100 Mpc. These figures demonstrate that, for the chosen combination of parameters, it is possible to explain the EUV and X-ray radiation of Coma by synchrotron radiation of electrons of "photonic" origin. VHE γ-ray luminosities 2×10^{45} and 3×10^{46} erg/s are requited to support the reported EUV and X-ray fluxes for the cases (a) and (b), respectively. Note that assuming a low energy cutoff in the γ-ray spectra below 100 TeV one may significantly reduce these energy requirements. From the point of view of energy requirements, the preference obviously should be given to the case of strong magnetic field, which also better agrees with the estimates derived from RM probes.

Actually, the case of low magnetic field, which predicts energy flux of inverse Compton TeV γ-rays on the level of $\sim 2 \times 10^{-10}$ erg/cm^{2}s, seems to be excluded by TeV observation. Such a flux hardly could be missed from the long-term observations of Coma by the HEGRA system of Cherenkov telescopes, even taking into account the extended character of this emission. The case of strong magnetic field predicts significantly reduced secondary γ-ray flux, which however remains sufficiently high to be probed by the next generation of Cherenkov telescope arrays.

Gamma-rays with energy above 100 TeV interact mainly with 2.7 K CMBR photons. The interactions with the diffuse FIR background radiation dominate at lower energies (see Fig. 1.4). Because at $E \leq 100$ TeV the mean free path of γ-rays for any FIR background model significantly exceeds 1 Mpc, the main contribution to the electrons produced within the central ≤ 1 Mpc regions of galaxy clusters comes from γ-rays with energy more than 100 TeV. Therefore, the results only slightly depend on the FIR background model.

The spectrum of electrons produced when γ-rays with a power-law distribution interact with a target photon field with a narrow distribution is described by Eq. (3.27). Starting from the minimum energy at $E_* = m_e^2 c^4 / 4\overline{h\nu} = m_e^2 c^4 / 12kT \simeq 65$ TeV, the electron spectrum sharply rises to $E_m \simeq 2.4 E_* \simeq 150$ TeV, and then at $E_e \gg E_m$ decreases as $E_e^{-(\Gamma+1)} \ln E$ (Γ is the photon index of injected γ-rays and $h\nu$ is the mean energy of target photons). The radiative cooling of electrons due to synchrotron and IC losses quickly leads to the formation of steady state spectra. Below E_* it has a standard E^{-2} type form, and correspondingly the synchrotron radiation has photon index 1.5. This spectrum continues down to the break point determined from the condition $t_{\rm rad} = t_0$, where t_0 is the age of the source. Although in this energy region the general spectral shape of the radiation is almost independent of the strength of magnetic field, the position of the break does depend on the magnetic field. For a source of age 10^7 yr and ICMF of 6μG, the break in the spectrum appears at frequencies around 30 GHz. For a lower field, $B = 3$ μG the break point is shifted to a frequency an order of magnitude higher. At energies above $E_{\rm m}$ the cooled electron spectrum depends strongly on the magnetic field. For $B = 6$ μG, the energy losses are dominated by synchrotron radiation, therefore the cooled electron spectrum is approximately proportional to E^{-4} (omitting the logarithmic term). The corresponding synchrotron radiation in this interval has a spectrum with photon index 2.5. The transition from the $\nu^{0.5}$ dependence to $\nu^{-0.5}$ dependence in the νF_ν ($\equiv E^2 f(E)$) presentation is clearly seen in Fig. 9.4.

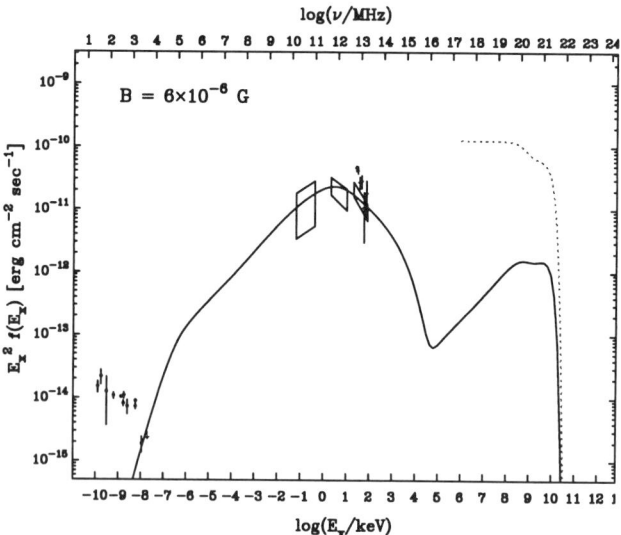

Fig. 9.4 Spectral Energy Distribution of synchrotron and IC radiation components of secondary electrons and positrons that are produced in the Coma cluster at interactions of "primary" γ-rays with the 2.7 K and diffuse extragalactic background photons (from Timokhin et al., 2003). It is assumed that "primary" γ-rays with high energy spectral cutoff at 700 TeV have been injected into ICM over the last 10^7 years with a rate 2×10^{45} erg/s. . The average magnetic field in the cluster up to 1 Mpc from the center is assumed to be 6μG. The radio, EUV and X-ray data are taken from Deiss et al. (1997), Lieu et al. (1999), and Fusco-Femiano et al. (1999), respectively.

For a magnetic field of $0.3\mu G$ the energy losses are strongly (by a factor of 100) dominated by IC losses on the 2.7 K CMBR. The synchrotron spectrum at low frequencies above the breaking point behaves in a similar manner to the previous case, $\nu F_\nu \propto \nu^{0.5}$ (see Fig. 9.5). The differences in the spectra of synchrotron radiation for low and high ICMFs become significant at highest, EUV and X-ray, frequencies. This is explained by the fact that in the case of a low magnetic field the energy losses of electrons responsible for the EUV and X-rays are dominated by the Compton scattering which proceeds in the Klein-Nishina regime. This makes the electrons spectrum *harder* compared to the injection spectrum.

Because of the large uncertainties, the current EUV and X-ray data do not give preference to synchrotron models with small or large intracluster fields. Hopefully, future detailed spectrometric and morphological studies in the broad energy interval from 0.1 keV to 100 keV will be able to provide observational tests between the large and small ICMF. The spatial distribution of synchrotron X-rays in the framework of this model depends on the energy distribution of γ-rays, radial distribution of the magnetic field, as well as on the spatial distribution of "primary" γ-ray sources.

In Fig. 9.6 the surface brightness distribution of synchrotron radiation

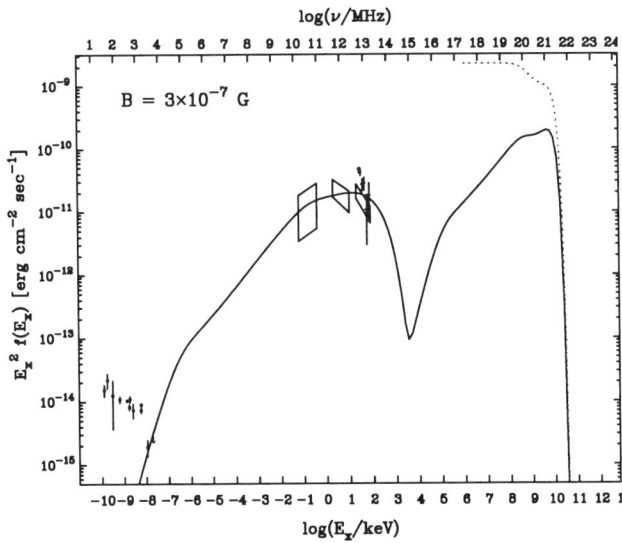

Fig. 9.5 The same as in Fig. 9.4, except for the strength of the intracluster magnetic field, $B = 0.3\mu G$, and the γ-ray injection rate 3×10^{46} erg/s.

is shown in different energy bands, assuming that the source(s) of primary γ-rays are concentrated in the central part of the cluster. The profiles are calculated for the same parameters as in Fig. 9.4. It is seen that with decrease of photon energy, the brightness distribution becomes broader. This reflects the reduction of the free path of γ-rays with energy. In this calculations we assume the average intracluster magnetic field of 6 μG. In reality, the intracluster magnetic field should, of course, decrease at larger distances from the center. This should lead to sharper profiles, especially at high energies. On the other hand, if sources of primary γ-rays are more or less homogeneously distributed in the cluster, we should expect quite flat brightness distributions at all photon energies.

The energy of a γ-ray photon interacting with background radiation fields is shared between the secondary electron and positron. However, the major fraction of the energy of the γ-ray photon is transfered to one of the electrons. Therefore the maximum of the synchrotron radiation of secondary electrons is expected at energy $h\nu \sim 10(B/1\,\mu G)(E_\gamma/10^{15}\,eV)^2$ keV. Thus, if the spectrum of γ-rays extends beyond 10^{17} eV, the maximum of synchrotron radiation will be shifted to the γ-ray domain. Therefore, the spectrum of synchrotron radiation dramatically depends on the position of

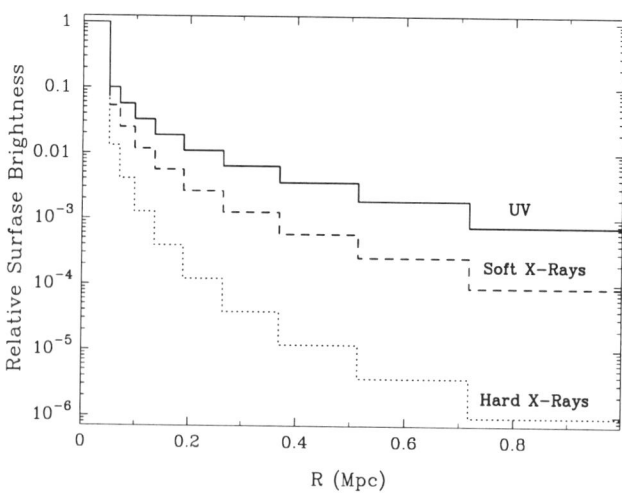

Fig. 9.6 Surface brightness distribution of the synchrotron radiation from Coma cluster in 3 spectral bands: 0.064 - 0.240 keV (EUV), 2 - 10 keV (soft X-rays) and 25 - 80 keV (hard X-rays), calculated for γ-rays produced in the central source (from Timokhin *et al.*, 2003).

the cutoff energy E_0 in the primary γ-ray spectrum. This effect is demonstrated in Fig. 9.7 assuming that primary γ-rays have differential energy distribution $\mathrm{d}N/\mathrm{d}E \propto \exp(-E/E_0)$ with (1) $E_0 = 10^{16}$ eV, (2) 10^{18} eV

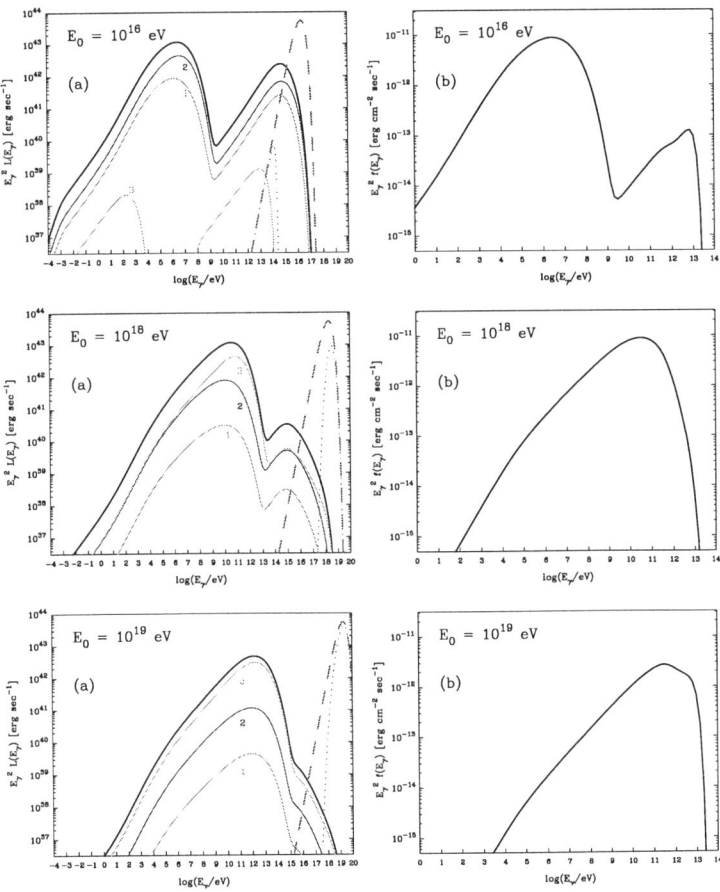

Fig. 9.7 Non-thermal luminosities (left panel) and fluxes (right panel) from a spherically symmetric galaxy cluster of radius 1.5 Mpc, assuming that γ-rays have been injected into ICM by a central source during the last 10^7 years. The spectra of the primary γ-ray source are shown by thick dashed lines. Dotted lines show the spectra of the primary γ-rays after propagating of 1.5 Mpc. The luminosity of the primary VHE source is assumed to be 10^{44} erg/sec, and the magnetic field $B = 1\,\mu$G. The luminosities of three different zones of the cluster are shown with thin solid lines: 1: $R = 0 - 1$ kpc, 2: $R = 10 - 40$ kpc, and 3: $R = 0.44 - 1.5$ Mpc. The overall luminosity of the cluster is shown by thick solid line. Expected energy fluxes of non-thermal radiation are calculated for a cluster at a distance of 100 Mpc, after correction for the intergalactic absorption.

and (3) 10^{19} eV. Note that these type γ-ray spectra can be formed by ultrahigh energy cosmic ray protons interacting with narrow-band radiation field. In this case E_0 is an order of magnitude less than the energy cutoff in the parent proton spectrum. On the other hand, the low energy part of the γ-ray spectrum (dN/dE = const) does not depend on the proton spectrum, but simply is result of the threshold of photomeson interactions.

In Figs. 9.7(a) are shown the luminosities of synchrotron radiation of secondary electrons. The spectra of primary γ-rays emitted by the central source, as well as the spectra of primary γ-rays after propagating of 1.5 Mpc (at the edge of the cluster), are also shown, by thick dashed and dotted lines, respectively. The area between the dashed and dotted curves indicates the amount of energy absorbed and then re-radiated in the cluster volume. It is seen that while in the case (1) the radiation of the secondary electrons peaks at MeV energies, in the case (2) and (3) the luminosity is dominated by GeV and TeV γ-rays, respectively. However, the very high energy γ-rays above 1 TeV suffer significant intergalactic absorption if sources are located beyond 100 Mpc. To demonstrate this effect, in Figs. 9.7b we show the expected γ-ray fluxes after correction for the intergalactic absorption, assuming that the source is located at a distance of 100 Mpc. For the assumed total luminosity in primary γ-rays of 10^{44} erg/s and distance to the source of 100 Mpc, the resulting GeV and TeV γ-ray fluxes can be probed with GLAST and forthcoming arrays of atmospheric Cherenkov telescopes. At the same time, because of limited sensitivity of γ-ray instruments in the MeV energy band, detection of the secondary synchrotron radiation initiated by 10^{16} eV primary γ-rays would be very difficult, unless the power of primary γ-rays significantly exceeds 10^{45} erg/s.

9.4 Nonthermal Radiation Components Associated with Very High and Extremely High Energy Protons

Beyond 100 Mpc, the Universe is opaque for extremely high energy (EHE) cosmic rays with energy $E \geq 10^{20}$ eV (see e.g. Berezinsky and Grigoreva, 1988; Aharonian and Cronin, 1994). Fortunately, observations of the electromagnetic radiation associated with these particles can compensate, at least partly, for this limitation, allowing an extension of the (indirect) study of EHE cosmic rays to cosmological scales.

The spatial and spectral characteristics of nonthermal emission initiated by interactions of protons with the 2.7 K CMBR and the ambient gas, depend on both the spectrum of the particles injected into the intracluster medium and the properties of the local environment. Below we discuss two different cases, assuming that (1) the source of EHE protons is surrounded by a rich galaxy cluster with B-field exceeding 1 μG, and (2) the EHE source is located in a low magnetic field environment.

9.4.1 High energy radiation from cores of clusters

One of the interesting implications of large intracluster magnetic fields is the relatively slow diffusion of cosmic rays in the ICM, allowing their effective confinement and accumulation in galaxy clusters over Hubble timescales $t \sim 1/H_0 \sim 10^{10}$ yr. The energy-dependent diffusion coefficient, together with the initial spectrum of particles injected into the ICM, determine the spectrum and the total energy content of protons in galaxy clusaters.

Fig. 9.8 shows two examples of contemporary proton spectra obtained within the "leaky-box" type approximation of particle propagation (Aharonian, 2002b) for the following model parameters:

(a) continuous injection of relativistic protons with a constant rate $L_p = 10^{45}$ erg/s during $\Delta t = 10^{10}$ yr; source spectrum of protons with power-law index $\Gamma_p = 2$ and exponential cutoff at $E_0 = 10^{20}$ eV; intracluster magnetic field $B = 3$ μG and gas density $n = 10^{-3}$ cm^{-3} within the central $R = 0.5$ Mpc region; escape time of protons from the high magnetic field region $\tau_{\rm esc} = 5 \times 10^7 E_{19}^{-1/2}$ yr;

(b) $L_p = 3 \times 10^{46}$ erg/s; $\Delta t = 10^8$ yr; $\Gamma_p = 1.5$, $E_0 = 10^{20}$ eV, $R = 0.5$ Mpc, $B = 5$ μG, $n = 10^{-3}$ cm^{-3}, $\tau_{\rm esc} = 2.0 \times 10^7 E_{19}^{-1/3}$ yr.

For the assumed magnetic field, B, and the cluster radius, R, the chosen escape times correspond to diffusion coefficients given in the power-law form $D(E) = D_0 E_{19}^\beta$ with D_0 and β compatible with the range of model parameters based on theoretical and phenomenological studies (see e.g. Völk et al., 1996, Berezinsky et al., 1997, Blasi and Colafrancesco, 1999). For the given diffusion coefficients, particle escape becomes an important factor in the formation of proton spectra above $E^* \propto (R^2/D_0 \, \Delta t)^{1/\beta} \sim 3 \times 10^{14}$ eV and 10^{17} eV, for the cases (a) and (b), respectively. As a result, respectively $E_p^{-2.5}$ and $E_p^{-1.83}$ type proton spectra are formed above these energies, before approaching the intrinsic (acceleration) cutoff at $E \sim 10^{20}$ eV.

Cases (a) and (b) correspond to two essentially different scenarios. Case (a) could be treated as a quasi-continuous injection of protons into the cluster during its age of about 10^{10} yr. In this case not only AGN jets, but also other sources of particle acceleration, e.g. large-scale (multi-hundred kpc) shock structures (Miniati et al., 2001) can contribute to the high energy proton production. The case (b) corresponds to the scenario in which the proton production is contributed by a single powerful AGN (see e.g. Enßlin et al., 1997) over its lifetime of about 10^8 yr. Since the escape time of protons with energy $E \geq 10^{18}\,\mathrm{eV}$ is less than 10^8 yr, the amount of highest energy protons confined in the cluster is effectively determined by recent accelerator(s) operating over the last $10^7 - 10^8$ years. Also, the assumed weak energy dependence of the escape time $\propto E^{-1/3}$ and the very hard proton spectrum $\propto E^{-1.5}$, with the injection rate $L_\mathrm{p} = 3 \times 10^{46}$ erg/s,

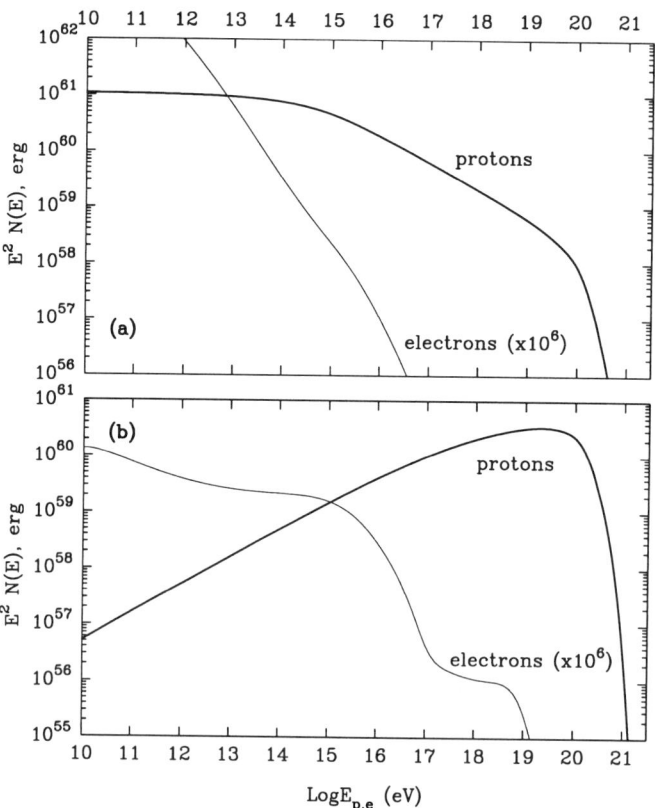

Fig. 9.8 The total energy content of protons and secondary electrons (multiplied by 10^6) inside a galaxy cluster (from Aharonian, 2002b). Calculations correspond to scenarios (a) and (b) (see the text for the assumed model parameters).

provide a much higher content of $\geq 10^{18}$ eV protons confined in the cluster for case (b) compared with case (a) (see Fig. 9.8). The assumed proton acceleration rate is large, but still acceptable. For example, the total kinetic energy of electron-proton jets in powerful radio sources can significantly exceed 10^{47} erg/s (Ghisellini and Celotti, 2001), provided that the kinetic energy power at the core is conserved along the jet up to ≥ 100 kpc scales (Celotti and Fabian, 1993). The numerical calculations show that, for the given duration and rate of proton injection, the total energy content of protons accumulated in the cluster is $W_\mathrm{p} = 1.4 \times 10^{62}$ erg and 2.1×10^{61} erg for cases (a) and (b), respectively.

Fig. 9.8 shows the spectra of secondary electrons produced at interactions of relativistic protons with the intracluster gas for $n = 10^{-3}$ cm^{-3}, and with the 2.7 K CMBR. These spectra consist of three components: (i) electrons from pp interactions, (ii) electrons from $p\gamma$ interactions due to Bethe-Heitler pair production and (iii) electrons from photo-meson processes. Due to radiative cooling, which in the case of electrons strongly dominates over escape losses at all energies, the first population electrons have a power-law spectrum $\propto E_\mathrm{e}^{-(\Gamma_\mathrm{p}+1)}$, while the spectra of the second and third electron populations are almost independent of the spectrum of primary protons, namely both cooled electron populations have standard E^{-2} type spectra with high-energy cutoffs at energies $\sim 10^{15}$ eV and $\sim 10^{19}$ eV, respectively. While in case (a) the second and third ($p\gamma$) electron components are strongly suppressed (Fig. 9.8a), in case (b) the features of all three components are clearly seen in Fig. 9.8b.

Although the energy content in secondary electrons is very small compared with the total energy budget, approximately half of the nonthermal energy of the cluster is eventually radiated away through the secondary electrons. Unlike primary (directly accelerated) electrons, which because of short lifetimes are concentrated in the proximity of their accelerators, the secondary electrons are homogeneously distributed over the entire cluster, and therefore their radiation has an extended (diffuse) character.

The nonthermal electromagnetic radiation initiated by interactions of high energy protons consist of five components — (1) synchrotron (marked as *e-synch*) and (2) inverse Compton (*IC*) photons emitted by secondary electrons, (3) synchrotron radiation of protons (p-synch), and π^0-decay γ-rays from (4) proton-proton (pp) and (5) proton-photon ($p\gamma$) interactions. The production rates of these radiation components are shown in Figs. 9.9a and 9.9b. It is seen that the spectral energy distributions of radiation characterising the (a) and (b) scenarios are essentially different.

In case (a) the nonthermal radiation is mainly contributed to, directly or via secondary electrons, by pp interactions. Since the pp interaction timescales in galaxy clusters with $n \leq 10^{-3}\,\text{cm}^{-3}$ exceed the source ages of about 10^{10} yr, the absolute fluxes of radiation are proportional to the product $nL_\text{p}\Delta t$. For the assumed index of accelerated protons, $\Gamma_0 = 2$, approximately the same fraction of the proton kinetic energy is released in π^0-decay γ-rays and $\pi^\pm$-decay electrons and positrons. On the other hand, the assumed magnetic field, $B = 3\ \mu\text{G}$, implies an energy density close to the density of the 2.7 K CMBR, therefore equal fractions of the electron energy are released through the synchrotron and inverse Compton channels. This results in the flat overall spectral energy distribution (SED) over a very broad frequency range from radio to multi-TeV γ-rays

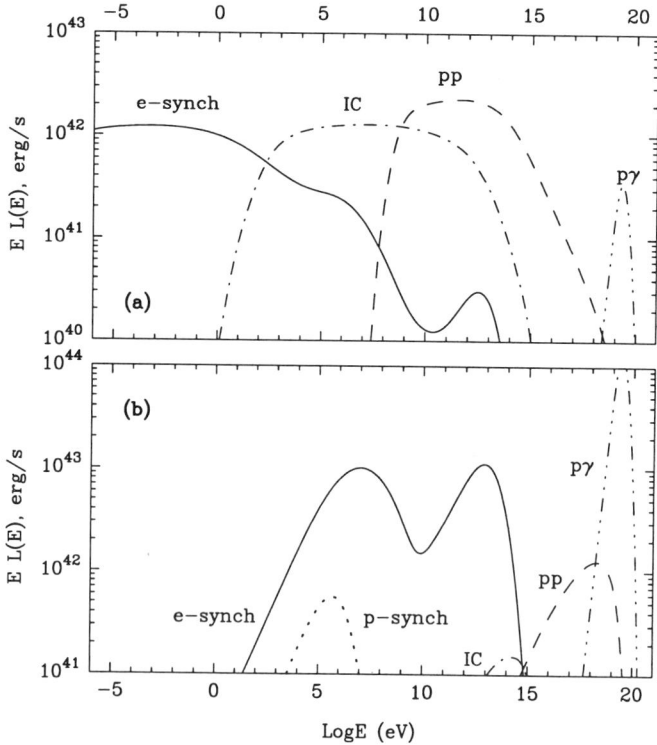

Fig. 9.9 Broad-band luminosities of nonthermal radiation initiated by protons in a galaxy cluster (from Aharonian, 2002b). The calculations correspond to two different scenarios of proton injection into the cluster (see the text for details).

(Fig. 9.9a). The energy domain below 1 keV is due to synchrotron radiation of electrons, while the interval between X-rays and low energy (≤ 100 MeV) γ-rays is contributed by inverse Compton mechanism. At higher energies the radiation is dominated by π^0-decay γ-rays produced at pp interactions. The local maximum at $10^{19} - 10^{20}$ eV in Fig. 9.9a, due to decays of π^0-photomesons, is relatively weak because of effective escape of protons from the cluster (see Fig. 9.8a)

For harder spectra of accelerated protons, e.g. with power-law index $\Gamma_p = 1.5$, the SED is strongly dominated by synchrotron radiation of secondary $p\gamma$ electrons with two prominent peaks at MeV and TeV energies (see Fig. 9.9b) corresponding to the radiation by electrons from Bethe-Heitler pair production and photo-meson production processes, re-

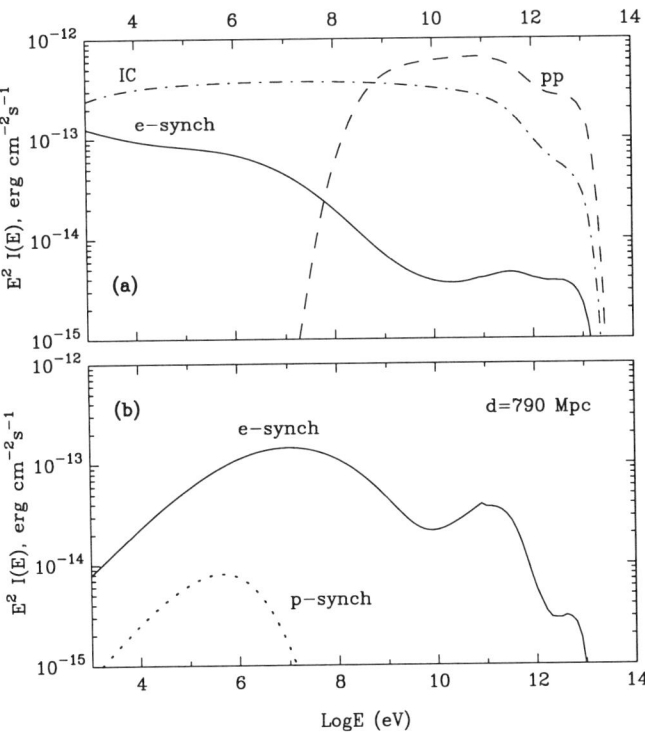

Fig. 9.10 Broad-band SED of nonthermal radiation of clusters corresponding to luminosities shown in Fig. 9.9a and Fig. 9.9b, and assuming that the source is located at $d = 175$ Mpc (a) and 790 Mpc (b), respectively (see the text for details and the model parameters used).

spectively. Note that the synchrotron radiation by primary (directly accelerated) electrons has an unavoidable, self-regulated cutoff below ~ 150 MeV, being defined by the balance between the synchrotron loss rate and the maximum-possible acceleration rate (see Chapter 3). Obviously in the case of synchrotron radiation of secondary electrons, there is no intrinsic limit on γ-ray energy. While at optical and radio frequencies the energy flux decreases as $\nu S_\nu \propto \nu^{0.5}$, at extremely high energies, $E \sim 10^{19} - 10^{20}$ eV, the prominent "$p\gamma$" peak dominates over the entire SED (Fig. 9.9b). However, this peak, as a part of the whole region of γ-rays above 10 TeV is not visible for the observer. Because of interactions with the diffuse extragalactic background radiation, these energetic γ-rays disappear during their passage from the source to the observer.

9.4.1.1 Detectability of gamma rays

In Fig. 9.10 we show the expected fluxes of radiation assuming that the sources with luminosities presented in Figs. 9.9a,b are located at a small distance (like Pictor A), 175 Mpc, and at a large distance (like 3C 273), 790 Mpc. These are two specific examples chosen to have a feeling for the detectability of radiation at different wavelengths. For the given source luminosities, the fluxes of high energy γ-rays are determined not only by the distance to the source ($F_\gamma \propto 1/d^2$), but also by the intergalactic photon-photon absorption. Note that the mean free path of γ-rays above 100 TeV is less than 1 Mpc, so strictly speaking we should include in calculations the radiation of next generation electrons produced inside the cluster. However, since the γ-ray luminosities sharply drop above 100 TeV (see Fig. 9.10), we can ignore these (3rd generation) photons without a significant impact on the accuracy of the calculations. Such an approximation, however, could be inappropriate for γ-rays with energy $\geq 10^{19}$ eV. The mean free paths of these γ-rays, interacting with the 2.7 K CMBR and extragalactic radio background photons, contain large uncertainties because of the lack of reliable information about the radio background at MHz frequencies (Berezinsky, 1970, Aharonian et al., 1992, Protheroe and Biermann, 1996, Coppi and Aharonian, 1997). Nevertheless it is likely that the extragalactic photon fields cannot prevent the $\geq 10^{19}$ eV γ-rays from travelling freely over distances more than 1 Mpc. Actually the interaction of the highest energy γ-rays with the cluster's own synchrotron radio emission could be a more important process. It is easy to show that the 10^{19} eV γ-rays can escape the central ~ 1 Mpc region of the cluster if the radio luminosity at frequencies ~ 10 MHz does not exceed 10^{42} erg/s. For comparison, the 10 MHz radio luminosity of the Coma cluster is about 10^{40} erg/s, but in most powerful clusters it could be significantly higher. For example, the power-law extrapolation (with spectral index 1.6) of the recently reported radio flux from the giant halo in Abell 2163 ($z = 0.203$) at 20 cm $S_\nu \simeq 0.15$ Jy, to lower frequencies (Feretti et al., 2001) gives for the luminosity of the cluster $\simeq 5 \times 10^{42}$ erg/s at 10 MHz. This implies that the $\geq 10^{19}$ eV γ-rays are absorbed inside this cluster, and thus we must deal with the scenario discussed in Sec.9.3. In other words, we have to include in the overall SED the synchrotron radiation of the secondary electrons. This would simply result in the linear increase (by a factor of two) of the flux of ≥ 1 TeV γ-rays (typical energy of synchrotron photons produced by 10^{19} eV electrons in the magnetic field $B \geq 1\,\mu$G) shown in Fig. 9.10b.

A quite different spectrum of γ-rays is expected from the "echo" of the original $\geq 10^{19}$ eV γ-rays if the free path of extremely high energy photons exceeds 1 Mpc. In this case the photon-photon interactions take place predominantly outside the cluster where the intergalactic magnetic field, $B_{\rm IG}$, is significantly smaller (most probably, $B_{\rm IG} \leq 10^{-8}$ G). Then we should expect an extended synchrotron emission from the pair-produced (and radiatively cooled) electrons with a hard differential spectrum $\propto E^{-1.5}$ up to the maximum energy in the SED at

$$E_{\rm max} \sim 5 (B_{\rm IG}/10^{-9}\,{\rm G})\,({\rm E_e}/10^{19}\,{\rm eV})^2\,{\rm GeV}\ . \tag{9.3}$$

The γ-ray spectrum beyond $E_{\rm max}$ depends on the shape of the proton spectrum in the region of the cutoff E_0, but in any case it is smoother (a basic feature of synchrotron radiation) than the proton spectrum in the corresponding region above E_0. The energy flux of radiation which peaks at $E_{\rm max}$ can be easily estimated from Fig. 9.9b, assuming that almost the entire 10^{44} erg/s luminosity in the primary $E \sim 10^{19} - 10^{20}$ eV γ-rays is re-radiated outside the cluster in the form of secondary synchrotron photons. Namely, we should expect an energy flux at the level $F_\gamma \sim 10^{-12} (d/1\,{\rm Gpc})^{-2}$ erg/cm^{2}s contributed mainly by the region around $E_{\rm max}$. It is interesting to note that although the secondary synchrotron radiation is produced outside the cluster, the angular distribution of this radiation seen by the observer would essentially *coincide with the size of the cluster*. Indeed, the ratio of the gyroradius to the mean synchrotron interaction path of electrons, which is estimated as $r_{\rm g}/\lambda_{\rm synch} \simeq 30\,E_{19}^2 (B_{\rm IG}/10^{-9}\,{\rm G})$, implies that the photo-produced electrons of energy $E_{\rm e} \sim 10^{19}$ eV will radiate synchrotron photons before they would significantly change their direction in the magnetic field $\geq 10^{-9}$ G, i.e. the synchrotron γ-rays produced outside the cluster will follow the direction of their "grandparents" – π^0-decay γ-rays produced inside the cluster. Below E_{max} the angular distribution of γ-rays could be quite different. In this energy region γ-rays produced by cooled electrons with steady state distribution $\propto E_e^{-2}$, have a hard power-law differential spectrum $\propto E^{-1.5}$. For $B_{\mathrm{IG}} \sim 10^{-9}$ G, the trajectories of $E \leq 10^{18}$ eV electrons will be curved significantly before they radiate synchrotron γ-rays. Therefore, the size of the low energy γ-rays, $E \leq 100$ MeV, produced by these electrons, is determined by the mean free path of primary $\sim 10^{19}$ eV γ-rays in the intergalactic photon field, and thus would exceed the angular size of the cluster. Detection of such radiation, e.g. by GLAST, would not only confirm the presence of the highest energy protons in the clusters of galaxies, but also could provide unique information about the intergalactic magnetic fields and the diffuse extragalactic radio background.

9.4.1.2 Detectability of X-rays

The "hadronic" nonthermal radiation luminosities shown in Fig. 9.10 depend on several model parameters like the magnetic field and the gas density of ICM, the total amount of protons confined in the cluster, *etc.* The spectral shape of radiation strongly depends on the spectrum of injected protons, as well as on the diffusion coefficient. However, since both the assumed ambient gas density and the proton injection rates are close to the maximum allowed numbers, the luminosities shown in Fig. 9.10 could hardly be dramatically increased. If so, the nonthermal X-ray fluxes of "hadronic" origin could be marginally detected only from relatively nearby clusters like Virgo, Coma, Perseus, the cluster surrounding Cygnus A, *etc.* The case of the Coma cluster is of a special interest because of the hard, most probably nonthermal, X-radiation recently reported from this source. As discussed in Sec.9.2.1, the most natural interpretation of this radiation by the inverse Compton scattering of radio electrons requires an intracluster field of $\sim 0.1\,\mu$G, noticeably smaller than the estimate of several μG deduced from Faraday rotation measurements. The synchrotron origin of EUV and hard X-rays from Coma seems a more attractive radiation mechanism, but in its conventional version, assuming that the electrons are directly accelerated in the ICM or are injected into the ICM by individual galaxies, this model cannot explain how these electrons could be accelerated to ≥ 100

TeV energies, and how they could produce diffuse X-radiation throughout the cluster, given their very short path-lengths compared to the cluster dimensions. Therefore it is worth exploring other models that assume a secondary origin of the multi-TeV electrons producing synchrotron X-rays. As discussed in Sec.9.3, an interesting possibility is that these electrons are produced by very high energy γ-rays interacting with the 2.7 K CMBR.

An alternative channel of production of secondary electrons in ICM could be, as discussed above, interactions of accelerated protons with the ambient gas or photon fields. In case (a), production of secondary electrons is dominated by interactions of protons with ambient gas. As shown in Fig. 9.9a, this model predicts an almost constant SED over a very large energy band from radio to TeV γ-rays, although it is contributed by three different radiation mechanisms – by synchrotron (at radio and optical wavelengths), IC (X and low-energy γ-rays) components radiated by secondary electrons, and by π^0-decay γ-rays. This model predicts a radio luminosity of $\sim 10^{42}$ erg/s which is higher by two orders of magnitude than the luminosity observed from Coma. Because of the large assumed magnetic field, $B = 3\,\mu$G, we face the same problem which arises in the standard inverse Compton models. This imposes an upper limit on the X-ray luminosity, $L_X \leq 10^{40}$ erg/s. However, this should not be treated as a robust constraint. In the "hadronic" models this limit can be easily removed by assuming a harder (than E^{-2}) power-law proton spectrum, and/or a low-energy spectral cutoff. But, of course, this would automatically also suppress the inverse Compton component of radiation produced by low energy secondary electrons. In contrast to the standard inverse Compton models, in the "hadronic" model we have an additional important channel for the production of X-rays - synchrotron radiation of multi-TeV secondary electrons. These electrons are produced more effectively at interactions of the accelerated protons with the 2.7 K CMBR than with the gas (see Fig. 9.9b).

In the "hadronic" model the synchrotron X-ray luminosity depends slightly on the strength of the magnetic field, thus in order to compensate for this deficit, we have to increase the proton injection rate by the same factor, i.e. up to $> 10^{47}$ erg/s. This uncomfortably high, at least for an object like Coma, acceleration power makes problematic, although it cannot completely exclude, the "hadronic" origin of X-rays. Decisive tests could be provided by new spectroscopic measurements (the hadronic model unambiguously predicts a hard X-ray spectrum with photon index of ~ 1.5 up to 1 MeV). Also, the model predicts large γ-ray fluxes both at MeV/GeV and TeV energies. These components have, however, different origins. The radiation up to 10 GeV is produced by "Bethe-Heitler" electrons, while the TeV γ-rays are due to the electrons from $\pi^\pm$ decays. While the TeV luminosity would be suppressed if the exponential cutoff in the proton spectrum occurs below 10^{20} eV, without any impact on the X-ray flux, the MeV/GeV γ-rays are tightly connected with X-rays. This implies that we should expect similar fluxes in ≤ 100 keV X-rays and ≥ 100 MeV γ-rays, i.e. if the "hadronic" origin of X-rays is correct, then γ-rays above 100 MeV from Coma should show up at the level of $\sim 10^{-11}$ erg/cm^{2}s.

Finally we note that the detection of "hadronic" X-rays from extended regions of clusters of galaxies is a rather hard task, even for instruments like Chandra and XMM-Newton, especially because of the presence of high local X-ray components like the thermal X-ray emission of the hot intra-cluster gas, nonthermal X-rays due to inverse Compton radiation of directly accelerated electrons, etc. Regardless of the details, the associated spectral band of ≥ 100 MeV γ-rays seems a more promising window to explore the "hadronic" processes with GLAST and, perhaps also with forthcoming 100 GeV threshold IACT telescope arrays.

9.4.2 Nonthermal radiation beyond the cluster cores

In regions of the intergalactic medium where the magnetic field is significantly lower than in the cores of rich clusters, the characteristics of nonthermal radiation are essentially different from the radiation features described in the previous section. The weak magnetic fields in such regions make the particle propagation faster, and prevent the dramatic synchrotron cooling of the highest energy (first generation) electrons. This allows an effective development of relativistic electron-photon cascades triggered by interactions of protons with the 2.7 K CMBR. The first stage of the cascade initiated by secondary electrons and γ-rays interacting with the same 2.7 K CMBR leads to formation of a standard γ-ray spectrum which can be approximated as $dN_\gamma/dE \propto E^{-1.5}$ at $E \leq 10$ TeV, and $dN_\gamma/dE \propto E^{-1.75}$ at $E > 10$ TeV with a sharp cutoff at $E \sim 100$ TeV. After the fast development in 2.7 K CMBR, the cascade enters the second (slower) stage. At this stage $E \leq 100$ TeV γ-rays produce $e^\pm$ pairs on the IR/O diffuse background radiation, while the Compton scattering of electrons is still dominated by 2.7 K CMBR photons. The second-stage cascade, which actually consists of 2 or 3 interactions, shifts the spectrum to lower (TeV and sub-TeV) energies, and broadens the angular distribution of emission arising from the deflections of $E < 100$ TeV electrons in the ambient chaotic magnetic field.

Unfortunately, the intergalactic magnetic fields and their fluctuations on very large ($\gg$ 1 Mpc) scales remain almost completely unknown, which does not allow us to reach definite conclusions concerning the expected characteristics of the radiation. Nevertheless, depending on the strength of intergalactic magnetic field, one of the following γ-ray emission components can be predicted.

9.4.2.1 Weak magnetic field

For a very weak intergalactic field, $B \leq 10^{-15}$ G, the cascade radiation arrives, because of the almost rectilinear propagation of primary protons and secondary pairs, from a direction centred on the source. Although quite speculative, such a large-scale intergalactic magnetic field cannot *a priori* be ruled out, in particular if an essential fraction of the Universe consists of huge, 100 Mpc scale voids (Einasto, 2001). In this case we may expect a point-like source of radiation, although the γ-rays are produced at distances ≥ 10 Mpc from the source. Indeed, for the given energy of a detected γ-ray photon $E_{\rm TeV} = E/1$ TeV, the emission angle is determined by the direction of electrons participating in the last interaction, namely by the deflection in the magnetic field of the parent electron of energy $E_{\rm e} = (E/4kT)^{1/2} m_e c^2 \simeq 17 E_{\rm TeV}^{1/2}$ TeV. The mean attenuation path of these electrons in the 2.7 K CMBR is about $\Lambda_{\rm e} \simeq 0.02 E_{\rm TeV}^{-1/2}$ Mpc, while the gyroradius is $r_{\rm g} \simeq 20\ E_{\rm TeV}^{1/2} (B_{\rm IG}/10^{-15}\,{\rm G})^{-1}$ Mpc. Correspondingly, the emission angle $\theta(\epsilon) \sim \Lambda_{\rm e}/r_{\rm g} \sim 10^{-3} E_{\rm TeV}^{-1} (B_{\rm IG}/10^{-15}\,{\rm G})$. Thus, for an intergalactic magnetic field of about 10^{-15} G, the cascade radiation at $E \geq 100$ GeV would be concentrated within an angle of $1°$. Since for sources at distances between 100 Mpc and 1000 Mpc, we expect a hard cascade spectrum with a cutoff between 100 GeV and 1 TeV, approximately half of the energy of $\geq 10^{20}$ eV protons (completely lost in interactions with the 2.7 K CMBR over distances ≤ 100 Mpc) would be released in this energy interval, and give a flux

$$F_\gamma \sim 5 \times 10^{-12} \left(\frac{L_{\rm p}(\geq 10^{20}\,{\rm eV})}{10^{45}\,{\rm erg/s}}\right) \left(\frac{d}{1\,{\rm Gpc}}\right)^{-2}\ {\rm erg/cm^2 s} \qquad (9.4)$$

At lower energies, the energy flux decreases ($\propto E^{1/2}$), and the angular size of emission increases ($\propto E$). Therefore searches for such emission from the directions of nonthermal extragalactic sources, in particular from AGN with powerful X-ray jets, can be done most effectively by 100 GeV threshold IACT arrays.

9.4.2.2 Intermediate magnetic field

If $B_{\rm IG} \geq 10^{-12}\,{\rm G}$, the cascade electrons are promptly isotropised. This leads to the formation of giant pair halos surrounding strong extragalactic TeV sources (see Chapter 11). The angular size of the extended γ-ray source depends on the photon energy. For a detected photon of energy $E_{\rm TeV}$, it is mainly determined by the mean free path of previous generation γ-rays of energy $E' \simeq 2E_{\rm e} \simeq 34 E_{\rm TeV}^{1/2}\,{\rm TeV}$. The free path of $E' \sim 10 - 15\,{\rm TeV}$ photons, which are responsible for the detected (last generation) 100 GeV cascade γ-rays, presently is poorly known, but, probably, does not exceed 50 Mpc. Thus the typical size of a 100 GeV halo radiation surrounding the EHE source at a distance 1 Gpc would be $\leq 3°$. The detection of pair halos presents a difficult experimental task, compared, in particular, with the detection of rectilinear cascade radiation discussed above. At the same time, the EHE sources can be revealed through their halo radiation independent of the orientation of AGN jets. It should be noticed in this regard that only in the case of an isotropically emitting source will the halo be centred on the source. If the relativistic outflow injecting EHE protons is directed away from the observer, the center of the halo would be displaced by an angle comparable to the typical angular size of the halo. The radiation characteristics of halos *initiated by EHE protons* primarily depend on the level of the diffuse extragalactic background, first of all at mid- and far- infrared wavelengths, but not on the intergalactic magnetic field, provided that the latter does not exceed $10^{-9}\,{\rm G}$. This question is discussed in Chapter 11.

9.4.2.3 Strong magnetic field

When considering electromagnetic cascades initiated by EHE protons in the intergalactic medium with magnetic field $B_{\rm IG} \geq 10^{-9}\,{\rm G}$, we must distinguish between two populations of secondary electrons. The electrons originating from the Bethe-Heitler pair production process are produced with typical energies $(m_{\rm e}/m_{\rm p})E_{\rm p} \sim 10^{15} - 10^{16}$ eV. For magnetic field $\leq 10^{-7}\,{\rm G}$, they cool mainly through inverse Compton scattering, and thus produce *faint* $({\rm e}^+, {\rm e}^-)$ halos in the fashion discussed above. Meanwhile, the electrons originating from the photo-meson production process, directly, via π^+-decays, or through interactions of π^0-decay γ-rays with the extragalactic diffuse radio background, have much higher energies, $\sim 1/10 E_{\rm p} \geq 10^{19}$ eV, taking into account that only $\geq 10^{20}$ eV protons interact effectively with the 2.7 K CMBR. Because of the Klein-Nishina effect, the energy losses of these electrons are dominated by synchrotron radiation, as long as the ambient magnetic field exceeds $\sim 10^{-9}$ G. This prevents cascade development, but instead provides another effective channel for production of high energy γ-rays. Almost the whole energy of $\geq 10^{20}$ eV protons is released, through the synchrotron radiation of secondary electrons, into the γ-rays with characteristic energy $\epsilon_{\max} \sim 50(B/10^{-8} \text{ G})(E_0/10^{20} \text{ eV})^2$ GeV. Because the gyroradius of 10^{20} eV protons in the magnetic field $B_{\text{IG}} \geq 10^{-9}$ G is comparable or less than their mean $p\gamma$ interaction path $\Lambda_{\text{p}} \sim 100$ Mpc, we should expect a diffuse radiation component emitted by huge intergalactic regions. This diminishes the chances for detection of this radiation component, especially from relatively nearby ($d \ll 1$ Gpc) objects. The situation is quite different for cosmologically distant objects. Because at distant cosmological epochs the 2.7 K CMBR was denser, $n_{\text{ph}} \propto (1+z)^3$, and hotter, $T_{\text{r}} \propto 1+z$, the mean free path of protons Λ_{p} has a strong z-dependence. At energies $E_{\text{p}} \leq 3 \times 10^{20}$ eV it can be approximated as $\Lambda_{\text{p}} \simeq 5.2(1+z)^3 \exp[3 \times 10^{20} \text{ eV}/(1+z)E]$ Mpc (Berezinsky and Grigoreva, 1988). For example, in the environments of quasars 3C 273 ($z = 0.158$) and PKS 0637-752 ($z = 0.651$), the mean free paths of 10^{20} eV protons are 44.6 Mpc and 7.1 Mpc, respectively. This implies that, if the X-ray emission from large scale jets of these powerful quasars has indeed a proton-synchrotron origin, we may expect an accompanying GeV γ-radiation component initiated by the highest energy escaping protons outside the jets, but still within $\sim 3°$ centred on 3C 273, and within 10 arcminutes centred on PKS 0637-752.

In summary, the powerful AGN jets, as well as the large-scale structures in galaxy clusters are potential sites of acceleration of protons and nuclei to energies $\geq 10^{19}$ eV. These particles interacting with the 2.7 K CMBR photons at the highest energies, and with the ambient gas at lower energies, initiate non-negligible nonthermal radiation components. The spatial and spectral distributions of these emission components extending from radio-wavelengths to high-energy γ-rays, contain unique information about both the highest energy protons and magnetic fields in the extended intergalactic environments. The γ-ray observations by GLAST and by the low-threshold IACT arrays may provide very sensitive probes of nonthermal phenomena in the cores and outskirts of galaxy clusters.

Chapter 10

TeV Blazars and Cosmic Background Radiation

Since the discovery of TeV γ-rays from Mkn 421 in 1992, many papers have been published with an unusual combination of two keywords - *Blazars* and *Cosmic-Infrared-Background* (CIB) radiation. The link between these two topics is not (initially) obvious. Indeed, while blazars constitute a sub-class of AGN dominated by highly variable (several hours or less) components of broadband (from radio to gamma-rays) *non-thermal emission* produced in relativistic jets, the CIB is a part of the overall diffuse extragalactic background radiation (DEBRA) dominated by *thermal emission* components produced by stars and dust, and accumulated over the entire history of the Universe. The scientific objectives of these studies are also different. While blazars may serve as ideal laboratories for the study of MHD structures and particle acceleration processes in relativistic jets, the CIB carries crucial cosmological information about the formation epochs and history of the evolution of galaxies.

Thus one may say that these two topics are relevant to quite independent areas of modern astrophysics and cosmology. Yet, the current studies of the CIB and blazars, more specifically the sub-population of blazars emitting TeV gamma-rays (*TeV blazars*), are tightly coupled through the intergalactic absorption of TeV radiation by CIB photons. Therefore, in this Chapter these two topics are discussed jointly, although with more emphasis on the questions related to the high energy processes in the small-scale jets of blazars. In the next Chapter we will return to the discussion of CIB related topics, but this time in the context of cosmological implications of very high energy γ-rays from distant extragalactic objects.

The astrophysical/cosmological importance of absorption was recognised in the 1960s (Nikishov, 1962; Gould and Schrèder, 1966; Jelley, 1966), but it became a really hot topic three decades later, after the detection of

TeV signals from several representatives of a specific class of active galaxies called BL Lac objects.

10.1 Cosmic Infrared Background Radiation

The CIB basically consists of two emission components produced by stars and partly absorbed/re-emitted by dust during the entire history of evolution of galaxies. Consequently, two distinct bumps in the spectral energy distribution (SED) of red-shifted radiation at near infrared (NIR) $\lambda \sim$ 1-2 μm and far infrared (FIR) $\lambda \sim$ 100-200 μm wavelengths, and a mid infrared (MIR) "valley" between these bumps are expected. Because of the heavy contamination caused by foregrounds of different origin, predominantly by the zodiacal (interplanetary dust) light, measurements of the CIB contain large uncertainties. Moreover, these results only conditionally can be treated as *direct measurements*, because their interpretation primarily depends on the modelling and removal of these foregrounds. Therefore, observations of the CIB generally allow derivation of the flux *upper limits* rather than detection of positive residual signals. In Fig. 10.1 we show the CIB fluxes based on the latest reports, and refer the reader to the review article by Hauser and Dwek (2001) for a comprehensive coverage of the subject concerning both the techniques of direct measurements and the interpretation of available observations.

Presently, the most reliable results are obtained from the COBE observations at NIR and FIR bands where the contribution of the zodiacal light becomes comparable with the CIB flux. The reported fluxes at 140 and 240 μm derived by three independent groups (Hauser et al., 1998, Schlegel et al., 1998, Lagache et al., 1999), and perhaps also at 100 μm (Lagache et al., 1999; Finkbeiner et al., 2000) are in a reasonable agreement with each other, but somewhat higher than the theoretical predictions. If the relation of these fluxes to the true extragalactic background is correct, this would imply that most of the star formation in the early Universe occurred in highly obscured, dusty environments.

Detections of the CIB are reported also in the NIR - at 2.2 μm and 3.5 μm wavelengths (Dwek and Arendt, 1998; Gorjian et al., 2000; Wright, 2001). These fluxes together with the upper limit at 1.25 μm (Wright, 2001) and the fluxes of optical light derived from the HST data below 0.8 μm (Bernstein et al., 2002; see, however, Mattila, 2003), agree with the recent theoretical calculations by Primack et al., (2001). An independent analysis of the COBE data by Cambresy *et al.*, (2001), as well as the results of the Japanese IRTS satellite (Matsumoto, 2000) are in agreement with the fluxes at 2.2 μm and 3.5 μm reported by Wright (2001), but at shorter wavelengths the claimed fluxes are noticeably higher compared to other measurements (see Fig. 10.1).

The "best guess" estimate of the SED at optical/NIR wavelengths of about 20-50 nW/m^{2}sr is comparable with the FIR flux of about 40-50 nW/m^{2}sr (Madau and Pozzetti, 2000). This indicates that an essential part of the energy radiated by stars is absorbed and re-emitted by cold dust. Our current knowledge of MIR, which carries information about the warm dust component, is quite limited. The only flux estimates in this band derived from ISOCAM source counts at 6 and 15 μm (Franceschini *et al.*, 2001) should be taken as lower limits.

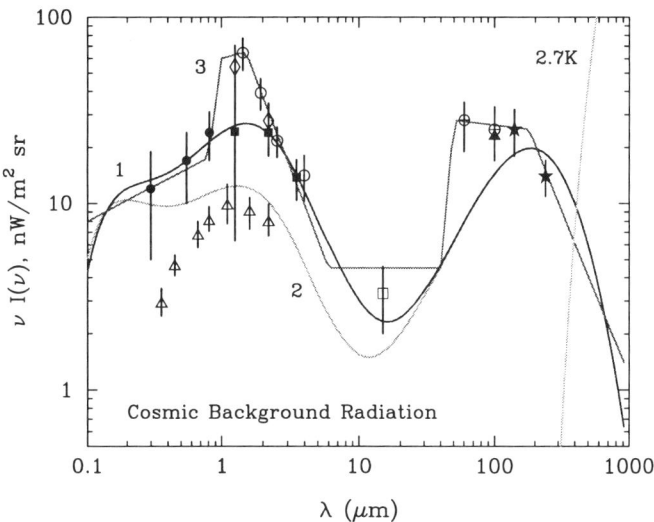

Fig. 10.1 Measured and modelled SEDs of the Cosmic Background Radiation. The reported fluxes at 140 and 240 μm (stars) are from Hauser *et al.* (1998), at 100 μm (filled triangle) from Lagache *et al.*, et al. (1999), and at 60 and 100 μm (open circles) from Finkbeiner *et al.* (2000). The point at 15 μm (open square) is derived from ISOCAM source counts (Franceschini *et al.*, 2001). The reported fluxes at NIR and optical wavelengths are shown by filled squares (Wright, 2001), open circles (Matsumoto, 2000), open diamonds (Cambresy *et al.*, 2001), and filled circles (Bernstein *et al.*, 2002). The open triangles from Pozzetti *et al.* (1998) correspond to lower limits. While the reference CIB models 2 and 3 should be considered as lower and upper limits, curve 1 may be treated as a representative of the "model of choice".

These data stimulated a renewed interest in theoretical studies of the CIB based on different phenomenological (backward, forward, cosmic chemical evolution, semi-analytical, *etc.*) approaches. In spite of certain achievements, the ability of current models to reproduce the CIB measurements should not be, however, overstated, especially because these models contain a number of adjustable parameters and in fact are "primarily designed for that purpose" (Hauser and Dwek, 2001).

In Fig. 10.1 three reference CIB models are shown. We *provisionally* assign them as "nominal" (1), "low" (2), and "extreme" (3) descriptions of the CIB. Curve 2 corresponds to the so-called "LCDM-Salpeter" model of Primack *et al.* (1999), but assuming a two times larger contribution of the dust component. Similar NIR spectra with a flux of about 10 nW/m^{2}sr at 1 μm have been assumed (Franceschini *et al.*, 2001). Curve 1 is close to the prediction by Primack *et al.* (2001) based on the so-called "Kennicut" stellar initial mass function (IMF). And finally, curve 3 is "designed" to match the extreme fluxes reported both at near and far infrared wavelengths.

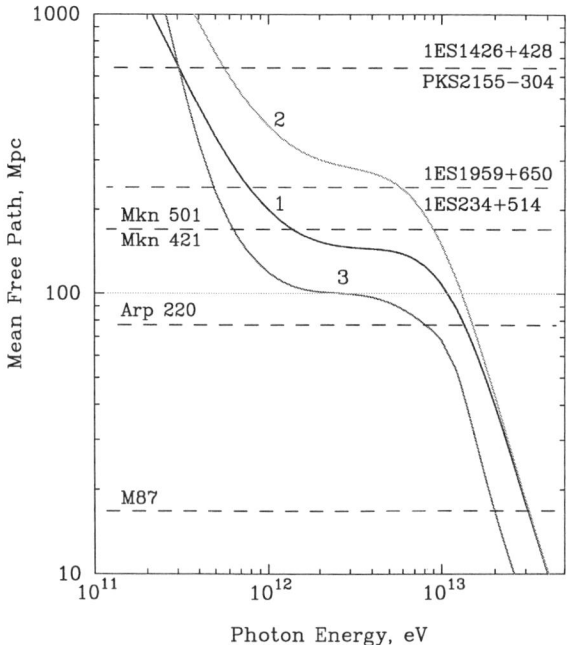

Fig. 10.2 Mean free-path of VHE γ-rays in the intergalactic medium calculated for the three CIB models presented in Fig. 10.1. (From Aharonian, 2001b).

10.2 Intergalactic Absorption of Gamma Rays

To calculate the mean free path of γ-rays, $\Lambda(E)$, in the intergalactic medium one must convolve the CIB photon number distribution, $n(\epsilon)$, with the pair production cross-section. Because of the narrowness of the latter (see Fig. 3.4), for broad-band photon spectra over half the interactions of a γ-ray photon of energy E occur with a quite narrow interval of target photons, $\Delta\lambda \sim (1 \pm 1/2)\lambda^*$, centered on $\lambda^* \approx 1.5(E/1\text{ TeV})$ μm. This gives a convenient approximation for the optical depth $\tau(E) = d/\Lambda(E)$ for γ-rays emitted by a source at a distance d:

$$\tau(E) \simeq 1 \left(\frac{u_{\text{CIB}}(\lambda^*)}{10\text{ nW/m}^2\text{sr}}\right) \left(\frac{E}{1\text{ TeV}}\right) \left(\frac{z}{0.1}\right) H_{60}^{-1}, \qquad (10.1)$$

where $u_{\text{CIB}} = \nu I(\nu) = \epsilon^2 n(\epsilon)$ is the SED of the CIB, and z is the source redshift. No deviation of the observed γ-ray spectrum from the intrinsic (source) spectrum, e.g. by a factor of ≤ 2, would imply $\tau(E) \leq \ln 2$, and consequently provide an upper limit on the CIB flux at $\lambda^*(E)$. For example, this condition for 1 TeV γ-rays from Mkn 421 or Mkn 501 ($z \simeq 0.03$) gives $u_{\text{CIB}} \leq 20$ nW/m^{2}sr around 1-2 μm. This estimate is quite close to the recent NIR measurements, therefore we may conclude that already at 1 TeV we should expect absorption features in the spectra of γ-rays from Mkn 421 and Mkn 501. This is demonstrated in Fig. 10.2 where we present accurate numerical calculations of the mean free path of γ-rays for the three CIB reference models shown in Fig. 10.1. The horizontal lines indicate the distances to the reported TeV blazars (see below), as well as to 2 nearby prominent extragalactic objects – the radiogalaxy M 87 and the ultraluminous starburst galaxy Arp 220. The mean free paths calculated for the "low" CIB model (curve 2 in Fig. 10.2), should be treated as an upper limit for the mean free path. This implies that we cannot ignore the intergalactic absorption of TeV γ-rays from sources located beyond 100 Mpc. On the other hand, this gives a unique chance to extract information about the CIB by detecting absorption features in TeV spectra of extragalactic objects.

Often such a possibility is reduced to a proposal to search for sharp cutoffs in the energy spectra of extragalactic TeV sources (e.g. Stecker et al., 1992). However, in many cases this could be a misleading recommendation. In fact, a cutoff in a γ-ray spectrum is not yet evidence for the intergalactic absorption and, vice versa, the lack of a cutoff cannot be interpreted as absence of intergalactic absorption. It is interesting to note that all CIB models with a conventional spectral shape between 1 and 10 μm predict an almost constant (energy-independent) mean free path of γ-rays, and correspondingly insignificant spectral deformation at energies between 1 and several TeV. The explanation of this effect is straightforward. If the spectrum of background photons in a certain energy interval has a power-law dependence, $n(\epsilon) \propto \epsilon^{-\beta}$, the mean free path in the corresponding γ-ray energy interval $\Lambda(E) \propto E^{1-\beta}$. Within $1\mu m < \lambda < 10\mu$ the CIB spectrum can be approximately described by a power-law, $u_{\rm CIB} \propto \lambda^{-1}$ (see Fig. 10.1), or $n(\epsilon) \propto \epsilon^{-1}$, therefore in the interval between 1 and several TeV the γ-ray mean free path only slightly depends on energy (Fig. 10.2).

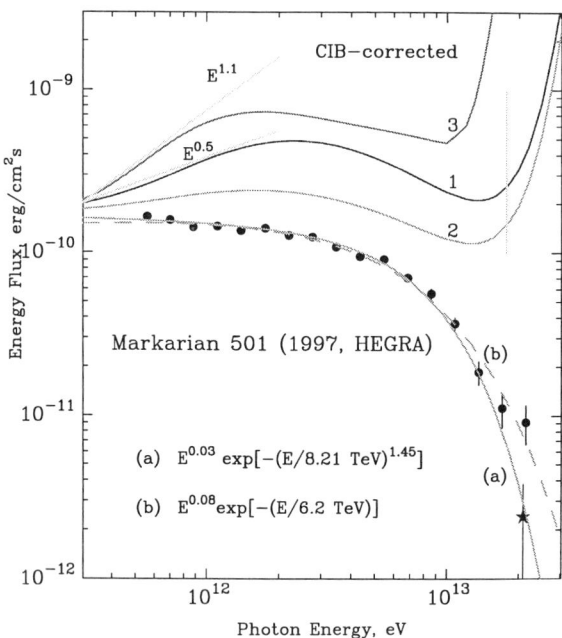

Fig. 10.3 The absorption-corrected SED of Mkn 501 reconstructed for the analytical presentation of the measured spectral points given by equation (a) and assuming the three different CIB models shown in Fig. 10.1. The experimental points and two different analytical fits correspond to the 1997 time-averaged spectrum of Mkn 501 detected by HEGRA (see Chapter 2). The experimental points are derived using a conservative technique for energy reconstruction with ≈ 20 per cent resolution, except for the lowest flux point at 21 TeV ("star"), which has been derived using a new method of energy determination of γ-rays with resolution of about 12 per cent. The spectrum (a) with a super-exponential cutoff better fits this point. The vertical line at 17 TeV indicates the edge of the spectrum measured with high statistical significance.

10.3 TeV Blazars

Blazars are AGN dominated by a highly variable component of non-thermal radiation produced in relativistic jets close to the line of sight (see e.g. Begelman et al., 1984, Urry and Padovani, 1995). The dramatically enhanced fluxes of the Doppler-boosted radiation, resulting from the fortuitous orientation of the jets towards the observer, make these objects ideal laboratories for studying the underlying physics of AGN jets through multi-wavelength studies of the temporal and spectral characteristics of radiation from radio to very high energy γ-rays. This first of all concerns the BL Lac objects, a sub-population of blazars of which 6 nearby representatives – Mkn 421 ($z = 0.031$), Mkn 501 ($z = 0.034$), 1ES 2344+514 ($z = 0.044$), 1ES 1959+650 ($z = 0.048$), PKS 2155-304 ($z = 0.116$), and 1ES 1426+428 ($z = 0.129$) – are established as TeV γ-ray emitters (see Chapter 2).

BL Lac objects are galaxies with extremely bright active cores dominating the radiation from the rest of the galaxy. The name of this source population originates from a specific galaxy called BL Lacertae, which on photographs looks like a star. The rapid variability of these objects in brightness and in polarization is explained by relativistic synchrotron jets. The TeV radiation of BL Lacs provides an independent, and perhaps the strongest, evidence in favour of this commonly accepted paradigm of blazars. Indeed, the enormous apparent VHE γ-ray luminosity of TeV blazars, reaching $L_{\rm app} = 4\pi d^2 f_\gamma \sim 10^{45}$ erg/s during the strongest flares of Mkn 421 and Mkn 501 can be reduced to more reasonable (from any point of view) level assuming that the radiation is produced in a sources moving with large Doppler factor, $L_{\rm int} \approx \delta_{\rm j}^{-4} L_{\rm app}$. This assumption also allows a larger (more comfortable) size of the emitter based on the observed variability timescales, $R \leq c\,\Delta t_{\rm var}\delta_{\rm j}$. Actually, the same applies to other components of nonthermal variable radiation as well. The uniqueness of γ-rays comes from their "fragility" due to interactions with ambient photon fields, and the assumption about the relativistic bulk motion appears to be unavoidable in order to overcome the problem of catastrophic internal γ-γ absorption. Indeed, assuming that the γ-radiation is emitted isotropically in the frame of a relativistically moving source, the optical depth at the *observed* γ-ray energy is estimated as

$$\tau_{\gamma\gamma} \simeq \frac{f_{\rm r} d^2 \sigma_{\rm T} \delta_{\rm j}^{-6}}{8 m_{\rm e}^2 c^3 \Delta t} \simeq 1.35 f_{-11} \Delta t_{\rm h}^{-1} \delta_{10}^{-6} (z/0.1)^{-2} H_{60}^{-2} E_{\rm TeV} \quad (10.2)$$

Here $\delta_{10} = \delta_{\rm j}/10$, $\Delta t_{\rm h} = \Delta t/1\,{\rm hour}$, $E_{\rm TeV} = E/1\,{\rm TeV}$, $d = cz/H_0 =$

$500(z/0.1)H_{60}^{-1}$ Mpc is the distance to the source with redshift $z \ll 1$, and $f_{-11} = f_r/10^{-11}\,\text{erg/cm}^2\text{s}$ is the observed energy flux at $h\nu \simeq 100\delta_{10}^2 E_{\text{TeV}}^{-1}$ eV which for an order of magnitude estimate is assumed to be constant at the observed ("blueshifted") optical to UV wavelengths that predominantly contribute to TeV γ-ray absorption. In particular, observations of multi-TeV γ-rays from Mkn 501, the spectrum of which in the high state extends to ~ 20 TeV, provide a very robust constraint on the jet's Doppler factor. Assuming that the observed optical/UV flux of this source in a high state, $f_{-11} \simeq 5$ (e.g. Pian et al., 1998), is produced in the jet, the catastrophic absorption of 20 TeV γ-rays can be avoided ($\tau_{\gamma\gamma} \leq 1$) only when $\delta_j \geq 10$. Finally, one may interpret the 1997 high state time-averaged differential spectrum of Mkn 501, $E^{-1.92}\exp(-E/6.2\,\text{TeV})$ (see Chapter 2), as a $E^{-1.92}$ pure power-law *production* spectrum modified by the internal γ-γ absorption with optical depth $\tau = E_{\text{TeV}}/6.2$. This would give an accurate determination of the jet's Doppler factor, taking into account the very weak dependence of δ_j on all relevant parameters, $\delta_j = 9.8\, f_{-11}^{1/6}\Delta t_{\text{h}}^{-1/6} H_{60}^{-1/3}$. Clearly, since there could be a number of other reasons for the steepening of the TeV spectrum, this estimate should be considered as a lower limit on δ_j.

As discussed in Chapter 2, the flux variability on different time-scales (and, plausibly, of different origin) is a remarkable feature of the TeV radiation from BL Lac objects. For Mkn 501 and Mkn 421, it ranges from the spectacular sub-hour flares to the extraordinary long (several months) high-activity states observed in 1997 and 2001, respectively. After the discovery of TeV γ-rays, Mkn 421 and Mkn 501 and some other BL Lac objects have been subject of intensive studies through multiwavelength campaigns. These observations revealed that the TeV flares correlate with X-rays on time-scales of hours or less. This is often interpreted as a strong argument in favour of the synchrotron-Compton jet emission models in which the same population of ultra-relativistic electrons is responsible for production of both X-rays and TeV γ-rays via synchrotron radiation and inverse Compton scattering, respectively. However, the very fact of correlation between X-ray and TeV γ-ray fluxes does not yet rule out other possibilities, including the so-called hadronic models which assume that the observed γ-ray emission is initiated by accelerated protons interacting with the ambient gas, low-frequency radiation, and magnetic fields.

Our poor current knowledge about the distortion to the initial γ-ray spectra caused by internal and intergalactic absorptions makes the identification of radiation processes extremely difficult. As long as the CIB remains highly uncertain at all relevant frequency bands, we cannot rely on the information contained in the *observed* γ-ray spectra. On the other hand, the spectral variability of TeV emission as well as the correlations of absolute TeV fluxes with other energy bands do not depend on the intergalactic absorption. Therefore, only detailed studies of spectral evolution of TeV γ-rays and their correlations with X-rays from several X-ray selected BL Lac objects in different states of activity, and located at different distances within 1 Gpc, will allow us to follow and resolve simultaneously the fluctuations predicted by different models on timescales close to the shortest ones likely in these objects.

The sensitivities of imaging atmospheric Cherenkov telescopes are nicely suited to search for short signals from TeV blazars. In particular, the HEGRA, CAT and VERITAS IACTs were able to follow TeV flares of Mkn 501, Mkn 521 and 1ES 1959+650 at the flux level of $\approx 10^{-11}$ erg/cm²s on timescales less than several hours, and thus were well-matched to the sensitivity and spectral coverage of X-ray detectors on RXTE and BeppoSAX for multiwavelength monitoring of flux variations. The sensitivity and energy coverage will be significantly improved once new and larger arrays like CANGAROO, H.E.S.S., MAGIC and VERITAS come online.

The best results so far in this regard became available recently, after the well coordinated multiwavelength observations of Mkn 421 in 2001. On several occasions, truly simultaneous observations by RXTE and TeV instruments with durations up to 6 hours per night were carried out. A nice example of such coverage, the 2001 March 19 flare detected by RXTE and VERITAS is shown in Fig. 2.17. This event demonstrates an impressive TeV/keV correlation. For unambiguous identification and deep understanding of acceleration and radiation mechanisms in the jets, the correlations of *absolute* γ-ray and X-ray fluxes are very important, but not yet sufficient. The key information seem to be contained in *spectral* variability in both energy bands on timescales comparable to the characteristic dynamical times of about 1 h or less. The strongest flares of Mkn 421 in 2001 provide us with such unique information. In particular, the spectral analysis of the 21/22 and 22/23 March 2001 flares detected by the HEGRA collaboration demonstrate a clear correlation of the hardness ratio with the absolute TeV flux as is shown in Fig.2.18.

10.4 Leptonic Models of TeV Blazars

The experimental studies of the spectral evolution of TeV γ-rays and their correlations with X-rays on sub-hour timescales not only imply a new standard in γ-ray astronomy, but also indicate that the quality of the TeV data is approaching the level which would allow us to follow and resolve simultaneously the predicted fluctuations in the putative synchrotron and IC emission components on timescales close to the shortest ones likely in these objects. This may have two important consequences (Coppi and Aharonian, 1999a):

• Matching the observed X-ray/TeV light curves (as opposed to simply fitting snapshot spectra obtained many hours and days apart) should provide a very stringent test of the so-called synchrotron-self Compton (SSC) model since we have two detailed handles on the *single electron* distribution responsible for both emission components. This test can rule out alternative *hadronic* models which are less attractive but still viable options for explanation of TeV emission.

• If the SSC model works, it will be possible to fix the key model parameters and calculate the blazar's intrinsic spectrum, comparing the observed variations of *absolute fluxes* and *spectral shapes* with the predictions of self-consistent, time-dependent numerical codes. This is a crucial point because with an estimate of the intrinsic TeV spectrum, then, and *only then*, it would be possible to estimate the intergalactic absorption effect by comparing the intrinsic and observed spectra, and thus to get information about the CIB.

The leptonic (or electronic) models of TeV blazars assume that both the X-ray and TeV emission components originate in relativistic jets due to synchrotron and IC radiation of the same population of directly accelerated electrons. The electronic models have two attractive features, First, the required TeV electrons in the jet can be readily explained through the (relatively) well understood shock acceleration mechanism. Secondly, both the synchrotron and inverse Compton radiation channels operate with very high efficiency. Indeed, for any reasonable set of model parameters, the characteristic cooling time of high energy electrons responsible for TeV γ-rays and synchrotron X-rays is comparable with typical dynamical times of these objects, of order $R/c \leq$ several hours. The IC cooling time of electrons $t_{\rm IC} \propto E_{\rm e}^{-1}$, and Compton scattering boosts the ambient photon energy proportionally to $E_{\rm e}^2$, so the characteristic time of γ-ray emission decreases with energy as $\propto E_\gamma^{-1/2}$. This predicts different variability timescales for GeV and TeV γ-rays. The relatively low energy electrons responsible for GeV radiation (as well as for synchrotron optical/UV photons) cannot promptly respond to changes of physical conditions in the source.

The electronic models are able to give a reasonable phenomenological description (Ghisellini *et al.*, 1998) of the observationally established sequence (Fossati *et al.*, 1998) in accordance of which less powerful objects have both synchrotron and IC peaks at a similar level of luminosity, and are located at higher (X-ray and TeV) energies. Meanwhile, in more powerful blazars the IC emission strongly dominates the emission, and both peaks are shifted towards lower (IR/optical and GeV) frequencies. Within the electronic models, this trend is clearly seen in Fig. 10.4, and can be explained by the energy cutoff in the electron spectrum that is determined by the balance between the energy loss rate and acceleration rate. In particular, assuming that the acceleration rates in different blazars are similar (for example, because of comparable strengths of relativistic shock waves), the high density of radiation fields in powerful blazars results in "early" energy cutoffs in the electron spectra compared to the less powerful objects. Consequently the synchrotron and IC peaks in powerful blazars will be shifted

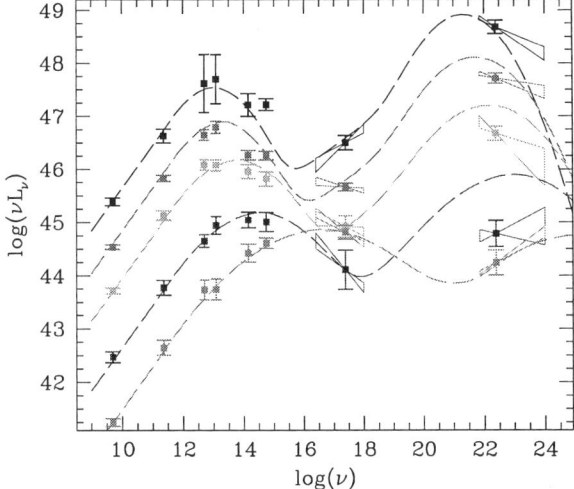

Fig. 10.4 The average SEDs of the "all blazars" sample combined according to radio luminosity, and irrespective of their original classification. The dashed curves are analytic representations obtained assuming that (1) the ratio of the synchrotron and IC peak frequencies is constant and (2) the amplitude of inverse Compton peak is proportional to the radio luminosity (from Fossati *et al.*, 1998).

towards lower energies.

BL Lac objects are relatively modest blazars without noticeable signs of broad emission lines and without thermal emission of the accretion disk (the so-called "blue bump"). In contrast, these are features commonly seen in powerful blazars. Therefore it is believed that in these objects, more specifically in the so-called High frequency Peak BL Lacs (HBL, as introduced by Padovani and Giommi, 1995), the synchrotron photons constitute the main target for the Compton scattering of electrons. If true, this would allow us to treat the radiation processes in these objects within the self-consistent SSC model. In its simplified one-zone, homogeneous version, this model assumes a spherical emission region filled with an isotropic electron population and a random magnetic field, moving towards the observer relativistically with constant Doppler factor δ_j. This model requires a minimum number of parameters, and provides conclusive predictions about the broad-band spectral and temporal characteristics.

10.4.1 *Constraints on the SSC parameter space*

Presently a general view dominates in the "blazar community" that the one-zone SSC model not only explains the X-ray/TeV correlations, but also gives satisfactory fits to the observed X-ray and TeV spectra of Mkn 421 and Mkn 501 . An example of the good spectral fits for the 1997 April flares of Mkn 501, obtained for fixed model parameters characterising the jet, and changing only the energy distribution of electrons, is shown in Fig. 10.5 (Tavecchio *et al.*, al. 2001).

It should be noted, however, that the γ-ray spectra shown are not corrected for the intergalactic absorption. As is seen in Fig. 10.3, this effect results in significant modification of the original γ-ray spectrum even for the "low" CIB model. It is clear that this effect cannot be neglected. At the same time, because of existing large uncertainties in the CIB flux, the effect of intergalactic absorption should be included in the modelling of the SEDs of TeV blazars as a free parameter.

The analysis of the spectral shape and variability of the synchrotron and IC components of non-thermal radiation of TeV blazars within the framework of a single-zone SSC model allows meaningful constraints on the parameter-space of the X-ray and γ-ray production region (Tavecchio *et al.*, 1998, Bednarek and Protheroe, 1999, Guy *et al.*, 2000). The key observables used in the derivation of the constraints on the jet parameters are the positions and amplitudes of the synchrotron and IC peaks, as well the CIB flux. Tavecchio et al. (1998) developed a convenient analytical approach, assuming that the electron spectrum is a broken power-law like the one described in the caption of Fig. 10.5. Later, Guy et al., (2000) incorporated the effect of intergalactic γ-ray absorption into this approach, and showed that this correction leads to a more consistent picture.

The homogeneous SSC model describes the emission of electrons in a single 'blob' with three basic model parameters: the radius R, magnetic field B, and Doppler factor of the bulk motion δ_j. Important information about the ratio R/δ_j is contained in the observed source variability time-scale, t_{var}. Thus, the constraints in the parameter-space for the given maximum

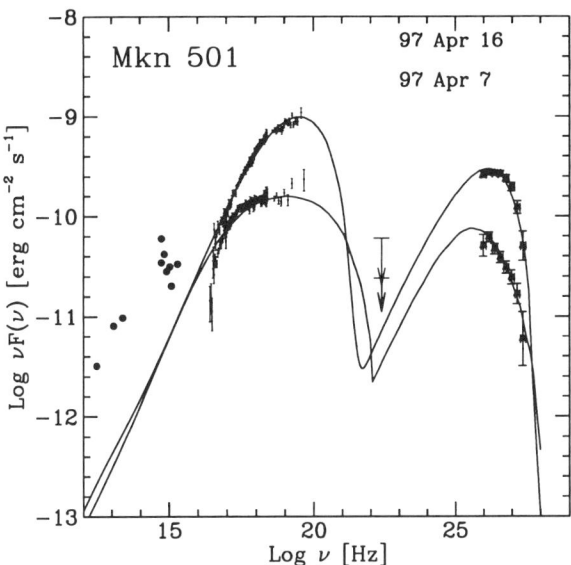

Fig. 10.5 Interpretation of the quasi-simultaneous X-ray (*BeppoSAX*) and TeV (CAT) observation of 1997 April 7 and 16 flares of Mkn 501 within the one-zone homogeneous SSC model (Tavecchio et al., 2001). It is assumed that the size $R = 1.9 \times 10^{15}$ cm, the Doppler factor $\delta_j = 10$ and the magnetic field $B = 0.32$ G of the 'blob' are unchanged. The spectral change of X-rays and γ-rays is attributed to the change in the energy distribution of electrons which is assumed to be a power-law with a low energy break γ_b and exponential cutoff at γ_{max}: $N(\gamma) = K_0 \, \gamma^{-n_1}(1+\gamma/\gamma_b)^{(n_1-n_2)} \exp{-(\gamma/\gamma_{max})}$ with $n_1 = 1.5$, $n_2 = 3$. It is assumed that the IC peak is formed in the Klein-Nishina regime and is mainly determined by the value of γ_b. For the flare of April 16 $\gamma_b = 7 \times 10^5$, while for the less intense April 7 flare $\gamma_b = 1.1 \times 10^5$, The amplitude of the electron distribution K_0 is assumed the same, thus the change of the overall luminosity both in X-ray and γ-rays is also determined merely by the change of γ_b. The intergalactic absorption of γ-rays is assumed to be negligible.

size, $R_\text{max} = t_\text{var} c \delta_\text{j}$, can be expressed in terms of two parameters on the ($\log B, \log \delta_\text{j}$) plane.

A distinct feature of the SSC model is the characteristic SED of radiation with two (synchrotron and Compton) bumps which, in the case of HBLs appear in the X-ray and the TeV γ-ray domains. The observations of X-rays require a population of relativistic electrons with a power-law spectrum broken at energy γ_b. The specific values of the indices are determined by the spectral shape of the synchrotron radiation below and above the synchrotron peak.

In the quiescent state, the synchrotron radiation of Mkn 501 is characterised by a synchrotron peak below 1 keV. During the April 16, 1997 flare the X-ray spectrum was exceptionally hard with a photon index of less than 2 at least up to 100 keV, indicating a dramatic shift of the synchrotron peak by at least two orders of magnitude (Pian et al., 1998). Due to the lack of statistics, both the synchrotron peak position and the synchrotron luminosity during the April 16 flare are not well defined.

The synchrotron spectrum of Mkn 501 in the high state can represented in the form of broken power-law with spectral indices $\alpha_1 = 0.5$ and $\alpha_2 = 1$ below and above the energy $\nu_\text{b} = 5.2 \pm 0.3 \times 10^{18}$ Hz (see Fig. 10.6). Additionally, a high energy cutoff at 300 keV is assumed, which could naturally be attributed to the cutoff in the electron acceleration spectrum. The index α_2 and the position of the cutoff are rather qualitative, but fortunately the final conclusions do not depend strongly on their exact values. The apparent synchrotron luminosity is then $\nu_s L(\nu_s) = 2.5 \pm 0.1 \times 10^{45}$ erg/s.

Generally, the SSC models predict a pronounced "Compton peak" in the SED of TeV blazars. However, as seen in Fig. 10.1, the observed γ-ray spectrum of Mkn 501 does not show any noticeable maximum. Remarkably, the γ-ray spectrum becomes closer to the theoretical predictions after being corrected for the intergalactic absorption. In particular, for the "low" CIB model in Fig. 10.1 with a scaling factor of 2.5, a clear "Compton peak" in the γ-ray spectrum appears around 2 TeV, with spectral indices ~ 0.5 and ~ 1.5 below and above the peak. Note that this is a feature predicted by the one-zone SSC model for the IC radiation, provided that the synchrotron X-radiation is described by a broken power-law with spectral indices $\alpha_1 = 0.5$ and $\alpha_2 = 1$ (Tavecchio et al., 1998, Krawczynski et al., 2000).

Three independent constraints in the ($\log B, \log \delta_\text{j}$) parameter plane are possible based on *(i)* the position of the synchrotron and Compton peak frequencies (area A), *(ii)* the synchrotron and Compton peak luminosities (area B), and *(iii)* the equilibrium between the radiative cooling and escape of electrons (area C).

The first condition leads to the following relation between B and δ_j

$$(A) \qquad B\,\delta_{10}^{-1} \approx 1.5 \frac{\nu_{s,19}}{\nu_{IC,26}^2} \qquad (10.3)$$

where the magnetic field B is in Gauss, $\delta_{10} = \delta_j/10$, $\nu_{s,19} = \nu_s/10^{19}\,\text{Hz}$, and $\nu_{IC,26} = \nu_{IC}/10^{26}\,\text{Hz}$.

The IC peak is very sensitive to the CIB absorption, and can be shifted by a factor of 3 (up to 10) towards higher frequencies for the absorption-corrected spectrum (Guy et al., 2000). Correspondingly, ignorance of the intergalactic γ-ray absorption would significantly overestimate the B/δ_j

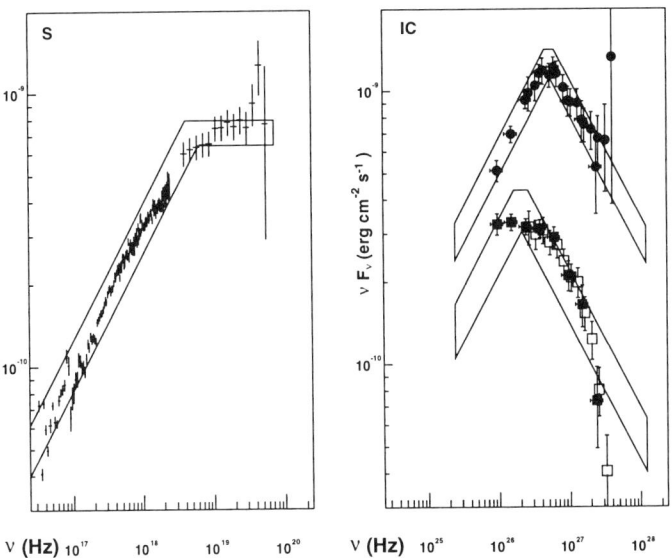

Fig. 10.6 The Spectral energy distribution of the April 16, 1997 flare of Mkn 501 obtained quasi-simultaneously in X-rays and TeV γ-rays by *BeppoSAX* and CAT (filled squares). The HEGRA 'time-averaged' spectrum obtained during the entire 1997 outburst of Mkn 501 and normalised to the flux around 1 TeV measured during the April 16 flare by CAT, is also shown (open squares). The filled dots correspond to the CIB absorption-corrected fluxes computed for the LCDM model of Primack et al. (1999), but rescaled (increased) by a factor of $SF = 2.5$. The 3 σ error boxes corresponding to the broken power-law model, are also shown (from Guy et al., 2000).

ratio. This is seen from comparison of the regions (A) in the the left and right panels of Fig. 10.7.

The comparison of the observed luminosities in the synchrotron and the IC peaks gives the second relation between the magnetic field and the Doppler factor:

$$(B) \quad B\delta_{10}^{2.5} \geq 0.5 \left(\nu_{s,19}\,\nu_{IC,26}\right)^{-1/4} \frac{(\nu_s L(\nu_s))_{45}}{(\nu_{IC} L(\nu_{IC}))_{45}^{1/2}} t_{var,h}^{-1}, \quad (10.4)$$

where $(\nu_s L(\nu_s))_{45}$ and $(\nu_{IC} L(\nu_{IC}))_{45}$ are the synchrotron and IC peak *apparent* luminosities in unites of 10^{45} erg s^{-1}, and $t_{var,h} = t_{var}/1\,h$ is the source variability time-scale. Here $t_{var,h}=10$ is assumed.

The region (B) shown in Fig. 10.7 is based on Eq.(10.4) with an allowed range of 3 σ uncertainties in the positions and apparent luminosities of the synchrotron and Compton peaks as described above. Note that ignorance of the CIB absorption of γ-rays would lead to the conclusion that the radiative cooling of electrons in the jet is well dominated by synchrotron losses. However, after correction for the intergalactic absorption, the IC luminosity in fact could be as high as (or even exceed) the synchrotron luminosity (see Fig. 10.6), implying that Compton scattering would become an equally important cooling channel.

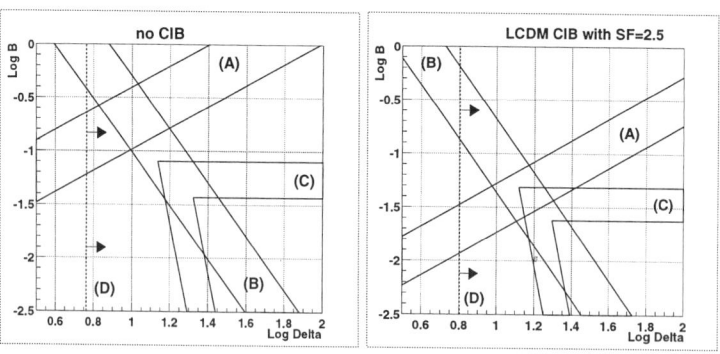

Fig. 10.7 Parameter-space for the April 16, 1997 flare of Mkn 501 (with $t_{var}=10$h) without (left panel) and with (right panel) correction for intergalactic γ-ray absorption (from Guy et al., 2000). The areas (A), (B), (C) and (D) are defined by Eqs. (10.3)-(10.7).

A relation between B and δ_j arises if one assumes that the break in the electron spectrum at γ_b is determined from the equilibrium between cooling and escape from the source (obviously, there could be other reasons for the break in the electron spectrum, *e.g.* connected with the character of the acceleration mechanism). This model assumption leads to the following constraints in the Compton-cooling dominated regime:

$$(C1) \quad B\delta_{10}^9 \geq 37.(\nu_s L(\nu_s)_{45})^2 \, t_{\text{var,h}}^{-2} \, \beta_{\text{esc}}^{-2} \nu_{\text{IC},26}^{-1} \tag{10.5}$$

and in the synchrotron-cooling dominated regime:

$$(C2) \quad B \geq 0.18 \, \beta_{\text{esc}}^{1/2} (t_{\text{var,h}} \, \nu_{\text{IC},26})^{-1/2} \tag{10.6}$$

Here β_{esc} is the electron escape velocity in units of the speed of light, i.e. a parameter describing the energy-independent escape time as $t_{\text{esc}} = \beta_{\text{esc}} R/c$. In Fig. 10.7 this rather uncertain parameter is allowed to vary from $1/3$ to 1.

To avoid strong absorption of TeV photons inside the 'blob' due to pair-production of γ-rays interacting with optical photons, a minimum value for δ can be computed; it is used to check the validity of the area defined by the intersection of the (A), (B) and (C) regions. For the Compton peak luminosity, we obtain

$$(D) \quad \delta > 6.6 \left(\frac{L(\nu_s)_{26}}{t_{var,h}} \sqrt{\nu_{IC,26} \nu_{s,19}} \right)^{0.2} \tag{10.7}$$

Fig. 10.7 shows that the inclusion of intergalactic absorption in the treatment results in more consistent relations between the different model parameters. In particular it moves region (A) into regions (B) and (C), and thus overcomes the incompatibility of different constraints on the principal model parameters. After correction for intergalactic absorption, all three regions (A), (B), and (C) in the ($\log B, \log \delta$) intersect around the point $B \simeq 0.05$ G, and Doppler factor $\delta \simeq 15$. This simple example shows that the intergalactic absorption of γ-rays should be taken into account in any realistic attempt to constrain self-consistently the parameter-space in future studies based on detailed modelling of temporal and spectral characteristics of TeV blazars.

10.4.2 Time-dependent SSC treatment

The results presented in Fig. 10.7 give important information about the range of principal parameters that can qualitatively describe the TeV observations of HBLs within the one-zone SSC model. However, this simplified and approximate procedure cannot supersede detailed numerical modelling of nonthermal processes in the jets. To test the SSC scenario for TeV blazars, a thorough theoretical treatment is needed even for the simplified one-zone version of the model. This concerns the time-dependent approach of the evolution of electron distributions as well as accurate calculations of the inverse Compton effect in the Klein-Nishina regime for both radiative losses of electrons and formation of γ-ray spectra. The key feature of this regime is that the Compton scattering transfers the energy of electron to the photon. This changes, for example, the mapping between the synchrotron and γ-ray emission components. Therefore in HBLs, the peaks of the synchrotron and Compton emission components generally are *not* produced by electrons of the same energy. Also, the Compton γ-ray flux at the highest energies tends to track the X-ray synchrotron flux only *linearly*, but not *quadratically* as expected in the Thomson regime. In some cases the Compton losses can strongly dominate over the synchrotron losses, and at the same time to proceed in the Klein-Nishina regime. In such conditions, we should expect very unusual combinations of X-ray and γ-ray spectra. This could be the case of 1ES 1426+428 (see below). Then approximations such as introductions a "Klein-Nishina cutoff" for the Compton cross-section, or using a continuous energy-loss approximation for solving of the electron energy distributions are quite dangerous and can easily lead to large errors in the spectra predicted for the peaks and tails of the Compton and synchrotron components (e.g., Coppi and Blandford 1990; Coppi, 1992; Coppi and Aharonian, 1999).

The second important requirement of SSC models (in fact, of any blazar model) is that they should treat self-consistently the time evolution of the electron population ensuring that the assumed electron energy spectrum is physically achievable from an initial acceleration spectrum, in contrast to (quite popular) parametric SSC fits which simply postulate the prompt electron spectra. The radiation of TeV blazars is extremely variable, and time-averaging obviously would discard significant information, thus making the interpretation of results rather ambiguous.

The potential of the time-resolved analysis to break model degeneracies has been demonstrated by Krawczynski *et al.* (2002) applying the time dependent SSC code of Coppi (1992) to the coordinated X-ray and γ-ray observations of Mkn 501 during the April-May outburst in 1997. It was assumed that freshly accelerated "external" electrons are injected into the emission generation region with a time-variable rate parameterised as function of the electron Lorentz factor γ, spectral index of particle acceleration s, amplitude $Q_0(t)$, minimum Lorentz factor $\gamma_{\min}$, and high energy cut-off

$\gamma_{\max}(t)$ as follows:

$$Q_e(\gamma, t) = Q_0(t)\gamma^{-s} \exp\left[-\gamma/\gamma_{\max}(t)\right]\Theta(\gamma - \gamma_{\min}) \qquad (10.8)$$

with $\Theta(x) = 0$ for $x < 0$ and $\Theta(x) = 1$ for $x \geq 0$. The canonical value of $s = 2$ expected for diffusive particle acceleration at strong shocks has been assumed.

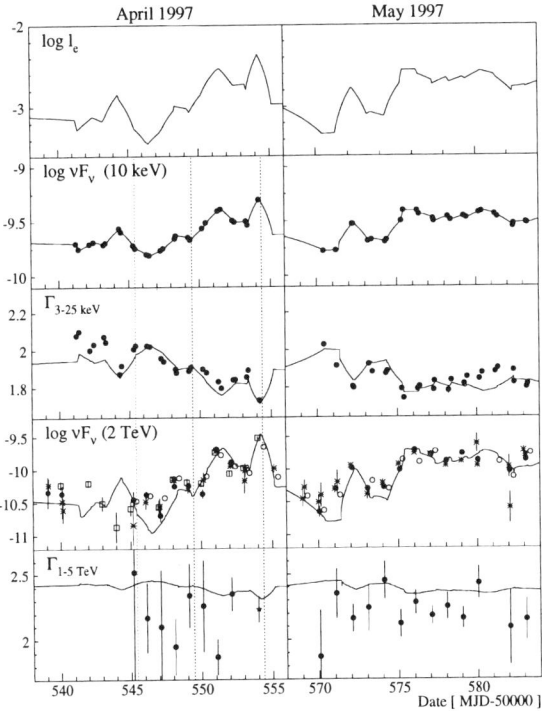

Fig. 10.8 Fit of a SSC model with two-emission components: (1) a quasi-stationary X-ray component, and (2) a time variable X-ray/TeV γ-ray component, assuming that flares are caused by the time-variable injection power of electrons, $Q_0(t)$. The panels, show from top to bottom: (i) the time-history of the injection power, (ii) the 10 keV X-ray energy flux (10^{-12} erg cm^{-2} s^{-1}), (iii) the X-ray 3-25 keV photon index, (iv) the 2 TeV energy flux (10^{-12} erg cm^{-2} s^{-1}), and (v) the 1-5 TeV photon index. The gamma-ray fluxes are from CAT (squares), HEGRA (solid points), and Whipple (open circles). The vertical dashed lines show the days with detailed spectroscopic observations by *BeppoSAX* (April 7, 11, and 16). The assumed model parameters are: $\delta_j = 45$, $R = 4.5 \times 10^{13}$ cm, $B = 1.1$ G, $t_{\text{esc}} = 10^4 (R/c)$, $\gamma_{\min} = 10^5$, $\gamma_{\max} = 1.4 \times 10^7$. (From Krawczynski *et al.*, 2002)

It is important to note that the low-energy cutoff in the spectrum of accelerated electrons γ_{min} is a critical model parameter. If the radiative cooling time of electrons with Lorentz factor γ_{min} is shorter than all the other characteristic timescales in the system, the main break of the electron spectrum occurs at γ_{min}. Thus, at high enough values ($\sim 10^5$), γ_{min} determines the energies at which the synchrotron and IC SEDs peak. Remarkably, all SSC fits require very large values for the low-energy cutoff (or break), $\gamma_{min} \geq 10^3$. The origin (theoretical ground) of γ_{min} remains unclear in contrast to the high energy cutoff γ_{max} which naturally arises from the

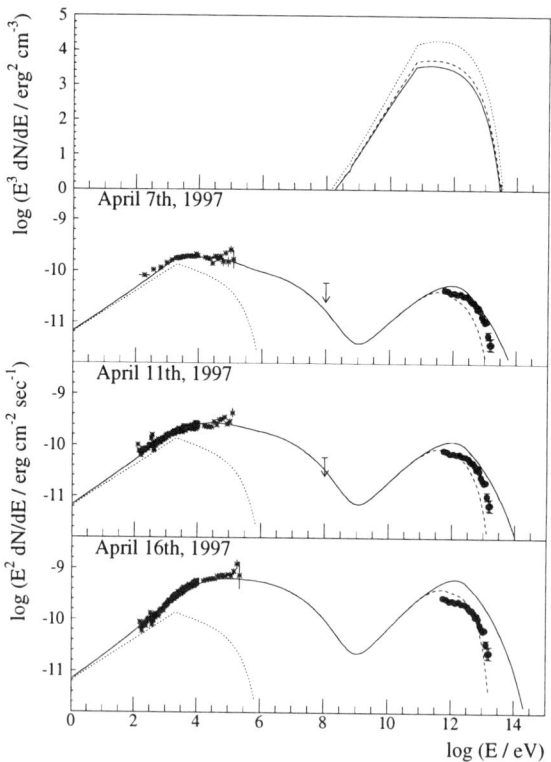

Fig. 10.9 For the model of Fig. 10.8, the upper panel shows the electron energy spectra ($E^3 \, dN/dE$, energy E in the jet frame) responsible for the observations of April 7 (solid line), April 11 (dashed line) and April 16 (dotted line) flares of Mkn 501 in 1997. The lower three panels compare the observed (points) with the modelled (solid line) SEDs. For illustrative purposes the dashed line shows the TeV gamma-ray energy spectra modified by extragalactic absorption. (From Krawczynski *et al.*, 2002).

balance between the electron acceleration and radiative cooling rates.

Seven free parameters of the model were assumed in the fitting procedure: the radius of the emission volume R, the jet Doppler factor δ_j, the mean magnetic field B, the escape time of relativistic electrons from the emission region t_{esc}, the electron acceleration rate Q_0, and the minimum and maximum Lorentz factors of accelerated particles γ_{min} and γ_{max}. The attempt was to fit the April and May flares of Mkn 501, including the RXTE 10 keV fluxes and 3-25 keV photon indices and the 2 TeV fluxes derived from CAT, HEGRA, and Whipple measurements. Given a hypothesis of what causes the flaring activity (a variable $Q_0(t)$, $\gamma_{max}(t)$, and/or $\delta_j(t)$), the fitting procedure consists of two steps:

(1) For a set of the postulated time-independent parameters (R, $\bar{\delta}_j$, B, γ_{min}, t_{esc}) the simplest possible time-dependent functions $Q_0(t)$ and $\gamma_{max}(t)$ which fit the X-ray flux amplitudes, are determined. These functions $Q_0(t_i)$, $\gamma_{max}(t_i)$ are derived iteratively by starting with a first guess, running the SSC model, and adjusting the relevant values until the X-ray fluxes are satisfactorily described.

(2) The time-independent parameters (R, $\bar{\delta}_j$, B, δ_j, t_{esc}) are varied to obtain the best fit to the observed X-ray photon indices and TeV gamma-ray flux amplitudes. Due to the large experimental uncertainties, fitting the TeV photon indices is not assigned a high priority. At the same time for 3 selected datasets (April 7, 11, and 16) for which relatively good spectral γ-ray measurements were available, the modelled photon indices were checked with the results from TeV observations.

The results of this study lead to quite interesting conclusions.

(i) One-component SSC models cannot fully describe the data. While, by construction, the models succeed in accounting for the temporal evolution of the X-ray fluxes they do not adequately explain the range of observed X-ray spectral indices, the broadband 0.1 keV-200 keV energy spectra, and the variation of TeV gamma-ray fluxes. Note, that this conclusion cannot be generalised, because it is applicable to the models in which only the rate of accelerated particles or only the maximum Lorentz factor of accelerated particles is assumed to be changed.

(ii) It is possible to get very good fits to the data, if one assumes that the X-rays originate from a superposition of a soft quasi-steady component and a hard rapidly variable component. Namely, postulating a quasi-steady X-ray component with a photon index of 2.2 and a 10 keV amplitude

$\nu F_\nu = 10^{-10}$ erg/cm^{2}s, the model assuming only a time dependent rate of accelerated particles $Q_0(t)$, can satisfactorily explain the absolute X- and γ-ray flux variations, as well as the spectral evolution of X-rays, for the following set of time-independent parameters: $\gamma_{\min} = 10^5$, $\gamma_{\max} = 1.7 \times 10^7$, $B = 1.1$ G, $R = 4.5 \times 10^{13}$ cm, $t_{\rm esc} = 10^4(R/c)$. This is demonstrated in Fig. 10.8. The above parameters are not quite similar to the "typical" parameters derived from the *parametric* SSC studies applied to the snapshot data (see Fig. 10.5). In particular, the required magnetic field is larger and the size is smaller than those assumed in the parametric SSC models. The results applied to the X-ray observations of April 7, 11, and 16 flares for which a nice spectral coverage was provided by *BeppoSAX*, give quite good agreement, achieved essentially assuming a very high value for the low-energy cutoff in the electron acceleration spectrum, $\gamma_{\min} = 10^5$. The results are presented in Fig. 10.9 where the modelled and observed TeV energy spectra are also shown. For all three days the time averaged 1997 HEGRA TeV gamma-ray energy spectrum normalised at 2 TeV to the mean flux (measured with all operational TeV telescopes) is used. The results shown in Fig. 10.8 and 10.9 generally prove that the time-dependent SSC model is indeed a powerful tool for further studies of applying electronic models to the flares of TeV blazars.

For illustrative purposes the dashed lines in Fig. 10.9 show the predicted TeV Gamma-ray energy spectra modified by extragalactic absorption, computed for the CIB model "LCDM, Salpeter Stellar Initial Mass Function" of Primack et al. (2001). The agreement of the predicted spectra with the data apparently is not very impressive. This may be interpreted as a result of a combination of different factors, in particular due to the simplifications in the proposed one-zone model, the lack of exactly simultaneous X-ray and TeV observations, uncertainties in the CIB fluxes, *etc.* More sophisticated modelling and better, simultaneously obtained spectrometric X-ray and TeV data are needed for further development of the theory. For example, we cannot ignore the fact that for CIB models predicting high NIR fluxes (as large as 20 nW/M^{2}s) the intrinsic γ-ray spectra of TeV blazars appear extremely hard. In particular, this is the case of Mkn 501 as it is shown in Fig. 10.3. Such spectra cannot be easily explained by the standard one-zone SSC model without invoking extreme jet parameters like Doppler factors $\delta_j \sim 50$, and very small magnetic fields $B \leq 0.1$ G, which significantly reduce the radiation efficiency. The situation is even more difficult for Mkn 421, the X-ray synchrotron cutoff of which never extends beyond 20 keV, and especially for the blazar 1ES 1426+428.

10.4.3 The case of 1ES 1426+428

Because of the relatively large redshift ($z = 0.129$), the TeV gamma radiation from the extreme BL Lac object 1ES 1426+428 (Costamante *et al.*, 2001) suffers severe intergalactic absorption. Therefore the discovery of TeV γ-rays from this source (see Chapter 2) arrived as a pleasant surprise. The deformation of the primary γ-ray spectrum due to the intergalactic absorption is characterised by the factor $\exp(-\tau)$, where $\tau(E)$ is the energy-dependent optical depth for a γ-ray photon of energy E emitted by a source at a distance d. The spectral deformation factors calculated for 3 essentially different CIB models show that above 300 GeV the intergalactic absorption leads to a strong steepening of the spectrum of γ-rays from 1ES 1426+428, but starting from 1 TeV to several TeV the spectrum is deformed only slightly, although the suppression of the absolute flux may be as large as a factor of 100 (see Fig. 10.10). Therefore the initial (source) spectrum of γ-rays, i.e. the γ-ray spectrum after correction for the intergalactic absorption, $J_0(E) = J_{\text{obs}}(E)\exp[\tau(E)]$, is expected to be quite different from the observed spectrum $J_{\text{obs}}(E)$.

An important implication of this approach of reconstruction of the source spectrum of 1ES 1426+428 is that the latter obtains a reasonable shape only for a rather limited class of models of CIB. Namely, any significant deviation of these models from the CIB model-1 shown in Fig. 10.1, especially at wavelengths between 1 and 10 μm, results in an unusually steep upturn in the reconstructed γ-ray spectrum. This is demonstrated in Fig.2.21 (Chapter 2). To some extent here we face a problem similar to the problem of possible existence of a pile-up in the source spectrum of Mkn 501 (see Fig. 10.3). Note that the sharp positive slopes of the γ-ray spectra ($E^2 dN/dE \propto E^3$) in Figs. 2.21b and 2.21c cannot be easily explained by conventional models invoking either a leptonic or a hadronic origin of γ-ray emission of BL Lac objects.

In principle, in the case of γ-ray production from monoenergetic protons interacting with ambient plasma (see Sec. 10.5.1), a $E^2 dN/dE \propto E^2$ type spectrum could be produced. However, the steeper ($\propto E^3$) rise of the reconstructed γ-ray spectrum unambiguously excludes this possibility. A sharper pile-up could be expected by speculating that the radiation is the result of bulk motion Comptonization of ambient low-energy thermal photons by a cold conical wind with bulk Lorentz factor exceeding 10^7 (see Sec. 10.6).

In any case, it is likely that we should be prepared for unusual conclusions concerning the acceleration and radiation processes in the small scale AGN jets. Note that even a "decent" power-law intrinsic TeV spectrum of 1ES 1426+428 ($E^2 dN/dE \propto E^{0.5}$), obtained for a model-1 type spectrum of CIB (see Fig. 2.21a), does not yet guarantee an easy explanation of the broad-band spectral energy distribution of this source. Independent of the details, it may require a significant revision of the current conceptual view according to which the synchrotron (X-ray) peak in the SED dominates over the inverse Compton (TeV) peak. Indeed, although the detected TeV flux is only few times 10^{-12} erg/cm^{2}s, after correction for the intergalactic absorption it exceeds 10^{-10} erg/cm^{2}s, i.e. it becomes at least an order of magnitude larger than the X-ray energy flux (Costamante et al., 2003). Thus, the TeV luminosity significantly exceeds the X-ray luminosity, and it is clear that this should have an impact on the TeV-blazar models. For the inverse Compton models, this would imply that the radiative cooling of electrons proceeds in the regime dominated by Compton losses, like in

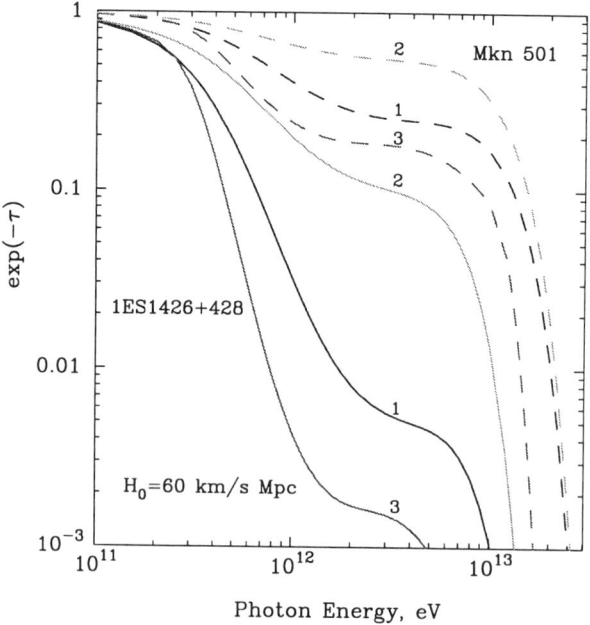

Fig. 10.10 Impact of the intergalactic absorption on the TeV spectra of Mkn 501 (dashed curves) and 1ES 1426+428 (solid curves). The curves marked as "1", "2", and "3" correspond to the absorption factors $\exp(-\tau)$ calculated for CIB models 1, 2, and 3 shown in Fig. 10.1.

the GeV blazars (or "red blazars"), but with the main energy release in the TeV domain, like in Mkn 421 and Mkn 501 ("blue blazars").

It should be noted that the two bumps in the spectral energy distributions of TeV blazars are always called "synchrotron" and "inverse Compton" peaks which tacitly implies that the leptonic origin of the TeV radiation is accepted as an undisputed fact, and all possible difficulties with the interpretation of data within this concept can be blamed on the imperfectness of the existing models. This currently popular view dominates over the so-called hadronic models. However, the TeV observations of blazars, in particular 1ES 1426+428, indicate that it is quite important to develop, parallel with improvement of leptonic models, alternative scenarios, attributing, for example, the TeV emission to the synchrotron radiation of protons, and the X-ray emission to the synchrotron radiation of primary (directly accelerated) or secondary (pair-produced) electrons (see Sec. 10.5.3.3).

10.5 Hadronic Models

Presently, the leptonic models, independent of the specific assumptions concerning the acceleration and radiation scenarios, represent the preferred concept for TeV blazars. Two important intrinsic features characterise these models: (i) electrons can be readily accelerated to TeV energies, e.g. through the shock acceleration mechanism, and (ii) they can radiate X-rays and TeV γ-rays with very high efficiency via synchrotron and inverse Compton channels.

The so-called *hadronic* models are generally lacking in these virtues. Although the associated acceleration of protons is expected with at least the same efficiency as that of the electrons (for most acceleration mechanisms), the hadronic models require proton acceleration to energies up to 10^{20} eV, otherwise they cannot offer efficient γ-ray production mechanisms in the jets. These models assume that the observed γ-ray emission is initiated by accelerated protons interacting with ambient mater (the so-called *mass-loaded* hadronic models; see e.g. Pohl and Schlickeiser, 2000), photon fields (photo-pion hadronic models; see e.g. Mannheim, 1993; 1996), magnetic fields ("pure" proton synchrotron model; Aharonian, 2000) or both magnetic and photon fields (Mücke and Protheroe, 2001; Mücke *et al.*, 2003).

10.5.1 Mass-loaded hadronic jet models

While the synchrotron and photo-pion hadronic jet models require acceleration of protons to extremely high energies, the mass-loaded hadronic models need protons of relatively modest energies. Yet, because of the low density of the thermal electron-proton plasma in the jet, the efficiency of this mechanism appears to be too low to explain the observed time variability and the high fluxes of the observed gamma-radiation (Aharonian et al., 1999c). Formally, the variability can be explained by invoking adiabatic losses caused by relativistic expansion of the emitting 'blob'. However, this assumption would imply, in fact, very inefficient γ-ray production with an intrinsic luminosity $L_\gamma^{(\rm int)} \simeq \dot{W}_{\rm p}(t_{\rm ad}/t_{\rm pp})$, where $\dot{W}_{\rm p}$ is the proton acceleration rate, $t_{\rm ad} \geq R/c \sim \Delta t_{\rm var}\delta_{\rm j}$ is the adiabatic cooling time, and $t_{\rm pp} \simeq 5 \times 10^{15}(n/1~{\rm cm}^{-3})^{-1}$s denotes the characteristic emission timescale for the production of π^0-decay γ-rays. Given that the proton acceleration rate cannot exceed the total power of the central engine, roughly the Eddington luminosity $L_{\rm Edd} = 1.3 \times 10^{45}(M/10^7 M_\odot)$ erg/s, and that $L_\gamma^{(\rm int)} \simeq 4\pi d^2 f_\gamma \delta_{\rm j}^{-4}$, where d is the luminosity distance, and f_γ is the detected γ-ray flux, one obtains

$$n \geq 4 \times 10^7 ~ f_{-10} ~ \Delta t_{\rm 3h}^{-1} ~ \delta_{10}^{-5} M_7^{-1}(d/1~{\rm Gpc})^2~{\rm cm}^{-3}~, \qquad (10.9)$$

where the following normalisations are used: $f_{-10} = f_\gamma/10^{-10}$ erg/cm²s, $\delta_{10} = \delta_{\rm j}/10$, $M_7 = M/10^7 M_\odot$, and $\Delta t_{\rm 3h} = \Delta t/3$ h is the radiation variability in the observer's frame in units of 3 hours - the characteristic time-scale observed from TeV blazars. For example, for Mkn 501 ($d \simeq 170$ Mpc) in a high state with parameters $f_{-10} \geq 1$, $\Delta t_{\rm 3h} \sim 1$, and $\delta_{10} \sim 1$, the baryonic density of plasma in the jet should exceed 10^6 cm^{-3}. This makes the relativistically moving 'blob' very heavy ($M \simeq 2.5 \times 10^{29}$ g) and the corresponding kinetic energy uncomfortably large, exceeding $E_{\rm kin} = Mc^2(\Gamma_{\rm j} - 1) \geq 2 \times 10^{51}$ erg.

Strictly speaking, this argument holds only against the *mass-loaded* hadronic models based on the assumption that γ-rays are produced *inside* small scale jets. However it does not exclude the 'pp' hadronic models in general. For example, an *external* scenario like the "relativistic jet meets target" (Morrison et al., 1984), i.e. assuming that γ-rays are produced in nearby dense gas clouds that (randomly) move across the jet (e.g. Dar and Laor 1997), remains an attractive possibility for the 'pp' hadronic models.

10.5.2 Photo-pion and synchrotron losses of protons

The models involving interactions of protons with radiation and magnetic fields require proton acceleration to extreme energies exceeding 10^{19} eV. This implies that the acceleration should proceed at the maximum (theoretically possible) rates. On the other hand, the necessary (for any TeV blazar model) condition of high efficiency of radiative cooling requires extreme parameters characterising the sub-parsec jets and their environments. But this should not prevent us from exploring hadronic models as a viable alternative to the leptonic models.

The efficiency of any γ-ray production mechanism is characterised by the ratio of the radiative cooling time to the typical dynamical time-scale of the source corresponding to the minimum variability of radiation in the jet's frame

$$t^* = \Delta t_{\text{var}} \, \delta_j \simeq 10^5 \, \Delta t_{3h} \delta_{10} \text{ s} \, . \tag{10.10}$$

Proton-synchrotron radiation becomes an effective mechanism of γ-radiation with characteristic cooling time $\leq 10^5$ s, only for protons with $E \geq 10^{19}$ eV in a strong magnetic field close to 100 G (see Eq.3.29). In this regime, the synchrotron losses may well dominate over non-radiative losses caused by adiabatic expansion or the escape of particles from the source.

It is interesting to compare the proton-synchrotron cooling time t_{sy} with the photo-pion cooling time, $t_{p\gamma}$. The energy flux of low-frequency radiation that originates in the jet can be presented in the following form

$$\nu S_\nu = 10^{-12} g_{\text{fir}} (h\nu/0.01 \text{ eV})^{-s+1} \text{ erg/cm}^2\text{s} \tag{10.11}$$

where g_{fir} is a scaling factor indicating the level of the FIR flux at $h\nu = 0.01$ eV ($\lambda \sim 100\,\mu\text{m}$) in units of 10^{-12} erg/cm^{2}s. The flux normalisation at FIR wavelengths, which play a major role in proton-photon interactions in the jet, makes the results of calculations quite insensitive to the choice of the spectral index s. Generally, for BL Lac objects, in particular for Mkn 421 and Mkn 501, $g_{\text{fir}} \leq 1$.

For a source with the *co-moving frame luminosity* $L'(\nu')\nu$, an observer at a distance $d = cz/H_0$ would detect a flux (Lind & Blandford 1985) $S_\nu = \delta_j^3 L'(\nu/\delta_j)/4\pi d^2$ The energy losses of protons in a low-frequency photon field with a broad power-law spectrum are dominated by photo-meson processes. For broad and flat spectra of target photons the cooling time of protons is estimated as $t_{p\gamma}(E) \simeq (c <\sigma_{p\gamma} f> n(\nu^*)h\nu^*)^{-1}$, thus for the flux given by Eq. (10.11) with s=0.5,

$$t_{p\gamma} \simeq 4.5 \times 10^7 \, \Delta t_{3h}^2 \, \delta_{10}^{5.5} \, (z/0.03)^{-2} \, g_{\text{FIR}}^{-1} \, E_{19}^{-0.5} \, \text{s}, \qquad (10.12)$$

where $<\sigma_{p\gamma} f> \simeq 10^{-28} \, \text{cm}^2$ is the photo-pion production cross section weighted by the inelasticity at a photon energy $\sim 300 \, \text{MeV}$ in the proton rest frame, and $h\nu^* \simeq 0.03 \, E_{19}^{-1} \, \text{eV}$. For both Mkn 501 and Mkn 421 the photo-pion cooling time cannot be less than $10^7 \, \text{s}$, unless we assume a very high ambient photon density. Formally this could be possible, for example, adopting a smaller blob size than follows from the observed flux variability $\Delta t_{3h} \sim 1$, and/or assuming a small Doppler factor of the jet, $\delta_j \ll 10$. However, the photon density in the source cannot be *arbitrarily* increased, otherwise it would result in a catastrophic absorption of TeV radiation inside the source.

In the field of ambient photons with a differential power-law spectrum $n(\nu) \propto \nu^{-(s+1)}$, the optical depth for photon-photon absorption is equal to $\tau_{\gamma\gamma}(\epsilon) = A(s)(\sigma_T/2) \, h\nu_0 n(h\nu_0) R$, where $h\nu_0 = 4(m_e c^2)^2/\epsilon$, and $A(s) = 7/12 \cdot 4^{s+1}(s+1)^{-5/3}/(s+2)$ (Svensson, 1987). For the spectral index $s = 0.5$,

$$\tau_{\gamma\gamma} \simeq 0.4 \, \Delta t_{3h}^{-1} \, \delta_{10}^{-5} \, g_{\text{fir}} \, (z/0.03)^2 \, (\epsilon/1 \, \text{TeV})^{1/2}. \qquad (10.13)$$

The optical depth $\tau_{\gamma\gamma}$ increases with energy increases $\propto \epsilon^{1/2}$, therefore at 1 TeV it should not exceed 1, otherwise the absorption of $\geq 10 \, \text{TeV} \, \gamma$-rays would become unacceptably large. For example, for Mkn 501, assuming that the low-frequency radiation in the blob is described by Eq. (10.11) with $s = 0.5$ and adopting a rather relaxed estimate for $g_{\text{fir}} \sim 0.3$ (e.g. Pian *et al.*, 1998), the condition $\tau_{\gamma\gamma}(1 \, \text{TeV}) \leq 1$ results in a robust lower limit on the blob's Doppler factor $\delta_j \geq 7$.

From Eqs.(10.12) and (10.13) we obtain a simple relation between the photo-pion cooling time of protons, $t_{p\gamma}$, and the optical depth $\tau_{1\text{TeV}}$:

$$t_{p\gamma} \simeq 1.8 \times 10^7 \, \Delta t_{3h} \, \delta_{10}^{1/2} \, \tau_{1\text{TeV}}^{-1} \, E_{19}^{-1/2} \, \text{s} \,. \qquad (10.14)$$

For a steeper spectrum of low-frequency radiation, e.g. a power-law with $s = 1$, and for the same normalisation to $\tau_{1\text{TeV}}$, the photo-pion cooling time is shorter:

$$t_{p\gamma} \simeq 10^6 \, \Delta t_{3h} \, \tau_{1\text{TeV}}^{-1} \, E_{19}^{-1} \, \text{s} \,. \qquad (10.15)$$

Even so, $t_{p\gamma}$ remains significantly larger than t^* given by Eq.(10.10), especially if we take into account that for $s = 1$ the optical depth depends more strongly on energy, $\tau_{\gamma\gamma} \propto \epsilon$.

The source transparency condition for multi-TeV γ-rays implies that in TeV blazars the photo-pion processes proceed on significantly larger time-scales compared with t^* and $t_{\rm sy}$ (see Fig. 10.11). In particular, for the power-law spectrum of low-frequency radiation with $s = 0.5$ we have $t_{p\gamma}/t^* \simeq 1.7 \times 10^2 \, \delta_{10}^{-0.5} \, \tau_{1{\rm TeV}}^{-1} \, E_{19}^{-0.5}$ and $t_{p\gamma}/t_{\rm sy} \simeq 4 \times 10^2 \Delta t_{3h} \, \delta_{10}^{0.5} \, B_{100}^{-2} \, \tau_{1{\rm TeV}}^{-1} \, E_{19}^{0.5} E$.

Thus, for any reasonable assumption concerning the geometry of γ-ray production region, as well as the spectral shape of low-frequency radiation in the blob, the detection of TeV γ-rays from any blazar would imply a low efficiency of the photo-pion processes in the jet, unless the energy of protons does not significantly exceed 10^{19} eV. In compact γ-ray production region(s) of the jet with a typical size $R \leq 3 \times 10^{15} \Delta t_{3h} \delta_{10}$ cm, proton acceleration to such high energies is possible only in the presence of a magnetic field $B \gg 10\,{\rm G}$. In such conditions, however, synchrotron radiation becomes a more effective channel for radiative cooling of protons. In principle, the difficulty with synchrotron losses could be overcome by adopting

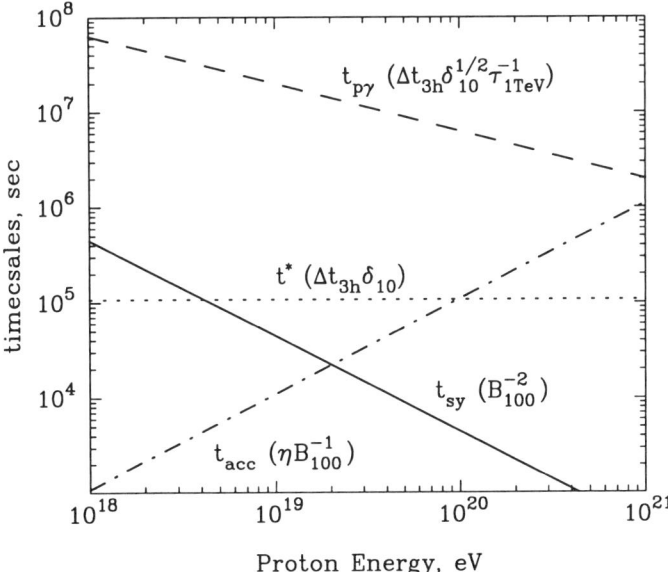

Fig. 10.11 The characteristic acceleration and energy loss (synchrotron, photo-meson and adiabatic) time-scales of protons in the blob. At each curve the scaling factors (the products of relevant physical parameters) are shown.

a weak ($B \leq 1$ G) magnetic field in the blob, but assuming that the EHE protons are accelerated outside of the blob, e.g. near the central compact object, and then transported along with the jet. However, this assumption does not solve the second problem connected with the low efficiency of the photo-pion processes in the jet imposed by the transparency condition for TeV γ-rays.

A regrettable consequence of this conclusion is the suppressed TeV neutrino flux. This however concerns only the objects seen in TeV γ-rays. The pair cascades initiated by secondary electrons and γ-rays from $p\gamma$ interactions may still remain a viable possibility for other AGN, in particular for the powerful GeV blazars detected by EGRET, where the radiation density is much higher than in the BL Lac objects, and, more importantly, the photo-pion cooling time of EHE protons is not constrained by the severe TeV γ-ray transparency condition.

10.5.3 *Proton synchrotron models*

The inner parts of AGN jets with magnetic fields at the level of $B \sim 10 - 100$ G are one of a few possible sites in the Universe for the production of cosmic rays to the highest observed energies of about 10^{20} eV (see Fig. 1.6), provided that the particle acceleration in the compact regions of inner jets proceeds at the rate $\eta r_g/c$ with $\eta \sim 1$, and that synchrotron or curvature radiation losses dominate over other radiative and non-radiative losses. If such an arrangement of favourable conditions indeed takes place in blazars, it seems quite natural to invoke the synchrotron radiation of protons for explanation of TeV radiation from blazars. Indeed, as discussed in Sec. 3.3.2, if protons are accelerated in the regime dominated by synchrotron losses, the spectral shape of the Doppler boosted γ-radiation in the observer's frame is determined essentially by the self-regulated synchrotron cutoff at $E_0 \simeq 3\eta^{-1}\delta_{10}$ TeV. Thus, if the proton acceleration takes place at the highest possible rate ($\eta \sim 1$), for typical Doppler factors $\delta j \geq 10$, the proton synchrotron radiation would extend well into the TeV domain. Remarkably, this process allowing very hard intrinsic γ-ray spectra, gives a reasonable explanation for the stable spectral shape of TeV emission as observed during the strong flares of Mkn 501 in 1997, and predicts significant spectral steepening in low states.

10.5.3.1 Fitting the TeV spectrum of Mkn 501

Fig. 10.12 demonstrates the ability of the proton synchrotron model to explain the spectra of TeV radiation of Mkn 501 both in high and low states. For the high state, the model spectrum is normalised, after correction for the intergalactic absorption (assuming CIB model no.2 in Fig. 10.1), to the measured flux at 1 TeV. This determines, for a given magnetic field, the required total energy in accelerated protons. For a proton energy distribution described by a "power-law with exponential cutoff" with $\alpha_p = 2$, the

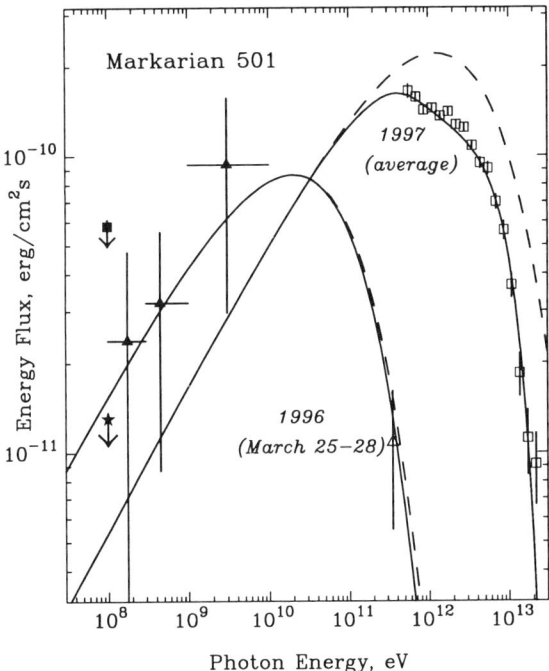

Fig. 10.12 The proton-synchrotron radiation of Mkn 501 in a high and low states. The dashed and solid curves correspond to the spectra of γ-radiation before and after corrections for intergalactic extinction, respectively. The time-averaged HEGRA data in the high state are the same as in Fig.10.3. The broad-band γ-ray data in the low state are obtained by the Whipple (open triangle) and EGRET (filled triangles) groups during the multiwavelength campaign in March 1996 (Kataoka et al. 1999). An "archival" upper limit on the 100 MeV flux based on the long-term observations of the source during the phase I period of EGRET is also shown (star). The high state of the source is fitted by $\alpha_p = 2$ and $E_0 = 1.3\,\text{TeV}$. The low state of the source is fitted by $\alpha_p = 2$ and $E_0 = 0.02\,\text{TeV}$ (from Aharonian, 2000).

spectrum of Mkn 501 in a high state is surprisingly well fitted assuming $E_0 = 1.3\,\text{TeV}$. If γ-rays are produced in the *synchrotron-loss-dominated* regime, from Eq. (3.35) we can easily estimate the ratio $\rho = \delta_{10}/\eta$ which is the most relevant parameter for determination of the position of the self-regulated synchrotron cutoff in a relativistically moving source – $\rho = 0.43$. Because the Doppler factor in Mkn 501 is believed to be close to 10, the conclusion could be drawn that during the entire 1997 outburst the acceleration of particles in Mkn 501 took place in the regime quite close of maximum acceleration rate ($\eta \sim 2$). If so, this explains in a rather natural way the essentially time-independent shape of the TeV spectrum observed during the strong flares in 1997. Comparing the minimum (energy-independent) escape time of particles $t_{\text{esc}} \sim R/c \simeq 3.3 \times 10^4\, R_{15}\xi$ s, with the synchrotron cooling time of the protons responsible for the production of γ-rays in the cutoff region, $t_{\text{sy}} \simeq 2.4 \times 10^4 (B/100\text{G})^{-3/2} \eta^{1/2}$ s, the following condition for formation of the self-regulated synchrotron cutoff should be satisfied

$$B \geq 80\ R_{15}^{-2/3} \eta^{1/3}\ \text{G}\ . \qquad (10.16)$$

If the particle acceleration takes place in the regime dominated by synchrotron losses, we may expect spectral variability caused by changes in the parameter ρ, i.e. due to variability of the acceleration rate and/or of the Doppler factor of the jet. Even a slight drop in the acceleration rate should lead to dramatic spectral changes of γ-rays. This effect can explain the detection by EGRET of a very flat spectrum at energies from 50 MeV to 5 GeV during the period of a low state of the source in March 1996. A rather surprising aspect of this report is that the energy flux of γ-rays at several GeV was significantly larger than the X-ray and TeV fluxes measured approximately at the same time by the ASCA and Whipple instruments (Kataoka *et al.*, 1999). In the framework of this model, the EGRET and Whipple fluxes could be explained with parameters $\alpha_p = 2$ and $E_0 = 0.02\,\text{TeV}$ (see Fig. 10.12), or, if the proton acceleration takes place in the regime dominated by synchrotron losses, $\rho \simeq 0.007$. This would imply a dramatic reduction of the acceleration efficiency ($\eta \simeq 140\delta_{10}$), provided that the jet's Doppler factor remains more or less constant. Note that this interpretation adopts a very small ρ parameter, but tacitly assumes that the proton acceleration takes place in the *synchrotron-loss-dominated* regime. Another, perhaps more realistic, scenario would be the source transition from the *synchrotron-loss-dominated* regime to the *particle escape-loss-dominated* regime, for example, due to a possible reduction of the magnetic field caused by the expansion of the blob.

The signature of the low state is a shift of the position of the spectral peak down to ≤ 100 GeV, with a hard GeV γ-ray spectrum, and very steep spectrum at TeV energies. While the hard GeV spectrum is indeed observed, the small TeV photon statistics do not provide adequate information about the spectrum at TeV energies. The γ-ray spectrum formed in the *escape-loss-dominated regime* depends on parameters characterising the production region, in particular the source size and the magnetic field. Therefore, at this stage of evolution, the source should show significant spectral variations at TeV energies, although in the GeV domain the flux variability should not necessarily be accompanied by spectral changes.

10.5.3.2 *X-rays from secondary electrons*

Within the proton synchrotron model, we may expect also synchrotron radiation produced by *directly accelerated* electrons - the counterparts of EHE protons. If the particle acceleration takes place in the synchrotron-loss-dominated regime, the self-regulated synchrotron cutoff of this component depends only on the parameter $\rho = \delta_{10}/\eta$, namely $E_0 \simeq 1.6\rho$ GeV, thus it correlates with the position of the cutoff in the proton synchrotron spectrum at $E_0 \simeq 3\rho$ TeV. The ratio of the energy fluxes of these two components is determined simply by the ratio of non-thermal energy channelled into the accelerated protons and electrons, $\dot{W}_\mathrm{p}/\dot{W}_\mathrm{e}$. In a magnetic field of about 100 G, the cooling time of electrons is shorter, almost at all relativistic energies, than the typical dynamical (e.g. light-crossing) times. This results in a well-established steady-state spectrum of electrons $\mathrm{d}N/\mathrm{d}E_\pm \propto E_\pm^{-(\alpha_0+1)}$, provided that the power-law index of acceleration spectrum $\alpha_0 \geq 1$. Consequently, a power law synchrotron spectrum with a photon index $(\alpha_0 + 2)/2$ would be formed. In particular, for $\alpha_0 = 2$, we should expect a flat synchrotron SED ($\nu S_\nu = const$) from optical/UV wavelengths to MeV/GeV γ-rays. The broad-band spectra of both Mkn 421 and Mkn 501 do not agree with such a pure power-law behaviour; in fact, the SED of both objects show pronounced synchrotron X-ray peaks. Therefore this (theoretically possible) population of directly accelerated electrons - counterparts of the EHE protons - cannot be responsible for the bulk of X-ray emission.

The X-ray emission could instead be due to electrons, produced in a different way, and/or in other region(s) of the jet. The X-ray light curves of some HBLs show so-called "soft" and "hard" lags (e.g. Takahashi *et al.*, 1999). A possible interpretation of this effect in terms of competing acceleration, radiative cooling, and escape timescales of synchrotron-emitting electrons (Takahashi et al. 1996; Kirk et al. 1998), requires that all these timescales are comparable with the light crossing time, $t = R/c \sim 10^4 - 10^5$ s. The cooling time of electrons responsible for a synchrotron photons of energy E_x is $t_{sy}^{(e)} \simeq 1.5 \times 10^3 (B/1\,\text{G})^{-3/2} (E_x/1\,\text{keV})^{1/2}$ s. Therefore in the X-ray production region the magnetic field cannot, independent of specific model assumptions, significantly exceed 0.1 G. Thus, the hypothesis of a proton-synchrotron origin of TeV radiation implies that the production regions of TeV γ-rays ($B \sim 100$ G) and synchrotron X-rays ($B \sim 0.1$ G) should be essentially different.

In the proton-synchrotron model of TeV radiation from BL Lac objects, two more components of X-radiation are expected. Indeed, in a field of about 100 G the accelerated protons of energy $E \sim 10^{15}$ eV *themselves* produce synchrotron X-rays. However, the contribution of this component to the observed X-ray flux is negligible. A much more prolific channel for X-ray production connected (indirectly) with the EHE protons, can be provided by electrons of non-acceleration origin, namely by secondary electrons produced at interactions of the primary TeV γ-rays with the ambient low-frequency radiation.

The appearance of secondary electrons in the jet results in production of a hard synchrotron X-ray component. It is clear that for an optical depth of $\tau_{\gamma\gamma} \sim 1$ the luminosity of this component would be comparable to to the luminosity of their "grandparents" - TeV γ-rays. The photoproduced electrons have a rather specific spectral shape, significantly different from that of directly accelerated particles. For example, the spectrum of electrons produced at interactions of high energy γ-rays with a photon index Γ and field photons with a narrow (e.g. Planckian) spectral distribution with a characteristic energy $\overline{h\nu}$, has the following characteristic form: starting from the minimum (allowed by kinematics) energy at $E_* = m_e^2 c^4 / 4\overline{h\nu}$, the electron spectrum sharply rises reaching the maximum at $E_m \simeq 2.4 E_* \simeq 0.15 (\overline{h\nu}/1\,\text{eV})^{-1}$ TeV, and then at $E \gg E_m$ it decreases as $q_\pm \propto E_\pm^{-(\Gamma+1)} \ln E_\pm$ (see Eq.(3.27) in Chapter 3).

The intense synchrotron losses quickly establish a steady-state broken power-law electron spectrum $\propto E_\pm^{-2}$ below E_m, and $\propto E_\pm^{-(\Gamma+2)}$ (if we ignore the weak logarithmic term) above E_m. Correspondingly, the synchrotron spectrum of the secondary pair-produced electrons is characterised by a smooth transition, from $\nu S_\nu \propto E_x^{0.5}$ to $\nu S_\nu \propto E_x^{-(\Gamma-1)/2}$, with a break point at $E_{x,b} \sim 100 (B/100\,\text{G})(\overline{h\nu}/1\,\text{eV})^{-2}$ keV.

Below we discuss the possibility of explaining both the X-ray and TeV γ-ray fluxes from 1ES 1426+428 within the 'pure' proton-synchrotron model.

10.5.3.3 Broad band SED of 1ES 1426+428 within the proton-synchrotron model

As discussed in Sec. 10.4.3, the detection of TeV γ-rays from 1ES 1426+428 may challenge the standard leptonic models of TeV blazars. The apparent luminosity of γ-rays after correction for the intergalactic absorption appears at least an order of magnitude larger than the synchrotron luminosity. Within the one-zone SSC model, this would imply a very low magnetic field, a very high density of relativistic electrons, and a correspondingly very strong (orders of magnitude) deviation from the equipartition condition. The external Compton models and the models of electromagnetic cascade in radiation-dominated environments allow more flexibility in modelling the unusual SED of this source.

Alternatively, the hard intrinsic TeV γ-ray spectrum of 1ES 1426+428 and high L_γ/L_X ratio (≥ 10) can be explained by the *proton-synchrotron* model, assuming that a small, approximately 1/10 fraction of the proton-synchrotron radiation suffers internal absorption. If so, the hard X-ray emission could be attributed to the synchrotron radiation of the secondary pair-produced electrons as discussed in the previous section. The detailed numerical calculations shown in Fig. 10.13 prove that indeed a simple proton-synchrotron model can reproduce the broad-band spectral energy distribution of 1ES 1426+428 quite well.

The calculations have been performed within the "leaky-box" approximation, assuming that all primaries (protons) and secondaries (electrons, photons) – are homogeneously distributed in the source, and that the probability of particles escaping the confinement region does not depend on their spatial coordinates. The escape time from a source with a characteristic size R is defined for all particles in the terms of free escape, $t_f = R/c \simeq 3.3 \times 10^4 R_{15}$ s. In addition, some calculations were performed for the regime of Bohm diffusion of charged particles with characteristic time $t_B \simeq 1.5 \times 10^5 B_{100} R_{15}^2 / E_{19}$ s. Note that these timescales are comparable with the synchrotron cooling time of *primary* (accelerated) extremely high energy protons, $t_{sy}^{(p)} \simeq 4.5 \times 10^4 B_{100}^{-2} E_{19}^{-1}$ s. At the same time, the synchrotron cooling time of electrons is much shorter than any of these timescales, $t_{sy}^{(e)} \simeq 0.04 B_{100} E_{12}^{-1}$ s ($B_{100} = B/100\mathrm{G}$, $E_{19} = E/10^{19}$ eV and $E_{12} = E/10^{12}$ eV). Thus, during the characteristic cooling time of $10^{19} - 10^{20}$ eV protons, the secondary electrons do not leave the source until they have cooled to MeV energies.

As long as the synchrotron cooling and escape times remain comparable, they play equally important roles in the formation of the proton spectra. For simplicity, it is assumed that both the magnetic field and the radius of the source do not change during the source evolution. Also, it is assumed that the protons are injected continuously into the blob with a rate $\dot{Q}_{\rm p}(E,t) = F(t) E_{\rm p}^{-\alpha_{\rm p}} \exp\left(-E_{\rm p}/E_0\right)$, i.e. it is assumed that the spectral shape of the injected protons, including the high energy cutoff E_0, does not change with time, while the absolute rate can be a function of time. The maximum acceleration energy E_0 is defined from the balance between the proton synchrotron cooling time and the minimum acceleration time $r_{\rm g}/c$, provided that the proton gyroradius does not exceed the source linear size. These are two extreme conditions, which cannot, most likely, be accommodated by the diffusive shock acceleration theory, even in its relativistic version. But this is the price we must pay if we want to invoke proton synchrotron radiation to explain the TeV γ-rays emission. This model unavoidably requires the maximum possible acceleration rates allowed by classical electrodynamics (Aharonian et al., 2002b).

Simplifications like the time-independent magnetic field and blob size would not always be the case, especially for the highly variable nonthermal phenomena related to X- and γ-radiation of the inner blazar jets. Nevertheless, they help to understand the general features of the model, in particular its ability to explain the observed *time-averaged* SED of 1ES 1426+428. Because of the lack of simultaneous X/TeV observations, as well as of any relevant information about the time variability of this source in X-ray and TeV γ-rays on short timescales, the construction of a more sophisticated model at this stage seems to be redundant. At the same time, it should be emphasized that the calculations do not postulate a snapshot spectra of protons, but rather treat the time evolution of the energy distribution of protons in the source adequately.

In Figs. 10.13 the spectra of 1ES 1426+428, performed for a certain combination of model parameters, are shown. Because presently only the time-averaged X-ray and TeV data of this source are available, the data are compared with the theoretical spectra calculated for the stage exceeding the characteristic energy loss and dynamical times of the system, thus these calculations can be interpreted as steady-state spectra. More specifically, the spectra are calculated at the stage 10^6 s after the continuous (in time and space) injection of protons into the γ-ray emission region starts.

The spectrum shown in Fig. 10.13 corresponds to a relatively compact blob of radius $R = 3 \times 10^{15}$ cm with magnetic field, $B = 100$ G, and Doppler factor $\delta_{\rm j} = 20$, and particle escape in the Bohm diffusion regime.

It is also assumed that the synchrotron TeV γ-rays are partly absorbed inside the source interacting with optical and EUV narrow-band emission components around 1 eV and 100 eV. In order to match the X-ray spectrum and absolute intensity, the optical depths were varied. For the fixed density n_0 and energy ϵ_0 of target photons, the optical depth is function of the energy and coordinates of production of the γ-ray photon. Therefore it is convenient to express the optical depth in terms of its value τ^* at the energy at which the cross-section approaches its maximum ($E^* = 4m_e^2 c^4/\epsilon_0 \simeq 1(\epsilon_0/1 \text{ eV})^{-1}$ TeV) and for the γ-ray photons produced at the center of the source, $\tau_* = n_0 \sigma_{\gamma\gamma}(E^*) R$. A good fit to both the X-ray and γ-ray spectra is achieved assuming $\tau^* \simeq 1$. Since the optical depth decreases at energies both below and above E^*, the internal γ-ray absorption in these two lines leads to quite an interesting deformation of the primary spectrum (Fig. 10.13) with maximum effects at energies around 100 GeV and 10 TeV (in the observer's frame). Further deformation of γ-ray spectrum takes place in the intergalactic medium. The calculations presented in Fig. 10.13 are performed for intergalactic absorption adopting the "nominal" (no. 1) CIB model shown in Fig. 10.1.

Obviously, the assumption of internal γ-ray absorption in two monochromatic optical/UV lines is a simplification. The calculations show that some other (more sophisticated) external photon fields can also reproduce the observed γ-ray spectra. On the other hand, the adoption of these two lines could be not far from the reality. Their energies in the host galaxy frame (obviously they cannot be linked to the jet) are at ϵ/Γ_j, i.e. ≈ 0.05 eV and ≈ 5 eV, and thus they can be contributed by the so-called dust torus and broad-line regions (see e.g. Celotti et al., 1998), albeit we do not have a clear evidence of the existence of such regions in 1ES 1426+428.

It is seen from Fig. 10.13 that the model spectra generally describe the X-ray and TeV γ-ray fluxes quite satisfactorily, although the calculated fluxes slightly exceed the reported upper limits at 100 MeV and 0.2 TeV, as well as pass below the detected fluxes above several TeV. Given (i) the large statistical and systematic errors, (ii) the uncertainties caused by the intergalactic absorption, and (iii) the fact that the reported γ-ray fluxes are derived from observations over time periods significantly exceeding the suspected variability timescales, we perhaps should not overemphasise the significance of perfect spectral fits. Otherwise, a better fit can be obtained by adjusting some of the model parameters. For example, a better agreement with the flux upper limits at low energies can be achieved assuming a flatter acceleration spectrum for protons or a faster escape of low energy particles.

The injection power of protons, $\dot{W}_p = \int E\dot{Q}_p(E)dE$, is defined from the requirement of explaining the intrinsic γ-ray luminosity of the source, $L_\gamma \approx 4\pi d^2 f_\gamma \delta_j^{-4} \approx 5 \times 10^{42} (\delta_j/10)^{-4}$ erg/s. Since in the presence of strong magnetic field the synchrotron radiation of γ-rays proceeds with high efficiency, a rather modest injection rate of protons of about $\dot{W}_p \approx 1.2 \times 10^{42}$ erg/s is sufficient. This is higher only by a factor of 4 than the intrinsic γ-ray luminosity (for the assumed Doppler factor $\delta_j = 20$). Correspondingly,

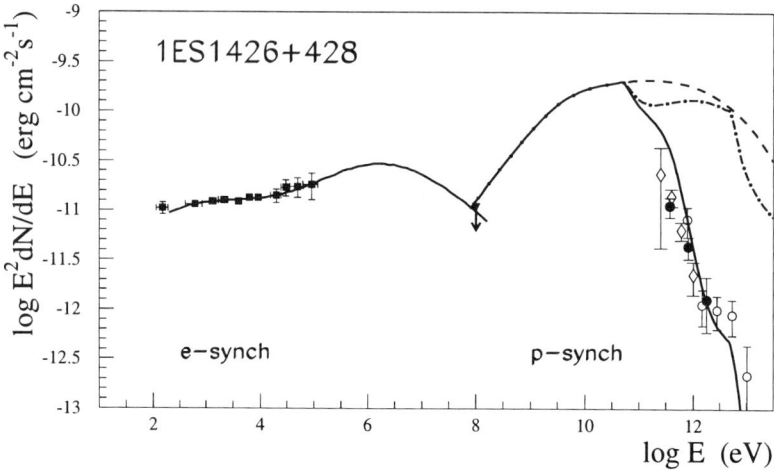

Fig. 10.13 Spectral Energy Distribution of 1ES 1426+428 calculated within the "pure" Proton Synchrotron model. While γ-rays originate from the synchrotron radiation of directly accelerated protons, X-rays are due to synchrotron radiation of secondary electrons. Photo-pion interactions of protons and inverse Compton scattering of electrons play a negligible role in the formation of electromagnetic radiation. The following model parameters are assumed: $B = 100$ G, $\alpha_p = 2$, $R = 3 \times 10^{15}$ cm, $\delta_j = 20$. The escape of charged particles proceeds in the Bohm diffusion regime. The dashed curve corresponds to the production spectrum of proton-synchrotron radiation, the dashed-dotted curve is the γ-ray spectrum after the internal absorption, and the solid curve represents the spectrum of γ-rays as seen by the observer, i.e. after the internal and intergalactic absorption. For parameters relevant to the internal and intergalactic absorption of γ-rays see the text. The measured time-averaged X-ray and TeV γ-ray fluxes are also shown. X-ray data are from Costamante et al., al. (2001), the TeV data are from the Whipple (Petry et al., 2002; filled points), HEGRA (Aharonian et al., 2002a; open circles), and the CAT(Djanati-Atai et al., 2002; diamonds) collaborations. The EGRET flux upper limit at 10^8 eV is also shown.

the steady-state energy density of protons is quite low, at the level of $w_{\rm p} \approx 2$ erg/cm^3. On the other hand, the density of magnetic energy is more than two orders of magnitude larger, $w_{\rm B} = B^2/8\pi \approx 400$ erg/cm^3. The huge total magnetic field of the blob, $R^3 B^2/6 \approx 5 \times 10^{49}$ erg formally would be sufficient to support observable TeV γ-ray fluxes for a long period. Namely, speculating that in such magnetized condensations an effective acceleration mechanism effectively transfers the magnetic energy to ultrarelativistic protons, a single blob ejected from the central source towards the observer could provide TeV γ-ray fluxes at the level shown in Fig. 10.13 during a time period exceeding 1 month.

The significant departure from the equipartition condition, when the magnetic pressure exceeds by two orders of magnitude the pressure due to the accelerated protons, seems to be a problem, in particular from the point of view of the stability of such a system. A more comfortable condition closer to the equipartition could be arranged assuming a smaller magnetic field. The model calculations show that with a 10 G field it is still possible to explain the X-ray and γ-ray data, although at the expense of a significant reduction of the radiation efficiency, and a significant increase in the injection power of protons to $\dot{W}_{\rm p} \approx 1.7 \times 10^{44}$ erg/s. The energy density accumulated at the stage 10^6 s in protons above 10^{18} eV is approximately 100 erg/cm^3, and in fact could be an order magnitude larger if the energy spectrum of protons continues to low energies. In this case, the energy density in protons exceeds, by more than two orders of magnitude, the energy density in the magnetic field. Therefore we may conclude that assuming an intermediate magnetic field between 10 and 100 G, we can explain the broad-band spectral features of 1ES 1426+428, and, at the same time, keep equipartition between the magnetic field and the relativistic protons. Note that within the proton synchrotron model the reduction of the magnetic field should be compensated by an adequate increase of the effective size of the particle accelerator in order to allow protons to be accelerated to highest energies. Important information about the blob size and the internal magnetic field, could be provided by future observations. In particular, detections of TeV flares from 1ES 1426+428 on timescales less than several hours would exclude small magnetic fields, $B \leq 30$ G and/or large blob sizes, $R \geq 10^{16}$ cm.

The described scenario couples X-rays to TeV γ-rays through synchrotron radiation of secondary electrons, therefore we must expect very tight X-ray/TeV correlations. The secondary electrons are promptly cooled in the strong magnetic field, so we may assume that the X-rays will arrive almost simultaneously together with the bulk of the TeV γ-rays. Nevertheless, more complicated correlations cannot be excluded, especially if the magnetic field drops significantly at the outskirts of the blob, so an essential part of the secondary electrons are produced and cooled in a low-field environment. This unfortunately makes the predictions about the X/TeV correlations less robust.

10.6 "IR background–TeV Gamma Ray Crisis"?

From the previous sections we may argue that the basic temporal and spectral characteristics of TeV blazars can be described, at least qualitatively, by the the current leptonic and hadronic models. At the same time all these models have difficulty explaining the increase or even pile-up that may appear at the end of the "reconstructed" spectrum of Mkn 501, if the reported FIR fluxes correctly describe the level of the CIB at far infrared wavelengths (see Fig. 10.1). Recently, motivated by such a non-standard spectral shape, several extreme hypotheses have been proposed to overcome the "IR background - TeV gamma-ray crisis" (see Chapter 11). In particular, Harwit *et al.* (1999) suggested the interesting idea that the HEGRA highest energy events are due to Bose-Einstein condensations interacting with the atmosphere. However, subsequently the HEGRA collaboration has demonstrated (Aharonian *et al.*, 2000c) that the detected shower characteristics are in fact in good agreement with the predictions for the events initiated by *ordinary* γ-rays. Another, even more dramatic hypothesis – violation of the Lorentz invariance – has been proposed to solve this problem (e.g. Kifune, 1999; Stecker and Glashow, 2001).

Below we discuss another non-standard, although, in our view, less dramatic hypothesis that can naturally accommodate pile-ups in the spectra of TeV blazars. The idea is based on a "non-acceleration" scenario for the production of γ-rays, assuming that the most energetic part of γ-ray spectrum of Mkn 501 above several TeV is due to *bulk motion Comptonization* of ambient low-frequency photons by a cold ultrarelativistic conical wind (Aharonian *et al.*, 2002).

Actually, relativistically moving plasma outflows in the forms of jets or winds are common for many astrophysical phenomena on both galactic or extragalactic scales. Independent of the origin of these relativistic outflows, the jet seem to be the only successful approach for understanding the complex features of nonthermal radiation of *Blazars*, *Microquasars* and *GRBs*.

The Lorentz factor of such outflows could be extremely large. In particular, in the Crab Nebula the Lorentz-factor of the unshocked pulsar wind is estimated to be between 10^6 and 10^7 (Rees and Gun, 1974). Meszaros and Rees (1997) argued that in the context of cosmological GRBs the magnetically dominated jet-like outflows from stellar mass black holes may attain extreme Lorentz factors exceeding 10^6. Lorentz factors of jets in the standard inverse Compton models of γ-ray blazars are rather modest, $\Gamma \sim 10$. However, there is no apparent theoretical or observation argument against bulk motion with a much larger Lorentz-factor (see, e.g., the discussion by Celotti et al., 1998). Because of existence of dense photons fields in the

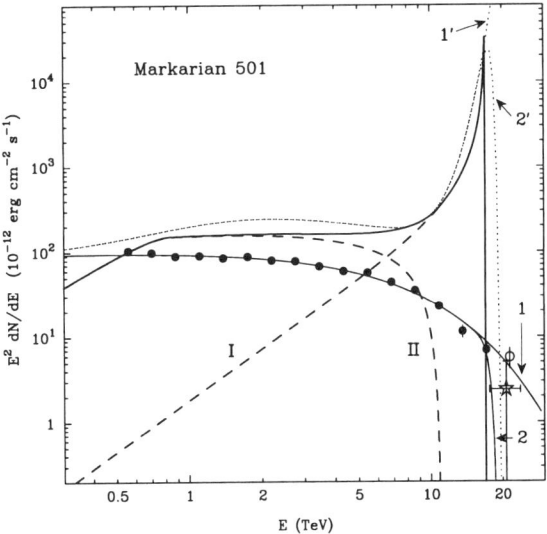

Fig. 10.14 Inverse Compton spectrum of a cold unshocked ultrarelativistic jet with a Lorentz factor of bulk motion $\Gamma = 3.33 \times 10^7$. The radiation component associated with Comptonization of ambient optical photons is shown by dashed curve I. The dashed curve II is the residual of the total TeV source emission (after subtraction of the unshocked wind component I), and could be attributed either to the Comptonization of the bulk motion on ambient FIR photons or to the radiation of 'blobs' in shocked jet. The heavy solid line represents the superposition of these two components. Fits to the observed flux of Mkn 501 are shown by thin solid lines. Curve 1 corresponds to the fit given by equation (b) in Fig. 10.3, and curve 2 – to the steepest possible spectrum above 17 TeV based on the most recent reanalysis of Mkn 501 HEGRA data (see Chapter 2). The intrinsic (absorption-corrected) spectra of Mkn 501, corresponding to the fits to the observed points "1" and "2", are shown by dotted lines "1'" and "2'", respectively (from Aharonian, Timokhin, Plyasheshnikov, 2001).

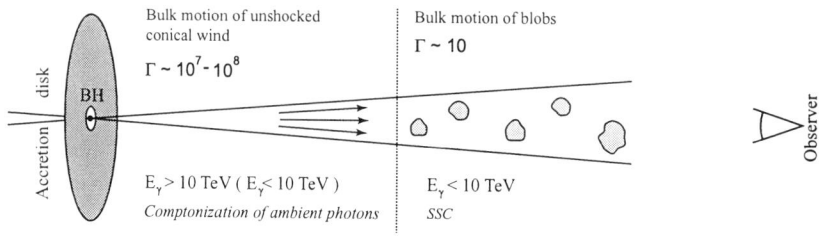

Fig. 10.15 Sketch of the two-stage γ-ray production scenario in a TeV blazar. At the first stage, the highest energy γ-rays above 10 TeV with a sharp spectral form ("pile-up") are produced due to Comptonization of the ambient optical photons by the cold ultrarelativistic jet with bulk motion Lorentz factor $\Gamma \sim 10^7 - 10^8$. Low energy γ-rays, $E_\gamma \leq 10$ TeV, can be also produced due to Comptonization of ambient FIR photons with a broad spectral distribution. At the second stage, relatively low energy ($E_\gamma \leq 10$ TeV) γ-rays are produced in the shocked jets ("blobs") moving with Lorentz factor of about $\Gamma \sim 10$ in accordance with the standard SSC model.

inner sub-parsec region of sources like Mkn 421 and Mkn 501, the Compton optical depth τ_C could be as large as 1. Obviously, for our model the most favourable value for τ_C lies between 0.1 (in order to avoid huge energy requirements for the outflow) and 1 (in order to avoid Compton drag). In particular, we have to assume that in the proximity of the black hole the outflow should be Poynting flux dominated, and that only at large distances from the central object, where the photon density is significantly reduced, the major part of the electromagnetic energy is transferred to the kinetic energy of bulk motion.

Due to the extremely large Lorentz factors, exceeding 10^7, Compton scattering on the ambient NIR/optical photons with energy more than 1 eV proceeds in deep Klein-Nishina regime; therefore the γ-radiation should have a very narrow distribution with energy $E \approx E_e = m_e c^2 \Gamma$. Meanwhile, the IC scattering on ambient far IR photons surrounding the central source, still takes place in the Thomson regime, and thus results in a smooth broadband spectrum.

Fig. 10.14 demonstrates that the overall *absorption-corrected* (for the CIB at FIR close to the extreme CIB model no.3 in Fig. 10.1). The spectrum of Mkn 501 can be satisfactorily explained in the terms of inverse Compton emission of the cold jet with a bulk Lorentz factor $\Gamma = 3.33 \times 10^7$, assuming an ambient radiation field with a narrow (Planckian) type radiation with temperature $kT = 2$ eV (dashed curve I). The dashed curve II, which formally is the residual from the subtraction of the unshocked wind component radiation from the intrinsic (reconstructed) TeV emission (solid heavy line), can be attributed to the bulk-motion Comptonization on ambient far IR photons produced e.g. by cold clouds surrounding the source. Alternatively, the "residual component" (curve II) can be referred to the SSC (or any other) radiation component of blobs in shocked jet. This two-stage (*pre-shock* plus *post shock*) scenario of formation of TeV γ-ray emission is schematically illustrated in Fig. 10.15.

The possibility of disentangling the multi-TeV emission with a characteristic sharp pile-up at the very end of the spectrum, $E_\gamma \simeq m_e c^2 \Gamma$, from the sub-10 TeV emission associated with the shocked structures (e.g. blobs) in the jet, not only solves the possible "IR background – TeV gamma-ray crisis" but also allows a more relaxed parameter space for the interpretation of X-rays and the *remaining* "low" energy (≤ 10 TeV) γ-rays within the conventional SSC scenario. Consequently, this offers more options for interpretation of X-ray/gamma-ray correlations. If the overall TeV radiation of Mkn 501 indeed consists of two, *unshocked* and *shocked*, jet emission components, we may expect essentially different time behaviours of these components. In particular, the "unshocked jet" (≥ 10 TeV) component should arrive earlier than the SSC components consisting of synchrotron X- and sub-10 TeV IC γ-rays. Therefore, an important test of the suggested two stage scenario of the TeV radiation of jets would be the search for correlations (or lack of such correlations) of ≥ 10 TeV radiation with both the low energy (e.g. 1-3 TeV) γ-rays and synchrotron X-rays. The low statistics of (heavily absorbed) γ-rays above 10 TeV makes the search for such correlations rather difficult, and requires ground-based instruments with huge, $\gg 0.1\,\mathrm{km}^2$ detection areas in this energy domain. The new generation IACT arrays should be able to perform such correlation studies.

Chapter 11

High Energy Gamma Rays — Carriers of Unique Cosmological Information

Extension of the γ-ray spectra of extragalactic sources to energies beyond 10 GeV opens a new exciting research area of gamma-ray astronomy – *observational gamma-ray cosmology*. The promise here is linked to the energy-dependent interactions of high energy γ-rays with the diffuse extragalactic background radiation (DEBRA), from UV to far infrared wavelengths, as well as to the effects caused by propagation of secondary (pair-produced) electrons in the intergalactic magnetic fields (IGMF). The absorption features in the spectra of primary (direct) γ-radiation, as well as the temporal and angular distributions of the secondary (cascade) γ-rays arriving from the directions of distant extragalactic objects with accurately measured redshifts, contain unique cosmological information about DEBRA and the IGMF.

11.1 Probing DEBRA Through γ-Ray Absorption Features

An observer looking within a narrow cone centered on a source at a distance $d = cZ/H_0$ will see an absorbed spectrum $J(E) = J_0(E) \exp\left[-\tau(E)\right]$, where $J_0(E)$ is the initial (source) γ-ray spectrum, and $\tau(E)$ is the optical depth characterising γ-γ pair production in the isotropically distributed extragalactic radiation fields. As discussed in Sec.10.2, because of the narrowness of the pair-production cross-section, for a broad-band background spectrum without very sharp features, over half the interactions of a γ-ray photon of energy E is contributed by a narrow interval of target photons. This allows a convenient approximation for the optical depth given by Eq.(10.1).

Usually, the proposals of studying DEBRA via gamma-ray astronomical means are formulated as an approach based on the search for cutoffs in the γ-ray spectra at energy E determined from the condition $\tau(E) = 1$ (e.g. Stecker et al., 1992). However, this is not always the case. Any realistic model of the Cosmic Infrared Background (CIB), with two bumps at NIR and FIR wavelengths, predicts a quite specific energy dependence for the intergalactic optical depth (see Fig. 10.2). Correspondingly, while at energies above 100 GeV and above 10 TeV we should expect significant spectral steepening (provided that the γ-ray source is located at a sufficiently large distance), at energies between 1 and 10 TeV the absorption does not necessarily imply strong modification of the primary spectrum. The reason of this effect is explained by the specific shape of the CIB spectrum $u_{\rm CIB}$ which in the region between near and mid infrared wavelengths behaves approximately as $\propto \lambda^{-1}$. The impact of this interesting effect can be seen in Fig. 10.3, where the SED of Mkn 501 is shown together with the absorption-corrected spectra reconstructed for 3 CIB models from Fig. 10.1. While at ≤ 3 TeV and ≥ 10 TeV the γ-ray spectrum is significantly modified, at intermediate energies the spectral change is less pronounced in spite of the strong suppression of the absolute flux.

On the other hand, the intergalactic modulation of primary γ-ray spectra at low energies (below several TeV) strongly depends on the flux and spectral shape of the CIB at wavelengths shorter than $\sim 2\mu$m. As is seen in Fig. 10.10, the steepening of γ-ray spectra at low energies is accompanied by a tendency of flattening ("recovery" of the primary spectrum) at energies above 1-2 TeV. This effect is especially strong in the case of distant sources with $z \geq 0.1$ for which the optical depth becomes large even at energies ~ 100 GeV. In Sec.10.4.3 we argued that this effect is already seen in the spectrum of 1ES 1426+428 located at $z = 0.129$. If true, this would imply that we are quite close to probing CIB at wavelengths from $\lambda \sim 0.3$ to 10 μm. However, this optimistic view does not yet guarantee fast and easy success. This can be accomplished only in the case of proper spectrometric measurements from several sources located at different distances between approximately 100 and 1000 Mpc, combined with an adequate understanding of the intrinsic γ-ray spectra. Fortunately, the current list of TeV blazars includes 6 objects – Mkn 421, Mkn 501, 1ES 2344+514, 1ES 1959+650, PKS 2155-304, 1ES 1426+428 – with redshifts ranging from 0.03 to 0.13, and thus nicely covering the required distance interval. The reported TeV fluxes from these blazars are significantly above the sensitivities of forthcoming 100 GeV energy threshold IACT arrays. This should allow not only high quality measurements of average γ-ray spectra in the energy interval between 100 GeV and 10 TeV, but, more importantly, detailed study of correlations with other non-thermal radiation components at lower frequency (radio, mm, optical X-ray and MeV/GeV) bands. This gives a certain optimism that eventually gamma-ray astronomers will be able to identify confidently the principal radiation mechanisms, to fix/constrain the relevant model parameter space, and reconstruct robustly the intrinsic γ-ray spectra based on multiwavelength studies of the spectral and temporal characteristics of blazars obtained on shortest possible (sub-hour) timescales. Then, it will be possible to estimate the effect of intergalactic absorption, and thus to infer information about the diffuse background at near and mid-infrared wavelengths.

At this stage of blazar/CIB studies, we face an ambiguity in interpreting the observations concerning both the intrinsic γ-ray spectra of TeV blazars and the flux of CIB. The only definite conclusion that can be drawn from

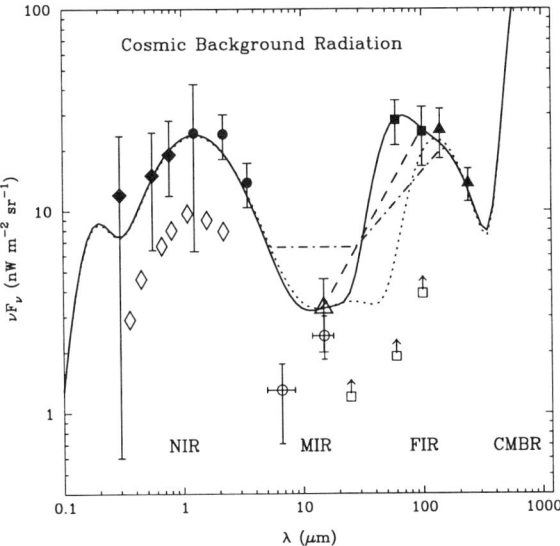

Fig. 11.1 Cosmic background radiation: available data and the "best guess" model spectra (from Aharonian et al., 2002a). The reported fluxes are shown with filled symbols: Bernstein et al. (2002) – diamonds, Wright et al. (2001) – circles, Finkbeiner et al. (2000) – squares, Hauser et al. (2001) – triangles. The low limits are shown by open symbols: Pozzetti et al. (2001) – diamonds, Biviano et al. (2001) – triangle, Franceschini et al. (2001) – circles, Hacking and Soifer (1999) – squares. Note that these fluxes are basically the same as in Fig. 10.1, but some more "lower-limit" points are added in this figure. The CIB models are shown by solid line – Model I, by dotted line – Model II, by dashed line – Model III, by dot-dashed line – Model IV.

available data, in particular from the best studied objects – Mkn 501 and Mkn 421, is that we are most likely detecting significantly absorbed TeV radiation from these objects. Remarkably, the absorption seems inevitable not only at energies above 10 TeV, for which the optical depth could be as large as 10, but also at sub-TeV energies (Coppi and Aharonian, 1999b; Guy et al., 2000). Note, however, that the analysis of intergalactic absorption at sub-TeV and multi-TeV energies leads to two different conclusions. The absorption-corrected γ-ray spectrum of Mkn 501 at low energies, based on the CIB fluxes reported at 2.2 and 3.5 μm and on the current theoretical predictions for NIR, has a reasonable shape that is in general agreement with both the leptonic and hadronic models. On the other hand, the corrections to the γ-ray spectrum at energies above 10 TeV based on the unexpectedly large CIB fluxes detected by COBE at 140 and 240 μm, result in an "unreasonable" source spectrum which sharply *curves up* above 10 TeV. Such a "non-standard" shape of the spectrum of intrinsic TeV radiation has been interpreted as an indication of "IR background – TeV gamma-ray crisis" (Protheroe and Meyer, 2000). Several dramatic assumptions, e.g. violation of the Lorentz invariance, have been proposed to overcome this "crisis". In Sec.10.6 we discussed a less drastic interpretation of the unusual intrinsic TeV spectrum, proposing that the pile-up in the source spectrum of Mkn 501 is a result of bulk motion comptonization of ambient optical radiation by a ultrarelativistic conical cold outflow (jet) with Lorentz factor $\Gamma_0 \geq 3 \times 10^7$

Two more proposals could be added to the list of "exotic" solutions: (i) speculating that Mkn 501 is located at a distance $\ll$ 100 Mpc, or (ii) assuming that TeV γ-rays from Mkn 501 are not direct representatives of primary radiation of the source, but are formed during the development of high energy electron-photon cascades in the intergalactic medium. At first glance, both ideas seem as relatively "neutral" proposals. But in fact they do contain dramatic assumptions. While the first hypothesis implies non-cosmological origin of the redshift of Mkn 501 (good news for the advocates of a non-cosmological origin of AGN and quasars), the second hypothesis requires an extremely weak intergalactic magnetic field at the level of 10^{-18} G or less.

Although fascinating, the appeal for revision of essentials of modern physics and astrophysics seems too premature. The nature of the FIR isotropic emission detected by COBE is not yet firmly established, and it is quite possible that the bulk of the reported flux, especially below 100 μm, is the result of a superposition of different local backgrounds. On the other hand, not only the flux at 60μm, but also the more reliable CIB fluxes reported at at longer, $\lambda \geq 100\mu$m, wavelengths are responsible (albeit in a less distinct form) for the appearance of the pile-up, unless we assume a very specific CIB spectrum in the MIR-to-FIR transition region. Therefore it is important to explore the impact of the CIB spectrum between mid and far infrared wavelengths on the deformation of the multi-TeV spectrum of primary γ-rays due to the intergalactic absorption.

For that reason in Fig.11.1 we show several model spectra of the CIB. Note that the spectral energy distribution (SED) of the CIB at optical/NIR wavelengths is comparable with the overall FIR energy flux. This indicates that an essential part of the energy radiated by stars is absorbed and re-emitted by dust in the form of thermal sub-mm emission. Currently the information at mid-infrared wavelengths is very limited. The only available measurement at 6 and 15 μm (see Fig. 11.1) derived from ISOCAM source counts (Franceschini et al., 2001) should be treated as lower limits. Therefore the flux estimate at the level of $\simeq 2 - 3 \mathrm{nW/m^2 sr}$, as well as the lower limits based on IRAS counts at 25-100 μm do not allow firm conclusions about the depth of the MIR "valley", which is dominated by radiation of the warm dust component. Consequently, it does not provide sufficient information for definite predictions regarding the slope of the spectrum in the most crucial (from the point of view of absorption of $\geq 10 \,\mathrm{TeV}$ γ-rays) MIR-to-FIR transition region.

As long as the available measurements of the CIB do not allow a quantitative study of the effect of absorption of γ-rays in the intergalactic medium, we can rely only on model predictions or on the "best guess" shape of the CIB spectrum. In this regard we note that the reported FIR fluxes present a common problem for all current CIB models. Therefore, if one assumes that the reported FIR fluxes have a truly diffuse extragalactic origin, an essential revision of the CIB models is needed in order to match the data.

In Fig. 11.1 several model spectra of the CIB are shown. Since we are interested, first of all, in the cosmic background fluxes at MIR and FIR, at shorter wavelengths we adopt a common approximation for all models which matches the reported optical and NIR fluxes. At wavelengths shorter than 10μm this approximation is quite close to model no.1 in Fig. 10.1. The template of the CIB spectrum in the principal MIR-to-FIR transition region, shown by dashed line (hereafter Model III) in Fig. 11.1, fits the reported fluxes including, within 2σ uncertainty, the 60 μm point. This idealised template has a rather flat shape in the MIR-to-FIR transition region, $\nu F_\nu \propto \lambda^s$ with $s \leq 1$. This results in short mean free paths of γ-rays above 10 TeV (≤ 50 Mpc) as can be seen in Fig. 11.2, and consequently in a pile-up in the reconstructed γ-ray source spectrum (see Fig. 11.3) defined as $J_0(E) = J_{\mathrm{obs}}(E) \exp[\tau(E)]$.

If we assume that the reported FIR fluxes correctly describe the level of the truly diffuse background radiation, only two ways are left for reduction of the effect of attenuation of ≥ 10 TeV γ-rays: (i) an *ad hoc* assumption of the CIB flux at wavelengths between 10 and 60 μm at the marginally acceptable (i.e. the ISOCAM low-limit) level, but with very rapid rise beyond 60 μm in order to match the reported fluxes at FIR, and (ii) adopting a Hubble constant of $H_0 \simeq 100$ km/s Mpc, i.e. assuming the smallest possible distance to Mkn 501, $d = cz/H_0 = 102$ Mpc ($z = 0.034$).

In Fig. 11.1 two other model spectra of the CIB are shown (solid line – Model I, dotted line – Model II). These models fit the data at NIR and FIR, but at the same time allow minimum intergalactic γ-ray absorption at $E \geq 10$ TeV, because both spectra are forced to be at the lowest possible level in the MIR-to-FIR transition region set by the ISOCAM lower limit at 15 μm. At wavelengths below 100 μm for both models the same spectral

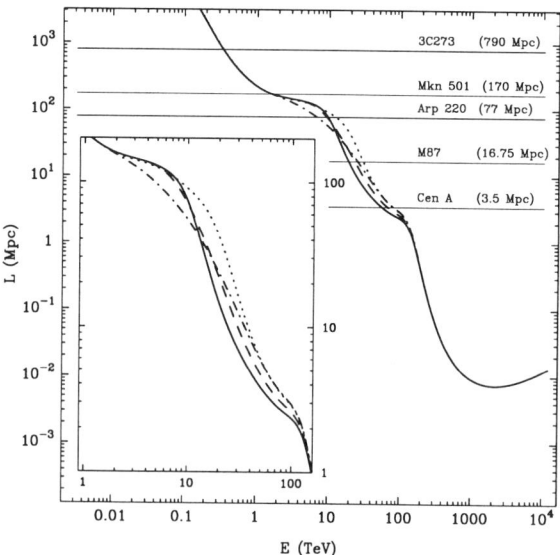

Fig. 11.2 Mean free path of γ-rays calculated for 4 different models of the CIB spectrum presented in Fig. 11.1. Solid line – Model I, dotted line – Model II, dashed line – Model III, dot-dashed line – Model IV. The horizontal lines indicate the distances to Cen A, M87, Arp 220, Mkn 501, and 3C 273 ($H_0 = 60$ km/s Mpc).

shape, described by a Planckian distribution, is assumed.

Model II marginally agrees with the 100 µm point but underestimates the flux at 60 µm by a factor of 5 compared with the flux reported by Finkbeiner *et al.* (2000). This model significantly suppresses the photon density between 50 and 100 µm, and correspondingly allows larger mean free paths for γ-rays with energies above several TeV (Fig. 11.2). Such a SED of the CIB results in almost E^{-2} type "reconstructed" spectrum of γ-rays from Mkn 501 at energies between approximately 2 and 20 TeV (Fig. 11.3). This has a simple explanation. For a constant SED of the CIB (the flat part of spectrum 2 in Fig. 11.1), the photon-photon pair-production optical depth is proportional to E, with an absolute value for the distance to Mkn 501 of 170 Mpc, $\tau(E) \approx 0.16(E/1 \text{ TeV})$. Therefore, in the absorption-corrected spectrum of γ-rays the correction factor e^τ compensates the exponential term of the observed spectrum of Mkn 501 (see Fig. 10.3). For conventional γ-ray production mechanisms, the E^{-2} type γ-ray source spectrum appears more natural than the spectra containing "pile-up" type sharp features. Therefore, Model II in Fig. 11.1 can be treated as the "favoured" one, although its flat spectral shape in the MIR-to-FIR transition region with a very fast increase after 60 µm does not agree with the current theoretical and phenomenological predictions.

Actually, all current CIB models have a problem to accommodate the reported flux at 60 µm. If this point, nevertheless, represents the level of the true diffuse extragalactic flux, we must assume, in order to minimise absorption effects, a spectrum close to Model I with a Wien type spectral shape below 60 µm, which perhaps is the steepest possible (i.e. physically justified) continuous spectrum. Even so, 60 µm photons themselves have sufficient energy for effective interaction with ≥ 15 TeV γ-rays. We can no longer suppress the severe γ-ray absorption at highest energies, and therefore the appearance of a sharp pile-up in the absorption-corrected spectrum above 10 TeV.

Finally, the CIB Model-IV (dot-dashed curve) shown in Fig. 11.1 assumes a significantly higher (by a factor of 2.5) flux at 15 µm compared to the reported ISOCAM lower limit, and passes through the low edges of the error bars of reported points at 100 and 140 µm. Remarkably, such a high MIR flux does not result in an unusual γ-ray source spectrum, but predicts an almost single power-law spectrum up to 10 TeV with photon index ≈ 1.5. Above 10 TeV we again observe a pile-up which however in this case is less pronounced than for Models I and III.

A γ-ray photon with energy E propagating through an isotropic photon field can interact, via electron-positron pair production, with ambient photons of energy $\epsilon \geq \epsilon_{\text{th}} = (m_e c^2)^2/E$ or wavelength $\lambda_{\text{th}} = 4.8(E_\gamma/1\text{TeV})$ μm. The latter relation is represented in Fig. 11.4 (heavy solid line). The cross-section of $\gamma\gamma$ interactions averaged over the directions of background photons peaks at $\lambda_{\max} \simeq \lambda_{\text{th}}/4$. Therefore, even for a broad-band spectrum of background photons, a significant fraction of the optical depth τ is contributed by a relatively narrow spectral band of the CIB centered on $\lambda \sim 1(E_\gamma/1\text{TeV})$ μm. In order to understand quantitatively the "γ-ray energy-CIB photon wavelength" relation, it is convenient to introduce the ratio $\kappa(E,\lambda) = \tau(E,\lambda)/\tau_0(E)$, where

$$\tau(E,\lambda) = \int_\lambda^{\lambda_{\text{th}}} \sigma_{\gamma\gamma}(E_\gamma,\lambda) n_{\text{CBR}}(\lambda)\,d\lambda\,, \qquad (11.1)$$

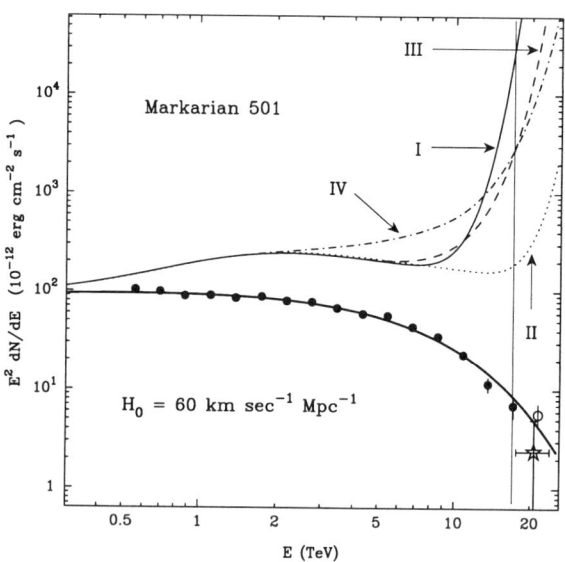

Fig. 11.3 Spectral Energy Distributions of Mkn 501 corrected for the intergalactic absorption (from Aharonian et al., 2002a). The experimental points (filled circles) correspond to the time-averaged spectrum of Mkn 501 measured by HEGRA during the flare in 1997; the star corresponds to the flux in the highest energy bin around 21 TeV obtained after reanalysis of the same data set, but with improved ($\sim 10\%$) energy resolution (see Chapter 2). The heavy line corresponds to the fit given by equation (b) in Fig. 10.3. The solid, dotted, dashed and dot-dashed lines represent the reconstructed (absorption-corrected) spectra of γ-rays for the CIB Models I,II, III and IV from Fig.11.1, respectively.

and $\tau_0 = \tau(E_\gamma, \lambda = 0)$ is the optical depth integrated over the entire spectral range of the CIB above the threshold λ_{th}.

In Fig. 11.4 the contour map of the function $\kappa(E, \lambda)$ is shown for 3 different models of the CIB in the spectral range of γ-rays from 1 to 100 TeV, and for the CIB photons from 1 to 500 μm. For the given γ-ray photon energy E_0, the spectral region of the CIB responsible for a fraction ξ of the total optical depth τ_0 is $[\lambda_\xi, \lambda_{\text{th}}]$, where λ_ξ and λ_{th} correspond to the points where the horizontal line $E = E_0$ intersects the corresponding level curve, and the threshold line, respectively. In Fig. 11.4 four levels for $\xi = 10, 50, 90$, and 99%, are shown. By definition, $\xi = 0$ corresponds to the threshold boundary $E_{\text{TeV}} = 0.21 \lambda_{\mu\text{m}}$. It is seen, for example, that at $E_0 = 20$ TeV approximately 50 per cent of the total optical depth τ_0 is contributed by background photons with wavelengths longer than 50μm.

From Fig. 11.4 it follows that the study of the CIB at wavelengths $\lambda \geq 50\mu$m can be best done with very high energy γ-rays, $E \geq 20$ TeV. Unfortunately, even the relatively close Mkn 421 and Mkn 501 cannot provide an unambiguous information about this important wavelength band;

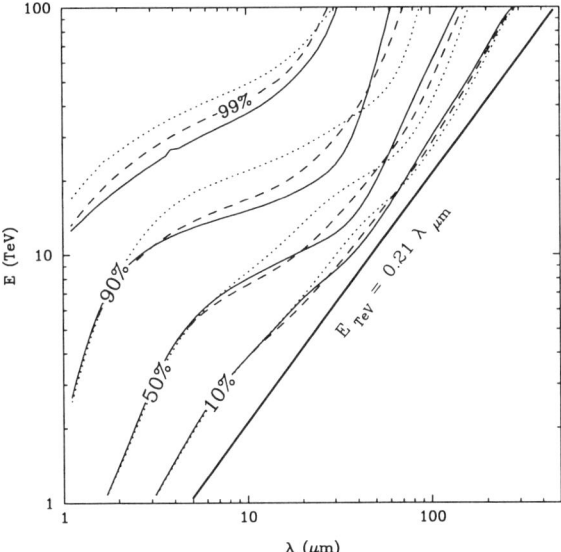

Fig. 11.4 The contour map of the function $\kappa(\lambda, E)$. The solid, dotted and dashed lines show the levels of $\kappa(\lambda, E)$ for the CIB Models I, II, and III from Fig. 11.1, respectively. The heavy solid line represents the threshold of $\gamma\gamma$ pair production.

the information is essentially lost due to the heavy intergalactic absorption of these energetic γ-rays. This task is rather contingent on discovery of nearby extragalactic multi-TeV γ-ray sources at distances $\ll 100$ Mpc, dictated by the condition that the mean free path of γ-rays effectively interacting with FIR should not significantly exceed the distance to the source, i.e. the optical depth at the relevant γ-ray energy $\tau(E) \sim 1$. Despite the lack of very close blazars, some other potential extragalactic γ-ray sources like the nearby radiogalaxies Centaurus A and M 87, and perhaps also the starburst galaxies Arp 220, M 82 and NGC 253 may (hopefully) provide us with multi-TeV γ-rays for such important studies. The intergalactic absorption factors for very high energy γ-rays from Mkn 501, M 87, and Centaurus A are shown in Fig. 11.5. The curves calculated for 3 different CIB models show that while Mkn 501 is best suited for the study of the CIB at wavelengths shorter than 20μm, the two nearby radiogalaxies M 87 and Centaurus A can be very helpful providing us with ≥ 20 TeV γ-rays for extraction of information about the CIB in the transition region from MIR to FIR, between 10 and 100μm.

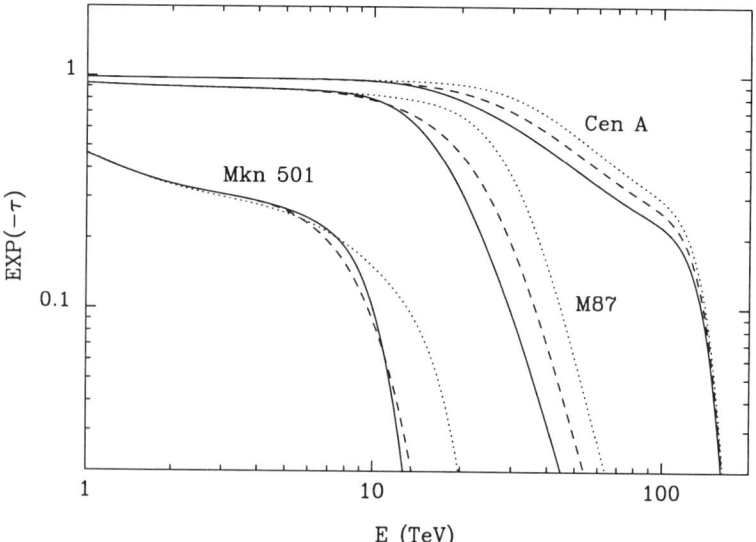

Fig. 11.5 Intergalactic absorption factor $\exp[-\tau(E)]$ for γ-rays from Mkn 501, M 87, and Centaurus A calculated for the three different CIB models from Fig. 10.1.

11.2 The Effect of Cascading in the CIB

Generally, the propagation of high energy γ-rays through a low frequency photon field cannot be reduced to the simple effect of γγ absorption. When a γ-ray is absorbed its energy is not lost. The secondary electrons and positrons create new γ-rays via inverse Compton scattering; the second generation γ-rays produce new (e^+, e^-) pairs, thus an electromagnetic cascade develops.

In the energy region $E \leq 100$ TeV, the γ-rays and electrons interact with photon fields of different origin. Although the energy density of the 2.7 CMBR greatly exceeds the density of other ("starlight" and "dust") components of CIB, because of the kinematic threshold of the reaction $\gamma\gamma \to e^+e^-$, γ-rays with energy less than several hundred TeV interact mostly with the infrared and optical background photons. The inverse Compton scattering of electrons does not have kinematic threshold, therefore the electrons interact mainly with the much denser CMBR photons.

If the cascade is initiated by primary photons with a hard (harder than E^{-2}) spectrum that extends to $E \geq 100$ TeV, at lower energies the secondary γ-rays may well dominate over the primary γ-rays. However, because of deflections of secondary electrons in ambient intergalactic magnetic fields, only a small fraction of cascade γ-rays may reach the observer synchronously with the low-energy primary photons. Depending on the strength of the IGMF, quite different temporal and spatial (angular) characteristics of the secondary γ-radiation are expected. For an IGMF exceeding 10^{-11} G, the secondary electrons are promptly isotropized, resulting in the formation of extended Pair Halos surrounding VHE sources. For a much weaker IGMF, which formally cannot be excluded, instead of detection of these extended and persistent structures, one should expect cascade radiation propagating close to the original direction of primary γ-rays. This effect has been invoked to explain the extension of the γ-ray spectrum from Mkn 501 beyond 10 TeV, in spite of the very large optical depth (Aharonian et al., 2002a). The idea behind this proposal is quite transparent. The highest energy primary γ-rays during their propagation through the intergalactic photon fields, initiate a cascade which at the absence of a strong ambient magnetic field provides an effective transfer of energy towards the observer or, in other words, "moves" the source closer to the observer.

In Fig. 11.6a the spectra of cascade γ-rays from Mkn 501are shown, corresponding to three different CIB models presented in Fig. 11.1, and calculated for a power-law spectrum of primary γ-rays with photon index $\Gamma = 1.9$. Within the experimental uncertainties, the cascade γ-ray spectrum corresponding to CIB Model II (light dotted line) agrees with the HEGRA points, while the γ-ray spectra calculated for CIB Models I and III (light solid and dashed curves, respectively) pass significantly below the measured fluxes at $E \geq 10$ TeV. However, at these energies the γ-ray spectra are very steep, therefore we must to take into account the limited instrumental energy resolution when comparing the theoretical predictions with the differential flux measurements. Namely, the experimental fluxes should be compared with the predictions after *convolving* the theoretical γ-ray spectra with the energy resolution function. The corresponding curves, assuming a Gaussian type instrumental energy spread function $G(E - E') = \exp(-(E - E')^2/2\sigma^2)/(\sqrt{2\pi}\sigma)$ with constant $\sigma = E/5$ (i.e. 20 per cent energy resolution) are shown by the heavy solid (Model I), dotted (Model II) and dashed (Model III) lines. It is seen that after this procedure the γ-ray spectra that correspond to the CIB Models I and III satisfactorily fit the observed γ-ray fluxes.

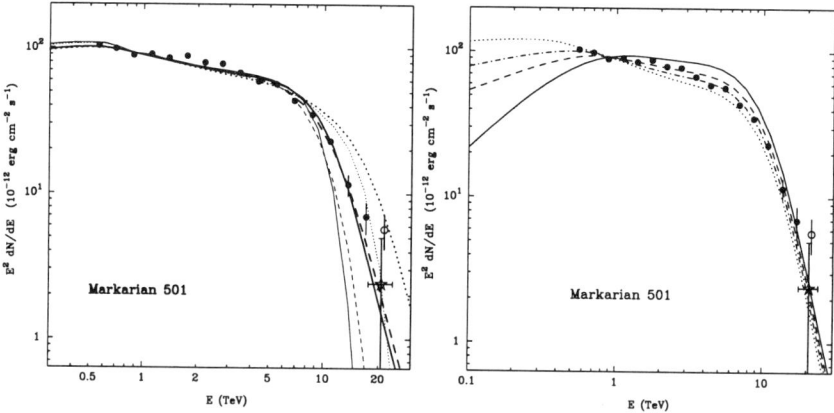

Fig. 11.6 Intergalactic cascade radiation spectra initiated by primary γ-rays from Mkn 501 calculated for the Hubble constant $H_0 = 60$ km/s/Mpc. The experimental points are same as in Fig. 11.3. **a** (left panel): the solid, dotted, and dashed curves are calculated for a photon index of primary γ-rays $\Gamma = 1.9$, and correspond to the CIB Model I, Model II, and Model III, respectively. The light and heavy lines represent the predicted cascade γ-ray fluxes before and after convolution with a Gaussian type energy spread function with 20% energy resolution. **b** (right panel): the solid, dashed, dot-dashed and dotted curves are the cascade γ-ray spectra calculated for the CIB Model III, and correspond to the the photon indices of the primary γ-rays $\Gamma=1.5, 1.8, 1.9$ and 2, respectively. All curves are obtained after the convolution of the cascade γ-ray spectra with the Gaussian type energy spread function with 20% energy resolution.

At high energies, the spectrum of cascade photons weakly depends on the photon index, Γ, and the maximum energy, $E_{\max}$, of primary γ-rays, provided that $\Gamma \leq 2$ and $E_{\max} \gg 10^2$ TeV. In Fig. 11.6b four cascade γ-ray spectra calculated for $\Gamma = 1.5, 1.8, 1.9, 2$ and $E_{\max} = 10^3$ TeV, are presented. All curves are obtained for CIB model III. The photon indices of primary γ-rays $\Gamma = 1.8, 1.9$, and 2 give quite similar cascade γ-ray spectra throughout the entire energy interval from 1 TeV to 20 TeV. They become distinguishable only at low energies, mainly because of contributions of primary (un-absorbed) γ-rays which dominate below 1 TeV.

An important feature of the well-developed cascade radiation is its standard spectral shape which is determined by propagation effects but not affected much by the details of the primary radiation from the source. Therefore, within this hypothesis it is possible to explain the quite stable spectral shape of Mkn 501 that was observed during the high state in 1997 despite the dramatic variation of the absolute flux on timescales less than several hours (Aharonian et al., 1999a,c). On the other hand, the possible spectral changes during the strong 1997 April flares reported by the CAT collaboration (Djannati-atai et al., 1999), as well as the noticeable steepening of the spectrum in a low state of Mkn 501 detected by the HEGRA telescope system (Sambruna et al., 2000; Aharonian et al., 2001b), do not contradict to this statement. These effects could be caused by variations of the ratio of the "cascade" component to the overall ("un-absorbed" plus "cascade") flux of γ-rays due, for example, to reduction of the maximum energy in the primary γ-ray spectrum at low or quiescent states.

The "intergalactic cascade" hypothesis requires extremely small intergalactic magnetic fields in the major fraction of the space between the source and the observer in order to avoid the significant time delays. Indeed, even a tiny intergalactic magnetic field of about 10^{-18} G leads to delays in arrival of TeV γ-rays, compared to the associated low energy photons, by more than several days (Plaga, 1995; Kronberg, 2001). This would obviously contradict the observed X-ray and TeV correlations, unless we assume that $B < 10^{-18}$ G. Although indeed very speculative, formally such small fields cannot be excluded, especially if we take into account that the typical scale of the so-called intergalactic voids, where the magnetic field could be arbitrarily small, is estimated as $\sim 120(H_0/100 \text{ km/s/Mpc})^{-1}$ Mpc (Einasto, 2001), i.e. quite comparable to the distance to Mkn 501.

The search for delays of high energy γ-rays arriving from distant extragalactic objects offers an interesting method to probe the primordial intergalactic magnetic fields down to 10^{-20} G. For practical realization of this interesting effect one must clearly distinguish the cascade time delays caused by the IGMF from the intrinsic time structures of radiation from compact variable sources. Moreover, due to the small but not negligible ($\theta \sim m_e c^2/E$) emission angles of the secondary particles produced in the $\gamma\gamma \to e^+e^-$ and $e\gamma \to e\gamma'$ reactions, we should expect a significant broadening and time delays of the cascade radiation even in the absence of a magnetic field, especially at low energies (Cheng and Cheng, 1996). The uncontrollable opening angle of pair production unfortunately reduces the applications of this interesting approach.

The "intergalactic cascade" hypothesis does not provide conclusive predictions for time correlations of TeV γ-rays with the nonthermal radiation components emitted by the central source at other energy bands. At the same time, this hypothesis allows robust predictions of the spectral characteristics of γ-radiation at different depths of the cascade development. In this regard, the inspection of this hypothesis based on the study and comparison of spectral characteristics of TeV γ-rays detected from sources at different redshifts is indeed worthwhile.

11.3 Pair Halos as Unique Cosmological Candles

Detection of "direct" intergalactic cascade radiation from distant extragalactic objects requires an extremely weak IGMF. Despite the recent achievements in radio measurements of intracluster magnetic fields, our knowledge of the IGMF on larger scales (i.e. beyond galaxy clusters) remains very poor (Kronberg, 2001; Grasso and Rubinstein, 2001). The method based on the microwave background anisotropy probes gives only upper limits on the primordial homogeneous field, $B \leq 3.4 \times 10^{-9}$ G (Barrow et al., 1997). But apparently it does not exclude the existence of huge "empty" regions – voids – with magnetic field as low as 10^{-18} G. If such voids would appear, just by chance, in front of some VHE emitters, then one should expect, as discussed above, "direct" cascade radiation towards these extragalactic objects. However, if the IGMF within a ≥ 10 Mpc radius around a VHE γ-ray source is relatively large, $B \geq 10^{-11}$ G, the mean free path of secondary TeV electrons exceeds the Larmor radius, and therefore the pair cascade would lead to production of an extended isotropic Pair Halo (Aharonian et al., 1994b). Actually, the formation of Pair Halos is unavoidable for all, even for highly beamed extragalactic γ-ray sources with spectra extending beyond 10 GeV. But only relatively compact Pair Halos produced around (multi) TeV sources can be detected. The Pair Halos are persistent structures with a monotonically increasing (with time) number density of non-thermalized electrons and positrons. Nevertheless, because the high energy γ-rays in Pair Halos are produced by TeV electrons the lifetime of which against Compton losses does not exceed 10^6 yr, the detectability of these structures requires γ-ray activity of the central (parent) source that has not stopped earlier than 10^6 yr ago. Detection of a Pair Halo at lower energies is very difficult, unless it is embedded in a strongly magnetised environment, e.g. in a galaxy cluster. In this case the energy of secondary electrons can be effectively released through synchrotron radiation in X-rays and perhaps even in γ-rays at detectable flux levels (see Chapter 9).

The formation and radiation of a Pair Halo is schematically demonstrated in Fig. 1.9. These hypothetical large-scale structures surrounding VHE sources can be revealed by their characteristic spectral and angular distributions. These features can be understood qualitatively from first principles. The energy E_γ and the arrival angle Θ of the detected photon from a source at a distance d are determined, on average, by the energy of the parent electron $E_e \approx (E_\gamma/4kT)^{1/2} m_e c^2$, and the distance l from the central source where this electron is produced. Because the mean free path of γ-rays, Λ, decreases with energy, the production site and the energy of the "parent" electron are defined, on average, by the "last γ-CIB interaction", rather independent of the history of previous generations of electrons and γ-rays. The energy of the leading electron varies between $E_e \approx (0.5-1)\, E_{\gamma,0}$, Therefore the energy E_γ and arrival angle Θ of a γ-ray photon seen by an observer contain information about their "grandparents" - γ-rays of energy $E_{\gamma,0} \sim 25(E_\gamma/1\ \mathrm{TeV})^{1/2}$ TeV, and of a mean free path $\Lambda(E_{\gamma,0}) = d \cdot \Theta$. Estimated in this way mean free path $\Lambda(E_{\gamma,0})$, based on the measured angular size Θ and the known distance to the source $d = cz/H_0$, tells us about the density of CIB around $\lambda \sim 1.2(E_{\gamma,0}/1\ \mathrm{TeV}) \simeq 30(E_\gamma/1\ \mathrm{TeV})^{1/2}$ μm.

Let's assume now, for an order-of-magnitude estimate, that the mean free path of 25 TeV γ-rays is between 10-20 Mpc (see Fig.11.2). Then the typical size of the halo in TeV γ-rays around a source at $z \sim 0.1$ is $\Theta \sim \Lambda(E_{\gamma,0})/d \sim 1°$ to $2°$. It is not difficult to guess also the approximate spectral shape of the radiation. It should have the characteristic spectral features of pair-cascades – a very hard energy distribution at low energies with photon index ≈ 1.5, and further gradual steepening to ≈ 2 in the transition region that precedes the cutoff.

Note that only in the case of an isotropically emitting γ-ray source will

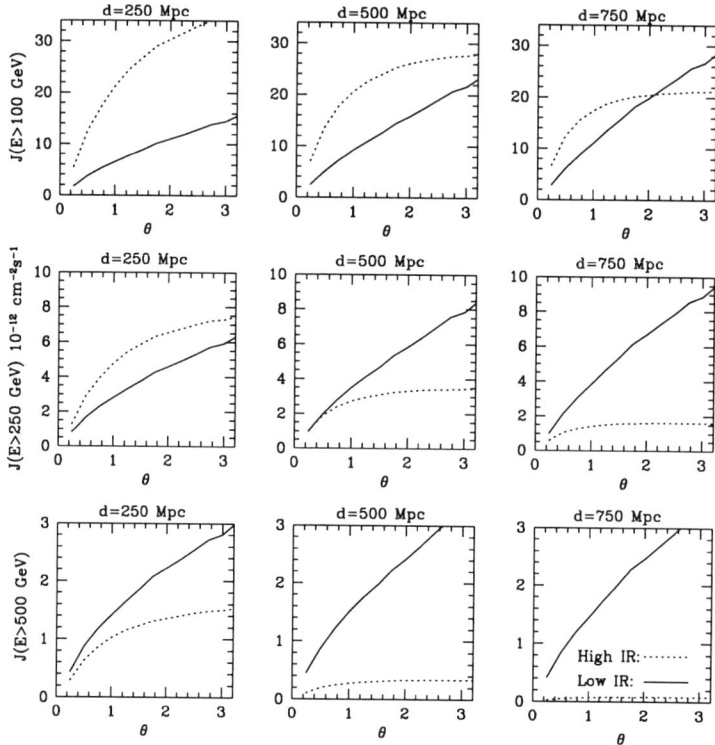

Fig. 11.7 The expected halo radiation fluxes at energies above 100 GeV, 250 GeV, and 500 GeV as a function of the detector opening angle for various combinations of source distance. The VHE luminosity of the central source above $E \gg 10\,\text{TeV}$ is normalised by the distance to the source as $L_0 = 10^{46}(d/1000\,\text{Mpc})^2\,\text{erg/s}$. For the CIB a flat (energy-independent) SED was assumed at two different levels: 4 nW/m²sr (low CIB) and 40 nW/m²sr (high CIB).

the Pair Halo be centered on the source. If the primary γ-ray emission is beamed, and directed away from us, the center of the halo is displaced by an angle comparable to the typical angular size of the halo. However the γ-ray radiation of the halo is emitted isotropically anyway, thus it can be recognised by its distinct variations in intensity with the angular size. These parameters depend weakly on details concerning the energy spectrum and geometry of the primary source, but they depend crucially on the flux of the background radiation, u_{CIB}, at mid- and far- infrared wavelengths.

The discovery of Pair Halos would provide us with unique "cosmological candles" to study of the time evolution of diffuse extragalactic background.

The γ-radiation from halos provides two *observables* – the angular and spectral distributions – and thus allows us to disentangle the two key *variables* – the CIB flux $u_{\rm CIB}$ and the distance to the source d – on a source-by-source basis.

The "Pair Halo" approach is complementary, to a certain extent, to the method based on the study of intergalactic absorption features as discussed in Sec.11.1. But it relies less critically on model-assumptions, and offers more interesting cosmological applications. On the other hand, the detection of Pair Halos is a more difficult task than the search for intergalactic absorption features in the direct component of radiation. Moreover, given the lack of information about source luminosities beyond 10 TeV, one cannot formulate with certainty the necessary conditions that would make the Pair Halos detectable. But in any case it is obvious that the limited sensitivities of both satellite-borne and ground-based γ-ray detectors for extended structures require very powerful VHE emitters. In this regard we note that the blazars, the representatives of the largest and brightest extragalactic γ-ray source population, for this task may not be sufficiently luminous. They are very bright because of the Doppler boosting (see Chapter 10), but, in fact, the intrinsic γ-ray luminosity of even the brightest blazars seems to be below 10^{46} erg/s.

The integral γ-ray fluxes above 100 GeV, 250 GeV and 500 GeV from Pair Halos at three different distances, calculated for isotropic central source γ-ray luminosities normalised to the source distance, $L_0 = 10^{46} \, (d/1000 \, {\rm Mpc})^2$ erg/s, are shown in Fig. 11.7. For the CIB, flat (energy-independent) spectral energy distributions at 4 and 40 nW/m²sr flux levels are assumed. These two values are lower and upper limits covering the whole range of uncertainties in the most crucial mid and far-infrared bands for formation of Pair Halos (see Fig. 11.1). The assumed the "VHE-luminosity – distance" relation keeps the total, integrated over all angles, halo flux at a constant level. It implies also that if there are objects of intrinsic VHE luminosity as large as 10^{46} erg/cm²s, the preferred distance range seems to be between several 100 Mpc to 1 Gpc in order to reduce the emission angle, and, at the same time, to have a reasonably high flux. Consequently, this compromise between the flux and the emission angle determines the energy range where the most effective studies can be performed. For probing the CIB at the present epoch, the energy interval between 100 GeV and 1 TeV seems to be optimal. The cosmological evolution of CIB can be studied with Pair Halos at $z \geq 1$, therefore the energy threshold of γ-ray detectors should be reduced to 10 GeV.

The γ-ray fluxes from Pair Halos shown in Fig. 11.7 can be (marginally) detected by 100 GeV threshold IACT arrays. The VHE γ-ray intrinsic luminosities assumed in these calculations are perhaps a bit optimistic, in particular for blazars taking into account that the Doppler boosting factors in these objects can be as large as 10^3. Nevertheless, there is a hope that the H.E.S.S. type Cherenkov telescope arrays equipped with large FoV high resolution cameras, should be able to provide the first sensitive searches for Pair Halos. In this regard, the TeV blazars 1ES 1426+428 at z=0.129 and PKS 2155-304 at $z = 0.116$, seem to be quite suitable objects to produce detectable Pair Halos at the H.E.S.S. sensitivity level.

In Fig.11.8a the spectral energy distributions of γ-rays expected within fixed intervals of angles centered on 1ES 1426+428 are shown. The angular distributions of arriving γ-rays at fixed energies are shown in Fig.11.8b. The calculations are performed (Eungwanichayapant, 2003) for the so-called "ΛCDM Cosmology+Salpeter IMF" CIB model (Primack, 2001) as shown in Fig. 11.9. The VHE emissivity of 10^{45} erg/s is assumed to be contributed by monoenergetic $E_0 = 100$ TeV γ-rays. The results of calculations do not, however, depend significantly on E_0, provided that $E_0 \gg 10$ TeV. Therefore the curves in Fig.11.8 are relevant to more realistic (broad-band) primary VHE γ-ray spectra as well. Fig.11.8a shows that the angular distributions become narrower as the energy of arriving γ-rays increases. It is seen also that the energy distributions of γ-rays within 0.1° to 0.3° and 0.3° to 1° intervals peak at 150-200 GeV, with an absolute flux around 0.5 eV/cm²s. These fluxes can be detectable by detectors like H.E.S.S. At somewhat higher fluxes, an adequate mapping of the halo also should be possible.

The visible parts of Pair Halos are determined by the density of the infrared background within ∼ 100 Mpc of VHE sources. Therefore we can probe directly the time evolution of the diffuse extragalactic background by observing γ-rays from Pair Halos surrounding sources with different, well established redshifts. The detection and identification of Pair Halos from cosmologically distant objects beyond $z = 1$ is possible effectively at energies below 30 GeV. It is important to note that whereas the spectral cutoffs in the radiation observed from the direction of these distant objects (both from the central source and from the surrounding halo) are caused by diffuse UV photons, the angular size of the halo is determined by the density of infrared photons, typically of 1 to $10\mu m$ wavelengths, at the epoch corresponding to the redshift of the central source z. This is a unique, unavailable by other astronomical means, cosmological channel of information

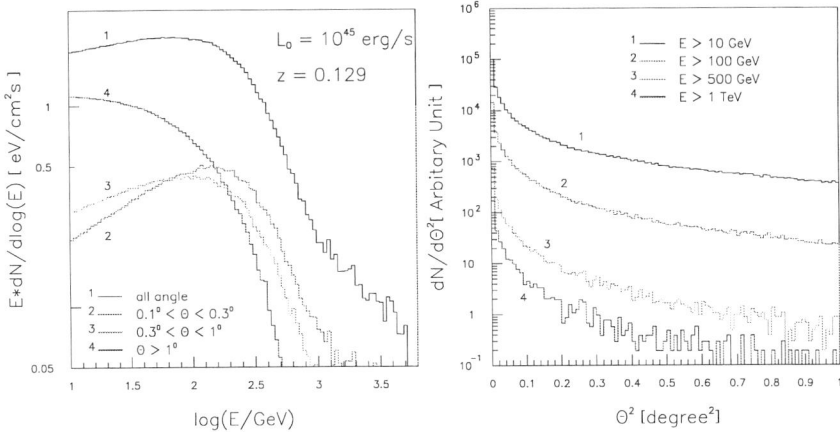

Fig. 11.8 Energy and angular distributions of γ-rays from a Pair Halo formed around a VHE source with redshift $z = 0.13$. The γ-ray luminosity of the source above 100 TeV is assumed to be 10^{45} erg/s. **a)** energy distributions within fixed angular bins (left panel); **b)** angular distributions for fixed γ-ray energies (right panel).

about the time evolution of the CIB, and, ultimately, about the epochs of galaxy formation and the character of their evolution. Such studies can be effectively performed by future major space- and ground based instruments like GLAST and 5@5.

Fig. 11.10 shows that more than 10 per cent of the primary VHE γ-ray luminosity is recycled through the Pair Halo and thus transferred to secondary γ-rays arriving within $1°$. Because the density of the CIB was significantly higher at epochs $z \geq 1$, the mean free path of γ-rays was correspondingly shorter. For example, the mean free path of a photon of energy 1 TeV at $z \geq 4$ does not exceed 10 Mpc (see Fig. 11.9). This implies that at cosmologically distant epochs even γ-rays of modest energy, ~ 1 TeV, emitted by the central object, can result in the formation of relatively compact Pair Halos. The chances of detection of such halos thus increases with z, provided that the central source emits γ-rays with adequate power.

In the case of detection of several such objects, we will be able to probe directly the cosmological evolution of the diffuse extragalactic background radiation. Also, these measurements can provide an independent test of the Hubble constant at large (and different !) redshifts. Finally, the intensities of the *primary* (compact, and possibly variable) and the halo (extended and steady) components of γ-ray emission from extragalactic sources tells

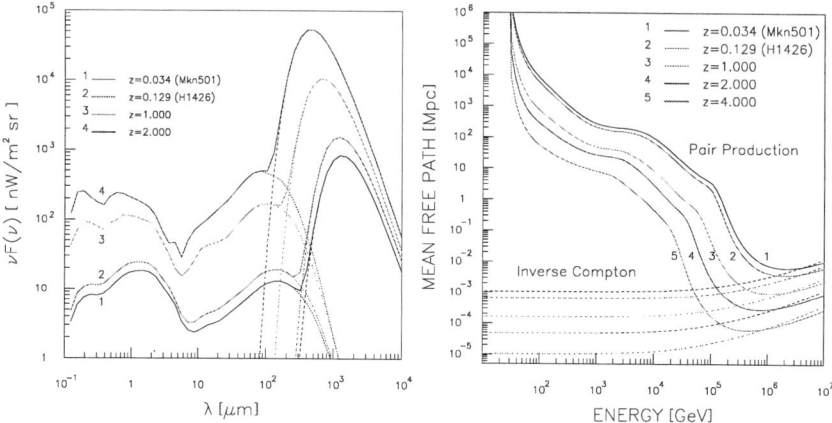

Fig. 11.9 Spectral Energy Distributions of the CIB at different cosmological epochs (from Primack, 2001) (left panel) and the corresponding mean free paths of γ-rays and electrons in the intergalactic radiation fields corresponding to the propagation lengths until the first interaction due to γ-γ pair-production and inverse Compton scattering, respectively (right panel)

us about the instantaneous and time integrated VHE powers of the central objects, and thus should allow us to estimate the total nonthermal energy released by individual extragalactic sources during their active phases. Another unique feature of Pair Halos is that they can be indicators of violent nonthermal events in past. In this case the Pair Halos represent relic structures around "retired monsters" which were very active in past, but presently are barely visible or even dead. And finally, Pair Halos can appear around powerful AGN with misaligned relativistic jets.

In Fig.11.11 we show the schematic images of Pair Halos around 3 different types of central objects - an isotropic source, a blazar, and a misaligned blazar. While in the case of an isotropic source of primary VHE γ-rays we expect a spherically symmetric image of the Pair Halo, in the case of a misaligned blazar with a double-jet structure we should expect two elongated images shifted from the central source. In the case of a blazar we again should see a symmetric image centered on the primary source. The luminosity of the central isotropic object gives a good estimate for the VHE power, and thus for such objects we may expect a quite bright halo emission. If the central source is a blazar, the high flux observed from the jet with a large but unknown Doppler factor, could lead to a misleading conclusion about the expected Pair Halo intensity. On the other hand, in

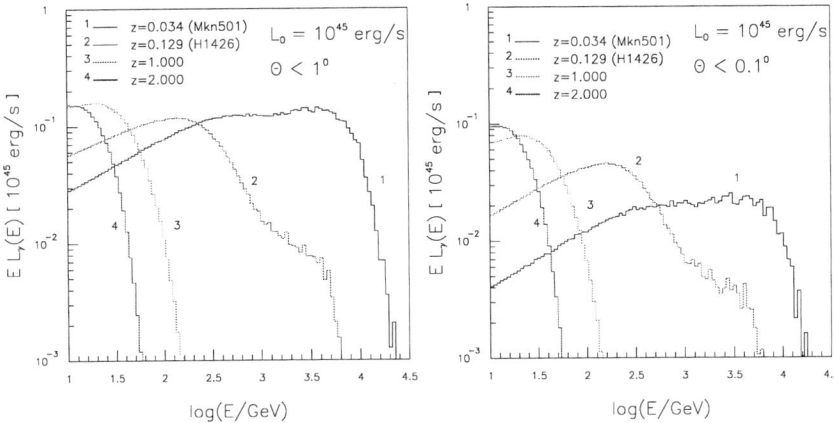

Fig. 11.10 Energy distributions of γ-rays from Pair Halos formed around sources at different z within 1° (left panel) and 0.1° (right panel). The calculations are performed for the time evolution of the diffuse background radiation shown in Fig. 11.9 at different cosmological epochs.

the case of a misaligned blazar, a pair of elongated halos can appear even without noticeable fluxes from the central object.

The two basic conditions for formation of Pair Halos are (i) the extension of the energy spectra of extragalactic γ-ray sources to TeV energies and (ii) relatively high magnetic field ($B \geq 10^{-11}$ G) environments in the ~ 10 Mpc neighbourhoods of these objects. The powerful jets of radiogalaxies and AGN seem to be the most promising candidates to create such large-scale structures. The possible production of very high (multi-TeV) and extremely high (PeV to EeV) energy γ-rays in both kpc and sub-pc AGN jets are discussed in Chapters 8 and 10. Because for formation of detectable Pair Halos we need γ-rays with energy well above 10 TeV (at least for a source with $z \ll 1$), the hadronic models of γ-ray production better match this requirement. Of particular interest is the scenario in which γ-rays are produced close to the central engines of powerful AGN due to interaction of protons, accelerated to extremely high energies, with ambient "hot" photon gas with a temperature of more than 10^4 K. Remarkably, for a certain combinations of model parameters not only can the photo-pion interactions proceed with high efficiency, but also the secondary γ-rays with energy exceeding 10^{15} eV can effectively escape from their production regions (see Fig. 8.10). As discussed in Chapters 8 and 9, these energetic photons can effectively power, through the synchrotron X-radiation of the secondary

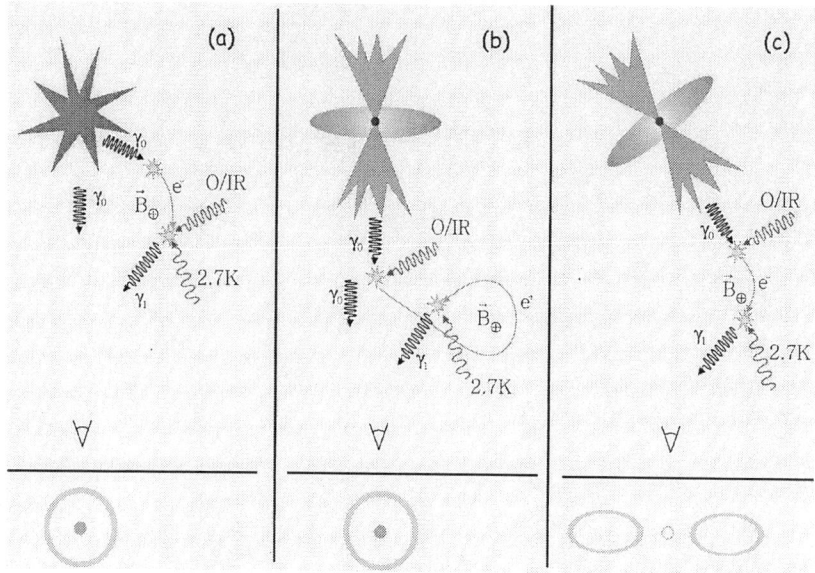

Fig. 11.11 Schematic images of Pair Halos around 3 different types of central objects - an isotropic source, a blazar, and a misaligned blazar.

pair-produced electrons, large (up to several hundred kpc) scale jets of AGN, as well as well as entire intracluster media. At the same time, lower energy γ-rays, $E \leq 100$ TeV, produced either at the base of the small-scale jet or during the cascade development in the large-scale jet can serve as ideal primary material for the creation of Pair Halos on ≥ 10 Mpc scales. In the case of galaxy clusters, these γ-rays can be contributed by several AGN, increasing the probability of a Pair Halo being detected.

Finally, as discussed in Chapter 9, if the energy spectrum of the central source extends to 10^{18} eV or even more, and the central AGN is located in a galaxy cluster with magnetic field as large as 0.1μG, we may expect bright high energy synchrotron (MeV/GeV) γ-ray cores inside the Pair Halos. The mean free path of these energetic γ-rays does not exceed 1 Mpc (see Fig. 1.4), thus they are absorbed within the galaxy cluster. The energy of secondary, pair-produced electrons is sufficient to produce synchrotron photons with characteristic energies extending from 100 MeV to several tens of GeV. Because the efficiency of conversion of the nonthermal power of the central engine to $\geq 10^{16}$ eV γ-rays can be as high as 1 per cent, and because the conversion of these EHE γ-rays to synchrotron photons in the intracluster medium proceeds with almost 100 per cent efficiency, this scenario can provide detectable γ-ray fluxes in a broad energy range depending on the high-energy cutoff in spectrum of primary γ-rays.

If the intracluster medium is powered by γ-rays from a single AGN with a jet away from the observer, then the observer can see only low energy γ-rays. Indeed, in order to see the synchrotron photons, the magnetic field should be sufficiently large to curve the electrons from their original direction which coincides with direction of parent γ-rays, and, therefore, with direction of the jet. Formally, this implies the following condition $l_{\text{synch}} > R_{\text{L}}$, where $l_{\text{synch}} = ct_{\text{synch}} \simeq 0.04 B_\mu^{-2} E_{17}^{-1}$ kpc is the mean free path of an electron of energy $E_{17} = E/10^{17}$ eV due to synchrotron losses in the field $B_\mu = B/1\mu\text{G}$, and $R_{\text{L}} \simeq 0.1\, E_{17} B_\mu^{-1}$ kpc is the Larmor radius. This immediately leads to the the following upper limit on the energy of the electron which can significantly change its direction before radiating a significant part of the kinetic energy, $E_{17} \leq 0.6\, B_\mu^{-1/2}$. For the given field B_μ and electron energy E_{17}, the characteristic energy of the synchrotron photon is $E_\gamma \simeq 0.5 B_\mu E_{17}^2$ GeV. Formally, one may expect synchrotron photons up to 1 TeV, if the spectrum of primary γ-rays extends beyond 10^{18} eV and the magnetic field exceeds 1 μG. However, given the above limit on the electron energy, and assuming that the primary EHE γ-rays are beamed away, the observer can see only relatively low energy synchrotron radiation, $\epsilon \leq 0.2$ GeV, independent of the magnetic field and the maximum energy of primary γ-rays. On the other hand, if the primary γ-rays are injected by a blazar, the observer can see not only very high energy photons, but also significantly enhanced radiation. Obviously, if the galaxy cluster is powered by several AGN with randomly oriented jets, these effects will be significantly smeared out.

Confirmation of such a scenario for the production of γ-rays in the intracluster medium would make galaxy clusters promising targets for searches of the accompanying γ-radiation of higher, multi-GeV to TeV energies from Pair Halos that should be unavoidably produced around these clusters on larger, 10 Mpc or more, scales by the relatively low energy, ≤ 100 TeV, parts of primary γ-ray beams of AGN.

11.4 Diffuse Extragalactic Background as Calorimetric Measure of the VHE Emissivity of the Universe

The intergalactic absorption of γ-rays prevents us from directly seeing most of the VHE emission in the Universe. Nevertheless, we can see lower energy photons associated with the secondary radiation produced in the electromagnetic cascades initiated by the absorbed γ-rays. Irrespective of the structure and strength of the IGMF, the ensemble of all VHE sources in the Universe produces isotropic cascade γ-radiation. This gives us a very effective cosmological tool to constrain or measure the VHE emissivity of the Universe by analysing the diffuse cascade background produced by all possible VHE phenomena that took place in the Universe over the Hubble timescale (Coppi and Aharonian, 1997). This includes, first of all, the ensemble of all (detected and potential) high energy γ-ray sources – from pulsars and SNRs in ordinary galaxies to most powerful extragalactic objects like AGN jets and galaxy clusters. Such cascades can be produced also by the highest energy cosmic rays triggered in interactions with 2.7 K CMBR photons, i.e. by the same process that leads to the so-called Greisen-Zatsepin-Kuzmin (GZK) cutoff (Greisen, 1966; Zatsepin and Kuzmin, 1966). Independent of the origin of these energetic particles, accelerated in individual astronomical objects and/or being result of more "exotic" cosmological scenarios like decaying GUT-scale particles or topological defects, the outcome is almost the same. All these scenarios produce a large γ-ray background (Aharonian et al., 1992; Protheroe and Stanev, 1996; Coppi and Aharonian, 1997; Sigl et al., 1997; Berezinsky et al., 1998). The flux of this radiation is rather insensitive to the spectrum of the original VHE radiation. Moreover, at low energies, $E \leq 100$ GeV, where the study of the isotropic γ-ray background-radiation is available by space-based instruments, the spectrum of the cascade radiation is weakly sensitive to the details of the spectrum and flux of DEBRA. Thus, cascade background radiation acts as a unique calorimeter, allowing us to measure the VHE energy input into the Universe based on the total level of the diffuse γ-ray background.

Indeed, using the spectral measurements of the extragalactic background radiation up to 100 GeV (see Fig. 1.3), we can estimate the maximum average VHE emissivity allowed in the Universe by equating the cascade background energy flux accumulated over a cosmological distance scale ($\frac{1}{4\pi}\dot{Q}_{e-m}d$ where $d \sim 1$ Gpc) with the observed EGRET energy flux above 100 MeV, $F_\gamma \sim 8 \times 10^3$ eV cm^{-2}sr^{-1}s^{-1}. Such a simple analysis gives $\dot{Q}_{\text{em}}^{\text{max}} \approx 5 \times 10^{-23}$ eV cm^{-3}s^{-1}. Detailed numerical modelling (Coppi and Aharonian, 1997), based on cascade calculations in the intergalactic radiation and magnetic fields, assuming certain cosmological evolution scenarios for both DEBRA and VHE source populations, gives a somewhat stronger upper limit. In particular, for any reasonable DEBRA model and for a moderately evolving source population (between $(1+z)^3$ and $1+z)^{9/2}$) the parameter $\dot{Q}_{\text{em}}^{\text{max}}$ appears in rather narrow limits, $\dot{Q}_{\text{em}}^{\text{max}} \simeq 3 \times 10^{-23}$ eV cm^{-3}s^{-1}. In a Hubble volume, $V_H \sim \frac{4\pi}{3}(c/H_0)^3$, this implies a maximum VHE source luminosity $\sim 3 \times 10^{50}$erg s^{-1}. Actually, this is not a large number compared, for example, with the bolometric luminosity of a single powerful quasar that could be as large as $\sim 10^{48}$erg s^{-1}.

This limit is rather insensitive to details of the IR/O background and applies to *any* cosmological population with significant VHE emission above ~ 1 TeV, for example, to any galaxy or cluster population with strong cosmic ray production at some stage in its history.

The majority of the identified EGRET sources are blazars, therefore it is straightforward to relate the bulk of the extragalactic background to this class of sources (Padovani *et al.*, 1993). The present estimates of contributions from unresolved (non-cascade) γ-ray blazars range from 25 per cent (Chiang and Mukherjee, 1998; Mücke and Pohl, 2000) to nearly all (Stecker and Salamon, 1996). If so, the above VHE cascade limits tighten and become even more interesting. In particular, depending on the IGMF and details of blazar beaming, they could imply that typical blazar AGN show an *intrinsic* spectral break (not due to DEBRA absorption) at ≥ 100 GeV – which would be an important constraint for blazar models. Also, they should be sufficient to definitely rule out the whole variety of currently popular *non-acceleration* ("top-down") scenarios for the origin of the highest energy cosmic rays.

At the same time, several AGN are already established as effective TeV emitters, and there is no shortage of ideas for producing VHE emission by other means (e.g., by decaying topological defects or annihilation of non-baryonic dark matter). Any residual background surviving a fluctuation/point source analysis could well be VHE cascade emission. Then, detection of a cascade background would provide combined information on the evolution of the underlying VHE source population. GLAST, with sufficient sensitivity to the diffuse γ-ray background up to 300 GeV, could set even tighter limits on VHE source populations, or hopefully, detect a cascade background.

Appendix A

Spherically symmetric diffusion from a single source

In the standard diffusion approximation (i.e. neglecting convection) the propagation of CRs is described by the familiar diffusion equation (Ginzburg and Syrovatskii, 1964) which in the spherically symmetric case reduces to the following form

$$\frac{\partial f}{\partial t} = \frac{D}{R^2}\frac{\partial}{\partial R}R^2\frac{\partial f}{\partial R} + \frac{\partial}{\partial \gamma}(Pf) + Q \tag{A.1}$$

Here $f(R, t, \gamma)$, with $\gamma = E/m_e c^2$, is the energy distribution function of particles at instant t and distance r from the source; $P(\gamma) = -d\gamma/dt$ is the continuous energy loss rate, and D denotes the energy-dependent diffusion coefficient $D(\gamma)$. It is assumed that D is independent of R, i.e. a homogeneous medium is supposed. For the δ-function-type initial distribution of particles both in space and time, i.e. for the Green's function with respect to R and t, Eq.(A.1) has a simple analytical solution for an arbitrary injection spectrum Q (Atoyan et al., 1995):

$$f(R, t, \gamma) = \frac{Q(\gamma_t)\,P(\gamma_t)}{\pi^{3/2}\,P(\gamma)\,R_{dif}^3} \exp(-\frac{R^2}{R_{dif}^2}) \ . \tag{A.2}$$

Here $\gamma_t \equiv g^{-1}(T-t)$, where $g^{-1}(T)$ is the inverse function of $g(\gamma)$, defined as

$$T = \int_{\gamma}^{\gamma_*} \frac{d\gamma_1}{P(\gamma_1)} = g(\gamma) \ . \tag{A.3}$$

The variable γ_t corresponds to the initial energy of particles which are cooled to a given γ during the time t, and

$$r_{dif}(\gamma, t) = 2\sqrt{\Delta u} \tag{A.4}$$

corresponds to the effective diffusion radius up to which the relativistic particles with energy γ propagate during the time t after their injection from the source. The function Δu is determined from the following equation:

$$\Delta u(\gamma, \gamma_t) = \int_\gamma^{\gamma_t} \frac{D(x)dx}{P(x)} \qquad (A.5)$$

Eq.(A.2) was obtained without any specification of the initial spectrum, the energy losses $P(\gamma)$, or the diffusion coefficient $D(\gamma)$. Since the function f in Eq.(A.2) depends on R in a simple exponential form, this solution is convenient for integration over spatially distributed sources, as well as over a finite particle injection time. For example, in the case of continuous injection of CRs from a stationary point source

$$f(r, \gamma) = \frac{1}{8\pi^{3/2} P(\gamma)} \int_\gamma^\infty \frac{Q(x)}{[\Delta u(\gamma, x)]^{3/2}} \exp\left(-\frac{r^2}{4\Delta u(\gamma, x)}\right) dx. \qquad (A.6)$$

In the case of stationary sources distributed uniformly in space at distances beyond some R_0 from us, we can substitute $Q(\gamma) \to q(\gamma)d^3r$ (q being the specific injection rate per unit volume) and integrate over the region $R > R_0$, which results in

$$f(\gamma) = \frac{1}{P(\gamma)} \int_\gamma^\infty q(x) \left[\mathrm{erfc}(a_0) + \frac{2 a_0}{\sqrt{\pi}} e^{-a_0^2}\right] dx, \qquad (A.7)$$

where $\mathrm{erfc}(z) = (2/\sqrt{\pi}) \int_z^\infty \exp(-x^2)\, dx$ is the error-function, and $a_0 = R_0/2\sqrt{\Delta u(\gamma, x)}$. Notice that, since $erfc\,(0) = 1$, in the limiting case $R_0 \to 0$ this latter equation gives the familiar result for the distribution of particles which are injected stationarily and uniformly into interstellar space, and suffer continuous energy losses. In this case the dependence on D disappears completely.

Appendix B

Evolution of relativistic electrons in an expanding magnetised medium

B.1 Kinetic equation

The formalism described below was developed by Atoyan and Aharonian (1999) to study the time evolution of the energy distribution of relativistic electrons in an expanding magnetised medium for an arbitrary time-dependent injection rate, taking into account both adiabatic and radiative energy losses, as well as losses caused by the *energy-* and *time-dependent* escape of electrons. The results of this convenient analytical approach can be applied to the study of the nonthermal radiation components of relativistic jets in AGN and microquasars.

The kinetic equation describing the evolution of the energy distribution function of relativistic electrons in a spatially homogeneous source $N \equiv N(\gamma, t)$ is the well known partial differential equation (e.g. Ginzburg and Sirovatskii, 1964):

$$\frac{\partial N}{\partial t} = \frac{\partial}{\partial \gamma}(PN) - \frac{N}{\tau} + Q. \qquad (B.1)$$

The Green's function solution to this equation in the case of time-independent energy losses and constant escape time $\tau(\gamma, t) = const$ was found by Syrovatskii (1959). However, in an expanding magnetised cloud under consideration we have to suppose that *all* parameters depend on both energy γ and time t, i.e. $Q \equiv Q(\gamma, t)$ is the injection spectrum, $\tau \equiv \tau(\gamma, t)$ is characteristic escape time of a particle from the source, and $P \equiv P(\gamma, t) = -(\partial \gamma / \partial t)$ is the energy loss rate.

Strictly speaking, Eq.(B.1) corresponds to a spatially homogeneous source where the energy gain due to in-situ acceleration of particles is absent. Actually, however, it may have wider applications. Indeed, in a general form the equation describing the evolution of the local energy distribution function $f \equiv f(\gamma, \mathbf{r}, t)$ of relativistic particles can be written (e.g. Ginzburg and Syrovatskii, 1964) as:

$$\frac{\partial f}{\partial t} = \mathrm{div}(D_r \mathbf{grad} f) - \mathrm{div}(\mathbf{u}_r f) + \frac{\partial}{\partial \gamma}(P_r f) - \frac{\partial}{\partial \gamma}(b_r f) + \frac{\partial^2}{\partial \gamma^2}(d_r f) \,, \quad \text{(B.2)}$$

where all parameters depend also on the radius-vector $\mathbf{r}$. The first two terms on the right-hand side of this equation describe the diffusive and convective propagation of particles, and the last two terms correspond to the acceleration of the particles through the first and second order Fermi mechanisms. If there are internal sources and sinks of particle injection (such as production and annihilation), then terms similar to the last two in Eq.(B.1) should be added as well.

Let us consider a source where the region of effective particle acceleration can be separated from the main emission region. This is likely to be the case of blobs in blazars and microquasars, where the probable site for in-situ acceleration of electrons is a relatively thin region around either the bow shock front formed ahead of the cloud, or possibly a *jet termination* shock formed behind the cloud. Meanwhile the main part of the observed flux should be produced in a much larger volume, V_0, of the post-shock region in the cloud, since the synchrotron cooling time of radio electrons is orders of magnitude larger than the dynamical times of the source. Since the acceleration efficiency (parameters b_r and d_r) should drop significantly outside the shock region, after integration of Eq.(B.2) over the volume V_0 the last two terms can be neglected.

The integration of the left side of Eq.(B.2) results exactly in $\partial N/\partial t$. Integration of the two propagation terms in the right side of Eq.(B.2) gives the net flux of particles, due to diffusion and convection, across the surface of the emission region. Obviously, these terms are expressed as the difference between the total numbers of particles injected into and escaping from the volume V_0 per unit time, so the last two terms of Eq.(B.1) are found (internal sources and sinks, if present, are also implied). Finally, integration of the energy loss term in Eq.(B.2) is reduced to the relevant term of Eq.(B.1), where P corresponds to the mean energy loss rate per particle of energy γ, i.e. $P = \int P_r f \mathrm{d}^3 r / \int f \mathrm{d}^3 r$. The volume V_0 of the cloud implies the region filled with relativistic electrons and *enhanced* magnetic field, where the bulk of the observed radiation is produced.

Thus, Eq.(B.1) is quite applicable to the study of sources with ongoing in-situ acceleration, as long as the volume V_0, where the bulk of nonthermal radiation is produced, is much larger than the volume ΔV of the region(s) in the source responsible for the effective acceleration of the electrons. Note that solutions for a large number of particular cases of the Fokker-Planck partial differential equation (which includes the term $\propto \bar{d}_r$ for stochastic acceleration), corresponding to different combinations of terms responsible for time-dependent adiabatic and synchrotron energy losses, stochastic and regular acceleration, were obtained long ago and have been qualitatively discussed by Kardashev (1962), assuming *energy-independent* escape of relativistic particles from the production region. However, in the jets of microquasars, and perhaps also blazars, the energy-dependent escape of electrons from the active zone may play an important role in spectral evolution of the resulting synchrotron and inverse Compton emission. Another important point is that the Fokker-Planck equation generally may contain a singularity, so transition from the solutions of that equation (if known), which are mostly expressed through special functions, to the case of $\bar{d}_r \to 0$ may not always be straightforward (for a comprehensive discussion of the problems related with singularities in the Fokker-Planck equation, as well as general solutions for *time-independent* parameters see Park and Petrosian (1995), and references therein). Meanwhile, substitution of the acceleration terms by effective injection terms in the regions responsible for the bulk of nonthermal radiation, allows us to disentangle the problems of acceleration and emission of the electrons, and enables analytical solutions to the first order equation Eq.(B.1) which are convenient both for further qualitative analysis and numerical calculations.

B.2   Time-independent energy losses

Suppose first that the escape time is given as $\tau = \tau(\gamma, t)$, but the energy losses are time-independent, $P = P(\gamma)$. The Green's function solution $G(\gamma, t, t_0)$ to Eq.(B.1) for an arbitrary injection spectrum $N_0(\gamma)$ of electrons implies δ-functional injection $Q(\gamma, t) = N_0(\gamma)\, \delta(t-t_0)$ at some instant t_0. At times $t > t_0$ it actually corresponds to the solution for the homogeneous part of Eq.(B.1), with initial condition $G(\gamma, t_0 + 0, t_0) = N_0(\gamma)$, while $G(\gamma, t_0 - 0, t_0) = 0$. Then for the function $F = PG$ this equation is reduced to the form

$$\frac{\partial F}{\partial t} = \frac{\partial F}{\partial \zeta} - \frac{F}{\tau_1(\zeta, t)}, \tag{B.3}$$

where instead of the energy γ, a new variable is introduced,

$$\zeta = g(\gamma) \equiv \int_{\gamma_*}^{\gamma} \frac{d\gamma_1}{P(\gamma_1)} \tag{B.4}$$

where γ_* is some fixed energy. Formally, ζ has the meaning of the time needed for a particle with energy γ to cool to energy γ_* (so for convenience one may suppose $\gamma_* = 1$). The function $\tau_1(\zeta, t) = \tau[\varepsilon(\zeta), t]$, where ε is the inverse function of $g(\gamma)$ which expresses the energy through ζ, i.e. $\gamma = \varepsilon(\zeta)$. The initial condition for $F(\zeta, t)$ reads

$$F(\zeta, t_0) = P[\varepsilon(\zeta)] N_0[\varepsilon(\zeta)] \equiv U(\zeta). \tag{B.5}$$

Transformation of Eq.(B.3) from variables (ζ, t) to $(s = \zeta + t, u = t)$ results in a partial differential equation over only one variable for the function $F_1(s, u) = F(\zeta, t)$:

$$\frac{\partial F_1}{\partial u} = -\frac{F_1}{\tau_1(s - u, u)}, \tag{B.6}$$

with the initial condition $F_1(s, u_0) = U(s - u_0)$ found from Eq.(B.5). Integration of Eq.(B.6) is straightforward:

$$F_1(s, u) = U(s - u_0) \exp\left[-\int_{u_0}^{u} \frac{du_1}{\tau_1(s - u_1, u_1)}\right]. \tag{B.7}$$

In order to return from variables (s, u) to (γ, t), it is useful to understand the meaning of the function $\varepsilon(s-x)$ which enters into Eq.(B.7) via Eq.(B.5) for U and the escape function $\tau_1 \to \tau$. Since ε is the inverse function of g, then for any z in the range of definition of this function, $z = g[\varepsilon(z)]$. Then, taking into account that $s = \zeta + t$ and $\zeta = g(\gamma)$, for $z = s - x$ one obtains $t - x = g[\varepsilon(s - x)] - g(\gamma)$. For the function g defined by Eq.(B.4), this equation results in

$$t - x = \int_{\gamma}^{\Gamma_x(\gamma, t)} \frac{d\gamma_1}{P(\gamma_1)} \tag{B.8}$$

where $\Gamma_x(\gamma, t)$ corresponds to $\varepsilon(s - x)$ after its transformation to the variables (γ, t). Thus, for a particle with energy γ at an instant t, the function $\varepsilon(s - x)$ is the energy $\Gamma_x \equiv \Gamma_x(\gamma, t)$ of that particle at time x, i.e. it describes the trajectory of individual particles in energy space.

Expressing Eq.(B.7) in terms of the Green's function $G = F/P$, the solution to Eq.(B.1) for an arbitrary $\tau(\gamma, t)$, but with time-independent energy losses, is found:

$$G(\gamma, t, t_0) = \frac{P(\Gamma_{t_0}) N_0(\Gamma_{t_0})}{P(\gamma)} \exp\left[-\int_{t_0}^{t} \frac{\mathrm{d}x}{\tau(\Gamma_x, x)}\right] . \tag{B.9}$$

Note that this is not a standard Green's function in the sense that the injection spectrum was supposed as an arbitrary function of energy $N_0(\gamma)$, and not necessarily a delta-function. Actually, it describes the evolution of relativistic particles with a given distribution $N_0(\gamma)$ at $t = t_0$. The solution for an arbitrary continuous injection spectrum is readily found after substitution $N_0(\gamma) \to Q(\gamma, t_0) \mathrm{d}t_0$ into Eq.(B.9) and integration over $\mathrm{d}t_0$:

$$N(\gamma, t) = \frac{1}{P(\gamma)} \int_{-\infty}^{t} P(\Gamma_{t_0}) Q(\Gamma_{t_0}, t_0) \exp\left[-\int_{t_0}^{t} \frac{\mathrm{d}x}{\tau(\Gamma_x, x)}\right] \mathrm{d}t_0 , \tag{B.10}$$

with the function Γ defined via Eq.(B.8). In the particular case of time and energy independent escape, $\tau(\gamma, t) = const$, this solution coincides with the one given in Syrovatskii (1959) in the form of double integral over t_0 and Γ, if Eq.(B.10) is integrated over energy with the use of general relations

$$\frac{\partial \Gamma_x}{\partial t} = -\frac{\partial \Gamma_x}{\partial x} = P(\Gamma_x) , \tag{B.11}$$

which follow from Eq.(B.8).

Some specific cases of Eq.(B.10) are worth brief discussion. Let the escape of particles be energy dependent but stationary, $\tau(\gamma, t) \to \tau(\gamma)$, and consider first the evolution of $N(\gamma, t)$ when energy losses are negligible, so $\Gamma_x \simeq \gamma$ for any t. Assuming for convenience that the form of injection spectrum does not change in time, i.e. $Q(\gamma, t) = Q_0(\gamma) q(t)$ with $q(t < 0) = 0$ (i.e. injection starts at t=0), Eq.(B.10) is reduced to

$$N(\gamma, t) = Q_0(\gamma) \tau(\gamma) \int_0^{t/\tau(\gamma)} q[t - \tau(\gamma) z] e^{-z} \mathrm{d}z . \tag{B.12}$$

For stationary injection $q(t \geq 0) = 1$ the integral results in the simple $(1 - e^{-t/\tau})$, so $N(\gamma, t) \simeq Q_0(\gamma) t$ until $t < \tau(\gamma)$, and then the escape of electrons modifies the particle distribution, compared with the injection spectrum, as $N(\gamma, t) \simeq Q_0(\gamma) \tau(\gamma)$. In the case of $\tau(\gamma) \propto \gamma^{-\Delta}$ it results in a power-law steepening of the injection spectrum by factor of Δ. This is the well known result of the so-called leaky-box model in cosmic ray theory. In the case of non-stationary injection, however, the modification of $Q(\gamma)$ is different. In particular, for an impulsive injection, $q(t) = \delta(t)$, it is reduced to a sharp cut-off of an exponential type above energies γ_t found from $\tau(\gamma_t) \simeq t$.

For a stationary injection of particles, Eq.(B.10) can be transformed to the form

$$N(\gamma,t) = \frac{1}{P(\gamma)} \int_\gamma^{\Gamma_0} Q_0(\Gamma) \exp\left[-\int_\gamma^\Gamma \frac{dz}{P(z)\tau(z)}\right] d\Gamma \qquad (B.13)$$

using Eq.(B.11). In the case of $\tau \to \infty$ (absence of escape) and large t, when $\Gamma_0 \equiv \Gamma_0(\gamma,t) \to \infty$, Eq.(B.13) comes to the familiar steady state solution for distribution of particles in an infinite medium. If the synchrotron (or IC) energy losses of electrons dominate, $P = p_2\gamma^2$, then $\Gamma_0 = \gamma/(1-p_2t\gamma)$. In this case $N(\gamma,t) \sim t\,Q_0(\gamma)$ until $p_2t\gamma \leq 1$, and then the radiative losses result in a quick steepening of a *stationary* power-law injection spectrum by a factor of 1. Meanwhile, in the case of impulsive injection the modification of the initial spectrum of electrons is reduced to an exponential cut-off at $\gamma \geq 1/p_2t$ (see Kardashev, 1962).

B.3 Expanding cloud

Energy losses of relativistic electrons in an expanding medium become time-dependent. The adiabatic energy loss rate is given as $P_{\rm ad} = v\gamma/R$, where R is the characteristic radius of the source, and v is the speed of spherical expansion. For electrons of higher energies, however, synchrotron losses may dominate. For the magnetic field we suppose a power-law dependence, $B = B_0(R_0/R)^{-m}$, where B_0 and R_0 are the magnetic field and the radius of the cloud at instant t_0. Thus,

$$P = \frac{\gamma}{R}\left(p_1 + p_2 \frac{\gamma}{R^\mu}\right), \qquad (B.14)$$

where $\mu = 2m - 1$. For adiabatic losses, $p_1 = v$, but the constants p_1 and p_2 are kept in parametric form in order to enable other losses with similar dependence on γ and R as well. Here we will suppose that the expansion speed $v = {\rm const}$, and consider evolution of the particles injected impulsively at instant t_0 with the spectrum $G(\gamma,t_0) = N_0(\gamma)$, as previously.

Since the energy losses depend on time via the radius $R(t) = R_0 + v(t-t_0)$, it is convenient to pass from the variable t to R. Then, for the function

$\Phi = \gamma G(\gamma, R)/R$, Eq.(B.1) reads:

$$R\frac{\partial \Phi}{\partial R} = \gamma \frac{\partial}{\partial \gamma}\left[\left(a_1 + a_2 \frac{\gamma}{R^\mu}\right)\Phi\right] - \left(1 + \frac{\gamma}{v\tau}\right)\Phi, \qquad \text{(B.15)}$$

where $a_1 = p_1/v$, $a_2 = p_2/v$, and for τ now we imply the function $\tau(\gamma, R)$. Transformation of this equation from variables (γ, R) to $(\psi = \ln(\gamma/R^\mu), \xi = \ln R)$ results in the equation

$$\frac{\partial \Phi_1}{\partial \xi} = \frac{\partial}{\partial \psi}[(\mu + a_1 + a_2 e^\psi)\Phi_1] - \left[1 + \frac{e^\xi}{v\tau_1(\psi,\xi)}\right]\Phi_1, \qquad \text{(B.16)}$$

where $\Phi_1 \equiv \Phi_1(\psi, \xi)$ and $\tau_1(\psi, \xi) = \tau(e^{\psi+\mu\xi}, e^\xi)$. The initial condition at $\xi_0 = \ln R_0$ reads $\Phi(\psi, \xi_0) = R_0^{\mu-1} e^\psi N_0(R_0^\mu e^\psi)$. Thus, we come to the equation formally coinciding with the one considered above, with 'time' (ξ) independent 'energy' (ψ) losses $P_*(\psi) = \mu + a_1 + a_2 e^\psi$, and arbitrary 'escape' function $\tau_*(\psi, \xi) = (1 + e^\xi/v\tau_1)^{-1}$. The solution to this equation is analogous to Eq.(B.9) :

$$\Phi_1(\psi, \xi) = R_0^{\mu-1} e^{\Psi_0} \frac{1 + c_* e^{\Psi_0}}{1 + c_* e^\psi} N_0(R_0^\mu e^{\Psi_0}) \exp\left[\xi_0 - \xi - \int_{\xi_0}^{\xi} \frac{e^z dz}{v\tau_1(\Psi_z, z)}\right], \qquad \text{(B.17)}$$

where $c_* = a_2/(\mu + a_1)$, $\Psi_0 \equiv \Psi_{\xi_0}$, and $\Psi_x \equiv \Psi_x(\psi, \xi)$ is the characteristic trajectory of a particle in the 'energy' space ψ, which is readily calculated from Eq.(B.8) for the given P_*:

$$\Psi_x(\psi, \xi) = -\ln[(c_* + e^{-\xi})e^{(\mu+a_1)(\xi-x)} - c_*] \qquad \text{(B.18)}$$

Returning now to the variables γ and R, the evolution of particles can be described by the function

$$G(\gamma, R, R_0) = \left(\frac{R_0}{R}\right)^{a_1} \frac{\Gamma_{R_0}^2}{\gamma^2} N_0(\Gamma_{R_0}) \exp\left[-\frac{1}{v}\int_{R_0}^{R} \frac{dr}{\tau(\Gamma_r, r)}\right]. \qquad \text{(B.19)}$$

The energy $\Gamma_r \equiv \Gamma_r(\gamma, R)$ corresponds to the trajectory of a particle with given energy γ at $r = R$ in the (Γ, r) plane, and can be represented as $\Gamma_r = \gamma \Lambda(\gamma, R, r)$, where

$$\Lambda(\gamma, R, r) = \frac{\left(\frac{R}{r}\right)^{a_1}}{1 + \frac{c_* \gamma}{R^\mu}\left[1 - \left(\frac{R}{r}\right)^{\mu+a_1}\right]}. \qquad \text{(B.20)}$$

In the formal case of $\mu + a_1 \to 0$, when $c_* = a_2/(\mu + a_1) \to \infty$, Eq.(B.20) tends to the limit $\Lambda' = (R/r)^{a_1}/[1 + a_2 \gamma R^{a_1} \ln(R/r)]$.

In the general case of continuous injection of relativistic particles with the rate $Q(\gamma,t) \to Q(\gamma,R)$, the evolution of their energy distribution during expansion of the cloud between radii R_0 and $R \geq R_0$ is found, using Eq.(B.19):

$$N(\gamma,R) = G(\gamma,R,R_0) + \qquad (B.21)$$

$$\frac{1}{v}\int_{R_0}^{R} \left(\frac{r}{R}\right)^{a_1} \Lambda^2(\gamma,R,r) Q(\Gamma_r,r) \exp\left[-\frac{1}{v}\int_r^R \frac{dz}{\tau(\Gamma_z,z)}\right] dr \, .$$

Here $G(\gamma,R,R_0)$ is given by Eq.(B.19). The substitution $R = R_0 + v(t-t_0)$ results in an explicit expression for $N(\gamma,t)$. If the source is expanding with a constant velocity v starting from $t = 0$, such a substitution results in formal changes $R \to t$, $R_0 \to t_0$, $r \to t'$, and $dr = vdt$ in Eq.(B.22). When only the adiabatic losses are important, i.e $a_1 = 1$ and $a_2 = 0$, Eq.(B.20) is reduced to a simple $\Lambda = t/t_0$, and

$$N(\gamma,t) = \frac{t}{t_0} N_0\left(\frac{t}{t_0}\gamma\right) \exp\left(-\int_{t_0}^{t} \frac{dx}{\tau(t\gamma/x,x)}\right) +$$

$$\int_{t_0}^{t} \frac{t}{z} Q\left(\frac{t}{z}\gamma,z\right) \exp\left(-\int_{z}^{t} \frac{dx}{\tau(t\gamma/x,x)}\right) dz \, . \qquad (B.22)$$

For the *energy-independent* escape, $\tau(\gamma,t) = \tau(t)$, a similar equation can be obtained from the relevant Green's function solution found by Kardashev (1962) for the case of "stochastic acceleration + adiabatic losses + leakage", if the acceleration parameter tends to zero.

It is seen from Eq.(B.22) that for a power-injection $Q(\gamma) \propto \gamma^{-\alpha_{\rm inj}}$ with $\alpha_{\rm inj} > 1$ the contribution of the first term quickly decreases, so at $t \gg t_0$ only the contribution due to continuous injection is important. This term is easily integrated assuming stationary injection and approximating $\tau = \tau_0 \gamma^{-\delta}(t/t_0)^s$, with δ and $s \geq 0$. In the case of $s < 1$, the energy distribution of electrons at $t \gg \tau(\gamma,t)$ comes to $N(\gamma,t) = Q(\gamma) \times \tau(\gamma,t)$, similar to the case of a non-expanding source. If $s \geq 1$, the condition $t \gg \tau(\gamma,t)$ can be satisfied only for sufficiently large γ, so only at these energies can the energy-dependent escape of particles from an expanding cloud result in a steepening of $N(\gamma,t)$.

Although Eq.(B.22) is derived under the assumption of a constant expansion speed v, it can be readily used in the numerical calculations for any profile of $v(t)$, by approximating the latter in the form of step functions with different mean speeds $\bar{v}_i$ in the succession of intervals (t_i, t_{i+1}).

Bibliography

Achterberg, A., Gallant, Y.A., Kirk, J.G., Guthmann, A.W. (2001) *MNRAS* **328**, 393.
Aharonian, F.A. (1991) *Astr. Space Sci.* **180**, 305.
Aharonian, F.A. (1993), in Proc. of Intern. Workshop on *Towards a Major Atmospheric Cherenkov Detector - II*, Ed. R.C. Lamb, Calgary, p.81.
Aharonian, F.A. (1995) *Nucl. Phys. B* **39A**, 193.
Aharonian, F.A. (1997), in Proc. of XVIII Intern. Symp. on *Lepton-Photon Interactions*, Eds. A. De Roeck and A. Wagner, Hamburg, p. 263.
Aharonian, F.A (1999) *Astropart. Phys.* **11**, 225.
Aharonian, F.A. (2000) *New Astronomy* **5**, 377.
Aharonian, F.A. (2001a) *Space Sci. Rev.* **99**, 187.
Aharonian, F.A. (2001b) in Proc. *ICRC 2001: Invited, Repporteur, and Highlight papers*, Hamburg, p. 250.
Aharonian, F.A. (2002a) *Nature* **416**, 797.
Aharonian, F.A. (2002b) *MNRAS* **332**, 215.
Aharonian, F. A. and Atoyan, A. M. (1981a) *Sov. Astron. Lett.* **7**, 395.
Aharonian, F. A. and Atoyan, A. M. (1981b) *Phys. Lett.* **99B**, 301.
Aharonian, F. A. and Atoyan, A. M. (1981c) *Astr. Space Sci.* **79**, 321.
Aharonian, F. A. and Atoyan, A. M. (1983) *Soviet Phys. - JETP* **58**, 1079.
Aharonian, F. A. and Sunyaev, R. A. (1984) *MNRAS* **210**, 257.
Aharonian, F.A. and Vardanian, V.V. (1985) *Astr. Space Sci.* **115**, 31.
Aharonian, F.A. and A.M. Atoyan, A.M. (1985) *Sov. Phys. - JETP* **62**, 189.
Aharonian, F.A. and Atoyan, A.M. (1991a) *ApJ* **381**, 220.
Aharonian, F.A. and Atoyan, A.M. (1991b) *J.Phys.G: Nucl. Part. Phys.* **17**, 1769.
Aharonian, F.A. and Cronin, J.W. (1994) *Phys. Rev. D* **50**, 1892
Aharonian, F.A. and Atoyan, A.M. (1995) *Astropart. Phys.* **3**, 275.
Aharonian, F.A. and Atoyan, A.M. (1996a) *A&A* **309**, 917
Aharonian, F.A. and Atoyan, A.M. (1996b) *Space Sci. Rev.* **75**, 357.
Aharonian, F.A. and Heinzelmann, G. (1996) *Nucl. Phys.* **B60**, 193
Aharonian, F.A. and Akerlof, C. (1997) *Ann. Rev. Nucl. Part Sci.* **47**, 273.
Aharonian, F.A. and Atoyan, A.M. (1998a) in Proc. "Neutron Stars and Pulsars: *Thirty Years after the Discovery*" (eds. M. Shibazaki *et al.*), Universal

Academy Press, Inc. (Tokyo), p.439.
Aharonian, F.A. and Atoyan, A.M. (1998b) *New Astron. Reviews* **42**, 579.
Aharonian, F.A. and Bogovalov, S. (1999) *Astron. Nachrichten* **320**, 332.
Aharonian, F.A. and Atoyan, A.M. (1999) *A&A* **351**, 330
Aharonian, F.A. and Atoyan, A.M. (2000) *A&A* **362**, 937.
Aharonian, F.A. and Völk, H.J [eds.] (2001) *AIP Conference Proceedings*, vol. 558, Melville, New York.
Aharonian, F.A. and Bogovalov, S.V. (2003)*New Astronomy* **8**, 85.
Aharonian, F.A. and Plyasheshnikov, A.V. (2003) *Astropart. Phys.*, **19**, 525.
Aharonian F.A., Kelner S.R. and Kotov, Yu.D. (1979), in Proc. *16th Intern. Cosmic Ray Conf., Kyoto*, vol. 1, p. 173; p.179.
Aharonian, F.A., Atoyan, A.M., Nagapetyan, A.M. (1983a) *Astrophysics* **19**, 187.
Aharonian, F. A., Kirillov-Ugryumov, V. G. and Kotov, Y. D. (1983b) *Astrophysics* **19**, 82.
Aharonian, F.A., Kririllov-Ugriumov, V.G., Vardanian, V.V. (1985) *Astr. Space. Sci.* **115**, 201.
Aharonian, F.A., Kanevsky, B., Vardanian, V.V. (1990) *Astr. Space. Sci.* **167**, 93.
Aharonian, F.A., Bhattacharjee, P. and Schramm, D.N. (1992) *Phys. Rev D* **46**, 4188.
Aharonian, F.A., Drury, L.O'C. and Völk, H.J. (1994a) *A&A* **285**, 645.
Aharonian, F.A., Coppi, P.S. and Völk, H.J. (1994b) *ApJ* **423**, L53.
Aharonian, F.A., Atoyan, A.M. and Völk, H.J. (1995) *A&A* **294**, L41.
Aharonian, F.A., Hofmann, W., Konopelko, A.K., Völk, H.J. (1997a) *Astropart. Phys.* **6**, 343; 369.
Aharonian, F.A., Atoyan, A.M. and Kifune,T. (1997b) *MNRAS* **291**, 162.
Aharonian, Konopelko, A.K., Völk, H.J., Quintana, H. (2001) *Astropart. Phys.* **15**, 335.
Aharonian, F.A., Timokhin, A.N., Plyasheshnikov, A.V. (2002a) *A&A* **384**, 847.
Aharonian, F.A., Belyanin A.A., Derishev, E.V., Kocharovsky, V.V., Kocharovsky, Vl.V. (2002b) *Phys.Rev.* **D66**, 023005.
Aharonian, F.A. et al. (HEGRA Collaboration) (1997) *A&A* **527**, L5.
Aharonian, F.A. et al. (HEGRA Collaboration) (1999a) *A&A* **342**, 69.
Aharonian, F.A. et al. (HEGRA Collaboration) (1999b) *A&A* **346**, 913
Aharonian, F.A. et al. (HEGRA Collaboration) (1999c) *A&A* **349**, 11.
Aharonian, F.A. et al. (HEGRA Collaboration) (1999d) *A&A* **350**, 75.
Aharonian, F.A. et al. (HEGRA Collaboration) (2000a) *ApJ* **539**, 317.
Aharonian, F.A. et al. (HEGRA Collaboration) (2000b) *A&A* **361**, 1073.
Aharonian, F.A. et al. (HEGRA Collaboration) (2000c) *ApJ* **543**, L39.
Aharonian, F.A. et al. (HEGRA Collaboration) (2001a) *A&A* **366**, 62.
Aharonian, F.A. et al. (HEGRA Collaboration) (2001b) *ApJ* **546**, 898.
Aharonian, F.A. et al. (HEGRA Collaboration) (2001c) *A&A* **366**, 74.
Aharonian, F.A. et al. (HEGRA Collaboration) (2001d) *A&A* **370**, 112.
Aharonian, F.A. et al. (HEGRA Collaboration) (2001e) *A&A* **373**, 292.
Aharonian, F.A. et al. (HEGRA Collaboration) (2001f) *A&A* **375**, 1008.
Aharonian, F.A. et al. (HEGRA Collaboration) (2002a) *A&A* **384**, L23.

Aharonian, F.A. et al. (HEGRA Collaboration) (2002b) *A&A* **393**, L37.
Aharonian, F.A. et al. (HEGRA Collaboration) (2002c) *A&A* **393**, 89.
Aharonian, F.A. et al. (HEGRA Collaboration) (2002d) *A&A* **395**, 803.
Aharonian, F.A. et al. (HEGRA Collaboration) (2003a) *A&A* **406** L9.
Aharonian, F.A. et al. (HEGRA Collaboration) (2003b) *A&A* **403** 523.
Aharonian, F.A. et al. (HEGRA Collaboration) (2003c) *A&A* **403**, L1.
Akhiezer, A.I., Berestetskii, V,B. *Quantum Electrodynamics* (1965), Interscience, New York.
Akhiezer, A.I., Merenkov, N.P., Rekalo, A.P (1994) *J. Phys. G: Nucl. Part. Phys.* **20**, 1499.
Allen, G.E., Berley, D., Biller, S. et al. (1995) *ApJ* **448**, L25.
Allen G.E., Keohane J.W., Gotthelf E.V., et al. (1997) *ApJ* **487**, L97.
Anchordoqui, L.A., Torres, D.F., McCauley, T.P., Romero, G.E., and Aharonian, F.A. (2003) *ApJ* **589**, 481.
Anderson M., Rudnick L., Leppik P., Perley R., Braun R. (1991) *ApJ* *373*, 146.
Anderson, M.C. and Rudnick, L. (1996) *ApJ* **456**, 234.
Anguelov, V. and Vankov, H. (1999) *J. Phys. G: Nucl. Part. Phys.* **25**, 1755.
Arons, J. (1996) *Space Sci. Rev.* **75**, 235.
Aschenbach, B. and Brinkmann, W. (1975) *A&A* **41**, 147.
Atoyan, A.M. and Aharonian F.A. (1996) *MNRAS* **278**, 525.
Atoyan, A.M. and Aharonian F.A. (1997) *ApJ* **409**, L149.
Atoyan, A.M. and Aharonian F.A. (1999) *MNRAS* **302**, 253.
Atoyan, A.M. and Völk, H.J. (2000) *ApJ* **535**, 45.
Atoyan, A.M. and Dermer, C.D. (2001) *Phys. Rev. Lett.* **87**, 221102.
Atoyan, A.M. and Dermer, C.D. (2003) *ApJ* **586**, 79.
Atoyan, A.M., Aharonian, F.A, and Völk, H.J. (1995) *Phys. Rev. D* **52**, 3265.
Atoyan, A.M., Tuffs, R., Aharonian F.A., and Völk, H.J. (2000a) *A&A* **354**, 915.
Atoyan, A.M., Aharonian F.A., Tuffs, R. and Völk, H.J. (2000b) *A&A* **355**, 211.
Atoyan, A.M., Aye, K.-M., Chadwick, P.M. et al. (2002) *A&A*, **383**, 864.
Baade, W. and Zwicky, F. (1934) *Phys. Rev.* **46**, 67.
Bacci, P. et al. (2002) *Astropart. Phys.* **17**, 151.
Bahcall, J. N., Kirhakos, S., Schneider, D. P. et al. (1995) *ApJ* **452**, L91.
Ball, L. and Kirk, J.G. (2000) *Astropart. Phys.* **12**, 335.
Ball, L. and Dodd, J. (2001) *PASA* **18**, 98.
Band, D.L. and Grindlay, J.H. (1986) *ApJ* **311**, 595.
Baring, M.G., Ellison, D.C., Reynolds, S.P., Grenier,I.A. and Goret, P. (1999) *ApJ* **513**, 311.
Baring, M.G. and Harding, A.K. (2001) *ApJ* **537**, 529..
Barrow, J.D., Ferreira, P.G., Silk, J. (1997) *Phys. Rev D* **78**, 3610.
Becker, W. and Trümper, J. (1997) *A&A* **326**, 682.
Bednarek, W. and Protheroe, R.J. (1997) *Phys. Rev. Lett* **71** 2616.
Bednarek, W. and Protheroe, R.J. (1999) *MNRAS* **302** 373.
Bednarek, W. (1997) *MNRAS* **285** 69.
Begelman, M.C. (2002) *ApJ* **568**, L97.
Begelman, M.C., Blandford, R.D., Rees, M.J. (1984) *Rev. Mod. Phys.* **56**, 255.
Beck, R. (1994) *A&A* **292**, 409.

Bell, A.R. (1997) *NNRAS* **179**, 573.
Bell, A. R. and Lucek, S. G. (2001) *NNRAS* **321**, 433.
Belyanin, A. A. and Derishev, E. V. (2001) *A&A* **379**, L25.
Benaglia, P., Romero, G.E. (2002) *A&A* **399**, 1121.
Berezhko, E.G. and Krymsky, G.F. (1988) *Sov. Phys.-Usp.* **31**, 27.
Berezhko, E.G. and Völk, H.J. (1997) *Astropart. Phys.* **14**, 201.
Berezhko, E.G. and Völk, H.J. (2000) *ApJ* **540**, 929.
Berezhko, E.G., Pühlhofer, G, and Völk, H.J. (2002), in Proc. *27th ICRC*, Hamburg, OG2.02.
Berezhko, E.G., Ksenofontov, L.T., and Völk, H.J. (2002) *A&A* **395**, 943.
Berezinsky, V. S. (1970) *Sov. J. Nucl. Phys.* **11**, 222.
Berezinsky, V.S. (1976) in Proc. *1976 DUMAND Summer Workshop*, Ed. A. Robert, FNAL, Batavia, p. 229.
Berezinsky, V. S. and Smirnov, A. Yu. (1975) *Astr. Space Sci.* **32**, 1975.
Berezinsky, V.S. and Prilutsky, O.F. (1978) *A&A* **66**, 325.
Berezinsky, V. S. and Grigoreva S.I. (1988) *A&A* **199**, 1.
Berezinsky, V.S., Bulanov, S.V., Ginzburg V.L., Dogiel V.A. and Pruskin, V.S. (1990) "Astrophysics of Cosmic Rays", North-Holland, Amsterdam.
Berezinsky, V. S., Gaisser, T.K., Halzen, F., Stanev, T. (1993) *Astropart. Phys.* **1**, 281.
Berezinsky, V.S., Kachelrieß, M. and Vilenkin, A. (1997a) *Phys Rev Lett.* **79**, 4302.
Berezinsky, V. S., Blasi, P. and Ptuskin, V. S. (1997b) *ApJ* **487**, 529.
Berezinsky, V.S., Blasi, P. and Vilenkin, A. (1998) *Phys Rev D* **58**, 103515.
Bernstein, R.A., Freedman, W.L., Madore, B.F. (2002) *ApJ* **571**, 107.
Bertsch D. L., Dame T. M., Fichtel C. E. et al. (1993) *ApJ* **416**, 587.
Bhattacharjee, P., Sigl, G. (2000) *Physics Reports* **327**, 109.
Biermann, P.L. and Strittmatter, P.A. (1987) *ApJ* **332**, 643.
Bignami,G.F., Hermsen, W. (1983) *ARA&A* **21**, 67.
Bignami, G.F., Caraveo, P.A., Lamb, R.C. (1983) *ApJ* **272**, L9.
Bildsten, L., Salpeter, E.E., Wasserman, I. (1992) *ApJ* **384**, 143.
Birkel, M. and Sarkar, S. (1998) *Astropart. Phys.* **9**, 297.
Blandford, R.D. and Cowie, L.L. (1982) *ApJ* **260**, 625.
Blandford, R.D. and Eichler, D. (1987) *Phys. Rep.* **154**, 1.
Blasi, P. (2000) *ApJ* **532**, L9.
Blasi, P. and Colafrancesco, S. (1999) *Astropart. Phys.* **12**, 169.
Bloemen, G.B.G.M. (1989) *Ann. Review of Astron. and Astrophys.* **29**, 469.
Bloemen, G.B., Dogiel, V.A., Dorman, V.L., Ptuskin, V.S. (1993) *A&A* **267**, 372.
Bloom, S.D. et al. (1997) *ApJ* **490**, L45.
Bloser, P.F., Andritschke, R., Kanbach, G., Schnfelder, V., Schopper, F., Zoglauer, A. (2002) *New Astronomy* **46**, 611.
Blumenthal, G.R. and and Gould, R.J. (1970) *Rev. Mod Phys.* **42**, 237.
Blumenthal, G.R. (1990) *Phys. Rev.* **D1**, 1596.
Bodo, J. and Ghisellini, G. (1995) *ApJ* **441**, L69.
Bogovalov, S.V. (1999) A&A **349**, 1017.
Bogovalov, S.V. and Aharonian, F.A. (2000) *MNRAS* **313**, 504.

Boldt, E. (1999) *Astroph. Lett & Communications*, astro-ph/9902040.
Bonamente, M., Lieu, R., Mittaz, J. (2001) *ApJ* **547**, L7.
Bonometto, S. and Rees, M.J. (1971) *MNRAS* **152**, 21.
Boone, L.M. et al. (2002) *ApJ* **579**, L5.
Borione, A. et al. (1994) *NIM* **A346**, 329.
Borione, A., Catanese, M.A., Chantell, M.C. et al. (1998) *ApJ* **493**, 175.
Bowyer, S., Lampton, M., Lieu, R. (1996) *Science* **274**, 1338.
Bowyer, S. and Berghöfer, T.W. (1998) *Astrophys. J* **506**, 502.
Böttcher, M. and Schlickeiser (1997) *A&A* **325**, 866.
Braun, R. (1987) *A&A* **171**, 233.
Brazier,K.T.S., Kanbach,G., Carraminana, A., Guichard, J. and Merck,M. (1996) *MNRAS* **281**, 1033.
Brecher, K. and Burbidge, G. R. (1972) *ApJ* **174**, 253.
Breitschwerdt, D., McKenzie, J. F. and Völk, H. J. (1991) *A&A* **245**, 79.
Breitschwerdt, D, Dogiel, V. A., Völk, H. J. (2002) *A&A* **385**, 216.
Brown, R.W., Hunt, W.F., Mikaelian, K.O., Muzinich, I.J. (1973) *Phys. Rev* **D8**, 3083
Brunetti, G. (2000) *Astropart. Phys.* **13**, 107.
Brunetti, G. (2002a), in Proc *"The Role of VLBI in Astrophysics, Astronomy and Geodesy"*, NATO Conf. Series, Eds, in press.
Brunetti, G. (2002b), in Proc. *"Matter and Energy in Clusters of Galaxies"*, Eds. S.Bowyer and C.-Y. Hwang, Taiwan, in Press.
Buckley, J. H., Akerlof, C. W.,Carter-Lewis et al. (1998) *A&A* **329**, 639.
Buckley, J., (2002) astro-ph/0201160
Bulik, T., Rudak, B. and Dyks, J. (2000) *MNRAS* **317**, 97.
Burbidge, G.R. (1956) *ApJ* **124**, 416.
Burbidge, G.R., Burbidge, E.M. and Sandage, A.R. (1963) *Rev. Mod. Phys.* **35**, 947.
Butt, Y.M., Torres, D.F., Combi, J.A., Dame, T., Romero, G.E. (2001) *ApJ* **562**, L167.
Butt, Y.M., Torres, D.F., Romero, G.E., Dame, T., Combi (2002) *Nature* **418**, 499.
Bykov, A. M. (2001) *Space Sci. Rev.* **99**, 317.
Bykov, A. M. and Toptygin, I. N. (2001) *Astron. Lett.* **27**, 625.
Cambresy, L. et al. (2001) *ApJ* **555**, 563.
Casse, M. and Paul, J.P. (1980) *ApJ* **237**, 236.
Catanese M., Bradbury S.M., Breslin A.C. et al. (1997) *ApJ* **487**, L143.
Catanese, M., Akerlof, C. W., Badran, H. M. et al. (1998) *ApJ* **501**, 616.
Catanese, M. and Weekes, T.C. (1999) *PASP* **111**, 1193.
Celotti, A. and Fabian, A. C. (1993) *MNRAS* **264**, 228.
Celotti, A., Fabian, A.C and Rees, M. J. (1998) *MNRAS* **293**, 239.
Celotti, A., Ghisellini, G. and Chiaberge, M. (2001) *MNRAS* **321**, L1.
Cesarsky, C.J. and Völk, H.J. (1978) *A&A* **70**, 367.
Cesarsky, C. (1992) *Nucl. Phys.* **28B**, 51.
Chadwick, P.M. et al. (1998a) *Astropart. Phys.* **9**, 131.
Chadwick, P.M. et al. (1998b) *ApJ* **503**, 391.

Chadwick, P.M. et al. (1999) *ApJ* **513**, 161.
Chandler, A.M., Koh, D.T., Lamb, R.C. et al. *ApJ* **556**, 59.
Chartas, G., Worrall, D.M., Birkinshaw, M. et al. (2000) *ApJ* **542**, 655.
Cheng, K.S., Ho, C. and Ruderman, M. (1986) *ApJ* **300**, 500.
Cheng, K. S. and Ruderman, M. (1989) *ApJ* **337**, L77.
Cheng, L.X., Cheng, K.S. (1996) *ApJ* **459**, L79.
Chi, X. and Wolfendale, A. W. (1991) *Journ. Phys. G: Nucl. Part. Phys.* **17**, 987.
Chiang, J., Mukherjee, R. (1998) *ApJ* **496**, 752.
Chudakov, A. E. et al. (1965) *Trans. Consult. Bureau* (P.N. Lebedev Institute, Moscow) **26**, 99.
Clarke, T. E., Kronberg, P. P. and Bhringer, H. (2001) *ApJ* **547**, L111.
Clayton, D.D. (1982) in Proc. *Essays in Nuclear Astrophysics*, Eds. C.A. Barnes et al., Cambridge University Press.
Combi, J. A., Romero, G. E. and Benaglia (2001a) *A&A* **333**, L91.
Combi, J. A., Romero, G. E., Benaglia, P. and Jonas, J. L. (2001b) *A&A* **366**, 1047.
Colbert, J.M. and Mushotzky, R.F. (1999) *ApJ* **519**, 89.
Coppi, P.S. and Blandford, R.D. (1990), *MNRAS* **245**, 453.
Coppi, P.S. (1992) *MNRAS* **258**, 657.
Coppi, P.S. and Aharonian, F.A. (1997) *ApJ* **487**, L9.
Coppi, P.S. (1998), in Proc. *High Energy Processes in Accreting Black Holes*, ASP Conf. Series, vol. 161, p. 375.
Coppi, P.S. and Aharonian, F.A. (1999a) *ApJ* **521**, L33.
Coppi, P.S. and Aharonian, F.A. (1999b) *Astropart. Phys.* **11**, 35.
Coroniti, F.V. (1990) *ApJ* **349**, 538.
Costamante, L., Ghisellini, G., Giommi, P. et al. (2001) *A&A* **371**, 512.
Costamante, L. and Ghisellini, G. (2002) *A&A* **384**, 56.
Costamante, L., Aharonian, F.A., Ghisellini, G., Horns D. (2003), in Proc. *Relativistic Jets in the Chandra and XMM Era*, in press.
Courvoisier, T. J.-L. (1998) *A&ARv* **9**, 1.
Cowsik, R. and Sarkar, S. (1980) *MNRAS* **191**, 855.
Cowsik, R. and Sarkar, S. (1984) *MNRAS* **207**, 745.
Cronin, J.W. (1999), *Rev. Mod. Phys* **71**, S165.
Cronin, J.W., Gibbs, K.G., Weekes, T.C (1993) *Ann. Rev. Nucl. Part Sci.* **43**, 883.
Crutcher, R.M. (1999) *ApJ* **520**, 706.
Dahlbacka, G.H., Chapline, G.P., Weaver, T.A. (1974) *Nature* **250**, 36.
Dame, T. M., Ungerechts, H., Cohen, R.S. et al. (1987) *ApJ* **322**, 706.
Dar, A. and Laor, A. (1997) *ApJ* **478**, L5.
Daugherty, J.K. and Harding, A.K. (1982) *ApJ* **252**, 337.
Daugherty, J.K. and Harding, A.K. (1996) *ApJ* **458**, 278.
Daum, A. et al. (1997) (HEGRA Collaboration) *Astrpart. Phys.* **8**, 1.
Davidson, F. and Fesen, R.A. (1985) *Ann. Rev. Astron Astrophys.* **23**, 119.
De Jager, O.C. and Harding, A.K. (1992) *ApJ* **396**, 161.
De Jager, O.C., Harding, A.K., Michelson, P.F., Nel, H.I., Nolan, P.L., Sreekumar, P. and Thompson, D. J. (1996) *ApJ* **457**, 253.

De Jager, O.C. and Harding, A.K. (1998) in Proc. "Neutron Stars and Pulsars: *Thirty Years after the Discovery*" (eds. M. Shibazaki *et al.*), Universal Academy Press, Inc. (Tokyo), p.483.

De Jager, O.C. and Stecker, F.W. (2001) *ApJ* **566**, 738.

de Naurois, M. et al. (2002) *ApJ* **566**, 343.

Deiss, B. M. , Reich W., Lesch H., Wielebinski R. (1997) *A&A* **321**, 55.

Derishev, E.V., Aharonian, F.A., Kocharovsky, V.V., Kocharovsky, Vl.V. (2003) *Phys.Rev.* **D68**, 043003.

Dermer, C. (1986) *A&A* **157**, 223.

Dermer, C.D. and Schlickeiser, R. (1991) *A&A* **252**, 414.

Dermer, C.D. and and Atoyan, A.M. (2002) *ApJ* **568**, L81.

Dermer, C.D., Strickman, M.S., Kurfess, J.D. [eds.] (1997) *AIP Conference Proceedings*, vol. 410, Woodbury, New York.

Dieckmann, M.E., McClements, K.G., Chapman, S.C., Dendy, R.O. and Drury, L.O'C. (2000) *A&A* **356**, 377.

Diehl, R. and Timmes, F.X. (1998) *PASP* **110**, 637.

Digel, S.W, Aprile, E., Hunter, S. D., Mukherjee, R. and Xu, F. (1999), *ApJ* **520**, 196.

Dingus, B. L. et al. (1988) *Phys. Rev Lett.* **61**, 1906.

Distefano, C., Guetta, D., Waxman, E., Levinson, A. (2002) *ApJ* **574**, 923.

Di Salvo, T. et al. (2001) *ApJ* **547**, 1024.

Djannati-Ataï, A., Piron, F., Barrau, A. et al. (1999) *A&A* **350**, 17.

Djannati-Ataï, A., Khelifi, B., Vorobiov, S. et al. (2002) *A&A* **391**, L25.

Dodson, R. and Golap, K. (2002) *MNRAS* **334**, L1.

Dogiel, V.L., Inoue, H., Masai, K., Schönfelder, V., Strong, A.W. (2002) *ApJ* **581**, 1061.

Dolag, K. and Enßlin, T. A. (2000) *A&A* **362**, 151.

Dorfi, E. A. (1991) *A&A* **251**, 597.

Dowdall, C. et al. (2002) *IAU Circular* No. 7903.

Dowens, A.J.B., Pauls, T. and Salter, C.J. (1986) *MNRAS* **218**, 393.

Drury L.O'C. (1983), *Rep. Prog. Phys.* **46**, 973.

Drury L.O'C., Aharonian, F.A. and Völk, H.J. (1994) *A&A* **287**, 959.

Drury, L. O'C., Duffy, P., Kirk, J. G. (1996) *A&A* **309**, 1002.

Drury, L.O.'C., Duffy, P., Eichler, D., Mastichiadis, A. (1999) *A&A* **347**, 370.

Drury L.O'C., Ellison, D.C., Aharonian, F.A. et al. (2001) *Space Sci. Rev.* **99**, 329.

Drury L.O'C. (2001) in: *AIP Conference Proceedings*, vol. 558 (eds. F. A. Aharonian and H. J. Völk), p. 71, Melville, New York.

Dwek, E. and Arendt, R.G. (1998) *ApJ* **508**, L9.

Eardley, D.M., Lightman, A.P., Shakura, N.I., Shapiro, S.L. and Sunyaev, R.A. (1978) *Comments Astrophys.* **7**, 151.

Edwards, P.G. and Piner, B.G. (2002) *ApJ* **579**, L67.

Eichler, D. and Wiita, P.J. (1978) *Nature* **274**, 38.

Eichler, D. and Vestrand, W. T. (1984) *Nature* **307**, 613.

Eikenberry, S.S. and and Fazio, G.G. (1997) *ApJ* **475**, L53.

Eilek, J. A. (1980) *ApJ* **236**, 664.

Einasto, J. (2001) *New Astronomy Reviews* **45**, 355.
Erber, T. (1966) *Rev. Mod. Phys.* **38**, 626.
Enomoto, R. et al. [eds.] (2003) Proceedings of the Symposium *The Universe Viewed in Gamma-rays*, Universal Academy Press, Inc. -Tokyo.
Enomoto, R., Tanimori, T., Naito, T. et al. (2002) *Nature* **416**, 823.
Enßlin, T., Biermann, P. L., Kronberg, P. and Wu, X.-P. (1997) *ApJ* **477**, 560.
Enßlin, T., Lieu, R., Biermann, P. (1999) *A&A* **344**, 409.
Enßlin, T. and Sunyaev, R.A. (2002) *A&A* **383**, 423.
Erlykin, A.D. and Wolfendale, A.W. (1997) *Journ. Phys. G: Nucl. Part. Phys.* **23**, 979.
Erlykin, A.D. and Wolfendale, A.W. (2000) *A&A* **356**, L63
Esposito,J.A., Hunter,S.D., Kanbach,G. and Sreekumar, P. (1996) *ApJ* **461**, 820.
Eungwanichayapant, A. (2003), PhD thesis, University of Heidelberg.
Fanslow, J. L., Hartman, R. C., Hildebrand, R. H. and Meyer, P. (1969) *ApJ* **158**, 771.
Fazio, G.G., Helken, G.H., Omongain, E., Weekes, T.C. (1972) *ApJ* **175**, L117.
Fender, R.P. et al. (1999) *MNRAS* **304**, 865.
Fender, R.P. (2001) *MNRAS* **322**, 31.
Fender, R.P. (2002) In: *"Relativistic flows in Astrophysics"*, Springer Verlag Lecture Notes in Physics **589**, 101.
Feretti, L., R. Fusco-Femiano, R., Giovannini, G., Govoni, F. (2001) *A&A* **373**, 106.
Fichtel, C.E. et al. (1975) *ApJ* **198**, 163.
Fichtel, C.E. and Trombka, J.I. (1997) "Gamma-Ray Astrophysics" (2nd ed), NASA Reference Publication 1386, Greenbelt.
Fierro, J.M., Michelson, P.F., Nolan, P.L. and Thompson, D.J. (1998) *ApJ* **494**, 734.
Finkbeiner, D.P., Devis, M. and Schlegel, D.J. (2000) *ApJ* **544**, 81.
Finley, J.P., Srinivasan, R., Saito, Y., Hiriyama, M., Kamae, T., Yoshida, K. (1998) *ApJ* **493**, 884.
Fossati, G., Maraschi, L., Celotti, A., Comastri, A., Ghisellini, G. (1998) *MNRAS* **299**, 433.
Franceschini, A. et al. (2001) *A&A* **378**, 1.
Fusco-Femiano, R., Dal Fiume, D., Feretti et al. (1999) *ApJ* **513**, L21.
Gabici,S. and Blasi, P. (2003a) *Astrphys. J* **583**, 695.
Gabici,S. and Blasi, P. (2003b) *Astropart. Phys.* **19**, 679.
Gaidos, J.A. et al. (1996) *Nature* **383**, 319.
Gaisser, T.K. (1990) *Cosmic Rays and Particle Physics*, Cambridge University Press.
Gaisser, T., Protheroe R.J. and Stanev, T. (1998) *ApJ* **492**, 219.
Gaisser, T.K. (2001) in: *AIP Conference Proceedings*, vol. 558 (eds. F. A. Aharonian and H. J. Völk), p. 27, Melville, New York.
Galbraith, W. and Jelley, J.V. (1953) *Nature* **171**, 349.
Geddes, J., Quinn, T. C., Wald, R.M. (1996) *ApJ* **459**, 384.
Gehrels, N. and Michelson, P. (1999) *Astropart. Phys.* **11**, 277.
Gehrels, N., Macomb, D. J., Bertsch, D. L., Thompson, D. J. and Hartman, R.

C. (2000) *Nature* **404**, 363.
Georganopoulos, M., Aharonian, F.A. and Kirk, J.G. (2002) *A&A* **388**, L25.
Ghisellini, G., Celotti, A., Fossati, G., Maraschi, L., Comastri, A. (1998) *MNRAS* **301**, 451.
Ghisellini, G. (1999) *Astron. Nachrichten* **320**, 232.
Ghisellini, G. and Celotti, A. (2001) *MNRAS* **327**, 739.
Giacani, E.B., Frail, D.A., Goss, W.M. and Vieytes, M. (2001) *Astron.J* **121**, 3133.
Ginzburg, V.L. and Syrovatskii, S.I. (1964) *Origin of Cosmic Rays*, New York: Macmillan.
Ginzburg, V.L. and Syrovatskii, S.I. (1965) *ARA&A* **3**, 297.
Ginzburg, V.L. and Ptuskin, V.S. (1984) *J. Ap. Astron.* **5**, 99.
Giovanoni, P.M. and Kazanas, (1990) *Nature* **345**, 319.
Gliozzi, M., Bodo, G. and Ghisellini, G. (1999) *MNRAS* **303**, L37.
Gold, T. (1968) *Nature* **218**, 731.
Gold, T. (1969) *Nature* **221**, 25.
Goldreich. P. and Julian, W.H. (1969) *ApJ* **157**, 869.
Gorjian, V., Wright, E.L., Chary, R.R. (2000) *ApJ* **536**, 550.
Gotthelf, E. V., Halpern, J. P., Dodson, R. (2002) *ApJ* **567**, L125.
Götting, N. et al. (HEGRA Collaboration) (2001), talk at the 27th ICRR, Hamburg.
Gould, R.J. (1965) *Phys. Rev. Lett.* **15**, 577.
Gould, R.J. and Schrèder, G.P. (1966) *Phys. Rev. Lett.* **16**, 252.
Gould, R.J. and Schrèder, G.P. (1967) *Phys. Rev.* **155**, 1408.
Gould, R.J. and Rephaeli, Y. (1978) *ApJ* **225**, 318.
Gralewicz, P., Wdowczyk, J., Wolfendale. A.W. and Zhang, L. (1997), *A&A* **318**, 925.
Grasso, D., Rubinstein, H.R. (2001) *Physics Reports* **348**, 163.
Greisen, K. (1966) *Phys. Rev. Lett* **16**, 748.
Grenier, I.A., Perrot, C. (1999), in Proc. *26th ICRC*, Salt Lake City, vol. 3, p. 476.
Grenier, I.A. (2000) *A&A* **364** L93.
Grenier, I.A. (2001) in: *AIP Conference Proceedings*, vol. 558 (eds. F. A. Aharonian and H. J. Völk), p. 191, Melville, New York.
Grindlay, J.E., Helmken, H.F., Hanbury Brown, R., Davis, J., Allen, L.R. (1975) *ApJ* **201**, 82.
Grove, J.E., Grindlay, J.E., Harmon, B.A. et al. (1997) in: *AIP Conference Proceedings*, vol. 410 (eds. C.D. Dermer et al.), vol. 1, p. 122, New York.
Grove, J. E., Johnson, W. N., Kroeger, R. A. et al. (1998) *ApJ* **500**, 899.
Guessoum, N., Kazanas, D. (1990) *ApJ* **358**, 525.
Guessoum, N., Jean, P. (2002) *A&A* **396**, 157.
Guilbert, P.W., Fabian, A.C., Rees, M.J. (1983) *MNRAS* **205**, 593.
Gurevich, A.V. and Istomin, I.N., (1985) *ZhETF*, **89**, 3.
Guy, J., Renault, C., Aharonian, F., Rivoal, M., Tavernet, J.-P. (2000) *A&A* **359**, 419.
Hacking, P.B., Sofier, B.T. (1991) *ApJ* **367**, L49.

Halzen, F. (2001), in: *AIP Conference Proceedings*, vol. 558 (eds. F. A. Aharonian and H. J. Völk), p. 43, Melville, New York.

Halzen, F. amd Hooper, D. (2002) *Rep. Prog. Phys.* **102**, 102.

Harding, A.K. (1996) *Space Sci. Rev.* **75**, 257.

Harding, A.K. et al. (1997), in Proc. *25th ICRC*, Durban, vol.3, p.325.

Harding, A.K. and and Muslimov, A. (1998) *ApJ* **508**, 328.

Harding, A.K. (2001) in: *AIP Conference Proceedings*, vol. 558 (eds. F. A. Aharonian and H. J. Völk), p. 115, Melville, New York.

Harding, A.K. amd Zhang, B. (2001) *ApJ* **548**, L37.

Harmon, B. A., Deal, K. J., Paciesas, W. S., Zhang, S. N., Robinson, C. R., Gerard, E., Rodriguez, L. F. and Mirabel, I. F. (1997) *ApJ* **477**, L85.

Harris, M.J. (1997) in: *AIP Conference Proceedings*, vol. 410 (eds. C.D. Dermer et al.), vol. 1, p. 418, Woodbury, New York.

Harris, D.E., Hjorth, J., Sadun, A.C., Silverman, J.D. and Vestergaard, M. (1999) *ApJ* **518**, 213.

Harris, D.E. (2001), in Proc: *Particles and Fields in Radio Galaxies* (eds. Laing R.A and Blundell K.M.), ASP Conf. Series, vol. 250, p. 204.

Harris, D.E. and Krawczynski, H. (2001) *Ap.J* **565**, 244.

Hartman,R. C., Bertsch, D. L., Bloom, S. D. et al. (1999) *ApJ (Suppl.)* **123**, 79.

Harwit, M. *ApJ* **510**, L83.

Harwit, M., Protheroe, R. J., Biermann, P. L. (1999) *ApJ* **524**, L91.

Hauser, M.G. et al. (1998) *ApJ* **508**, 25.

Hauser, M.G and Dwek, E. (2001) *ARA&A* **39**, 249.

Hayakawa, S. (1966) *Prog. Theor. Phys. Suppl.* **37**, 594.

Hayakawa, S. (1969) *Cosmic ray physics. Nuclear and astrophysical aspects*, New York, Wiley-Interscience.

Heinz, S. and Sunyaev, R.A. (2002) *A&A* **390**, 751

Heitler, W. (1954) *The quantum theory of radiation*, Oxford, Clarendon Press.

Helfand, D.J., Gotthelf, E.V. and Halpern, J. P. (2001) *ApJ* **556**, 380.

Henri, G., Pelletier, G., Petrucci, P.O. and Renaud, N. (1999) *Astropart. Phys.* **11**, 347.

Herterich, K. (1974) *Nature*, **250** 311.

Hester, J.J., Scowen, P.A., Sankrit, R. et al. (1995) *ApJ*, **448** 240.

Hewish, A., Bell, S.J., Pilkington, J.D.H., Scott, P.F. and Collins, R.A. (1968) *Nature* **217**, 709.

Hillas, A.M. (1984a) *ARA&A* **22**, 425.

Hillas, A.M. (1984b) *Nature* **312**, 50.

Hillas, A.M. (1985), in Proc. *19th ICRC*, La Jolla, vol. 3, p. 445.

Hillas, A.M. (1986) *Space Sci. Rev.* **75**, 17.

Hillas, A.M. et al. (1998) *ApJ* **503**, 744.

Hirotani, K. (2001) *ApJ*, **549** 495.

Hirotani, K. and Shibata, S. (2002) *ApJ* **564**, 369.

Hjellming, R.M. and Johnston, K.J. (1988) *ApJ* **328**, 600.

Hoffman, C.M., Sinnis, C., Fleury, P., Punch, M. (1999) *Reviews of Modern Physics* **71**, 897.

Hofmann, W. (HESS collaboration) (2003) in Proc. "The Universe Viewd in

Gamma Rays", Universal Academy Press, Inc.-Tokyo), 359.
Hofmann, W. et al. (2000) *Astropart Phys.* **12** 207.
Horan, D. et al. (2002) *ApJ* **571**, 753.
Horns, D. et al. (HEGRA Collaboration) (2003) in Proc. *28th ICRC*, Tsukuba, vol.4, p. 2373.
Horns, D., Costamante, L., Götting, N. et al (HEGRA Collaboration) (2002) in Proc. "The Universe Viewd in Gamma Rays", Universal Academy Press, Inc. (Tokyo), 273.
Hoyle, F., Fowler, W.A., Burbidge, G.R. and Burbidge, E.M. (1964) *ApJ*, **139** 909.
Hunter, S. D., Bertsch, D. L., Catelli, J. R. et al. (1997a) *ApJ* **481**, 205.
Hunter, S. D., Kinzer, R. L. and Strong A. W. (1997b), in Proc. "4th Compton Symposium", *AIP Conference Proceedings*, vol. 410 (eds. C. D. Dermer *et al.*,) , AIP Conf. Proceedings (New York), p. 192, Melville, New York.
Hunter, S. D. (2001) in: *AIP Conference Proceedings*, vol. 558 (eds. F. A. Aharonian and H. J. Völk), p. 171, Melville, New York.
Itoh, C. et al. (2002) *A&A* **396**, L1.
Itoh, C. et al. (2003a) *ApJ* **584**, L65.
Itoh, C. et al. (2003b) *A&A* **402**, 443.
Iyudin (2000) *Nucl Phys. B* **85**, 263.
Ivanenko, I.P. *Electromagnetic Cascade Processes* (1968), Moscow State University Press (in Russian).
Ivanenko, I.P. and Lagutin, A.A. (1991) in *Proc. 22nd ICRC*, Dublin, vol. 1, p. 121.
Jackson, J.D. (1975) *Classical Electrodynamics*, John Wiley and Sons Inc., New York.
Jauch, J.M. and Rohrlich, F. (1955)*The Theory of Photons and Electrons*, Addison-Wesley, Cambridge, MA.
Jelley, J.V. and Porter, N.A. (1963) *MNRAS* **4**, 275.
Jelley, J.V. (1966) *Phys. Rev. Letters* **16**, 479.
Jester, S., Röser, H.-J., Meisenheimer, K., Perley, R. (2002) *A&A* **385**, 27.
Jokipii, J.R. (1966) *ApJ* **146**, 480.
Jokipii, J. R. (1976) *ApJ* **208**, 900.
Jokipii, J. R. and Morfil, G.E. (1985) *ApJ* **290**, L1.
Jones, F.C. (1968) *Phys. Rev.* **167**, 1159.
Jones, F.C. and Ellison, D.C. (1991) *Spacs Sci. Rev.* **58** 259.
Jorstad, S,G., Marscher, A.P., Mattox, J.R., Wehrle, A. E., Bloom, S.D., Yurchenko, A.V. (2001) *ApJS* **134**, 181.
Kafatos, M., Shapiro, M. M. and Silberberg, R. (1981) *Comments Astrophys.* **9**, 179.
Kaiser, C.R., Sunyaev, R. and Spruit, H.C. (2000) *A&A* **356**, 975.
Kalashev, O.E., Kuzmin, V.A. and Semikoz, D.V. (2001) *Mod. Phys. Lett. A* **16**, 2505.
Karakula, S. and Tkaczyk, W. (1993) *Astropart. Phys* **1**, 229.
Kardashev, N.S. (1962) *Sov. Astron.* **6**, 317.
Karle, A. et al. (1995) *Astropart. Phys* **3**, 321.

Kataoka, J. et al. (1999) *ApJ* **514**, L138.
Kaufman Bernado, M.M., Romero, G.E. and Mirabel, I.F. (2002) *A&A* **385**, L10.
Kennel, C.F. and Coroniti, F.V. (1984) *ApJ* **283**, 694; 710.
Kifune, T. et al. (1995) *ApJ* **438**, L91.
Kifune, T. (1999) *ApJ* **518**, L21.
Kifune, T. (2001) in: *AIP Conference Proceedings*, vol. 558 (eds. F. A. Aharonian and H. J. Völk), p. 594, Melville, New York.
King, A.R., Davies, M.B., Ward, M.J., Fabbiano, G. and Ellvis, M. (2001) *ApJ* **552**, L109.
Kinzer, R.L., Purcell, W.R., Johnson, W.N. et al. (1996) *A & A Suppl Ser.* **120**, 317.
Kirk, J.G. and Mastichiadis, A. (1989) *A&A* **213**, 75.
Kirk, J.G. (1997), in Proc: *Proc. Relativistic Jets in AGNs* (eds. Ostrowski M. et al.), Cracow, p. 145.
Kirk, J.G., Rieger, F.M., Mastichiadis, A. (1998) *A&A* **333**, 452.
Kirk, J.G., Ball, L., Skjæraasen, O. (1999) *Astropart. Phys.* **10**, 31.
Kirk, J.G. and Dendy, R.O. (2001) *Journ. Phys. G: Nucl. Part. Phys.* **27**, 1589.
Knödlseder, J. (2000) *A&A* **360**, 539.
Kolikhalov, P.I., and Sunyaev, R.A. (1979) *Soviet Astr.* **56**, 338.
Konopelko, A. et. (1999) *Astropart. Phys.* **10**, 275.
Koyama, K., Petre, R., Gotthelf, E.V. et al. (1995) *Nature* **378**, 255.
Koyama, K., Kinugasa, K., Matsuzaki, K. et al. (1997) *Publ. of the Astronomical Society of Japan* **49**, L7.
Koyama, K. (2001) in: *AIP Conference Proceedings*, vol. 558 (eds. F. A. Aharonian and H. J. Völk), p. 82, Melville, New York.
Kozlovsky, B., Murthy, R.J., Ramaty, R. (2002) *ApJ (Suppl.Ser.)* **141**, 523.
Körding, E., Falcke, H. and Markoff, S. (2002) *A&A* **382**, L13.
Krawczynski, H., Coppi, P. S., Maccarone, T., Aharonian, F. A. (2000) *A&A* **353**, 97.
Krawczynski, H., Coppi, P. S., Aharonian, F. A. (2002) *MNRAS* **336**, 721.
Krennrich, F. et al. (2002) *ApJ* **575**, L9.
Kronberg P.P. (2001) in: *AIP Conference Proceedings*, vol. 558 (eds. F. A. Aharonian and H. J. Völk), p. 451, Melville, New York.
Kronberg, P. P., Dufton, Q. W., Li, H., Colgate, S. A. (2001) **560**, 178.
Kushida, J., Tanimori, T., Kubo, H. et al. (2003), in Proc. "The Universe Viewd in Gamma Rays", Universal Academy Press, Inc. - Tokyo, 291.
Lagage, P.O. and Cesarsky, C.J. (1983) *A&A* **125**, 249.
Lagache, G. et al. (1999) *A&A* **344**, 322.
Lamb, R.C. and Macomb, D.J. (1997) *ApJ* **488**, 872.
Laming, J. M. (2001) *ApJ* **546**, 1149.
Landau, L.D. (1938) *Nature* **141**, 333.
Landau, L.D. and Rumer, G. (1938) *Proc. R. Soc* **A166**, 213.
Laurent, P. and Titarchuk, L.G. (1999) *ApJ* **511**, 289.
Le Bohec, S. et al. (2000) *ApJ* **539**, 209.
Lerche, I. and Schlickeiser, R. (1980) *A&A* **239**, 1089.
Leventhal, M. et al. (1978) *ApJ* **225**, L11.

Levinson, A. and Blandford, R. (1996) *ApJ* **456**, L29.
Levinson, A. (2000) *Phys. Rev. Lett* **85**, 912.
Liang, H., Dogiel, V. A., Birkinshaw, M.(2002) *MNRAS* **337**, 567.
Lichti, G. G., Balonek, T., Courvoisier, T. J.-L. et al. (1995) *A&A* **298**, 711.
Lieu, R., Mittaz, J.P., Bowyer, et al. (1996) *Science* **274**, 1335.
Lieu, R., Ip, W.-H., Axford, W. I. and Bonamente, M. (1999) *ApJ* **510**, L25.
Lind, K.R. and Blandford, R.D. (1985) *ApJ* **295**, 358.
Lipari, P. (2002) *Astropart. Phys.* **16**, 295.
Lloyd-Evans, J.R.N., Coy, A., Lampert, J. et al. (1983) *Nature* **305**, 784.
Loeb, A., Waxman, E. (2000) *Nature* **405**, 156.
Longair, M. (1992), "High energy astrophysics" (2nd ed.), Cambridge University Press.
Lucarelli, F. et al. (2003) *Astropart. Phys.* **19**, 399.
Lucek, S. and Bell, A.R. (2000) *MNRAS* **314**, 65.
Lyubarsky, Yu.E. (1995), in: Sunayev R.A. (ed.) Physics of pulsars. Astrophysics and Space Physics reviews, Harwood Academic Publishers.
Lyubarsky, Y. and Kirk, J. G. (2001) *ApJ* **547**, 437.
Madau, P. and Pozzetti, L. (2000) *MNRAS* **312**, L9.
Mahadevan, R, Narayan, R. and Krolik, J. (1997) *ApJ* **486**, 268.
Makishima, K. et al. (2000) *ApJ* **535**, 632.
Malkan, M.A. and Stecker, F.W. (2001) *ApJ* **555**, 641.
Malkov, M.A. and Völk, H.J. (1995) *A&A* **300**, 605.
Malkov, M.A. (1997) *ApJ* **485**, 638.
Malkov, M.A. (1999) *ApJ* **511**, L53.
Malkov, M.A. and Drury, L.O'C. (2001) *Rep. Prog. Phys.* **64**, 429.
Malkov, M.A., Diamond, P. H. and Jones, T. W. (2002) *ApJ* **571**, 856.
Manchester, R.N. and Taylor, J.H. (1977) *Pulsars*, W.H.Freeman and company, San Francisco.
Mannheim, K., Krülls, W.M. and Biermann, P.L. (1991) *A&A* **251**, 723.
Mannheim, K. (1993) *A&A* **269**, 67.
Mannheim, K. (1996) *Space Sci. Rev.* **75**, 331.
Markoff, S., Falcke, H. and Fender, R.P. (2001) *A&A* **372**, L25.
Marsden, P.L. et al. (1984) *Ap.J.* **278**, L29.
Marshall, H.L., Harris, D.E., Grimes, J.P. et al. (2001) *ApJ* **549**, L167.
Mastichiadis, A. (1991) *MNRAS* **253**, 235.
Mastichiadis, A. (1996) *A&A* **305**, 53.
Mastichiadis, A. and De Jager, O.C. (1996) *A&A* **311**, L5.
Mastichiadis, A., Protheroe, A.P., Szabo, A.P. (1994) *MNRAS* **266**, 910.
Mastichiadis, A. and Kirk, J.G. (1997) *A&A* **320**, 19.
Mathis, J. S., Mezger, P. G. and Panagia, N. (1983) *A&A* **128**, 212.
Matsumoto, T. (2000) *The ISAS Science Report* **SP No. 14**, 179.
Mattila, K. (2003) *ApJ* **591**, 124.
McConnell, M.L. and Ryan, J.M. [eds.] (2000) *AIP Conference Proceedings*, vol. 510, Melville, New York.
McConnell, M.L. et al. (2000) *ApJ* **543**, 928.
McConnell, M.L. et al. (2002) *ApJ* **572**, 984.

McKay, T. A., Borione, A., Catanese, M. et al. (1993) *ApJ* **417**, 742.
McLaughlin, M.A. and Cordes, J.M. (2000) *ApJ* **538**, 818.
Meisenheimer, K., Yates, M. G. and Roeser, H.-J. (1997) *A&A* **325**, 57.
Melrose, D. (1980) *Plasma Astrophysics*, vol. 1, Gordon & Breach, New York.
Melrose, D. and Crouch, A. (1997) *PASA* **14**, 251.
Meszaros, P. and Rees, M.J. (1997) *ApJ* **482**, L29.
Mezger, P.G., Tuffs, R.J., Chini, R., Kreysa, E. and Gemuend, H.-P. (1986) *A&A* **167**, 145.
Michel, F.C. (1982) *Rev. Mod. Phys.*, **54**, 1.
Michel, C. (1998), in Proc. "Neutron Stars and Pulsars: *Thirty Years after the Discovery*" (eds. M. Shibazaki *et al.*), Universal Academy Press, Inc. (Tokyo), p.263.
Mignani, R.P., Caraveo, P.A. and Bignami, G.F. (1999) *A&A* **343**, L5.
Miniati, F., Ryu, D., Kang, H., Jones, T. W., Cen, R. and Ostriker, J. P. (2001) *ApJ* **542**, 608.
Miniati, F. (2002) *MNRAS* **337**, 1009.
Miniati, F. (2003) *MNRAS* **342**, 199.
Mirabel, I.F. and Rodriguez, L.F. (1994) *Nature* **371**, 46.
Mirabel, I.F. et al. (1998) *ApJ* **330**, L9.
Mirabel, I.F. and Rodriguez, L.F. (1999) *Annu. Rev. Astron. Astrophys.* **37**, 409.
Mioduszewski, A.J., Rupen, M.P., Hjellming, R.M., Pooley, G.G. and Waltman, E.B. (2001) *ApJ* **553**, 766.
Montmerle, T. (1979) *ApJ* **231**, 95.
Morfill G.E., Forman M. and Bignami, G. (1984), *ApJ* **284**, 856.
Mori, M. (1997) *ApJ* **478**, 225.
Mori, M., Bertsch, D. L., Dingus, B.L. et al. (1997) *ApJ* **476**, 842.
Morrison, P., (1957) *Nuovo Cimento* **7**, 858.
Morrison, P., Sadun, A. and Roberts, D. (1984) *ApJ* **280**, 483.
Morselli, P., (2002) *Frascati Phys. Ser.* **XXIV**, 363.
Moskalenko, I.V. and Strong, A.W. (1998) *ApJ* **476**, 842.
Mukherjee, R. (2001) in: *AIP Conference Proceedings*, vol. 558 (eds. F. A. Aharonian and H. J. Völk), p. 324, Melville, New York.
Mukherjee, R., Bertsch, D.L., Bloom, S. D. et al. (1997) *ApJ*, **490**, 116.
Muraishi, H. Tanimori, T, Yanagita, S. et al. (2000) *A&A* **354**, L57.
Mücke, A., Rachen, J.P., Engel, R., Protheroe, R.J. and Stanev, T. (1999) *PASA* **16**, 160.
Mücke, A., Pohl, M. (2000) *MNRAS* **312**, 177.
Mücke, A. and Protheroe, R.J. (2001) *Astropart. Phys.* **15**, 121.
Mücke, A., Protheroe, R.J., Engel, R., Rachen, J.P. and Stanev, T. (2003) *Astropart. Phys.* **18**, 593.
Nagano, M. and Watson, A.A. (2000) *Rev. Mod. Phys.* **72**, 689.
Nagirner, D.I. and Putanen, J. (1993) *A&A* **275**, 325.
Naito,T. and Takahara,F.(1994) *Journ. Phys. G: Nucl. Part. Phys.* **20**, 477.
Nelson, R., Hirayama, H., Rogers, D. (1985) *Preprint SLAC-265*, Stanford University.
Neronov, A., Semikoz, D., Aharonian, F. and Kalashev, O. (2002) *Phys. Rev*

Lett. **89**, 051101-1.
Neronov, A. and Semikoz, D. (2002) *Phys. Rev.* **D66**, 123003.
Neshpor, Yu.I et al. (1998) *Astron. Lett.* **24**, 134.
Neshpor, Yu.I et al. (2001) *Astronomy Reports* **45**, 249.
Nikishov, A.I. (1962) *Sov. Phys. JETP* **14**, 393.
Nishimura, J. (1967) *Handbuch der Physik* **Bd.XLVI/2**, 1.
Nishimura, J., Fujii, M., Taira, T. et al. (1980) *ApJ* **238**, 394.
Nishimura, J., Kobayashi, T., Komori, Y. and Yoshida, K. (1997) *AdSpR* **19** 767
Nishiyama, T. et al. (1999), in Proc. *26th ICRC*, Salt Lake City, vol. 3, p. 370.
Nolan, P.L., et al. (1981) *Nature* **293**, 275.
Nolan, P.L. et al. (1993) *ApJ* **409**, 697.
Ong, R. (1998) *Phys. Rep.* **305**, 93.
Ormes, J.F., Ösel, M.E. and Morris, D.J. (1988) *ApJ* **334**, 722.
Ostrowski, M. (1998) *A&A* **335**, 134.
Owen, F.N., Hardee, P.E. and Cornwell, T.J. (1989) *ApJ* **340**, 698.
Owens, A. J. and Jokipii, J. R. (1977) *ApJ* **215**, 685.
Ozernoy, L.M., Prilutskii, O.F., Rosental, I.L. (1973) *High Energy Astrophysics* (in Russian), Atomizdat, Moscow.
Ozernoy, L.M. and Aharonian, F.A. (1979) *Astr. Space Sci.* **66**, 497.
Pacini, F. (1968) *Nature* **216**, 567.
Pacini, F. (1971) *ApJ* **163**, L17.
Padovani, P., Ghisellini, G., Fabian, A.C., Celotti, A. (1993) *MNRAS* **260**, L21.
Padovani, P. and Giommi, P. (1995) *ApJ* **444**, 567.
Parizot, E. (2000) *A&A* **362**, 786.
Paredes, J.M., Marti, J., Ribo, M. and Massi, M. (2000) *Science* **288**, 2340.
Paragi, Z. Fejes, I., Vermeulen, R.C., Schilizzi, R.T., Spencer, R.E. and Stirling, A.M. (2001) in Proc IAU Symp. 205 *Galaxies and their constitutions at the highest angular resolutions*, ASP, p.112.
Park, B. T. and Petrosian, V. (1995) *ApJ* **446**, 699.
Paul, J. (2001) in: *AIP Conference Proceedings*, vol. 558 (eds. F. A. Aharonian and H. J. Völk), p. 183, Melville, New York.
Pelletier, G. (2001) in: *AIP Conference Proceedings*, vol. 558 (eds. F. A. Aharonian and H. J. Völk), p. 289, Melville, New York.
Petre, R., Allen, G.E., Hwang, U. (1999) *Astron. Nachrichten* **320**, 199.
Petrosian, V. (2001) *ApJ* **557**, 560.
Petry, D. et al. (2002) *ApJ* **580**, 104.
Pian, E., Vacanti, G., Tagliaferri G., et al. (1998) *ApJ* **492**, L17.
Plaga, R. (1995) *Nature* **374**, 430.
Plyasheshnikov, A. V. and Bignami, G. F. (1985) *Nuovo Cimento* **8C**, 39.
Plyasheshnikov, A.V., Aharonian, F.A., Völk, H.J. (2000) *J. Phys. G: Nucl. Part. Phys.* **26**, 183.
Plyasheshnikov, A.V. and Aharonian, F.A. (2002) *J. Phys. G: Nucl. Part. Phys.* **28**, 267.
Pohl, M. (1996) *A&A* **307**, L57.
Pohl, M. and Esposito, J. A. (1998) *ApJ* **507**, 327.
Pohl, M. (1998) *A&A* **339**, 587.

Pohl, M. and Schlickeiser, R. (2000) *A&A* **354**, 395.
Pollock, A.M.T. (1985) *A&A* **150**, 339.
Porter, T.A. and Protheroe, R.J. (1997) *Journ. Phys. G: Nucl. Part. Phys.* **23**, 1765.
Pozzetti, L. et al. (1998) *MNRAS* **298**, 1133.
Prilutsky, O.F., Rozental, I.L. (1970) *Acta Phys. Hung.* **29** 51.
Primack, J.R., Bullock, J.S., Somerville, R.S., MacMinn, D. (1999) *Astropart. Physics* **11**, 93.
Primack, J.R., Somerville, R.S., Bullock, J., Devriendt, J.E.G. (2001) in: *AIP Conference Proceedings*, vol. 558 (eds. F. A. Aharonian and H. J. Völk), p. 463, Melville, New York.
Prosch C., Feigl, E., Plaga, R. et al. (1996) *A&A* **314**, 275.
Protheroe, R.J. (1986) *MNRAS* **221**, 769.
Protheroe, R.J. and Stanev, T. (1993) *MNRAS* **264**, 191.
Protheroe, R.J. and Stanev, T. (1996) *Phys. Rev. Lett.* **77**, 3708.
Protheroe, R.J. and Biermann, P.L. (1996) *Astropart. Phys.* **6**, 45.
Protheroe, R.J. and Stanev, T. (1999) *Astropart. Phys.* **10**, 185.
Protheroe, R.J. and Meyer, H. (2000) *Phys. Letters B* **493**, 1.
Ptuskin, V. S., Völk, H.J., Zirakashvili, V.N. and Breitschwerdt, D. (1997) *A&A* **321**, 434.
Punch, M. et al. (1992) *Nature* **358**, 477.
Purcell, W.R. et al. (1997) *ApJ* **491**, 725.
Purcell, W. R., Grabelsky, D. A., Ulmer, M. P. et al. (1993) *ApJ* **413**, L85.
Rachen, J.P. and Bierman, P.L. (1993) *A&A* **272**, 161.
Ramana Murthy, P.N. and Wolfendale, A.W. (1993) "Gamma-ray astronomy", Cambridge University Press.
Ramaty, R., Kozlovsky, B. and Lingenfelter, R.E. (1979) *ApJ Suppl Ser.* **40**, 487
Reed, J. E., Hester, J. J.,Fabian, A. C.. and Winker, P. F. (1995) *ApJ* **440**, 706.
Rees, M.J and Gunn, J.E. (1974) *MNRAS* **167**, 1.
Rees, M.J. and Meszaros, P. (1994) *ApJ* **430**, L93.
Reimer, O. and Pohl, M. (2002) *A&A* **390**, L43.
Rephaeli, Y. (1977) *Astrophys. J.* **212**, 608.
Rephaeli, Y. and Gruber, D. (2002) *Astrophys. J* **579**, 587.
Ritz, S., Gehrels, and Shrader, C. [eds.] (2001) *AIP Conference Proceedings*, vol. 587, Melville, New York.
Roberts, T.P and Warwick,R.S. (2000) *MNRAS* **315**, 98.
Roberts, T.P et al. (2001) *MNRAS* **325**, L7.
Roberts, M.S.E., Romani, R.W., Kawai, N. (2001) Astroph.J. Suppl. Ser., **133**, 451.
Romero, G., Benaglia, P. and Torres, D.F. (1999) *A&A* **348**, 868.
Romani R.W. (1996) *ApJ* **470**, 469.
Romani, R.W. and Yadigaroglu, I.A. (1995) *ApJ* **438**, 314.
Rossi, B. and Greisen, K. (1941) *Rev. Mod. Phys.* **13**, 419.
Rowell, G.P. (2002), in Proc. : "*Relativistic flows in Astrophysics*".
Röser, H.-J., Meisenheimer, K., Neumann, M., Conway, R.G., Perley, R.A. (2000), *ApJ* **252**, 458.

Ruderman, M.A. and Sutherland, P.G. (1975) *ApJ* **196**, 51.
Quinn, J. et al. (1996) *ApJ* **456**, L83.
Safi-Harb, S. and Ögelman, H. (1997) *ApJ* **483**, 868.
Sambruna, R.M., Urry, C.M.,Tavecchio, F., Maraschi, L., Scarpa, R., Chertas, G., Muxlow, T. (2001) *ApJ* **549**, L161.
Sambruna, R.M., Aharonian, F.A., Krawczynski, H. et al. (2000) *ApJ* **538**, 127.
Samorski, M., Stamm, W. (1983) *ApJ* **268**, L17.
Sams, B.J., Eckart, A. and Sunyaev, R. (1996) *Nature* **382**, 47.
Sarazin, C.L. and Kempner, J.C. (2000) *ApJ* **533**, 73.
Sazonov, S.Y. and Sunyaev, R.A. (2000) *ApJ* **543**, 28.
Schatz, G. (2002) *Astropart. Phys.* **17**, 13.
Schlegel, D.J., Finkbeiner, D.P. and Devis M. (1998) *ApJ* **500**, 525.
Schönfelder, V. et al. (2000) *Astron. Astrophys. Suppl. Ser.* **143**, 145.
Schwartz, D.A., Marshall, H.L., Lovell, J.E. et al. (2000) *ApJ* **540**, 69.
Scott, J.S. and Chevalier, R. (1975) *ApJ* **197**, L5.
Seo, E.S. and Ptuskin, V.S. (1994) *ApJ* **431**, 703.
Shklovskii, I.S. (1959) *Sov. Astron.* **4**, 243.
Sigl, G., Lee, S., Schramm, D. N., Coppi, P. *Physics Letters B* (1997) **392**, 129.
Sikora, M., Begelman, M.C., Rudak, B. (1989) *ApJ* **341**, L33.
Sikora, M. (2001), in Proc: *Blazar Demographics and Physics* (eds. Padovani P. and Urry C.M.), ASP Conf. Series, vol. 227, p.95.
Sikora, M. and Madejski G. (2001) in: *AIP Conference Proceedings*, vol. 558 (eds. F. A. Aharonian and H. J. Völk), p. 275, Melville, New York.
Sikora, M., Blazejowski, M., Moderski, R., Madejski, G.M. (2002) *ApJ* **577**, 78.
Simpson, J.,A. (1983) *Ann. Rev. Nucl. Part Sci.* **33**, 323.
Skibo, J.G., Ramaty, R. and Purcell, W.R. (1996) *A & A Suppl Ser.* **120**, 403.
Slane, P., Gaensler, B,M., Dame, T.M. et al. (1999) *ApJ* **525**, 357.
Spencer, R.E. (1998) *New Astron. Reviews* **42**, 653.
Sreekumar, P., Bertsch, D.L., Dingus, J.A. et al. (1996) *ApJ* **464**, 628.
Sreekumar, P., Bertsch, D.L., Dingus, J.A. et al. (1998) *ApJ* **494**, 523.
Stecker, F.W. (1969) *Phys. Rev. Lett* **80**, 1816.
Stecker, F.W. (1971) *Cosmic Gamma-Rays*, NASA-SP249, Washington D.C.
Stecker, F.W., De Jager, O.C. and Salamon, M.H. (1992) *ApJ* **390**, L49.
Stecker, F.W. and Salamon, M.H. (1996) *ApJ* **464**, 600.
Stecker, F.W. and Glashow, S.L. (2001) *Astropart. Phys.* **16**, 97.
Stepanian, A.A. et al. (1983) *Izv. Krym. Astrophyz. Obs.* **66**, 234.
Stirling, A.M., Spencer, R.E., de la Force, C.J. et al. (2001) *MNRAS* **327**, 1273.
Strong, A. W., Diehl, R., Schönfelder V., et al. (1997), in Proc. "4th Compton Symposium", *AIP Conference Proceedings*, vol. 410 (eds. C. D. Dermer et al.), AIP Conf. Proceedings (New York), p. 1198, Melville, New York.
Strong, A. W., Moskalenko, I. V. and Reimer, O. (2000) *ApJ* **537**, 763.
Sturner,S.J. and Dermer,C.D. (1995) *A&A* **293**, 17.
Sturner,S.J., Skibo, J.G., Dermer, C.D. and Mattox, J.R. (1997) *ApJ* **490**, 619.
Sturrock, P. (1971) *ApJ* **164**, 529.
Sugiho, M., Kotoku, J., Makishima, K., Kubota, A., Mizuno, T., Fukazawa, Y. and Tashiro, M. (2001) *ApJ* **561**, L73.

Sunyaev, R.A. and Titarchuk, L.G. (1980) A&A **86**, 121.
Sunyaev, R.A. and Revnivtsev, M. (2000) A&A **358**, 617.
Svenson, R. (1987) MNRAS **227**, 403.
Swordy, A. (2001) Space Sci. Rev. **99**, 85.
Syrovatskii, S.I. (1959) Sov. Astron. **3**, 22.
Takahashi, T., Madejski, G., Kubo, H. (1999) Astropart. Phys. **11**, 177.
Takahashi, T. et al. (2000) ApJ **542**, L105.
Tanimori, T. (2003), in Proc. "The Universe Viewd in Gamma Rays", Universal Academy Press, Inc. -Tokyo, 37.
Tanimori, T., et al. (1998a) ApJ **492**, L33.
Tanimori, T., et al. (1998b) ApJ **497**, L25.
Tatischeff, V. (2002), in *Stades Ultimes de l'Evolution Stellaire*, EAS Pub. Series.
Tavecchio, F., Maraschi, L., Ghisellini (1998) ApJ **509**, 608.
Tavecchio, F., Maraschi, L., Sambruna, R.M. and Urry, C.M. (2000) ApJ **544**, L23.
Tavecchio, F. et al. (2001) ApJ **554**, 725.
Taylor, J.H. and Cordes, J.M. (1993) ApJ **411**, 674.
Thompson, D. J., Bailes, M., Bertsch, D. et al. (1999) ApJ **516**, 297.
Thompson, D.J. (2001) in: *AIP Conference Proceedings*, vol. 558 (eds. F. A. Aharonian and H. J. Völk), p. 103, Melville, New York.
Timokhin, A.N., Aharonian, F.A., Neronov, A. (2003) A&A, submitted.
Torres, D.F., Butt, Y.M. and Camilo, F. (2001) ApJ **560**, L155.
Torres, D.F., Romero, G.E., Dame, T.M., Combi, J.A., Butt, Y.M. (2003) Physics Reports **382**, 303.
Tuffs, R.J., Drury, L. O'C., Fischera, J. et al. (1997) in Proc. 1-st ISO Workshop on Analytical Spectroscopy (ESA SP-419), p.177.
Uchaikin, V.V. and Ryzhov, V.V. (1998) *The Stochastic Theory of Transport of High Energy Particles*, Novosibirsk, Nauka (in Russian).
Uchiyama, Y., Takahashi, T., Aharonian, F.A. and Mattox, J.,R. (2002a) ApJ **571**, 866.
Uchiyama, Y., Takahashi, T. and Aharonian, F.A. (2002b) PASJ **54**, L73.
Uchiyama, Y., Aharonian, F.A. and Takahashi, T. (2003) A&A, **400**, 567.
Urry, C.M. and Padovani, P. (1995) PASP **107**, 803.
Usov, V.V. and Melrose, D.B. (1995) Ast. J. Phys. **48**, 571.
Valinia, A. and Marshall, F.E. (1998) ApJ **505**, 134.
Valinia, A., Tatischeff, V., Arnaud, K., Ebisawa, K. and Ramaty, R. (2000), ApJ **543**, 733.
van der Laan, H. (1966) Nature **211**, 1131.
van der Meulen, R.D., Bloemen, H., Bennett, K., Hermsen, W., Kuiper, L., Much, R.P., Ryan, J., Schonfelder, V. and Strong, A. (1998) A&A **330**, 321.
Vestrand,W.T., Sreekumar,P., Mori, M. (1997) ApJ **483**, L49.
Völk, H. J., Aharonian, F. A. and Breitschwerdt, D. (1996) Space Sci. Rev. **75**, 279.
Völk, H.J. (2001), in Proc. *High Energy Astrophysical Phenomena*, Les Arcs.
Völk, H.J. and Atoyan, A.M. (2000) ApJ **541**, 88.

Völk, H. J., Berezhko, E. G., Ksenofontov, L. T., Rowell, G. P. (2002) *A&A* **396**, 649.
Waxman, E. and Coppi, P. (1996) *ApJ* **75**, L76.
Webber, W. R., Simpson, J. A. and Cane, H. V. (1980) *ApJ* **236**, 448.
Weekes, T.C. and Turver, K.E. (1977), in Proc *12th ESLAB Symposium*, Frascati, p. 279.
Weekes, T.C. et al. (1989) *ApJ* **342**, 379.
Weekes, T.C. (1992) *Space Sci. Rev.* **59**, 315.
Weekes, T.C., Aharonian, F.A., Fegan, D.J., Kifune, T. (1997) in: *AIP Conference Proceedings*, vol. 410 (eds. C.D. Dermer et al.), vol. 1, p. 361, Woodbury, New York.
Weisskopf, M.C., Hester, J.J., Tennant, A.F., et al. (2000) *ApJ* **536**, L81.
Willingale, R., West, R.G., Pye, J.P. and Swewart, G.C. (1996) *MNRAS* **278**, 749.
Wilson, A.S., Young, A.J. and Snopbell, P.L. (2000) *ApJ* **544**, L27.
Wilson, A.S., Young, A.J. and Snopbell, P.L. (2001) *ApJ* **547**, 740.
Winkler, P., F. and Long, K.S. (1997) *ApJ* **491**, 829.
Wolfendale, A.W. (1993) "The Hess Memorial Lecture", in Proc. 23rd ICRC (Calgary), *Invited, Rapporteur & Highlight papers*, World Scientific, p. 143.
Wright, E.L. (2001) *ApJ* **553**, 538.
Yamauchi, S., Kawai N. and Aoki, T. (1994) *PASJ* **46**, L109.
Yoshida, T. and Yanagita, S. (1997), In: Proc 2nd INTEGRAL Workshop on the Trtansparent Universe, SP-382, ESA, Paris, p.85.
Yoshikoshi,T., Kifune, T., Dazeley, S. et al. (1997) *ApJ* **487**, L65.
Zatsepin, G.T., Kuzmin, V.A. (1966) *JETP Lett* **4**, 78.
Zdziarski, A.A. (1988) *ApJ* **289**, 514.
Zdziarski, A.A. and Lightman, A.P. (1985) *ApJ* **294**, L79.
Zirakashvili, V. N., Breitschwerdt, D., Ptuskin, V. S. and Völk, H. J. (1996) *A&A* **311**, 113.

Index

γ-γ pair-production, 112, 117, 443
π^0-decay γ-rays, 106, 141, 184, 196, 199, 205, 218, 222, 275
1ES 1426+428, 67, 76, 399, 415, 417, 427, 454
1ES 1959+650, 67, 71, 399
1ES 2344+514, 67, 399
2.2 MeV line emission, 111
3C 273, 27, 34, 340, 346
3C 279, 27, 34
3rd EGRET Catalog, 33

AGN, 1, 9, 10, 12, 321, 323
annihilation radiation, 179, 182, 187

BATSE, 23
BeppoSAX, 401
BHCs, 295
binary pulsar, 265
BL Lac objects, 67, 394, 399
black holes, 12, 16, 314, 356, 433
blazars, 13, 393, 399
Bohm diffusion, 209, 213, 219, 241, 342, 355, 427
bottom-up scenario, 138
bremsstrahlung, 101, 160, 166, 181, 187, 188, 195, 206, 275
bulk-motion Comptonization, 17, 435

CANGAROO, 44, 46, 50, 52, 206
Cas A, 61, 231, 237
CAT, 44, 85, 401

Centaurus A, 27, 33, 80
Chandra, 230, 240, 256, 323
charged pions, 107
Cherenkov radiation, 83
CIB, 393, 394, 398, 438, 439
CMBR, 6, 9, 16, 147, 177, 205, 206, 284, 323, 447
cold wind, 17, 415
COMPTEL, 23, 25, 273
Compton GRO, 1, 23, 31
concept 5at5, 93, 94
continuous source, 151, 152
COS B, 32, 221, 233
cosmology, 7, 22, 437
Crab Nebula, 12, 47, 243, 268, 270, 278
curvature radiation, 130
Cyg X-1, 25, 26, 295, 297
Cyg X-3, 32, 41, 64

Dark Matter, 21
DEBRA, 393, 437
diffuse γ-ray emission, 173, 182, 185, 195
Diffuse Extragalactic Background, 19, 460
diffusion coefficient, 149, 158, 161, 170, 174
diffusive escape, 175, 185, 213
diffusive shock acceleration, 140, 142, 152, 201, 208, 236
Doppler factor, 14, 399, 404, 408

EGRET, 23, 33, 144, 145, 148, 179, 202, 244, 280, 309, 422
electromagnetic cascade, 6, 16, 131, 460
electromagnetic window, 1
electron-positron annihilation, 105
electronic models, 217
equipartition, 215, 231, 331, 335, 360, 361, 427, 431
extragalactic cosmic rays, 9

galactic cosmic rays, 7, 135
Galactic Disk, 136, 138, 139, 147, 164, 165, 173, 178, 185, 191, 194
galaxy clusters, 9, 359
gamma-ray astronomy, 1, 4, 7, 23, 83, 87, 106, 437
gamma-ray pulsars, 38, 54
GeV Blazars, 34
GeV bump, 183, 185, 196
GLAST, 2, 10, 90
Gould Belt, 42
grammage, 136
GRBs, 3, 10, 12
GRS 1915+105, 294, 295, 301

H.E.S.S., 46, 87
hadronic cascades, 122
hadronic models, 217, 219, 228, 355, 417
HEGRA, 44, 45, 48, 50, 61, 64, 76, 80, 85, 401
hot spots, 350, 351, 357
Hubble constant, 455

IACT arrays, 2, 10
IC γ-rays, 187, 205, 210, 218, 235
imaging, 84
impulsive source, 149, 171
INTEGRAL, 31
intergalactic absorption, 371, 397, 406, 430
intergalactic magnetic field, 363, 386, 389, 437, 449
internal shocks, 298, 300
interstellar medium, 137, 149

intracluster magnetic fields, 359, 368, 379
intracluster medium, 359, 379
inverse Compton scattering, 113

KED winds, 253, 254
Klein-Nishina regime, 114, 434
knee, 136, 138, 142
knots, 338, 346, 357

large scale jets, 321, 323, 330, 347
Leaky-Box, 175
leptonic models, 355, 402, 417
light cylinder, 244, 253, 259, 263
local CRs, 147, 161, 189
Lorentz, 415
Lorentz factor, 10, 254, 323, 339
Lorentz invariance, 432

M 87, 80
MAGIC, 46, 87
MHD jets, 16
MHD model, 268, 270, 273, 279
microblazars, 311
microquasars, 12, 293, 295, 297, 301, 309, 313, 316
millisecond pulsars, 252
Mkn 421, 67, 68, 393, 399, 401, 404, 440
Mkn 501, 67, 68, 399, 404, 422, 440
molecular clouds, 147, 156, 164, 191

neutron stars, 12, 243
nuclear γ-ray line emission, 110

OB associations, 40
OSSE, 23, 26
outer gap, 246, 247, 250, 258

Pair Halos, 447, 450, 451, 457
pair-production, 101
PeV SNRs, 238, 241
photodisintegration, 121
photomeson production, 121
PKS 2155-304, 67, 79, 399, 454
plasmons, 299, 300

plerions, 51, 243, 281
polar cap, 246, 247, 251
Poynting flux, 12, 253, 327, 356, 434
proton bremsstrahlung, 188, 196
proton synchrotron radiation, 126, 331, 342, 417, 422
proton-neutron bremsstrahlung, 111
proton-proton interactions, 107
pulsar magnetospheres, 244, 255, 265
pulsar winds, 12, 16, 243
pulsars, 12, 243, 246, 249, 250

relativistic outflows, 12, 267, 300, 432
rotation powered pulsars, 245
RX J1713.7-3946, 58, 206, 222, 225
RXTE, 401

SAS-2, 32
sea of GCRs, 136
Sedov phase, 201, 209
self-regulated synchrotron cutoff, 129
SN 1006, 58, 206

SNOBs, 40
SNRs, 40, 140, 141, 199
SSC, 402, 404, 409
star formation regions, 144, 161
stereo imaging, 85
synchrotron radiation, 123, 209, 240

TeV blazars, 67, 399, 402, 417, 432
TeV J2032+4130, 65
Thomson regime, 115, 434
top-down scenario, 138
two-temperature plasma, 111

unidentified sources, 29, 39
unshocked wind, 16, 50, 253, 280

VERITAS, 46, 87, 401

Whipple, 44, 47, 68, 76, 85

X-ray binaries, 293, 314